Treated Wastewater in Agriculture

Use and Impacts on the Soil Environment and Crops

Edited by

Guy J. Levy, Pinchas Fine and Asher Bar-Tal
Institute of Soil, Water and Environmental Sciences
Agricultural Research Organization,
The Volcani Center, Israel

A John Wiley & Sons, Ltd., Publication

This edition first published 2011

Blackwell Publishing was acquired by John Wiley & Sons in February 2007. Blackwell's publishing programme has been merged with Wiley's global Scientific, Technical, and Medical business to form Wiley-Blackwell.

Registered office
John Wiley & Sons Ltd, The Atrium, Southern Gate, Chichester, West Sussex, PO19 8SQ, UK

Editorial offices
9600 Garsington Road, Oxford, OX4 2DQ, UK
The Atrium, Southern Gate, Chichester, West Sussex, PO19 8SQ, UK
2121 State Avenue, Ames, Iowa 50014-8300, USA

For details of our global editorial offices, for customer services and for information about how to apply for permission to reuse the copyright material in this book please see our website at www.wiley.com/wiley-blackwell.

Library of Congress Cataloging-in-Publication Data

Treated wastewater in agriculture : use and impacts on the soil environment and crops / edited by Guy J. Levy, Pinchas Fine and Asher Bar-Tal. – 1st ed.
p. cm.
Includes bibliographical references and index.
ISBN 978-1-4051-4862-7 (hardback : alk. paper) 1. Water reuse. 2. Sewage irrigation–Environmental aspects. 3. Water in agriculture. I. Levy, Guy J. II. Fine, Pinchas. III. Bar-Tal, A. (Asher)
TD429.T735 2011
631.5'8–dc22

2010011216

A catalogue record for this book is available from the British Library.

This book is published in the following electronic formats: ePDF (9781444328578); Wiley Online Library (9781444328561)

Set in 10/12pt Times by Thomson Digital, Noida, India
Printed and Bound in Singapore by Markono Print Media Pte Ltd

1 2011

Contents

Preface

Irrigated agriculture produces one-third of the world's crop yield and half the return from global crop production. Yet, in many parts of the world, especially in semiarid and arid regions the future of irrigated agriculture is threatened by existing or expected shortages of freshwater. These shortages result mainly from the ever-increasing demand put upon water resources by the world's rapidly growing population and their improving standard of living. The constant rise in population and in water use per capita leads also to an ever-growing volume of municipal sewage water which requires to be disposed. Water recycling and the use of treated municipal sewage effluents, (herein referred to as treated wastewater (TWW)) for agriculture, industry and non-potable urban and environmental applications can afford a highly effective and sustainable strategy to exploit a water resource in areas afflicted by water scarcity. Irrigation with TWW can contribute a significant quantity of nutrients, and hence can contribute to the conservation of diminishing resources.

However, irrigation with TWW is not free of risk both to crop production and the soil environment. Potential risks include reduction in yield due to elevated salinity and specific ion toxicity, migration of pollutants towards surface- and groundwater, and deterioration of soil structure. It is important, therefore to understand the way in which parameters such as quality of TWW, irrigation management practices, and soil and crop characteristics affect processes occurring in the irrigated field.

The central role that irrigated agriculture plays in food production, coupled with the increasing need to utilize TWW for irrigation, motivated this attempt to assemble relevant core knowledge and recent advances in research on irrigation with TWW in the form of a comprehensive book. Our goal was to prepare a volume that consolidates the state-of-the-art knowledge on the various aspects of irrigation with TWW and analyzes the possible impacts (either positive or negative) of such irrigation water, both from the agricultural and the environmental perspectives.

The book is divided into 14 chapters arranged in two parts. The first part includes four chapters that cover technical, regulatory, and economic aspects of TWW reuse. The first chapter takes the reader step by step through the multitude of processes available for the treatment of municipal sewage effluents, from extensive, low-tech processes such as lagooning and constructed wetlands to enhanced tertiary and quaternary processes, all aimed at providing TWW that comply with quality criteria set by local and international regulators.

Treatment of sewage effluents should lead to the effective control of health hazards associated with the use of TWW and safeguard the farming community, the consumers of crops, and the population at large from exposure to pathogenic microorganisms that are originally present in the treatment stream. The second chapter deals with the question of what constitutes sufficient protection of the public welfare and introduces the idea that it might be socially and morally justified to allow exposure of the population to a predetermined and well-regulated level of risk to be associated with the reuse of TWW. The author endorses an approach to treating sewage effluents for agricultural use based on the assessment and characterization of the associated risk rather than adoption of the best available technology to treat the sewage effluents to an excessive, unnecessarily high level.

Chapter 3 presents the most updated regulations and guidelines for TWW use in agriculture embraced by various countries. The different basic philosophies adopted to protect public health and the environment are discussed in this chapter.

The last chapter in the first part (Chapter 4) highlights economic considerations involved in the use of TWW for irrigation. It presents a basic approach to pricing and cost allocation associated with the use of TWW by both large and small farming communities. This chapter also lends support to the role of government as a regulator and an arbitrator regarding the strong external interests and complex economic issues involved in the use of TWW.

The second part of the volume covers the impact of irrigation with TWW on the agricultural ecosystem. The agricultural and environmental aspects of the presence of organic and mineral forms of major nutrients (nitrogen and phosphorus) in TWW are discussed in Chapter 5 (5.1 and 5.2). The fate of organic N and ammonium in soil, including chemical transformations, mobility, uptake by plants, and the risk of groundwater contamination by excessive leaching, as well as emission of greenhouse gases and other losses to the atmosphere, are described and discussed in Chapter 5 (5.1). Examples of experiments and observations in which TWW was used are presented and compared with freshwater. In Chapter 5 (5.2) the fate of inorganic and organic P in soil, including its chemical transformations, mobility, uptake by plants, and the risk of excess P accumulation in agricultural soils and potential contamination of surface water bodies by runoff loaded with P are reviewed and discussed. Results of laboratory studies, field experiments and surveys of plots in which TWW was used are presented and compared with the results of control runs with freshwater. An additional section of this Chapter 5 (5.3) focuses on a subject that thus far has received little attention, namely the chemistry of calcium and of carbonates in TWW-irrigated soils. This topic is important from the environmental point of view because TWW contains relatively high concentrations of carbonates (and in particular bicarbonates), and their accumulation in soils and groundwater may be an important pathway of carbon sequestration.

The inorganic constituents, the concentrations of which in TWW are frequently higher than in freshwater, are discussed in Chapters 6, 7, and 8. Chapter 6 focuses on two elements (boron and chloride) that may reach toxic levels in TWW-irrigated soil. Boron reactions and interactions in soils and its uptake, transport, and distribution in plants are reviewed in Chapter 6 (6.1), as is the issue of boron toxicity resulting from the relatively high boron concentration often encountered in TWW. The specific toxicity of the chloride ion to plants is discussed in Chapter 6 (6.2), where a summary of the role of the chloride ion in plant physiology, as well as some examples of the effect on plants of chloride added through irrigation with TWW, are also given. Chapter 7 discusses the fate of heavy metals in TWW-irrigated soils. It highlights the role of two main factors, pH and Eh, that affect the behavior and fate of metals due to these parameters' strong influences on the solubility of metals and organic matter and on the properties and stability of the surfaces of the soil's solid components. It is argued that, in many cases, especially when less advanced methods for TWW production and higher irrigation rates are employed, it is the effect of the TWW on metals already present in the soil rather than on the metals contained in the irrigation water themselves that will govern the mobility and availability of metals in the TWW–soil–plant system.

The fact that TWWs are appreciably more saline than the freshwater from which they originated deserves attention. The salinity of the water increases throughout the long path of TWW formation. In Chapter 8, general aspects of the effect of salinity on crop production are presented, as well as some specific examples of the effect on irrigated trees. Innovative management methods to counteract the damage that may be caused by salinity are highlighted.

The potentially adverse effects on the stability of soil-structure and on the soil's hydraulic properties, which are associated with the elevated levels of certain organic and inorganic constituents in TWW as compared with freshwater are evaluated in Chapter 9. The reviewed literature reveals that reports on the effects of TWW on soil's physical characteristics are inconsistent and that existing knowledge is insufficient to support reliable modeling efforts for predicting the soil's response to irrigation with TWW.

The fate of organic matter and organic contaminants present in TWW and their effect on irrigation systems and the soil environment are discussed in Chapters 10, 11, 12, and 13. Biofilm buildup and its role in clogging irrigation systems, an issue that was often overlooked in the past, is discussed in Chapter 10. Special emphasis is put on the presentation of up-to-date knowledge of the mechanisms of biofilm formation and of state-of-the-art practical information on the effects of biofilm buildup on irrigation systems. Some useful schemes to minimize problems associated with biofilm formation are presented. Chapter 11 reviews the impact of various components of TWW, including microorganisms, on the soil's microbial population and activity. High levels of mineral solutes, as well as of dissolved organic carbon, detergents, pharmaceuticals, pollutants such as pesticides and other organic chemicals and trace metals, may affect the diversity, structure, and functioning of microbial communities, and hence also affect soil fertility and structure. The authors conclude that current knowledge on the effects of TWW on the soil microflora is insufficient. Especially lacking are data on the effect of the various components of TWW, separately or in combination, on the composition of the microbial community in soils.

Chapter 12 reviews the long-term risks posed by the potential interactions between the dissolved organic matter in TWW and anthropogenic chemicals present both in the soil and in the TWW. Special emphasis is put on the binding of anthropogenic chemicals to TWW-originated dissolved organic matter and the resultant possibility of enhanced transport of pollutants to groundwater. The characteristics of the organic matter in TWW and its influence on the soil organic matter are discussed in Chapter 13. Based on available knowledge, it is concluded that addition of dissolved or particulate organic matter through irrigation with TWW can ultimately result in either an increase or a decrease in the soil's organic matter content, depending on site-specific soil properties and conditions and microbial activity.

Transport of water and solutes in the soil profile as affected by irrigation with high sodium adsorption ratio (SAR) water is described in Chapter 14. High SAR values are common in TWW. The results of the flow and transport simulations discussed in Chapter 14, suggest that on the field-scale, under realistic flow conditions and over an extended period of time, the adverse effects of low solute concentration and a relatively high SAR on the flow and the transport are smaller as compared with the effects measured in laboratory systems in which the transport obeys the classical Darcy equation. These systems (unlike the conditions in the field) consist of a one-dimensional vertical, spatially homogeneous flow domain. This finding has practical implications regarding the use of TWW for irrigation. The data presented may be used for water quality classification as related to soils of different textures and as a tool for water and soil management.

In as much as this book covers a wide range of topics related to the use of TWW, it may serve as a reference book for scientists, agronomists, engineers, ecologists, and students. Hopefully, this volume will contribute to the continuation of capacity building in the many areas related to the use of TWW for irrigation and to optimizing the impact of irrigation with TWW on the agricultural ecosystem and the environment at large.

Contributors list

Andreas N. Angelakis	National Agricultural Research Foundation (NAGREF), Institute of Iraklio, Iraklio, Greece
Shmuel Assouline	Institute of Soil, Water and Environmental Sciences, Agricultural Research Organization, The Volcani Center, Bet Dagan, Israel
Asher Bar-Tal	Institute of Soil, Water and Environmental Sciences, Agricultural Research Organization, The Volcani Center, Bet Dagan, Israel
Bnayahu Bar-Yosef	Institute of Soil, Water and Environmental Sciences, Agricultural Research Organization, The Volcani Center, Bet Dagan, Israel
Alon Ben-Gal	Institute of Soil, Water and Environmental Sciences, Agricultural Research Organization, Gilat Research Center, Israel
Yona Chen	Department of Soil and Water Sciences, Robert H. Smith Faculty of Agriculture, Food and Environment, The Hebrew University of Jerusalem, Rehovot, Israel.
Carlos G. Dosoretz	Division of Environmental, Water and Agricultural Engineering, Faculty of Civil and Environmental Engineering and Grand Water Research Institute, Technion-Israel Institute of Technology, Haifa, Israel
Gil Eshel	Soil Erosion Research Station, Ministry of Agriculture & Rural Development, State of Israel, Ruppin Institute, Israel
Pinchas Fine	Institute of Soil, Water and Environmental Sciences, Agricultural Research Organization, The Volcani Center, Bet Dagan, Israel
Zev Gerstl	Institute of Soil, Water and Environmental Sciences, Agricultural Research Organization, The Volcani Center, Bet Dagan, Israel

David Giraldi Department of Civil Engineering, University of Pisa, Pisa, Italy

Ellen R. Graber Institute of Soil, Water and Environmental Sciences, Agricultural Research Organization, The Volcani Center, Bet Dagan, Israel

Amir Hass Agricultural and Environmental Research Station, Gus R. Douglass Land-Grant Institute, West Virginia State University, Institute WV, USA, and USDA-ARS, Appalachian Farming Systems Research Center, Beaver, West Virginia, USA

Efrat Hadas Investment Financing Department, Ministry of Agriculture and Rural Development, Bet Dagan, Israel

Renato Iannelli Department of Civil Engineering, University of Pisa, Pisa, Italy

Elizabeth Jüeschke Institute of Geography, Ruhr University, Bochum, Germany

Uzi Kafkafi The Robert H. Smith Institute of Plant Sciences and Genetics in Agriculture, The Robert H. Smith Faculty of Agriculture, Food and Environment, The Hebrew University of Jerusalem, Rehovot, Israel

Rachel Karyo Sackler Faculty of Medicine, Tel Aviv University, Israel

Ilan Katz Division of Environmental, Water and Agricultural Engineering, Faculty of Civil and Environmental Engineering and Grand Water Research Institute, Technion-Israel Institute of Technology, Haifa, Israel

Elisha Kenig Irrigation and Soil Division, Extension Service, Ministry of Agriculture and Rural Development, Bet Dagan, Israel

Rami Keren Institute of Soil, Water and Environmental Sciences, Agricultural Research Organization, The Volcani Center, Bet Dagan, Israel

Yoav Kislev Department of Agricultural Economics and Management, Faculty of Agricultural, Food and Environmental Quality Sciences, Hebrew University of Jerusalem, Rehovot, Israel

Guy J. Levy Institute of Soil, Water and Environmental Sciences, Agricultural Research Organization, The Volcani Center, Bet Dagan, Israel

Nico E. Marcar	CSIRO Sustainable Ecosystems, Canberra, Australia
Bernd Marschner	Institute of Geography, Ruhr University, Bochum, Germany
Uri Mingelgrin	Institute of Soil, Water and Environmental Sciences, Agricultural Research Organization, The Volcani Center, Bet Dagan, Israel
Dror Minz	Institute of Soil, Water and Environmental Sciences, Agricultural Research Organization, The Volcani Center, Bet Dagan, Israel
Nikolaos V. Paranychianakis	Department of Environmental Engineering, Technical University of Crete, Greece
David Russo	Institute of Soil, Water and Environmental Sciences, Agricultural Research Organization, The Volcani Center, Bet Dagan, Israel
Miquel Salgot	Water Research Institute, University of Barcelona, Barcelona, Spain
Hillel Shuval	Hadassah Academic College-Jerusalem and The Hebrew University of Jerusalem, Israel
Michael J. Singer	Land Air and Water Resources Department, University of California Davis, CA, USA
Daryl P. Stevens	Arris Pty Ltd, Richmond, Victoria, Australia
Jorge Tarchitzky	Department of Soil and Water Sciences, Robert H. Smith Faculty of Agriculture, Food and Environment, The Hebrew University of Jerusalem, Rehovot, Israel
Tivi Theiveyanathan	CSIRO Sustainable Ecosystems, Canberra, Australia
Uri Yermiyahu	Institute of Soil, Water and Environmental Sciences, Agricultural Research Organization, Gilat Research Center, Israel

Part I

General Aspects

Chapter 1
Sources and composition of sewage effluent; treatment systems and methods

Renato Iannelli and David Giraldi

1.1 Sources of usable wastewater

From an ideal point of view, all kinds of wastewater can be reused if they undergo appropriate reclamation treatments. At present, available technologies allow removal of almost all detectable contaminants from wastewaters, making them suitable for every use, despite their original pollution levels. However, selection of the usable wastewater source is the first, and most important, aspect of every reclamation project.

Quality, quantity and location are three important characteristics for the possible use of wastewater. The quality of wastewater defines the required treatment level and related costs. Quantity considerations are strictly related to scale economies for reclamation costs and returns; but, are also related to the comparison between the available wastewater to be reclaimed and the demand for usable water. The location of the source is an important factor that affects the costs related to transport of wastewater from the source to the reclamation plant and then to the final reuse destination. This can be a reason to opt for a reclaimed wastewater source rather than primary water to be transported from a distant location.

Comparison between the cyclic behaviors of potentially usable wastewater and water demand is another aspect of significant relevance in terms of required exploitation costs. As neither produced wastewater nor water demand are usually constant in time, the assessment should include a comparison of cyclic variations of available wastewater with that of water demand. If the two variation shapes match, the construction of compensation tanks/reservoirs can be avoided or significantly reduced, with remarkable cost savings. Conversely, non-matching shapes result in a requirement for compensation tanks/reservoirs, the volume of which depend on the differences between the shapes. Specifically, cycles with long periods (as, for instance, the annual cycle of the agricultural water demand, which is required only during the irrigation season) require the construction of storage reservoirs of extremely large volumes if the production of usable wastewater is constant all year round, as in the case of urban wastewater.

Treated Wastewater in Agriculture, First Edition, edited by Guy J. Levy, Pinchas Fine and Asher Bar-Tal

The main sources of usable wastewater can be basically classified into domestic, industrial, and a combination of the two, as often found in urban sewer systems.

1.1.1 Domestic and municipal wastewater

Domestic wastewaters are discharged from residential areas, commercial areas (offices, hotels, restaurants, shopping centers, theaters, museums, airports, etc.), and institutional facilities (schools, hospitals, old people's homes, prisons, etc.). The contaminants in domestic wastewaters are almost the same all over the world, although some differences can be found between developed and underdeveloped countries, particularly related to chemicals used in personal care and housekeeping products. The average concentration of contaminants mainly depends on the water supply per capita, which varies with water availability and climatic regions. According to different conditions, we can have weak, medium or strong sewage effluent. Moreover, the quantity of domestic wastewater depends also on the water supply per capita. The temporal variability of both wastewater flow and concentration of contaminants is due to the habits of community residents and seasonal conditions.

Municipal wastewater often includes both domestic and industrial wastewaters that are collected in the same sewer system. The variable, and sometimes partially unknown, incidence of the industrial component can result in significant variations of wastewater composition, with relevant effects on complexity, effectiveness, and reliability of the reclamation treatment.

1.1.2 Municipal, combined, and dedicated stormwater sewers

For combined sewer systems, municipal wastewater also includes urban stormwater. Both flow rate and pollution level can increase significantly during storm events. If urban stormwater is collected in dedicated sewer systems, it can also be reused. The amount of urban stormwater runoff has progressively increased in recent decades as a result of urban expansion; large areas of vegetated and forested land have been replaced by impervious surfaces. In the past, stormwater was commonly collected for direct water reuse, especially in the countryside, but recently it has been recognized that urban stormwater can be significantly polluted, especially during the first flush, thus requiring specific reclamation processes. Certain land uses and activities, sometimes referred to as stormwater "hotspots" (e.g., commercial parking lots, vehicle service and maintenance facilities, and industrial rooftops), are known to produce high loads of pollutants such as metals and toxic chemicals. Table 1.1 presents the principal pollutants found in urban stormwater and typical pollutant sources.

1.1.3 Industrial wastewater

Most scientific papers, technical guidelines (e.g. the U.S. Environmental Protection Agency (EPA) guidelines for water reuse, EPA, 1992), and published case studies deal with the use of municipal wastewaters. Nevertheless, around 20% of worldwide water production is used in the industrial sector, compared to 7% in the municipal sector

Table 1.1 Principal pollutants found in urban stormwater and typical pollutant sources (adapted from CDEP, 2004)

Stormwater pollutants	Potential sources
Excess nutrients	
Nitrogen, phosphorus (soluble)	Animal waste, fertilizers, failing septic systems, landfills, atmospheric deposition, erosion and sedimentation, illicit sanitary connections
Sediments	
Suspended, dissolved, deposited, sorbed pollutants	Construction sites, stream bank erosion, wash off from impervious surfaces
Pathogens	
Bacteria, viruses	Animal waste, failing septic systems, illicit sanitary connections
Organic materials	
Biochemical oxygen demand (BOD), chemical oxygen demand (COD)	Leaves, grass clippings, brush, failing septic systems
Hydrocarbons	
Oil and grease	Industrial processes; commercial processes; automobile wear, emissions, and fluid leaks; improper oil disposal
Metals	
Copper, lead, zinc, mercury, chromium, aluminum (soluble)	Industrial processes, normal wear of automobile brake linings and tires, automobile emissions and fluid leaks, metal roofs
Organic micropollutants	
Pesticides, VOCs, SVOCs, PCBs, PAHs (soluble)	Residential, commercial, and industrial application of herbicides, insecticides, fungicides and rodenticides; industrial processes; commercial processes
Deicing constituents	
Sodium, calcium, potassium chloride, ethylene glycol, other pollutants (soluble)	Road salting and uncovered salt storage. Snowmelt runoff from snow piles in parking lots and roads during the spring snowmelt season or during winter rain on snow events

PAHs, polycyclic aromatic hydrocarbon; PCBs, polychlorinated biphenyls; SVOCs, soluble volatile organic carbon; VOCs, volatile organic carbon.

(Kretschmer et al., 2002), therefore the potential of using industrial wastewater for irrigation should be carefully evaluated.

Industrial wastewater can come from a single industrial plant or from industrial districts, and its quantity and characteristics are highly variable with respect to the industrial processes involved. The temporal variability of industrial wastewater is mainly due to the phases of the industrial processes (startup, production, cleaning and cyclic maintenance). Moreover, the presence of internal water recycling can affect both the amount and composition of the wastewater.

Wastewater from food processing industries probably presents the highest potential for use in agriculture, as this is often rich in nutrients useful for crop development. Moreover, factories are often located near crops, thus resulting in lower transportation costs for the reclaimed water. Förtser and colleagues (1988, cited in Kretschmer et al., 2002) have studied the impact on soil, plants, and crops irrigated with wastewater from food

Table 1.2 Treatment and composition of wastewater from selected food industries (Kretschmer et al., 2002)

Source of wastewater	Pretreatment	Contaminants	N mg/l	P mg/l	K mg/l
Distilleries	Mechanical purification, neutralization	Alkali, acids, soda, chlorine-compounds	25	1	20
Brewery malting	Mechanical purification, neutralization	Yeast, carbohydrates, settleable solids	40	5	50
Fish processing	Mechanical purification, fat separation, dilution, chlorination, desodoration	Scale, fats, oils, organic acids, salt, H_2O_2	500	—	—
Potato flour	Mechanical purification	None	550	140	95
Canning	Mechanical purification neutralization, desodoration	Salts, organic acids, detergents, corrosive substances	60	10	35
Dairy	Mechanical purification	Disinfectants	35	10	20
Starch	Mechanical purification, neutralization, dilution	Salts, acids	300	45	415
Cider	Mechanical purification, neutralization, precipitation	Detergents	870	160	—
Sugar	Mechanical purification	Strontium, tar, prussic (cyanic) acid	50	10	—

processing industries. Table 1.2 lists some food processing industries from which wastewater may be potentially employed. There is also some evidence of other industrial sources for wastewater use (Into et al., 2004; Chen et al., 2005; Gerhart et al., 2006; Qin et al., 2006; Wang et al., 2006; Galil and Levinsky, 2007).

Compared to municipal wastewater, industrial wastewater presents a higher variability of required reclamation treatments, ranging from the peculiar reuse problems of unpolluted cooling water to the very specific treatments required for heavy polluted and strongly time-varying waters used inside the industrial processes and in washing cycles.

Finally, industrial wastewater is more case-specific, thus dedicated studies for choosing the best reclamation technology are often necessary, which, in turn, increase the complexity and reduce the generality of reclamation projects and concepts.

1.2 Main characteristics of usable wastewater

Typical compositions of untreated domestic wastewater (weak, medium, strong) are reported in Table 1.3. These data can also represent untreated municipal sewages, if the industrial component is not relevant. Table 1.4 provides average pollutant concentrations in urban stormwater. Conventionally treated wastewater presents a lower contamination level; however, secondary effluents can still contain contaminants that are of particular concern for reuse applications. Table 1.5 summarizes the most important quality parameters of wastewater and their significance, especially with regard to agricultural use. As reported in the table, some parameters appear to be beneficial for irrigation purposes (organic matter, nutrients), therefore they should not be removed from wastewater over a certain level. However, their presence in the reclaimed water should be carefully controlled, as they can become harmful for agricultural reuse if not

Table 1.3 Typical composition of untreated domestic wastewater (after Tchobanoglous et al., 2003)

		Concentration		
Contaminants	**Unit**	**Weak**	**Medium**	**Strong**
Solids, total (TS)	mg/l	390	720	1200
Dissolved, total (TDS)	mg/l	270	500	860
Fixed	mg/l	160	300	520
Volatile	mg/l	110	200	340
Suspended solids (SS)	mg/l	120	210	400
Fixed	mg/l	25	50	85
Volatile	mg/l	95	160	315
Settleable solids	mg/l	5	10	20
Biochemical oxygen demand, 5-days, 20°C (BOD_5)	mg/l	110	190	350
Total organic carbon (TOC)	mg/l	80	140	260
Chemical oxygen demand (COD)	mg/l	250	430	800
Nitrogen (total as N)	mg/l	20	40	70
Organic	mg/l	8	15	25
Free ammonia	mg/l	12	25	45
Nitrites	mg/l	0	0	0
Nitrates	mg/l	0	0	0
Phosphorus	mg/l	4	7	12
Organic	mg/l	1	2	4
Inorganic	mg/l	3	5	10
Chlorides	mg/l	30	50	90
Sulfate	mg/l	20	30	50
Oil and grease	mg/l	50	90	100
Volatile organic compounds	µg/l	<100	100–400	>400
Total coliforms	no/100 ml	10^6–10^8	10^7–10^9	10^7–10^{10}
Fecal coliforms	no/100 ml	10^3–10^5	10^4–10^6	10^5–10^8
Cryptosporidum oocysts	no/100 ml	10^{-1}–10^0	10^{-1}–10^1	10^{-1}–10^2
Giadria lamblia cysts	no/100 ml	10^{-1}–10^1	10^{-1}–10^2	10^{-1}–10^3

Table 1.4 Average pollutant concentrations in urban stormwaters (CDEP, 2004)

Constituent	**Units**	**Concentration**
Total suspended solids	mg/l	54.5
Total phosphorus	mg/l	0.26
Soluble phosphorus	mg/l	0.10
Total nitrogen	mg/l	2.00
Total Kjeldahl nitrogen	mg/l	1.47
Nitrite and nitrate	mg/l	0.53
Copper	mg/l	11.1
Lead	mg/l	50.7
Zinc	mg/l	129
Biological oxygen demand	mg/l	11.5
Chemical oxygen demand	mg/l	44.7
Organic carbon	mg/l	11.9
Polycyclic aromatic hydrocarbon	mg/l	3.5
Oil and grease	mg/l	3.0
Fecal coliform	colonies/100 ml	15000
Fecal strep	colonies/100 ml	35400
Chloride (snowmelt)	mg/l	116

Table 1.5 Main water quality parameters and their significance (adapted from Kretschmer et al., 2002)

Parameter	Significance
Total suspended solids (TSS)	TSS can lead to sludge deposition and anaerobic conditions. Excessive amounts cause clogging of irrigation systems. Measures of particles in wastewater can be related to microbial contamination and turbidity. Can interfere with disinfection effectiveness
Organic indicators	Measure of organic carbon.
Total organic carbon Degradable organics (chemical oxygen demand, biological oxygen demand)	Their biological decomposition can lead to depletion of oxygen. For irrigation only excessive amounts cause problems. Low to moderate concentrations are beneficial
Nutrients N, P, K	When discharged into the aquatic environment they lead to eutrophication. In irrigation they are beneficial, nutrient source. Nitrate in excessive amounts, however, may lead to groundwater contamination
Stable organics (e.g. phenols, pesticides, chlorinated hydrocarbons)	Some are toxic in the environment, accumulation processes in the soil
pH	Affects metal solubility and alkalinity and structure of soil, and plant growth
Heavy metals (Cd, Zn, Ni, etc.)	Accumulation processes in the soil, toxicity for plants
Pathogenic organisms	Measure of microbial health risk due to enteric viruses, pathogenic bacteria and protozoa
Dissolved inorganic compounds, total dissolved solids (TDS), electrical conductivity (EC) and sodium adsorption ratio (SAR)	Excessive salinity may damage crops. Chloride, sodium and boron are toxic to some crops, extensive sodium may cause permeability problems

properly supplied. Other contaminants are always detrimental; for some others (e.g. stable trace organic contaminants) negative effects are only suspected and further investigations are required.

1.2.1 Potentially beneficial substances

The main beneficial components for agricultural reuse are nutrients such as nitrogen, phosphorus, potassium, and other elements in minimal concentrations. A low to moderate concentration of biodegradable organic carbon can also be beneficial in controlled situations, specifically in degraded soils. As these components are not required to be removed by treatments above a certain level, the cost for wastewater reclamation can be reduced, with benefits for economic sustainability of reclamation projects.

Nitrogen is found in domestic sewage as a product of human or animal metabolic activity; nitrogen production can be roughly estimated in the range of 9.4–13.8 gN/person/d (Beccari et al., 1993). Untreated domestic sewage contains nitrogen, mainly in organic and ammonia forms; nitrates are found in secondary effluents only after conventional

Table 1.6 Qualitative estimation of nitrogen and phosphorus production by homogeneous industrial processes (Beccari et al., 1993)

Industrial production	N_{org}	NH_3	NO_3	PO_4
Steel		+		+
Cars		+		+
Food		+		+
Bottling				+
Iron			+	+
Cannery	+			
Dairy	+			
Organic and inorganic chemicals				+
Galvanic				+
Plastic materials	+	+	+	+
Paper		+		+
Plaster, asbestos			+	+
Electric power				+
Fertilizers	+	+	+	+
Animal feed				+
Explosives			+	
Nuclear			+	
Slaughterhouses, sausage factories	+			
Corn starch	+			
Aluminum				+
Oils	+			+
Distilleries		+		+
Textiles (synthetic fibers)		+		

oxidation treatments. Some industrial processes can also be a source of nitrogen. Table 1.6 provides a qualitative estimation of organic, ammonia, and nitrate nitrogen by homogeneous industrial processes. However, the typical variability of industrial sources does not allow a numerical estimation of nitrogen production.

Despite the fact that nitrogen is the main nutrient for crop development, an excessive amount may delay maturity and adversely affect harvest quality and quantity. This phenomenon affects some crops more than others and, in some cases, specific periods of their growing cycles (such as delayed ripening in citruses and sugar cane). However, this is not usually a concern for municipal reclaimed water, as nitrogen concentration is commonly insufficient to produce satisfactory crop yields, leading to requirement for supplemental fertilization (EPA, 1992). In any case, it is recommended to continuously control the overall supply of nitrogen so as to match the crop requirements.

Nitrogen mass applied with reclaimed water should not exceed the amount required for crop production, as excess nitrate can rapidly leach below the root zone (Feigin et al., 1991; also see Chapter 5 (5.1), thus contaminating underlying groundwater. Presently, the increasing presence of nitrates in groundwater exploited for human consumption is a cause of great concern.

Phosphorus, like nitrogen, is mostly found in domestic sewage as a product of human and animal metabolic activity, or as a consequence of the use of detergents. Phosphorus production can be roughly estimated within the range of 2.2–4.9 gP/person/d. Domestic sewage contains phosphorus mainly as orthophosphate, but the balance between orthophosphate and polyphosphate can greatly vary during a single day. The presence of organic phosphorus in sewage is very limited (approximately 10% of total P), as it is easily degraded to inorganic forms. Industrial processes are another possible source of phosphorus (Table 1.6), but their contributions are usually difficult to assess because most industrial processes produce phosphorus concentrations and loads that can vary significantly with time, making estimations particularly difficult. Phosphorus in reclaimed water is often lower than crop requirements, and there is no evidence of negative effects on the crop coming from excessive phosphorus in wastewater used for irrigation (Chapter 5 (5.2).

1.2.2 Harmful substances in sewage effluent

Below is a summary of the types of particular contaminants that can be found in domestic and industrial wastewater, with a brief description of their possible adverse effects on crops, soil, irrigation systems, workers, consumers, and groundwater. These issues are covered in detail in various chapters of this volume.

Pathogenic microorganisms (viruses, bacteria, protozoa, and helminth ova) represent the most common threat to the agricultural use of wastewater, both for workers and end consumers of crops. These microorganisms are found in wastewater as a result of the excreta from infected populations. Water disinfection treatments can minimize the pathogenic threat, but disinfection byproducts should be carefully controlled as they can be themselves a risk to both animal and human life. In addition, in the agricultural use of wastewater, a proper irrigation system provides a complementary treatment that significantly limits associated health risk (Chapter 2).

Suspended solids in wastewater occur as a result of the anthropic use and hydrological cycle of water. The main problem of these solids for agricultural reuse of water is the clogging of irrigation systems, related to both the amount and particle size of the suspended solids and the type of the irrigation system (surface irrigation, sprinkler irrigation, surface drip or trickle irrigation, subsurface drip, etc.). Suspended solids, particularly sand, can produce abrasion in the impellers of pumps. When leaves, flowers, and fruits are directly irrigated (e.g. sprinkler irrigation), surface deposits of suspended solids can aesthetically depreciate the production of the crop. Also, because of the possible wind drift of water droplets or aerosols from sprinkler systems, their use with reclaimed water requires special measures like the establishment of buffer zones around the irrigated area or the dropping of the sprinkler nozzles closer to the ground. In addition, the use of sprinkler irrigation of crops destined for raw human consumption is restricted by some regulatory agencies, as it results in the direct contact of reclaimed water with the crops (EPA, 1992).

The presence of salts in wastewater is related to the natural salinity of feed waters; the level of salinity can increase in wastewater due to the contribution of salts by human activities or to evapotranspiration (ET) effects. Compared with irrigation using potable water, the higher salinity of wastewater (mainly from Na^+, Cl^-, and bicarbonate, and also from K^+, Ca^{2+}, ammonium, and sulfate) increases soil solution salinity, lowering its osmotic potential, thus progressively causing the wilting of plants and higher levels of leaching of the soil (Chapter 8). The production of crops is reduced in proportion to the salinity level of the reclaimed water. The higher relative concentration of sodium (and the ensuing sodium adsorption ratio) can produce negative effects on soil structure and permeability, and on seed emergence and plant development. In general, the progressive increase of soil salinity is considered a reason for concern, especially in areas subjected to the risk of desertification, due to the tendency of progressive reduction of soil fertility.

Chloride, sodium and boron can present immediate specific toxicity effects for plants. Phosphorus and heavy metals (Zn, Cu, Ni, Mn, Cd) can have a long-term effect due to their increasing concentration in the soil and bioaccumulation of the toxic compounds and detrimental effects on the soil itself. Accumulation of contaminants mainly occurs where ET is higher, causing leaves to suffer necrosis. Heavy metals can also be toxic for the end consumers of agricultural products (animals and humans): bioaccumulation of heavy metals can occur in fruits and vegetables. Heavy metals, together with nitrates, can pollute the groundwater, if the soil is not able to retain or treat these elements during percolation. Toxic effects depend on concentration of contaminants, crop species, uptake rates by plants, and irrigation systems. Heavy metals are toxic only if their presence in wastewater is related to industrial discharge; at low concentrations, as found in domestic wastewater, heavy metals may even be useful for crops.

Boron is an essential microelement for crops that has a small concentration window between deficiency and toxicity. Its affinity to soil particles is higher than many other anions, thus it is not easily leached from soils. Boron in wastewater comes from several human uses, such as glass and ceramics production, detergents, bleaches, fire retardants, disinfectants, alloys, specialty metals, preservatives, pesticides, and fertilizers. The main concern from boron-related loss of fertility of agricultural soils is from irrigation with brackish water; however, wastewater use in arid and semiarid soils also involves risk from toxic levels of boron accumulation in soil. This was dealt with in Israel by a unique agreement signed between the Ministry of Environmental Protection and the detergent industry to gradually phase out the use of boron (and sodium) in detergents. The outcome was a marked decrease of boron concentrations in wastewater observed in recent years (Table 1.7).

Stable trace organic contaminants include chemicals with proven endocrine-disrupting effects, such as polycyclic aromatic hydrocarbons, alkylphenols, organotins, brominated flame retardants, etc. Similar harmful contaminants are pharmaceuticals, plasticizers, pesticides, and degradation products of some detergents. As biological processes do not do

Table 1.7 Boron measured in the sewage of three main cities in Israel (data derived from Weber and Juanicó, 2004)

	Boron in sewage		
Year	**Haifa ppm**	**Tel Aviv ppm**	**Western Jerusalem ppm**
1998		0.650	
1999	0.530	0.590	0.655
2000	0.535	0.575	0.630
2001	0.410	0.436	0.357
2002	0.283	0.215	0.300

much to remove these contaminants, they can be found both in untreated wastewater and secondary effluents. Their presence in wastewater is due to both industrial and domestic activities.

Finally, as mentioned above, even beneficial components such as nitrogen, phosphorus, and organic matter, if occurring in excess, can be harmful to crops, soil, surface water, and groundwater.

1.2.3 Quantity considerations

Managing and allocating reclaimed water supplies may significantly differ from management of traditional sources of water. For example, groundwater or surface impoundments are, at the same time, both a water source and a storage facility that can be exploited by water demands. However, wastewater is continuously generated and, if not immediately used, needs to be stored or disposed of. Depending on the volume and pattern of projected water demands, seasonal storage, translocation, and pumping requirements can become a significant design consideration, and have a substantial impact on the capital cost of the system (Fine et al., 2006). Seasonal storage systems will also impact operational expenses. This is particularly true if the quality of the water is degraded in storage by algae growth and pretreatment is required to maintain the desired or required water quality. Alternatively, economic restrictions (which are associated also to the actual need for water) may prevent construction of designated reservoirs, confining the reclamation operation to the irrigation season only, and disposing of the wastewater for the rest of the year.

The need for seasonal storage in reclaimed water programs generally results from one of the two following requirements. First, storage may be required during periods of low demand for subsequent use during higher demand periods. Second, storage may be required to reduce or eliminate the discharge of excess reclaimed water into surface water. This can happen in particular situations, for instance, when the agricultural use of wastewater is adopted also to control the disposal of nutrients to eutrophic water bodies. These two needs for storage are not mutually exclusive, but different parameters are considered when developing an appropriate design for each. Where resource management is the primary consideration, rather than pollution abatement, reclaimed water supply and user demands must be calculated, and the most cost-effective means of allocating that resource must be determined.

When reclaimed water is viewed as a resource or commodity, the users' needs must be anticipated and accommodated in a similar manner to potable water supplies. In short, the supply must be available when the consumer demands it. The so-called concept of safe yield expresses the idea that the provider must guarantee the availability of the water source. It is commonly applied to surface water bodies in assessing available potable water supplies. Conversely, in the case of reclaimed water use, typical schemes of agreement adopted to now between providers and users avoid a guarantee of continuous delivery. This is done primarily to allow for interruption of service in the event of treatment plant upsets, but allowances for shortages have also been included, in some cases, to protect the providers against obligations that can become excessively onerous.

Presently this concept is changing, primarily in developed countries. As water reuse is assuming a greater role in conserving potable supplies, reclaimed water is becoming a commodity, and water reuse systems are emerging as a new utility, the considerations of safe yield become necessary also for reclaimed water considered as a source (EPA, 1992). The concept of "safe yield", then, is being progressively introduced in such situations for reclaimed waters.

Where water reuse is being implemented to reduce or eliminate wastewater discharges to surface waters, state or local regulations usually require that adequate storage be provided to retain excess wastewater under a specific return period of low demand. In some cold climate areas, storage volumes may also be specified according to projected non-application days due to freezing temperatures.

1.3 Wastewater treatments

Hereafter, we will discuss different methods for treating wastewater, with special attention to reclamation purposes for agricultural use. For a comprehensive description of various treatment technologies refer to, for example, Tchobanoglous et al. (2003), whereas biological treatments are discussed in detail in Grady et al. (1999).

A traditional classification of wastewater treatment stages (which is presented in the following sections) catalogs them as preliminary, primary, secondary, and tertiary, with a further quaternary class that has been introduced recently. The scheme is centered on the main treatment, the secondary one, which is, in most cases, biological. The secondary treatment is typically necessary to meet the standards, set in Europe and most developed countries worldwide, to discharge into surface water. Hence, in developed areas, most urban areas, industries, and farms are already served by wastewater treatment plants (WWTPs) including at least preliminary, primary, and secondary treatment stages.

As the reclamation of wastewater to be reused is a relatively new concept in Europe, almost all medium- and large-scale schemes for wastewater reclamation are derived from upgrading conventional secondary treatment processes, including nutrient removal and further treatments. Some authors (Bixio et al., 2006) anticipate that, as has occurred in many similar situations in the past, EU environmental regulations will be taken as reference by other countries in the future and will become a sort of de facto world reference. Hence, they assume that the configuration of reclamation typical to the EU will become common elsewhere too.

New alternatives to the traditional biological-centered scheme may also become viable in the future, but most authors consider this possibility feasible only in the long term (de Koning et al., 2008).

Finally, there are also some cases where existing processes required for effluent disposal can be avoided when the wastewater is reused instead; this typically happens with the removal of nutrients in the agricultural use of reclaimed water, where the nutrient content of the reclaimed water can assume, after a proper assessment, a beneficial role for the irrigated crops.

1.3.1 Classification of treatment stages

As mentioned above, the possible stages of wastewater treatment are commonly classified as preliminary, primary, secondary and tertiary, but more recently, the increased utilization of membrane treatment to obtain very high-quality reusable water has led to introduction of the concept of quaternary treatments. Although it is inadequate to classify the wide range of treatments that modern techniques have made available, this scheme should be mentioned because of its consolidated use in the description of the treatment train for wastewater disposal or use.

The preliminary treatments include mechanical operations aiming to easily remove materials that can interfere with subsequent stages, like sand, grit, oils, greases, and other non-decomposable solids or liquids separable with raw screening, filtration, sedimentation, or flotation processes.

The term "primary" is reserved for primary sedimentation, which separates the primary sludge to be treated by means of a specific (aerobic or anaerobic) stabilization stage. Whereas the preliminary treatments separate inert or quite stable material from wastewater, primary sedimentation separates sludge with a high content of biodegradable organics, which requires a specific aerobic or anaerobic stabilization stage. The reason for having separate phases for preliminary and primary treatments is that these provide the opportunity to have separated fluxes of inert and decomposable sludges. This is particularly relevant when considering that anaerobic stabilization of primary sludge, being able to produce energy, is considered an additional resource rather than simply a required additional disposal treatment.

The secondary treatments include core removal operations aiming to reach, in most cases, the quality standards to return used water to the natural environment. As this classification was originally introduced for urban wastewater, the secondary stage typically includes biological treatments such as oxidation ponds, activated sludge, or trickling filters. However, in some situations, for example industrial treatments or temporary or compact alternatives for domestic wastewater treatments, chemico-physical treatments such as coagulation-flocculation-clarification, are mentioned as alternative options to biological secondary treatment. The secondary clarification is commonly considered part of the secondary stage.

The tertiary treatments include subsequent operations necessary to further increase removal levels obtained by biological secondary treatments.

When the plant treats wastewater to dispose into natural bodies, typical tertiary treatments include a disinfection stage, and, in some cases, removal of nitrogen and phosphorus. For removal of nitrogen, the most popular options (e.g. the anoxic-aerobic activated sludge process) are actually performed biologically by way of specific enhancements of the secondary biological stage rather than by a proper tertiary treatment. For phosphorus, biological removal, although possible, is not common, instead chemical-physical removal can provide both an enhancement of the secondary stage or a proper tertiary treatment, the latter option being preferred when a higher efficiency is required. As stricter discharge standards that require efficiency and reliability of the removal of solids higher than that achievable by the mere secondary clarifier are being introduced in the most advanced countries, tertiary filtration and/or chemical-assisted sedimentation are also tending to occur.

In most wastewater reclamation projects, enhancement of the tertiary stage of existing WWTPs is the most popular treatment option. Depending on the quality standards required for the reclaimed water, chemical-assisted sedimentation, filtration, activated carbon adsorption, microfiltration or ultrafiltration (UF) treatments can be included, often coupled with a stricter disinfection stage.

When very high-quality reclaimed water is required, the introduction of high-density membranes has led to the development of quaternary treatments that allow removal of almost all the compounds that can possibly occur in wastewater, including small ions. However, such treatments (including nanofiltration and reverse osmosis) require a thorough pretreatment of the wastewater to inhibit fouling and guarantee more stable operation of the membranes. It became evident that the traditional tertiary sand or activated carbon filtrations, even when followed by fine cartridge filtration, are insufficient pretreatments for nanofiltration (NF) or reverse osmosis (RO). Hence, traditional filtration is progressively being substituted by "dual membrane" treatments, that couple a first stage of larger pore membrane filtration (microfiltration, MF, or UF) followed by NF or RO.

1.3.2 Preliminary and primary treatments

The preliminary and primary treatments aim to remove materials that can be easily separated from raw wastewater and disposed of, such as fats, oils, grease, sand, gravel, stones, grit in general, larger settleable solids, and floating materials.

1.3.2.1 Removal of large objects from influent sewage

In mechanical treatment, the influent wastewater is strained to remove all large objects that are deposited in the sewer system, such as rags, sticks, condoms, sanitary towels (sanitary napkins) or tampons, cans, fruit, etc. This is most commonly done with a manual or mechanically automated raked screen. This type of waste is removed because it can damage or clog the equipment in the sewage treatment plant.

The present tendency is towards reduction of the characteristic screening cut-off (from the current 5 mm to 2 mm or less), using special mechanical equipment like vibrating or rotating sieves with high pressure back-washing.

1.3.2.2 Sand and grit removal

Primary treatments typically include a sand or grit channel or chamber in which the velocity of the incoming wastewater is carefully controlled (by hydraulic control, mechanical mixing, or air injection) to allow sand, grit, and stones to settle, while keeping the majority of the suspended organic material in the water column. This equipment is called a detritor or sand catcher. Sand grit and stones need to be removed early in the process to avoid damage to pumps and other equipment in the following treatment stages. Sometimes there is a sand-washer (grit classifier) followed by a conveyor that transports the sand to a container for disposal.

1.3.2.3 Flotation

Low-density materials like oil, grease, and fats should also be removed to avoid problems in the subsequent treatment phases. The treatment is usually gravimetric, but involves skimming the material accumulated at the surface, instead of removing the solids settled at the bottom. Often this treatment is coupled with grit removal (especially in the case of air injection, which significantly improves the upwards motion of low-density substances) using skimming equipment at the surface and grit removal at the bottom of the same tank.

Maximum flotation efficiency is obtained by injection of very small air bubbles. In the case of dissolved air flotators (DAF), this is obtained by injection of a secondary high-pressure water flux saturated with air. Preliminary chemical coagulation-flocculation can further enhance flotation efficiency. The high efficiency of the DAF system in removing suspended solids also allows it to be used as a tertiary treatment.

1.3.2.4 Primary sedimentation

Many plants incorporate a primary stage in large tanks, commonly called primary "clarifiers", "settlers" or "sedimentation tanks", where the very low flowing speed allows fecal solids to settle and light material to float.

Prior to an activated sludge secondary stage, the main purpose of primary sedimentation is to reduce the organic load on the secondary stage itself (thus extending its overall capacity by reducing the retention time), and produce a homogeneous and solid-free wastewater suitable for biological treatment and a sludge flux that can be separately treated or processed. Prior to an attached growth secondary stage (see Section 1.3.3), the main function of the primary settler is to remove solid particles to avoid the risk of early clogging of the support media of the secondary reactor. There are also cases where primary settlement is present as the only (partial) treatment; in this case, sometimes accepted for very small treatment plants, the settler is coupled with a tank for storing and stabilizing the separated sludge (Imhoff tank).

Primary settlers are usually equipped with mechanically driven scrapers that continually drive the collected sludge towards a hopper in the base of the tank, and the floating material to a skimmer on the surface from where they can be pumped to further sludge treatment stages.

As biological treatment usually includes a further sedimentation stage (the secondary clarifier), primary sedimentation is omitted in some cases. Specifically, the activated sludge process, because of its relative insensitivity to clogging, does not strictly require primary sedimentation, and, in the case of biological nitrogen and/or phosphorous removal enhancements, primary settling may even prove a disadvantage. On the other hand, primary sedimentation is often included in activated sludge plants when the produced sludge is being stabilized by anaerobic digestion with biogas production for energy recovery. This is because sludge from primary settlers produces much more biogas than sludge from secondary clarifiers.

1.3.3 Secondary biological treatments

Biological treatments (usually termed secondary stage) are designed to substantially degrade the biological content of sewage deriving from human and food waste, soaps, and detergents.

Even if some chemical alternatives are possible, biological processes eminently represent the most common treatment for civil and for many industrial and intensive livestock breeding wastewaters, when they have to be disposed back into the environment after use. The biological process usually requires some finishing treatments to be reused, even in agriculture.

Biological treatments can be implemented by way of technological (mechanical) facilities, aimed at reducing as much as possible the required area for the plant. This is the most traditional approach, presented in this section and the following subsections. An alternative, that is becoming more and more attractive, forsakes the goal of minimizing the plant area, pursuing the goal of being more environmentally friendly instead. This class of treatments, which is usually considered best suited for small to medium size plants, is presented in section 1.3.8.

The mechanical-biological processes can be aerobic or anaerobic: the latter are sometimes used, usually followed by a final aerobic finishing stage, in wastewater with a very high organic load, providing an opportunity to produce biogas that is exploited for energy production. However, anaerobic treatments are mostly reserved for biological stabilization of primary and secondary sludge as an alternative to aerobic stabilization. In most cases biological treatment of wastewater is aerobic, sometimes completed with anoxic or anaerobic stages to enhance nutrient removal.

1.3.3.1 Secondary mechanical-biological process: suspended and attached growth options

All biological treatments rely on a biomass, mostly composed of saprophyte bacteria, that is concentrated in a reaction tank by recirculating a settled sludge (suspended growth processes) or allowing a biofilm to grow on specific solid supports of various types and shapes (attached growth processes). The biomass eats the organic substrate of the fed effluent, thus removing it. The eaten substrate is then metabolized partly to the final stable form of CO_2, which is freed into the atmosphere, partly to grow the biomass itself. A small part of biomass is finally removed from the process by a final settler, filter or membrane (depending on the specific process). A number of different techniques are also used to feed the large quantity of oxygen required by the high-concentration biota, with various process options, as well as techniques (stirring, percolation, filtration), to allow the best contact between the bacteria and the substrate to be treated.

The most common treatment, to date, is the so-called activated sludge process, belonging to the suspended growth class, whereas several fixed film alternatives are usually reserved for specific situations.

Because the list of aerobic mechanical-biological process alternatives is extremely long, a complete discussion is impossible in this chapter, thus only the principal treatments are briefly introduced.

Activated sludge

Activated sludge plants encompass a variety of mechanisms and processes that use dissolved oxygen to promote the growth of biological flocks that substantially remove organic biodegradable matter and, also, by ancillary processes like flocculation and adsorption, lower fractions of non-degradable organic and even inorganic compounds.

In its basic form, an activated sludge plant is composed of an aeration tank, where the wastewater is mechanically aerated and mixed, and a final settler that separates the treated clarified effluent from a sludge flux that is mainly returned by a pump to the aeration tank

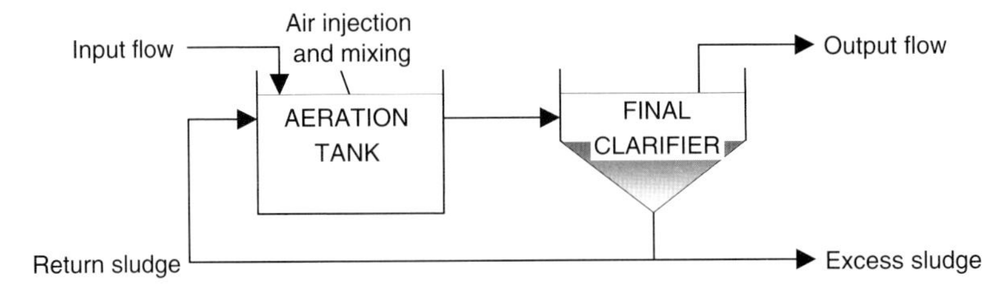

Figure 1.1 Basic configuration of an activated sludge plant.

itself (Fig. 1.1). A small fraction of the sludge (excess sludge) is extracted daily as a subproduct of the process and disposed of after specific stabilization and dewatering. The excess sludge from domestic plants usually has properties that allow, in controlled conditions, its use as an agricultural soil amendment.

Several shapes of aeration tank are possible, as well as alternative equipment for aeration and mixing, with widely ranging efficiency, compactness, and cost. Many enhancements are available as well, usually involving multiple stages, also including anoxic and anaerobic phases, aiming to increase efficiency or enhance removal of nutrients. These enhancements have the drawback of being higher complexity, making them especially suitable for large plants (Fig. 1.2).

Another variant receiving increasing interest today is the so-called membrane biological reactor (MBR), in which a low-density membrane filtration stage is set as a substitute of the final settler, with several advantages in terms of increased efficiency and compactness. This technique is discussed later, together with the other membrane filtration processes (Section 1.3.7.3).

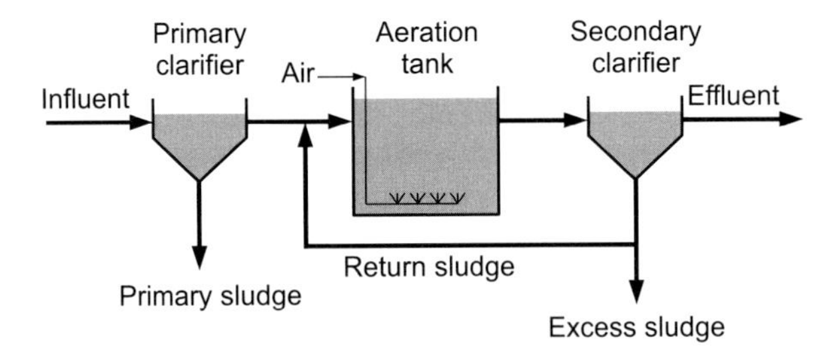

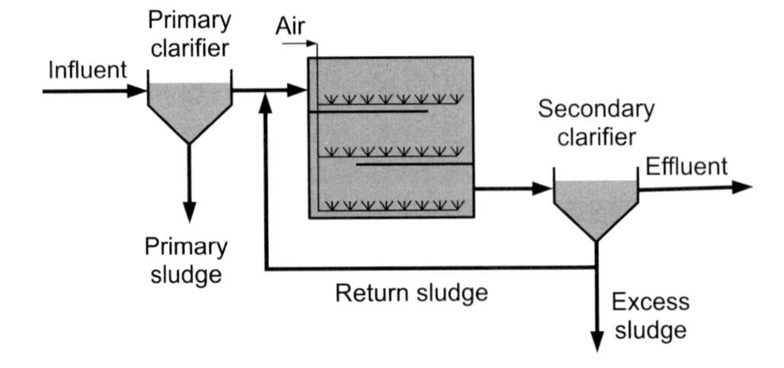

Figure 1.2 Description of activated-sludge processes for biochemical oxygen demand (BOD) removal and nitrification (classification from Tchobanoglous et al., 2003).

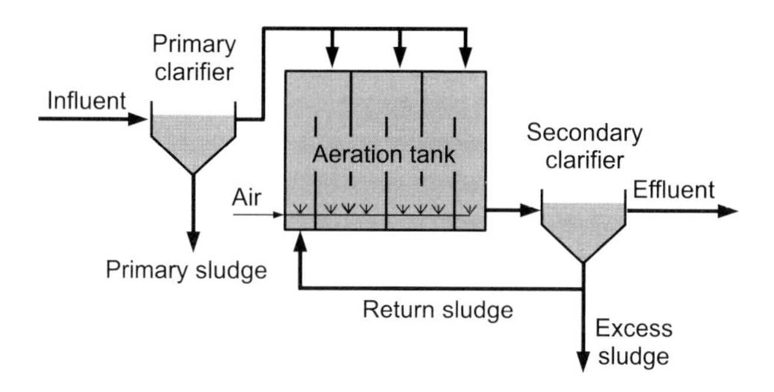

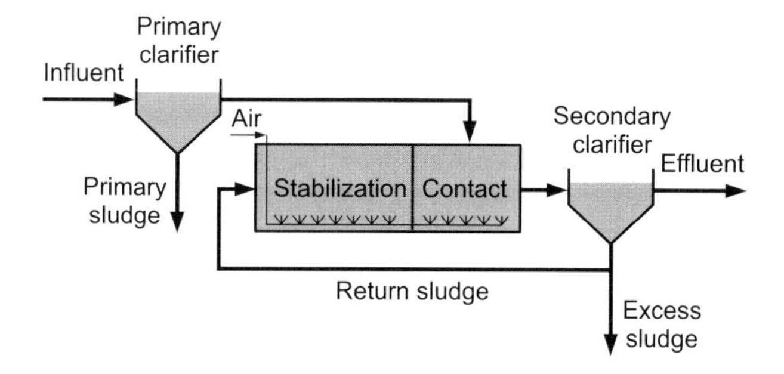

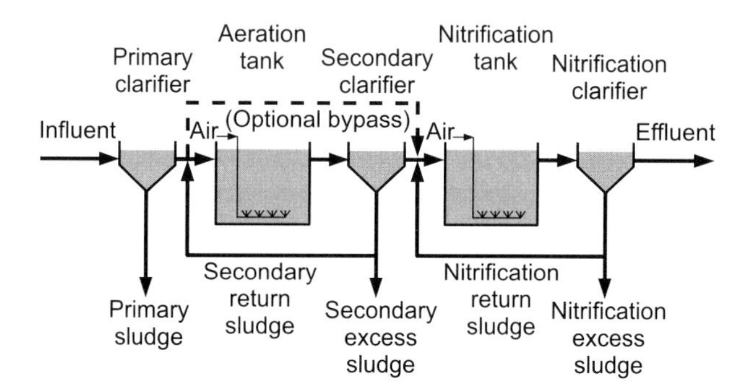

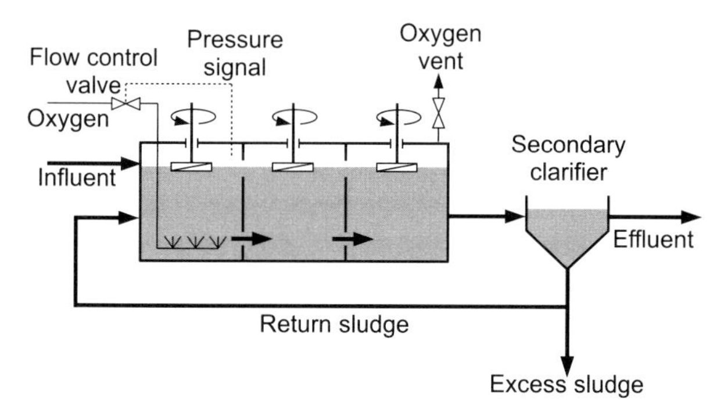

Figure 1.2 (*Continued*)

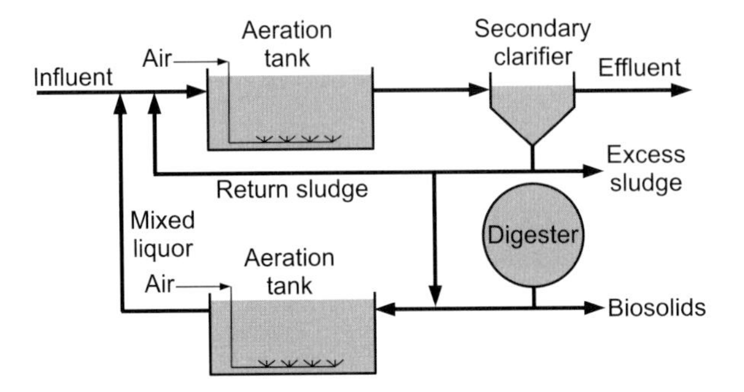

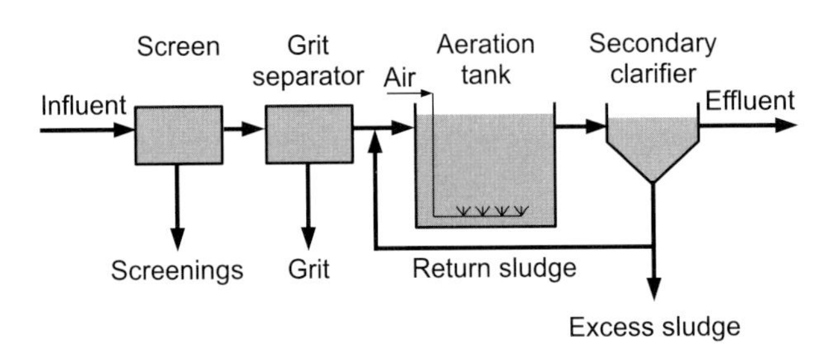

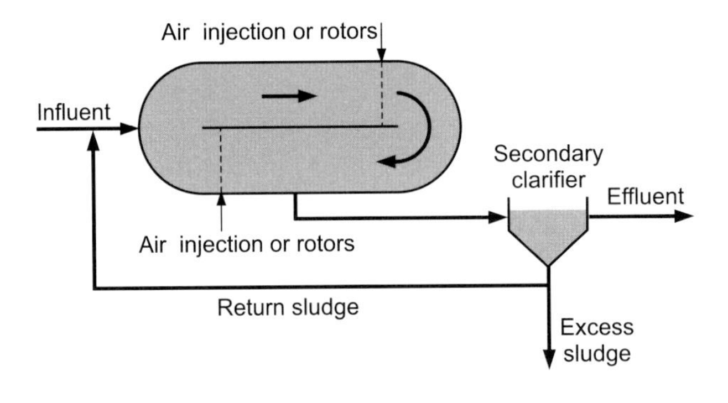

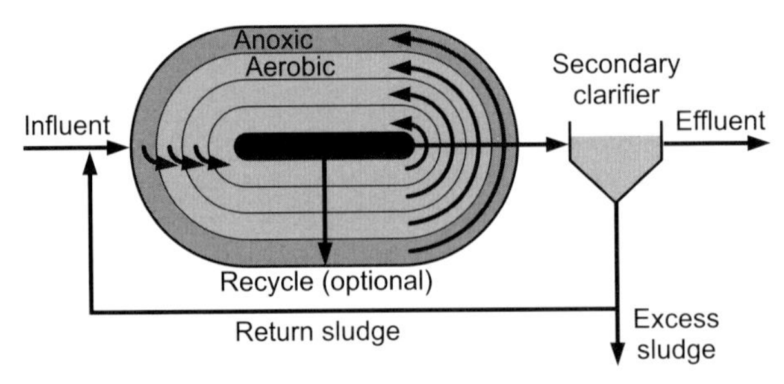

Figure 1.2 *(Continued)*

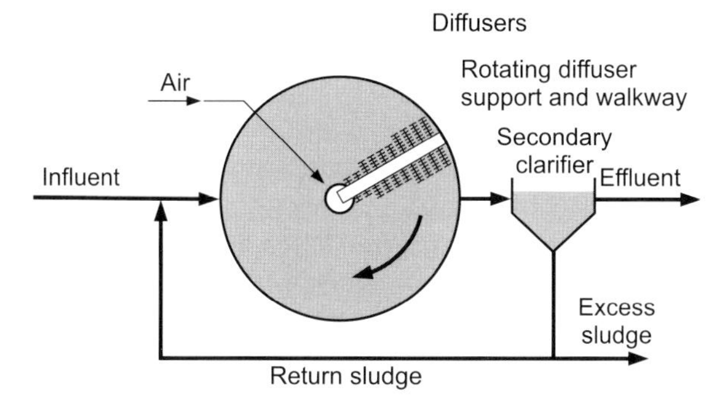

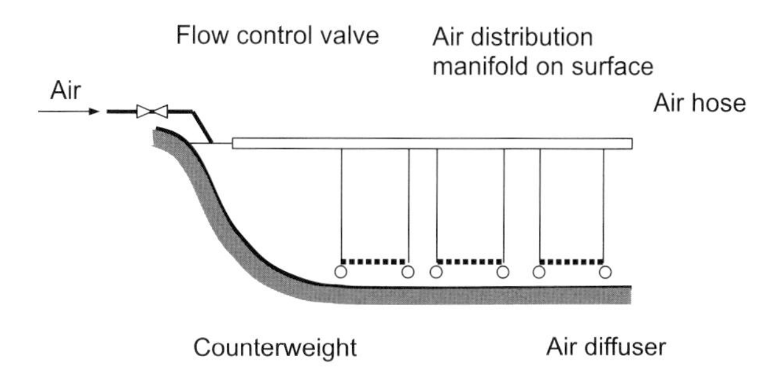

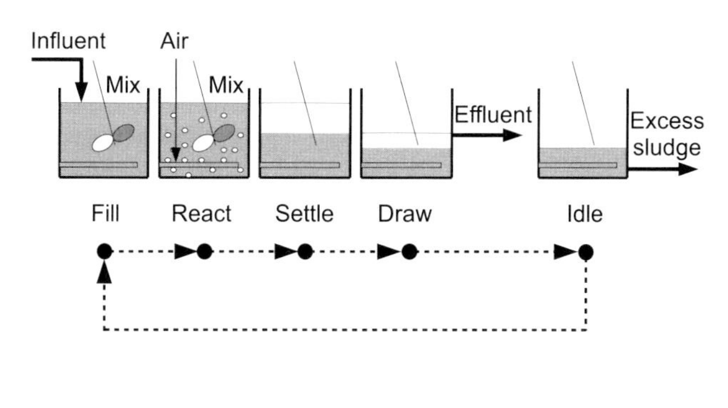

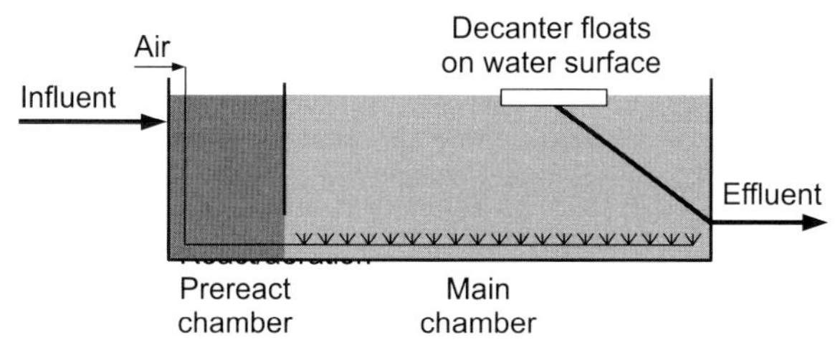

Figure 1.2 *(Continued)*

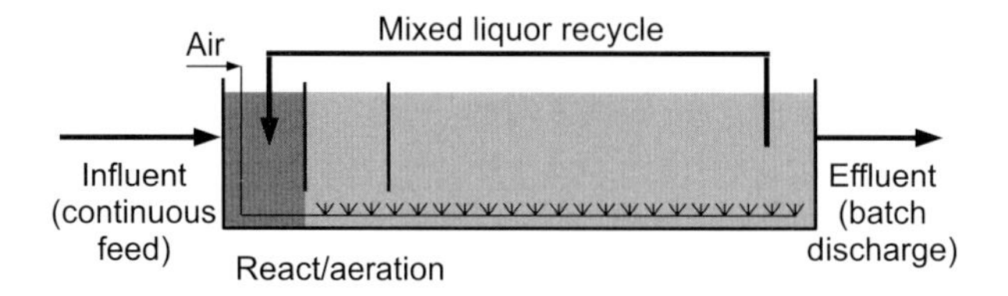

Figure 1.2 *(Continued)*

As with all biological oxidation processes, the activated sludge method is sensitive to temperature, with the rates of biological reactions increasing with temperature in the range between 0°C and 40°C. Most surface aerated vessels operate at between 4°C and 32°C.

The efficiency of the activated sludge process is recognized to be generally higher than the alternative attached growth processes presented in the following sections. This is mainly because this process is able to select bacterial species that can form flocks of biological sludge (which are called "activated sludge", giving the name to the process itself), which present, in normal conditions, high settling speed. These flocks are able to exercise some ancillary processes of flocculation and adsorption, which increase the removal efficiency of the process even for non-biodegradable organics or inorganic substances. In particular, this process is effective in the removal of colloidal solids, which are difficult to remove using the alternative attached growth processes.

The main weakness of this process is in the relative instability of the flock formation, which involves, in the presence of several anomalies that are sometimes difficult to investigate, the establishment of sludge illnesses (like the so-called bulking, rising, and pin-point effects). These reduce the settleability of the flocks of sludge, causing significant reductions in the overall efficiency of the process. To avoid this risk, accurate and competent management is required, making this process unsuitable for very small plants.

Trickling filters

The most popular attached growth treatment, especially in older plants, is the so-called trickling filter (Fig. 1.3), where the settled sewage water is spread onto the surface of a deep bed made up of cobbles (average diameter of 6–8 cm), or specially fabricated plastic

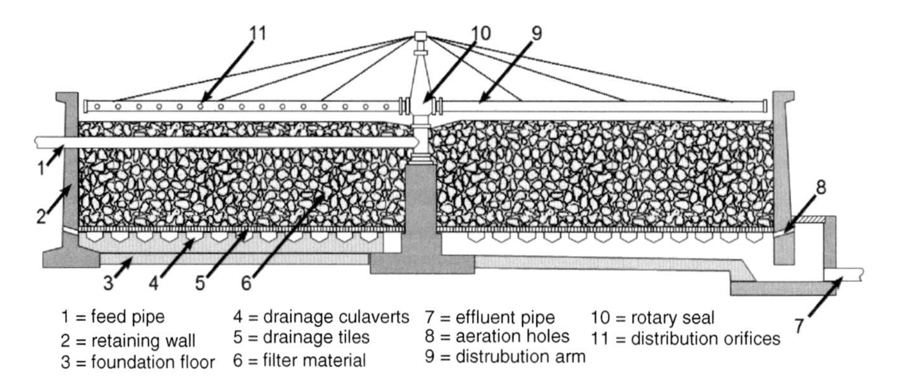

Figure 1.3 Traditional trickling filter.

media with high surface areas to support the biofilm that forms. The water is distributed through perforated pipes, sometimes installed on radial branches rotating along a central pivot. The distributed water trickles through this bed and is collected in drains at the base. These drains also provide a source of air that flows up through the bed, keeping it aerobic. A biological film of bacteria, protozoa, and fungi grows on the media surfaces, feeding on the biodegradable organic content of the spread liquor, thus removing it. In normal operation, the biofilm grows continuously and is detached after reaching a certain thickness. This detached material is transported by the wastewater flux and is finally separated by the subsequent secondary settler-clarifier. Organic overloading of beds or insufficient hydraulic flux can excessively increase the thickness of the film, leading to clogging of the filter media and ponding on the surface (EPA, 2000).

Trickling filters are less efficient than activated sludge, especially for colloidal components. Their process operation cannot be adjusted as easily as for activated sludge in response to loading variations. In particular, the nutrient (nitrogen and phosphorus) control in these systems is limited. On the other hand, trickling filters present lower operative costs because natural aeration allows for saving on mechanical aeration costs. Another advantage is the general avoidance of the sludge settling decay problems described for the activated sludge treatment.

An interesting field of application is the pretreatment of high organic load industrial wastewater (Eckenfelder, 2000). The trickling filter pretreatment allows for savings on costs for mechanical aeration, due to the natural aeration typical of this system, and a final activated sludge treatment allows exploitation of the higher overall removal efficiency of that treatment.

Rotating biological contactors

Rotating biological contactors (RBCs) are mechanical secondary treatment systems, which are robust and capable of withstanding surges of organic load. The system is constituted of horizontal packs of thin plastic disks slowly rotating around a horizontal axis, with 40% of their diameter immersed in the treated wastewater. Bacteria and microorganisms present in the sewage can become attached to the surface of the disks, breaking down and stabilizing organic pollutants. The necessary oxygen is obtained from the atmosphere as the disks rotate. As the microorganisms grow, they build up on the media until they are sloughed off due to shear forces provided by the rotating disks in the sewage. Effluent from the RBC is then passed through final clarifiers where the microorganisms in suspension settle as sludge. The sludge is withdrawn from the clarifier for further treatment. Compared with activated sludge, RBCs have the same advantages and disadvantages as trickling filters, with much lower hydraulic head losses. However, this system incurs higher costs because of the strict requirement of covered tanks for proper operation.

Biofilters

Biological Aerated and Anoxic Filters (also called Biofilters or BAF) combine filtration with biological carbon reduction, nitrification, and denitrification. Biofilters usually include a reactor filled with a filter media. The media are either in suspension or supported by a gravel layer at the foot of the filter. The dual purpose of these media is to support the highly active biomass attached and to filter suspended solids. Carbon reduction and ammonia conversion occur in aerobic mode and are sometimes achieved in a single reactor, whereas nitrate conversion occurs in anoxic mode. Biofilters are operated either in up-flow or down-flow configuration, depending on the design specified by the manufacturer. Unlike other biologic treatments, this system does not require a final settler as the sludge is retained in the granular media and periodically removed by back-washing. A preliminary settler is required instead, that also can be used as thickener for sludge from back-washing. In combination with lamellar primary settlers, this system allows very compact installations, at the cost of higher requirements of mechanical equipment.

Moving bed bioreactors
Moving bed bioreactors (MBBR) combine the advantages of suspended growth and attached film systems, adding, in conventional activated sludge tanks, plastic support media that allow specific biofilm to grow attached to them. These media are retained in the tanks by screens, allowing specialized biofilm to grow in specific phases of the process without passing through final settler and return-sludge pumping. This yields higher efficiencies, especially when carbon, nitrogen, and occasionally phosphorus removal are joined in multiple stages. This system is also suited to retrofit activated sludge for increased organic loads.

1.3.3.2 Secondary sedimentation

The final step of most secondary treatments is to settle out the biological activated sludge or detached biofilm, so as to produce a clarified effluent with very low levels of organic material and suspended solids. This is done in the secondary clarifier-settler. In the activated sludge process, it also assumes the very important role of thickening the return sludge to be recirculated to the reactor to guarantee the high operative concentration of activated sludge.

1.3.3.3 Multi-stage treatments

The previously mentioned treatments can be combined in multiple stages to increase overall removal efficiency. For the activated sludge system, the single sludge multi-stage combination presents a single final settler, whereas the multiple sludge version requires a specific settler for each stage, obtaining increased efficiency at the cost of a more complicated plant. In attached growth systems, a single settler is required in any case, thus allowing simpler multi-stage configuration with enhanced efficiencies. For this reason, a three-stage process is considered standard for RBC systems.

Combinations of trickling filters, RBC, or even anaerobic systems with a final activated sludge stage are often proposed for high organic load effluents to reach a high overall efficiency at lower operative costs.

1.3.3.4 Package plants and batch reactors

In order to use less space, treat difficult waste, deal with intermittent flow, or achieve higher environmental standards, a number of hybrid treatment plant types have been designed. Such plants often combine all, or at least two of the three, main treatment stages into one combined stage. In cases where a large number of sewage treatment plants serve small

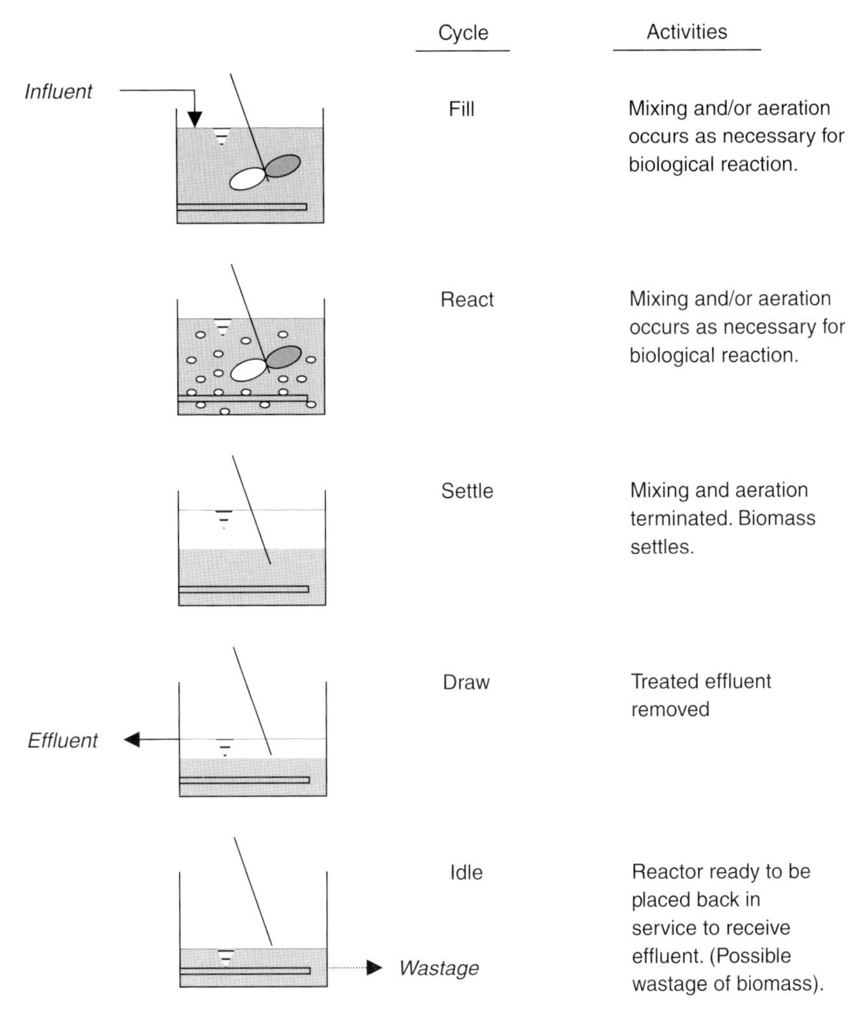
Cycle
Activities
Influent
Fill
Mixing and/or aeration occurs as necessary for biological reaction.
React
Mixing and/or aeration occurs as necessary for biological reaction.
Settle
Mixing and aeration terminated. Biomass settles.
Draw
Treated effluent removed
Effluent
Idle
Reactor ready to be placed back in service to receive effluent. (Possible wastage of biomass).
Wastage

Figure 1.4 Sequence of treatment phases in a sequencing batch reactor (adapted from Grady et al., 1999).

populations, package plants are a viable alternative to building discrete structures for each process stage.

In the sequencing batch reactors (SBR), the treatment stages found in the activated sludge system (aeration, anoxic mixing, anaerobic mixing, sedimentation) are accomplished as sequencing batch phases in a single tank by activating and deactivating specific electromechanical equipment under computer-assisted timing control (Fig. 1.4). These plants are typically best suited for small installations (the US EPA factsheet for the SBR technology indicates a typical maximum input flow rate of 18900 m^3/d; EPA, 1999), where better performance can be achieved than in traditional activated sludge. However, some large-scale applications are also present worldwide. The typical drawback of sensitivity to timing errors is progressively decreasing with the increasing scientific knowledge of the process and availability of reliable monitoring instrumentation at affordable costs.

1.3.4 Nutrient removal

As discussed in Section 1.2.1, the agricultural use of wastewater is often considered a way to rationally exploit its nutrient content for beneficial fertilization of irrigated crops. Thus, specific removal of nitrogen and phosphorus is not usually considered a mandatory treatment in wastewater reclamation for agricultural use.

Nevertheless, the great concern about the risks related to overfertilization in sensitive water basins requires a conscious comparative analysis of nutrients available in reused water and required by the irrigated crops.

In some cases, control is advisable over nutrients in specific forms. Ammonia nitrogen removal may be required not only for eutrophication control, but also because of its toxicity to aquatic organisms, or for the relatively high biological oxygen demand. Conversely, nitrate control may be necessary where reclaimed water reaches potable water supply aquifers or for agricultural irrigation of certain crops, such as sugar cane and corn, during specific periods of their growing cycle (EPA, 1992).

Moreover, some specific requirements in terms of nutrient control come from the sensitivity of specific crops, the seasonally changing irrigation water requirements, or the need of simple disposal of wastewater in certain seasons when irrigation is not required. In such cases, nutrient control in treated wastewater can become necessary, especially in eutrophication-sensitive basins, for specific periods of the year.

Based on these considerations, a discussion of the most common nutrient removal techniques follows, with a special focus on their capability to easily undergo periodical startup and stop transients.

1.3.4.1 Nitrogen removal

For removal of nitrogen, biological removal techniques are largely prevalent over chemical methods, for both efficacy and cost reasons. Biological removal can be performed by mechanical-biological techniques, usually implemented as specific enhancements of the activated sludge process, or by natural treatments, like lagooning or constructed wetlands, usually implemented as tertiary treatment stages.

The most common mechanical-biological method for nitrogen removal from wastewater is effected in two stages, through the biological oxidation of nitrogen from ammonia to nitrate with the intermediate step of nitrite (nitrification), followed by its reduction to nitrogen gas (denitrification) to be released to the atmosphere and thus removed from the water. Other forms of nitrogen in wastewater do not usually require specific removal treatments as they are easily converted to ammonia (organic nitrogen) or nitrates (nitrite nitrogen) during standard biological processes, or removed in particulate form by sedimentation.

The first of the two stages, nitrification, can be performed by way of many suspended and attached growth treatment processes designed to foster the growth of nitrifying bacteria. In the activated sludge process, nitrification is accomplished by designing the process to operate at a solid retention time that is long enough to allow the full development of the slow-growing nitrifying bacteria. Nitrification will also occur in trickling filters that operate at low biochemical oxygen demand (BOD)/total Kjeldahl nitrogen (TKN) ratios of the influent, either in combination with BOD removal, or as a separate advanced process, following any type of secondary treatment. Usually, multi-stage processes reach a much higher efficiency. A well designed and operated nitrification process will produce an effluent containing $\leq$1.0 mg/L ammonia nitrogen.

The second stage, biological denitrification, is coupled with nitrification to complete the process converting nitrates to gaseous N_2, which escapes into the atmosphere. During this process, nitrates are used by a variety of heterotrophic anaerobic-facultative bacteria as the terminal electron acceptor in the absence of dissolved oxygen (anoxic conditions). A carbonaceous food source is also required by the bacteria in these processes.

Biological denitrification can be performed by a number of processes. These include specific enhancements of most suspended growth and some attached growth treatment processes, provided they are designed to create the proper conditions. Specifically, the treatment line must include a stage containing nitrate nitrogen, a carbon source, and facultative-anaerobic heterotrophic bacteria, in the complete absence of dissolved oxygen (DO). This kind of reactor is called anoxic. As an alternative, a single reactor with low DO concentration or alternate aerobic-anoxic conditions allows simultaneous nitrification-denitrification.

As nitrification consumes alkalinity, thus resulting in pH decrease, whereas denitrification produces alkalinity, coupling of the two stages can, in most cases, avoid a specific chemical pH control of the overall process. Biological denitrification processes can be designed to achieve effluent nitrogen concentrations between <2 mg/L and 12 mg/L nitrate nitrogen. The outlet total nitrogen will be somewhat higher depending on the concentration of suspended and soluble (<0.45 μm) organic nitrogen.

Ammonia nitrogen also can be removed from effluents by several chemical or physical treatment methods, such as air stripping, liming, ion exchange, RO and break-point chlorination, whereas the options for nitrates are less extensive, being practically limited to the single RO process, which involves significantly higher costs. All these methods have proven to be presently uneconomical or too difficult to operate in most municipal applications (Tchobanoglous et al., 2003).

Other possible nitrogen control options range in the field of natural treatments (see discussion in Section 1.3.8) include tertiary aerobic-anoxic lagoons or constructed wetlands with floating or rooted macrophytes. Complete natural secondary-tertiary treatments based on multi-stage horizontal-vertical subsurface and surface flow constructed wetlands are sometimes proposed, especially for small size installations (e.g. Kadlec and Knight, 1996).

From an operational point of view, as chemical treatments have shorter startup periods, they could theoretically be optimally managed for seasonal usage (e.g. when a treatment is required to dispose the treated wastewater into the environment out of the irrigation period), but the high costs of chemical nitrogen removal make this option feasible only in very specific cases. The mechanical-biological treatments, as well as the natural treatments discussed in Section 1.3.8, can also reach this goal, although with some problems of long transients due to the requirement for adaptation of the involved biota. However, some successful experiments of seasonal management of biological nitrogen removal are described in the literature (e.g. Hatziconstantinou and Andreadakis, 2003).

1.3.4.2 Phosphorus removal

Phosphorus can be removed from wastewater by either chemical or biological methods, or by a combination of the two. Unlike for nitrogen, chemical methods are presently largely prevalent over biological processes, mainly for reliability reasons. Nevertheless, enhanced biological phosphorus removal (EBPR) techniques are increasingly being exploited, mainly to pursue significant savings on operative costs.

Chemical phosphorus removal is done by precipitating the phosphorus from solution by the addition of iron, aluminum, or calcium salts. Several options are available to include the chemical precipitation of phosphorus in the line of the activated sludge process (Tchobanoglous et al., 2003); the most common are presented in Figure 1.5.

Enhanced biological phosphorus removal relies on culturing specific bacteria known as polyphosphate accumulating organisms (PAOs) that, under anaerobic conditions, are able to take up volatile fatty acids (VFAs) and convert them to intracellular poly-β-hydroxyalkanoates (PHAs). According to the current understanding of PAO metabolism, PAOs gain the energy required for VFA uptake and conversion to PHA through hydrolysis of their intracellularly stored polyphosphate (poly-P) to soluble orthophosphates (Smolders et al., 1994; Mino et al., 1998). Under aerobic conditions, PAOs take up orthophosphate to recover poly-P level, grow, and replenish their glycogen storage gaining the required energy by oxidation of PHA.

Several different techniques, some proprietary, are available to exploit the described mechanism. They are commonly classified into two groups: mainstream processes, when the phosphorus segregation happens on the main wastewater treatment line, and sidestream processes, where P is segregated in the return sludge line. Among the various available processes are (for further details see Tchobanouglos et al., 2003; Beccari et al., 1993):

- mainstream processes without N treatment: Phoredox, A/O™;
- mainstream processes with N treatment: A^2/O™, modified Bardenpho, standard UCT, modified UCT, VIP, Johannesburg;
- sidestream processes: Phostrip™;
- the SBR process (see Section 1.3.3.4), which can also be enhanced to allow nitrification, denitrification and biologic phosphorus removal.

All the listed processes rely on the addition of an anaerobic stage to the standard aerobic stage. If anaerobic, anoxic, and aerobic stages are provided, the removal of nitrogen (by nitrification-denitrification) and phosphorus is possible at the same time.

In all cases, the phosphorus is removed from the process by taking the poly-P-rich waste sludge extracted immediately after the aerobic stage and exposing it to anaerobic conditions (absence of DO and nitrates). Orthophosphates are then released in dissolved form and can be extracted with the supernatant from a settler. As this supernatant presents a very high concentration of orthophosphates, they can be segregated by treatment with lime, aluminum salts, or iron salts with higher efficiency and less consumption of chemicals than standard on-stream methods of chemical P removal. Alternatively, the high phosphate concentration also allows some crystallization processes, including struvite controlled crystallization, which produces a slow release fertilizer that can be used in agriculture (Pastor et al., 2008).

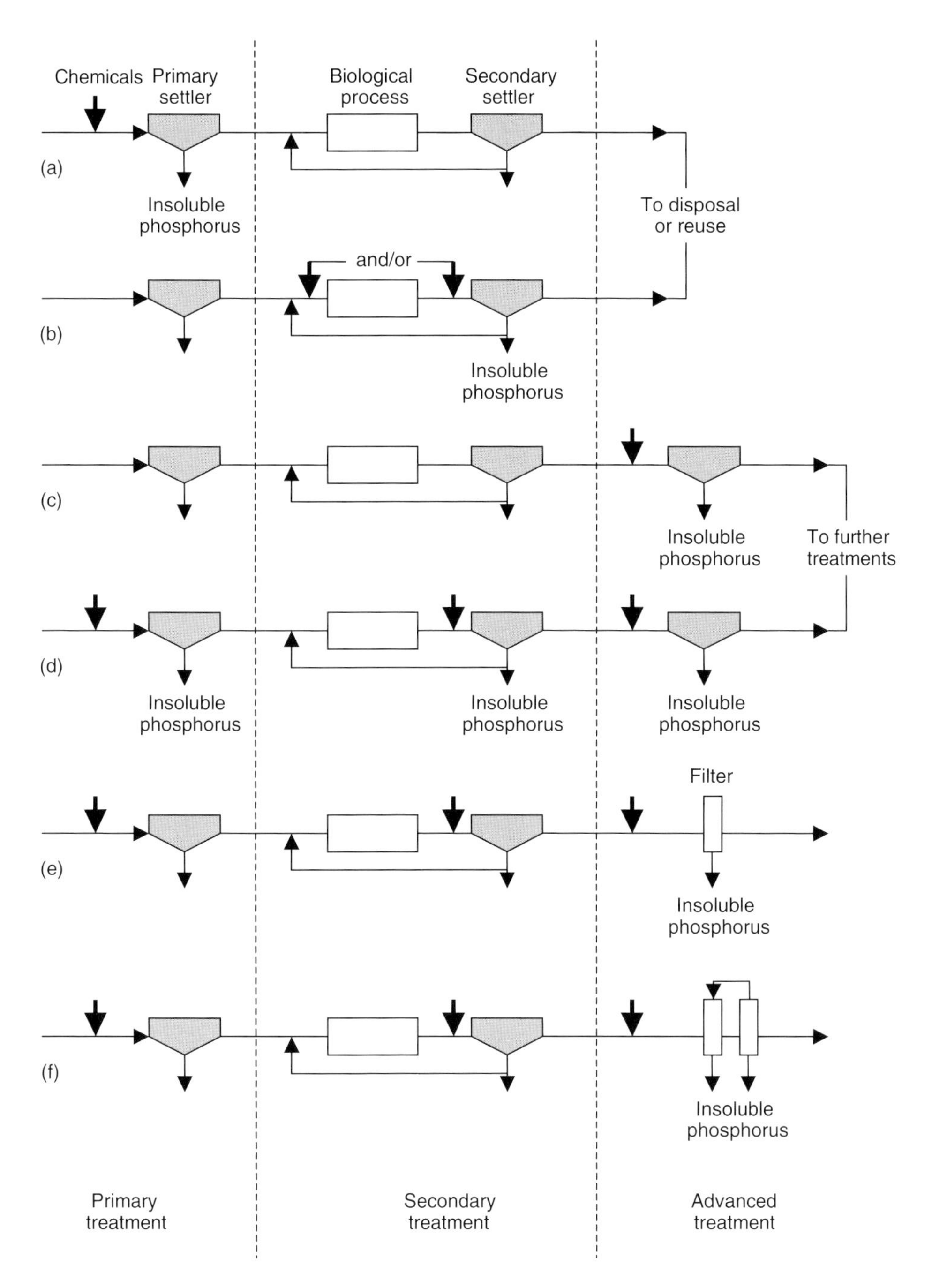

Figure 1.5 Alternative options for chemical phosphorus removal in the activated sludge process: (a) pretreatment, (b) simultaneous treatment, (c) post-treatment, (d–f) several alternatives of the "split treatment" (adapted from Tchobanoglous et al., 2003).

The following compares the advantages of chemical and biological P removal methods.

Advantages of chemical P removal:

- reliability;
- low sensitivity to environmental factors including possible feed instability;
- lower final attainable P concentration (post-treatment and split treatment);
- less construction costs and plant compactness (simultaneous treatment).

Advantages of biological P removal:

- lower operation costs due to smaller use of chemicals;
- lower sludge production;
- better reusability of the produced sludge in agriculture.

1.3.5 Disinfection

The purpose of disinfection of wastewater is to substantially reduce the number of pathogenic microorganisms in the water that is discharged back into the environment, recycled, or reused. As the hygienic safety of wastewater is a major issue for possible use, the disinfection stage represents one of the most important aspects of all the possible options of use. With regards to agricultural applications, different levels of disinfection are required in different countries, especially when unrestricted use is planned. Thus, this stage is usually the principal adaptation of the reclamation treatment according to the intended disposal/recycling/use planned for the wastewater.

Often, further tertiary treatments are carried out, with the main purpose being to increase the general quality of wastewater to allow a more effective and reliable disinfection with reduced side effects. Specifically, a polishing treatment stage such as filtration, coagulation-flocculation-sedimentation, membrane filtration, constructed wetland or oxidation pond is often carried out prior to disinfection to reduce the suspended solids. This is particularly important for ultraviolet (UV) radiation treatment, where a low suspended solids content is the main provision for a successful disinfection; but it is also important for chlorination and, in general, for all chemical disinfection methods, to reduce the risk of byproduct formation. A continuous monitoring of turbidity before the disinfection stage is also advisable to maintain high disinfection performance and reduce the risk of byproduct formation.

A very important question, still partially unresolved, concerns the analytical techniques required to monitor the efficacy of the disinfection systems. To guarantee a predefined level of health risk related to the residual content of pathogen microorganisms after disinfection, it is necessary to evaluate the content of pathogens belonging to all classes of potential pathogen microorganisms present in wastewater. Unfortunately, this range is very extended, and includes microorganisms of significantly varying dimensions and resistance to the various kinds of treatments (viruses, bacteria, protozoa, fungi, worms and the latent-life forms of some, such as protozoa cists, bacteria spores, or worm eggs). More research is being undertaken to develop cost-effective analytical techniques to accurately evaluate the residual infection risk, but affordable technical solutions are still limited to some simple pathogen-indicator classes mainly related to bacteria. Specifically, as fecal coliforms are not defined taxonomically, their count is being progressively substituted by *Escherichia coli*, and some authors (Brodsky, 1997; Mossel, 1997) have suggested that the term "fecal coliforms" should be excluded from microbiology and the term "Thermotolerant coliforms" (TTC) should be considered as a more appropriate description of these organisms (Baylis and Petitt, 1997; Chapter 2).

Unfortunately, the most monitored classes (first of all bacteria) are often the easiest to be abated from water, and on-line monitoring is still too problematic and expensive to be systematically performed.

The simplest classification of the several treatment possibilities available for disinfection of treated wastewater, is into chemical, physical, and natural-biological treatments.

1.3.5.1 Chlorine-Based chemical treatments

Among chemical disinfectants, chlorine-based treatments (mainly chlorine gas, hypochlorite, or chlorine dioxide) are probably still the most common worldwide due to their low cost and long-term history of effectiveness. Chloramine, which is sometimes used for drinking water, is not used in wastewater treatment because of its lower immediate efficacy and higher persistence. The main disadvantage is that chlorination of residual organic material can generate chlorinated-hydrocarbons, which can be carcinogenic or harmful to the environment. These include trihalomethanes and N-nitrosamines, including the potent carcinogen N-nitrosodimethylamine (NDMA) (Mitch et al., 2003). Moreover, chlorine can react with residual ammonia or nitrate in the effluent, thus forming different types of chloramines, which are less efficient disinfectants.

An important characteristic of chlorination is the presence of a longer term residual disinfection activity after treatment, due to residues of active chlorine or formed chloramines directly related to overstoichiometric dosages. This overdosage is given for various reasons:

- to increase the disinfection efficiency without increasing the dimension of the contact tank (to obtain compacter and cheaper plants);
- for setup and control reasons, as a real time adjustment of the dosage as a function of continuously varying flux and pollutant concentration is very difficult to obtain;
- to obtain a controlled residual disinfection activity (specially in cases where a long network of pipes and/or reservoirs is interposed between the treatment plant and the water use location).

The residual disinfection activity, although useful in some cases, may also be harmful if the effluent is disposed of into a natural aquatic environment. Furthermore, as residual chlorine is toxic to aquatic species, the treated effluent would need to be chemically dechlorinated, increasing the complexity and cost of the treatment.

As a basic comparison of the three mentioned chlorination techniques, the following "rule of thumb" can be stated:

- Chlorine gas is the least expensive solution and probably the most widely used. However, it is also the most disadvantageous for the following reasons: the gaseous reagent presents storage and handling risks as it is toxic and corrosive, there is possible formation of several disinfection byproducts (DBP, including trihalomethanes) due to the high reactivity with a wide range of organic and inorganic compounds possibly present in wastewater or in the receiving water body, the disinfection is very efficient with respect to bacteria, but not as much with other microorganisms, such as viruses and protozoa. Hence, this treatment is progressively being phased out.

- Sodium or calcium hypochlorite present similar features to chlorine gas, but in liquid form, which significantly simplifies handling and storing. The main drawback in comparison with chlorine gas is its moderately higher cost. For this reason hypochlorites are often being used as backup, mainly in cases of emergency in WWTPs that do not disinfect continuously. All other considerations are similar to chlorine gas.
- Chlorine dioxide (ClO_2) chlorination presents different properties: it is a highly reactive gas that must be produced on-site by reacting two liquid compounds. Usually hydrochloric acid and sodium chlorite are reacted in the following simple and fast reaction: $5NaClO_2 + 4HCl \rightarrow 4ClO_2 + 5NaCl + 2H_2O$. Chlorine dioxide is more selective than other chlorine forms (among others, it does not react with ammonia). In fact, its common use is as a drinking water disinfectant to destroy water impurities in order to avoid trihalomethane formation upon exposure to free chlorine. It is a more effective disinfectant against water-borne pathogenic microbes such as viruses, bacteria, and protozoa. Because it requires more complex handling, it is used more often in larger WWTPs.

1.3.5.2 Other chemical alternatives

Some alternative chemical disinfectants are proposed with the aim to increase efficacy and reduce drawbacks in terms of toxicity of disinfection byproducts to human health. The most interesting chemical alternatives are ozone and peroxy-acetic acid (PAA), although other oxidants (e.g. hydrogen peroxide, potassium permanganate, and lime) are also being used in a few cases.

Ozone, O_3, is a gas produced on-site by passing a flux of dried air or pure oxygen through a high voltage potential. This allows transformation of a high fraction of the O_2 molecules contained in the treated gas to their allotropic form O_3. The water to be disinfected is then treated by passing it through one or more serially connected contact chambers, where the produced gas is injected by gas blowers of different types (bubblers, venturi-pipes, turbines).

Ozone is a very unstable and reactive gas and oxidizes most organic materials it comes in contact with, thereby destroying many pathogenic microorganisms. Ozone is considered to be safer than chlorine. Whereas chlorine has to be stored on-site (and is highly poisonous in the event of an accidental release), ozone is generated on-site upon demand. Ozonation also produces fewer disinfection byproducts than chlorination. Inasmuch as *N*-nitrosodimethylamine might form by ozonation of precursors that occur in wastewater, it has been shown that ozonation oxidizes them more thoroughly compared with chlorinating agents (e.g. ClO_2) (Lee et al., 2007). Recently, the formation of NDMA from tolylfluanid (a wide spectrum fungicide) has been shown in ozonated water (Pesticides Safety Directorate, 2007). This prompted the suspension of sale and supply of products containing tolylfluanid (see 2007/322/EC European Union Commission Decision of 4 May 2007).

The main disadvantages of ozone disinfection are the high cost of the ozone-generation equipment, the high requirement for electrical energy, and the requirement of highly skilled operators. Finally, the fast decomposing rate of ozone can be seen as an advantage or a disadvantage depending on the specific applications, as in the following examples:

- When the treated wastewater is directly discharged into the environment, the completed decay of the ozone disinfection activity is an advantage as it avoids any sort of negative effects of the residual disinfectant on the receiving ecosystem.
- If, conversely, a distribution network of the reclaimed water is provided, a residual disinfection activity could be desirable to prevent possible recontaminations along the pipes of the network. In these cases, ozonation can be coupled with a final low-concentration chlorine dosage to provide the desired residual disinfection effect.

Peroxy-acetic acid (CH_3COOOH or $C_2H_4O_3$ or PAA) is an interesting liquid oxidant that has been used in the treatment of water in recent years. It can easily and effectively substitute traditional hypochlorite disinfection without any variation in storing and dosing equipment, with a slightly higher disinfection effectiveness, and with a significantly lower risk of DBP production. It presents a wide disinfection efficiency on bacteria (less efficacy on viruses) that does not significantly decrease in the presence of higher concentrations of suspended solids (Masotti and Verlicchi, 2005). Promising results have been also obtained using PAA in combination with UV irradiation (Caretti and Lubello, 2003). Its persistency is lower than that of chlorine (Masotti and Verlicchi, 2005). Its industrial production cost is presently about double that of hypochlorites, but prices are steadily decreasing.

1.3.5.3 Radiation treatments

Among physical treatments, disinfection with UV light radiation represents a solid, efficient, and cheap alternative to the previously mentioned chemical treatments. As no chemicals are used, the treated water has less adverse effects on the organisms that later consume it. Ultraviolet radiation causes damage to the genetic structure of bacteria, viruses, and other organisms, making them incapable of reproduction.

The key disadvantages of UV disinfection are: the need for frequent lamp maintenance and replacement, and the need for a high-quality effluent to ensure that the targeted microorganisms are not shielded from the UV radiation, as any solids present in the effluent might protect the microorganisms from the UV light. Also, some regrowth phenomena after UV disinfection were observed in several cases. Thus, disinfection efficiency and persistence can improve by coupling radiation with a chemical treatment such as chlorination, peroxide, or peroxy-acetic acid (Koivunen, 2005).

In some developed countries (e.g. the UK), UV radiation is becoming the most common means of disinfection, as a result of concerns regarding the impact of chlorine on chlorinating organics in treated wastewater and receiving water bodies (Tree et al., 1997).

Other radiation disinfection treatments are presently used for water disinfection, or are foreseen as promising options for the near future, but, they are probably more suitable for drinking water treatment rather than for agricultural wastewater use. Among others, are photo-catalytic disinfection based on TiO_2 induced catalysis of UV lamp radiation or natural solar radiation, which was also proposed for disinfection of small quantities of drinking water in rural areas of developing countries (Malato et al., 2007).

1.3.5.4 Mechanical separation

A pure physical disinfection technique can be based on mechanical separation of microorganisms by membranes of suitable pore size. However, several studies have pointed out that bacteria are able to pass through porous membrane structures, even when nominal pore size is smaller than bacterial size (Kobayashi et al., 1998; Shinde et al., 1999; Causserand et al., 2003). For this reason, reliable disinfection can be attained only with membranes of pore size significantly lower than the average dimension of the microorganisms to be removed (bacteria and viruses), specifically UF or NF membranes (average pore size up to 10^{-3} μm). Furthermore, complementary chemical disinfection is necessary in any case to ensure complete hygienic safety due to possible infiltration of microorganisms through irregularities in the membranes themselves, or in the ancillary mechanical equipment and containing vessels. A good and reliable pure membrane disinfection (without the need for final chemical treatment) is still not available.

It should, however, be emphasized that the above limitations are particularly relevant in the case of disinfecting potable water. As far as effluents are concerned, even for unrestricted agricultural use, guaranteed complete microorganism removal is often not necessary and not required by most legislative standards (Chapter 2).

1.3.6 Other tertiary chemico-physical treatments

1.3.6.1 Sand filtration

Sand filtration removes much of the residual suspended matter, thus drastically reducing suspended solids and turbidity of reclaimed water. Also, most microorganisms, and particularly the largest ones such as protozoa, are removed with a significant removal efficiency, especially when sand filtration is enhanced with a previous coagulation-flocculation stage. Therefore, sand filtration is commonly considered as the most traditional and effective prerequisite for efficient and reliable disinfection, especially for UV disinfection (the effectiveness of which is drastically limited by the content of suspended solids) and chemical disinfection (where a high content of solids increases the consumption of chemicals and the risk of formation of disinfection byproducts).

The process of filtration is performed by flowing the wastewater under treatment through a bed of porous, granular material such as sand. Depending on the filtration velocity, sand filters are classified as slow filters – with filtration velocity ranging from 0.2 to 1 m/h – and rapid filters that present filtration velocity in the range from 2 to10 m/h.

Rapid filters act only as a mechanical barrier, effective mainly on suspended solids. Conversely, slow filters – which require a much larger surface area – also allow the presence of a biota that adds to the pure mechanical separation a certain biological removal of biodegradable colloids and dissolved organics, especially if oxygen transfer into the bed is enhanced by discontinuous operation of the inflow.

During their operation, both kinds of filters are progressively clogged by the particles removed from the treated wastewater, causing an increase of hydraulic losses in the filter. When the maximum acceptable limit is reached, the efficiency of filtration starts to reduce, and the filters must be cleaned. The cleaning of rapid filters is obtained by periodically back-washing them with part of the treated water, which is stored in a dedicated tank.

During back-washing, this water (sometimes mixed with air) is forced upward through the media, expanding the filter bed slightly and carrying away the impurities in the wash-through water. The frequency of back-washing cycles and the consumption of treated water for back-washing depend on the size of the filter and the solid contents of the treated wastewater. The frequency can range from a few hours to a few days, and the consumption of treated water ranges from 3% to 4% up to 10% to 15%. The use of compressed air during back-washing can significantly reduce the consumption of treated water.

Two kinds of rapid filters are available: gravity-flow filters and pressure filters. The former consists of a cylindrical tank completely open at the top, where the flux under treatment flows from a layer of free surface water superimposed on the filtration layer. The latter consists of a closed cylindrical steel tank where water is pumped through the filter under pressure. Pressure filters can accept higher head losses, thus reducing the frequency of back-washing, at the cost of higher consumption of energy.

Slow filters usually require a reduced cleaning frequency, ranging from months to years depending on the media size, the hydraulic load, and the efficiency of the pretreatments. The cleaning is operated manually, removing solids from the surface, and, in some cases, replacing the first layer of sand. The complete replacement of the bed is required only every several years.

Slow filters can be operated also with plants on their surface, and in this case they are referred to as "subsurface constructed wetlands" (see discussion in Section 1.3.8.2). Planting has several positive effects on the process, including a more environmentally friendly appeal.

For reuse applications, sand filtration is established as part of the traditional tertiary treatment stage, as the importance of suspended solids removal is second only to disinfection.

Rapid filtration is often preferred over slow filtration because of its higher compactness. However, for small communities, the simplicity of operating and maintaining slow filters can be a factor in final selection of the filtration treatment.

1.3.6.2 Coagulation-flocculation

The coagulation-flocculation treatment is a two-step process that improves removal efficiency of the smallest particles from wastewater. It is part of the traditional tertiary treatment train for unrestricted irrigation.

Specifically, particles with average size in the range 10^{-3}–1 μm are commonly classified as colloids, and are usually the main cause of turbidity of water. The removal of colloids is difficult with mechanical methods:

- gravity separation methods (sedimentation and flotation) require very long detention times, thus requiring tanks of sizes unaffordable in most cases;
- sand filtration presents excessive pore size for effective removal of colloids;
- membrane filtration results in early clogging, where water has a significant concentration of colloids.

Coagulation-flocculation improves removal of the colloids by allowing the aggregation of small particles into larger agglomerates. This is done by neutralizing the surface electric charge that hinders aggregation of particles otherwise made possible by van der Waals attraction forces.

The coagulation stage consists of dosing electrolytes such as alum or ferric chloride, which are adsorbed on the surface of particles, thus neutralizing the opposite electrostatic charge that is on the surface. Alternatively, some long-chain polymers can be dosed, which are adsorbed on the surface of several particles at the same time, binding them together despite the electrostatic forces that make them repel from each other. Coagulation typically requires about 25–60 s of retention time in a fast mixed tank. The wastewater is then transferred into a slowly stirred tank (flocculation stage), where small flocks are allowed to aggregate, growing to an optimal size for subsequent sedimentation, flotation, or filtration. This process usually takes about 5–20 min and polyelectrolytes can be added at this stage to support and improve flocculation.

Traditionally, the flocks are then separated in a designated settler, in a sand filter, in a flotator or in a combination of these treatments. Several improvements of this process are possible, for example the combination of flocculation and settling stages in a single specially shaped treatment tank, where the settled sludge is continuously recycled to minimize the consumption of chemicals and maximize the efficiency of removal. Also micro-sand (20–100 μm) is sometimes dosed into the settling tank to increase the weight of the flocks, increasing the sedimentation speed. The sand is subsequently separated from the flocks by a hydrocyclone and recycled. Further details and comments on possible combined processes concerning the selection of the optimal treatment train are given in Section 1.4.

The coagulation-flocculation treatment, as in other chemical processes, generates chemical sludge that needs to be treated and disposed of. The disposal is often more problematic than with sludge from biological processes, due to the higher quantity produced and the presence of metals. Another drawback of this treatment is that the added chemicals can introduce into the reclaimed water low quantities of substances that can be harmful for reuse purposes.

1.3.6.3 Activated carbon adsorption

Adsorption over activated carbon is a simple treatment able to remove a wide selection of soluble contaminants from water, both organic (aromatic solvents, polynuclear aromatics, chlorinated aromatics, phenols, aromatic and aliphatic amines, surfactants, soluble organic dyes, fuels, chlorinated solvents, aliphatic and aromatic acids, pesticides and herbicides) and inorganic (nitrogen, sulphites, heavy metals, chlorine). Activated carbon is characterized by a vast system of pores and molecular sizes within the carbon matrix, resulting in the formation of a material with extensive internal surface area.

Activated carbon is available in two different forms: Granular Activated Carbon (GAC) and Powdered Activated Carbon (PAC). Whereas PAC is mainly dosed in the wastewater flow and removed by filtration after an appropriate contact time, GAC is used in fixed or moving bed filters or columns. Initial cost of PAC is lower as it is usually produced by fine grinding the byproduct of GAC production. However, GAC can be regenerated, saving more than 60% of the initial cost, whereas regeneration of PAC is not economically feasible.

1.3.6.4 Advanced oxidation processes

Advanced oxidation processes (AOPs) of wastewater effluent are innovative treatment technologies specifically developed to remove new emerging pollutants and contaminants that present low reaction rates with common oxidants. Many of these contaminants are unregulated and currently are being studied. They can be grouped in various classes: pharmaceuticals (e.g. antibiotics, analgesics, X-ray contrast agents, etc.), steroids and hormones (e.g. contraceptives), personal care products (e.g. fragrances, sun-screen agents, insect repellents, etc.), antiseptics, surfactants, flame retardants, industrial additives, gasoline additives, disinfection byproducts, plasticizers, and pesticides. Some have been already recognized as carcinogenic and endocrine disruptors, and their presence in reclaimed water is particularly hazardous. Therefore, AOPs become more and more important in reclamation treatments.

Most AOPs aim to produce hydroxyl radicals, as free radicals oxidize a wide range of compounds at a fast rate; additional oxidizing agents could be present in the process, often as a side effect of the production of free radicals. Examples of AOPs are (de Koning et al., 2008): ultrasounds, photolysis by UV, Fenton reaction, pulsed power plasma or electro-hydraulic discharge reactors, electrochemical oxidation, and radiation processes. Advanced oxidation processes have a conspicuous consumption of energy and/or chemicals so they are sequenced after traditional processes that remove the more easily oxidizable compounds.

1.3.6.5 Other chemical processes

Ion exchange removes ions from wastewater by sorption of cations (mercury, trivalent chromium, copper, ammonia, iron, etc.) or anions (chromate, phosphate, boron etc.) onto the exchange medium. Ion exchange media usually consist of resins made from synthetic organic materials (or inorganic or natural polymers) that possess ionic functional groups, which are saturated with easily exchangeable ions. After the resin exchange capacity has been exhausted, resins can be regenerated for reuse. Various factors affect the applicability and effectiveness of this process: oil and grease can clog the exchange resin; suspended solids content greater than 10 mg/l can cause resin blinding; the pH of the influent water can affect the ion exchange selectivity; and oxidants in the influent water can damage the ion exchange resin. The wastewater formed during the regeneration step requires special treatment and disposal.

The control of pH in wastewater is commonly carried out to guarantee a pH value suitable for water disposal or use. The pH is also sometimes controlled to create optimal chemical conditions for other tertiary chemical processes, such as chemical precipitation. The pH is regulated by the appropriate injection of acids (HCl, H_2SO_4, HNO_3, H_2CO_3, H_2SiO_3) and/or bases (NaOH, KOH, $Ca(OH)_2$, NH_4OH, $Fe(OH)_2$). Note that carbon dioxide is commonly used to neutralize basic wastewater exploiting the reaction of formation of carbonic acid in aqueous solution $CO_2 + H_2O = H_2CO_3$.

Chemical precipitation is mainly used to remove soluble heavy metals and phosphorus. The solid precipitates formed can then be separated from the liquid portion. Precipitation can be obtained either by converting the contaminant into an insoluble form or by changing the composition of the wastewater to diminish the solubility of the contained substances.

The most common chemicals used for precipitation are lime (calcium oxide), ferrous sulfate, alum (aluminium sulfate), ferric chloride and anionic, cationic or non-ionic polymers. The specific approach taken depends on the contaminants to be removed, the pH and alkalinity of wastewater, the phosphate level, the point of injection, and the mixing modes. Although chemical precipitation is a well-established treatment, the most recent research concentrates on combining chemical precipitation with other treatments such as photochemical oxidation, reverse osmosis, and biological methods to optimize its performance (EPA, 2000a).

1.3.6.6 Dilution

Although not properly a treatment system, the dilution of wastewater with freshwater can lead to compliance with required standards for reuse. This is not a common practice in traditional WWTPs – aiming to treat wastewater prior to discharge in water bodies – for two main reasons: (i) dilution cannot reduce the overall load of pollutants on receiving water bodies; and (ii) dilution leads to pollution of the freshwater used to dilute the wastewater. However, the situation can be reversed in the case of wastewater use, especially in agriculture. Compounds like nitrogen and phosphorus turn into nutrients, rather than pollutants, when their concentrations are not excessive for the irrigated crops; the pollution of the freshwater used for dilution saves the resting freshwater that is replaced by the used wastewater. Dilution can be adopted, for example, in the case of salty wastewater, as salinity is a major drawback for agricultural use. Control of salinity can be achieved only by very expensive processes such as RO.

1.3.7 Membrane technologies and quaternary processes

Among the wide range of treatment methods available, membrane filtration is probably developing at the fastest rate. In fact, membrane technology currently enables the treatment of most kinds of effluents for practically all levels of treatment (Wintgens et al., 2005). But, from a financial point of view, even for the most advanced countries, low-quality raw water and the high-quality levels required involve construction and operation costs that are still unaffordable in most situations. The rapid development of these technologies, however, is expected to change this situation. This would allow, on the one hand, the use of membranes to replace other more traditional technologies, and on the other, the reuse of previously unaffordable wastewater resources.

High-porosity membranes (MF, UF) are increasing in terms of general dependability, reduction of scaling and fouling phenomena, with increased cyclic duration and endurance to physical and chemical aggression. Thus, these technologies are serious candidates for replacing traditional treatments (coagulation-flocculation-sedimentation, sand filtration, active carbon adsorption). They are able to control not only suspended solids and turbidity, but also pathogen contents to replace chemical disinfection. They can also be used as a pretreatment for low-porosity membrane treatments (dual membrane concept, Del Pino and Durham, 1999).

Low-porosity membranes (NF and RO) are being continuously improved for salt removal in brackish water (reduction of required transmembrane pressure, TMP, and increased fouling and scaling resistance) and saline water (increased endurance to high TMP, resistance to high saline aggression and efficiency of energy recovery techniques).

Lastly, the coupling of MF or UF membranes with biological treatments in MBR applications enables traditional secondary treatments to be replaced by more compact and simpler plants that can achieve tertiary quality levels in a single secondary stage.

1.3.7.1 Membrane filtration

Although the feasibility of agricultural wastewater use projects is generally a mere question of cost, the application of membrane treatments is continuously increasing for the following reasons:

- The continuous development of MF and UF membranes is lowering their cost and increasing their efficiency and reliability in terms of the reduction of scaling and fouling, the reduction of required TMP and pretreatments and an increase in efficacy as a pathogen barrier. This development makes these membranes an effective competitor for more traditional treatments such as coagulation-flocculation-clarification, sand filtration, and chemical disinfection.
- The increasing demand for, and scarcity of, water worldwide are changing the financial breakeven point, as cheaper alternative water resources are continuously decreasing. The need for salt removal treatments for agricultural water is increasing worldwide.
- An awareness of the threat posed by the rising inorganic salt content in agricultural soil and groundwater due to the excessive salinity of irrigation water is increasing worldwide (Bouwer, 2000). Hence, the need for salt control on irrigation water is growing.
- A comparison between wastewater, natural brackish water, and seawater as a source of water for agriculture usually points to wastewater as the cheapest solution, even in cases when the used wastewater presents a salt content requiring high-density membrane treatments for proper irrigation use. This is because the total cost of the membrane treatment rises significantly with the increase of salt content of the water to be treated.

For the above reasons, opportunities are increasing for using low-density and high-density membranes to treat wastewater for agriculture use.

With regard to MF and UF membranes:

- conductivity and dissolved solids contents remain unaffected by both MF and UF treatments;
- decoloration due to UF is more noticeable than decoloration due to MF;
- filtration can eliminate detergents and phenol concentrations by 40%. Fe, Zn, Al, Cr, Cu and Mn can also be significantly eliminated not only by direct precipitation with, for example, hydroxides or phosphates, but also through the association of metals to suspended matter and macromolecules;
- in terms of pathogen controls, MF and UF membranes look promising for agricultural reuse. See discussion in 1.3.5.4;
- coupled with powdered activated carbon, UF and MF can be used to remove contaminants (including natural precursors of disinfection byproduct);

- UF treatment, coupled with powdered activated carbon, provides water that is generally suitable for unrestricted irrigation purposes, as it is high in nutrients (N and P are practically insensitive to filtration), low in micropollutants and microorganics, and exhibits favorable inorganic ratios.

With regard to dense membranes, in wastewater treatment and reclamation, RO systems are typically used as polishing processes that have a significant impact on bulk parameters, such as 65–80% and 85–99% total organic carbon (TOC) removal with NF and RO, respectively. Reverse osmosis systems are effective in removing various contaminants, including base neutral compounds, dissolved metals, and pathogens. The use of dense membranes to control the content of inorganic salts in agricultural water reclamation applications is increasing.

It is not uncommon for RO membranes in water reclamation applications to experience an average annual flux decline of 25–30%, even with frequent membrane cleaning. It should be noted that membrane rejection properties are susceptible to change after cleaning.

1.3.7.2 Dual membrane and direct membrane filtration concepts

Despite the continuous improvement in membrane performance and tolerance to contaminants, the control of NF and RO membrane fouling continues to be a major challenge in wastewater reclamation. Thorough pretreatment has been repeatedly stressed as the first step to control membrane fouling and ensure success (Kim et al., 2002).

After secondary biological treatments, RO pretreatment in wastewater reclamation uses articulate sequences of conventional stages such as flocculation, lime or alum clarification, recarbonation, settling, filtration, and activated carbon adsorption. A simplified process used in smaller systems involves in-line flocculation followed by pressure filtration (direct filtration), producing water of a slightly lower quality, with a performance largely dependent on feed water quality, hydraulic loading rates, and a good control over alum and cationic polymer dosages.

Following various successful demonstrations such as Water Factory 21 (Orange County, CA, USA) (Reith and Birkenhead, 1998), the industry is moving from lime clarification towards MF and UF as a pretreatment. For instance, investigation of UF pretreatment for RO recycling of secondary effluents from a refinery demonstrated that UF can consistently remove over 98% of suspended solids and colloids and 30% of chemical oxygen demand, irrespective of influent water quality and operating conditions (Teodosiu et al., 1999). The simplicity of membrane operations makes it an attractive option for wastewater reclamation.

Another newly emerging concept is direct membrane filtration, where a low-density membrane treatment follows simple preliminary pretreatments such as screening, sedimentation, or DAF with possible flocculants or powdered activated carbon dosing. This technology, which is still at an experimental stage, has mostly been aimed at reuse projects, where there are interesting advantages such as compactness and simplicity of treatment (van Nieuwenhuijzen et al., 2000; Ravazzini et al., 2004; de Koning et al., 2008).

Table 1.8 Membrane bioreactor (MBR) performances versus conventional processes (adapted from Wisniewski, 2007)

	Raw water				Treated water			
Processes	**TSS mg/l**	**COD mg/l**	**Turbidity NTU**	**Germs /100 ml**	**TSS mg/l**	**COD mg/l**	**Turbidity NTU**	**Germs /100 ml**
Trickling bed	200	700	120	10^8	35	125	10	10^6
Activated sludge	200	700	120	10^8	30	80	5	10^6
Physical-chemical	200	700	120	10^8	60	130	20	10^7
MBR	200	700	120	10^8	0	20	<2	$<10^2$

COD, chemical oxygen demand; NTU, nephelometric turbidity units; TSS, total suspended solids.

1.3.7.3 Membrane biological reactors

Membrane biological reactors combine an activated sludge treatment with a membrane liquid–solid separation process. The membrane component utilizes low pressure MF or UF membranes and eliminates the need for clarification and tertiary filtration, achieving better results than filtration for suspended solids and pathogens (Table 1.8).

The membranes can be simply immersed in the aeration tank or alternatively bundled in a specific sidestream tank connected to the aeration tank by a recirculation line. Whereas the former has the simplest structure, the latter makes higher TMP possible, allowing the use of lower porosity membranes. However, very low effluent standards are achieved even with MF membranes in immersed configuration, and the lower TMP of newer crossflow membranes also enables the use of UF membranes in immersed configuration.

One of the key benefits of MBRs is that they overcome the limitations associated with the poor settling of sludge in conventional activated sludge (CAS) processes. Membrane bioreactors use a considerably higher concentration of mixed liquor suspended solids (MLSS) than CAS systems, which are limited by sludge settling.

The MBR process is typically operated at MLSS in the range of 8000–12 000 mg/l, whereas CAS is operated in the range of 2000–3000 mg/l. The elevated biomass concentration in the membrane enables very effective removal of both soluble and particulate biodegradable materials at higher loading rates. Thus, increased sludge retention times (SRTs) – usually exceeding 15 days – ensure complete nitrification even in extreme cold weather.

The cost of building and operating a MBR is usually higher than conventional wastewater treatments; however, as the technology is becoming more and more popular and is gaining wider acceptance throughout the industry, the life-cycle costs are steadily decreasing. The increased use of MBR technology is also enhanced by the definition of dedicated design criteria (Iannelli and Giraldi, 2004). In addition, MBR facilities are already preferable in particular situations such as developed urban areas with limited land for the plant or also in unrestricted agricultural reuse applications. This is due to the ability to substitute a treatment train (activated sludge, flocculation, filtration, disinfection) with just one stage.

For all these reasons MBR processes are considered as one of the most promising technologies for agricultural use of civil wastewater.

1.3.8 Natural treatments

1.3.8.1 Lagooning

Lagoons and stabilization ponds provide settlement and further biological treatment via storage of wastewater in large man-made basins. The depth of the water determines the processes, which are classified as aerobic for shallower ponds (up to 80 cm), whereas deeper basins are called facultative or even anaerobic. The shallow aerobic ponds are mostly used as tertiary treatments, and are practically only of interest for wastewater reclaiming. Colonization by native macrophytes, especially reeds, is often encouraged. Small filter feeding invertebrates, such as Daphnia and some species of Rotifera, greatly assist in the treatment by removing fine particulates. As aerobic ponds are effective in the control of suspended solids and pathogens and for lamination and homogenization, their use in reclamation projects is greatly encouraged.

Sometimes excessive algal or weed growth can occur, leading to an increase in suspended solids in the outlet. Additional filtration treatments are thus required, which are often carried out in a final subsurface-flow constructed wetland.

1.3.8.2 Constructed wetlands

Constructed wetlands (CWs) include engineered reedbeds and a range of similar methodologies, all providing a high degree of aerobic biologic improvement. They are often used instead of secondary treatments for small communities.

Natural wetlands should not really be considered as treatment systems for water reuse as inflows, outflows, and reclamation processes are not usually monitored and controlled. However some reclamation processes naturally occur in wetlands, and water withdrawal can take place, especially for agricultural purposes. Therefore, wetlands would seem to be natural treatment stages for possible indirect reuse. On the other hand, there are several CWs used for wastewater reclamation and use. Rousseau et al. (2008) reported various reuse options for reclaimed wastewater by CWs: restricted and unrestricted irrigation of agricultural crops, watering of green areas such as gardens, golf courses, and public parks, flushing toilets, cleaning purposes, cooling water, water supplies for natural wetlands and nature reserves, and groundwater replenishment. Constructed wetlands are also considered as a possibility for nutrient reuse via plant and animal biomass production. The same authors highlighted that small-scale applications of reclamation by CWs are most popular in Europe, whereas large-scale applications are widespread in Australia and the USA.

Several types of CWs can be used, from the simple free water surface to the more complex subsurface vertical flow system (EPA, 2000; Kadlec et al., 2000). For water reuse they are usually operated as tertiary treatments in combination with other wastewater treatment units, but for some applications (e.g. restricted irrigation) they can also be used as stand-alone treatments. An international survey of relevant studies on tertiary free water surface CWs worldwide (Ghermandi et al., 2007) found that these systems can meet reuse quality standards for a variety of applications, in particular for no-contact to medium-quality uses.

Constricted wetlands for reclamation purposes present the classical benefits of CWs: simple maintenance and operation, buffering capacity, creation of wildlife habitats, low environmental impact, and public acceptance. Drawbacks include large surface requirements, mosquito and muskrat problems, odor, and clogging. Evaporation could be a major drawback for reclamation purposes and needs a careful evaluation. Giraldi and Iannelli (2007) highlighted that the reduction in outflow can become a significant reason for limitation of CW use in reclamation projects. First of all, a reduction in water flow could lead to a notable increase in the dissolved compounds that do not undergo purification processes inside the wetland. This means that more expensive treatments would be required in order to reach reuse effluent standards. In addition, salts could increase to unacceptable levels for water reuse in agriculture and several other uses. Lastly, variations in ET rates with seasonal and daily parameters, such as climatic conditions and vegetation development, could further reduce the quantity of water that can normally be relied on for reuse.

To limit ET, and hence to safeguard wastewater use, subsurface flow systems – especially vertical ones – are preferable to free water flow CWs as they present lower ET losses. Protecting CW systems from strong winds and, if possible, from exposure to direct sunlight (e.g. setting the systems behind buildings) could further reduce ET losses.

1.3.8.3 Wastewater storage reservoirs

Similar to ponds, wastewater storage reservoirs (WWSRs) differ from other extensive treatments as they bring together reclamation with seasonal water volume regulations for agricultural or landscape irrigation. Wastewater from municipalities or food industries is accumulated in WWSRs after a primary and/or secondary treatment; the inflow is generally constant during the year whereas water withdrawal only occurs during the irrigation season. Wastewater storage reservoirs require long residence times and large areas, but they present low energy and maintenance requirements.

Complex physical-chemical and biological processes occur in WWSRs, as is typical of hypertrophic water bodies with a slow water turnover. Therefore, their reclamation efficiency could vary greatly from case to case as a consequence of stored wastewater, climatic conditions, ecosystem characteristics, design features, and operation modalities (Cirelli et al., 2008). Different operation schemes could be adopted (Cirelli, 2003) (Fig. 1.6): a single reservoir with continuous inflow (a) is the simplest, but water quality problems could occur when irrigation starts, as the retention time drastically decreases. These problems could be partially solved with two reservoirs in series (b) using the fresher water from the first reservoir, only for restricted reuse. Another solution is to use two or more reservoirs in parallel (c) each of them with batch inflow and a minimal retention time.

1.3.8.4 Soil aquifer treatment

Soil aquifer treatment (SAT) is a natural and economical process for wastewater reclamation. Secondary effluents are infiltrated into a confined groundwater aquifer. During infiltration in the vadose zone, various biologic and physical processes are activated to purify wastewater (biologic degradation, filtration, adsorption, ion exchange, etc.). In order to maintain aerobic conditions in the soil aquifer, infiltration can be carried out by alternate flood and percolating phases. The groundwater aquifer then acts as storage for the treated wastewater, which is withdrawn from recovery wells surrounding the recharging area; the reclaimed water is mainly used for irrigation. Soil aquifer treatment

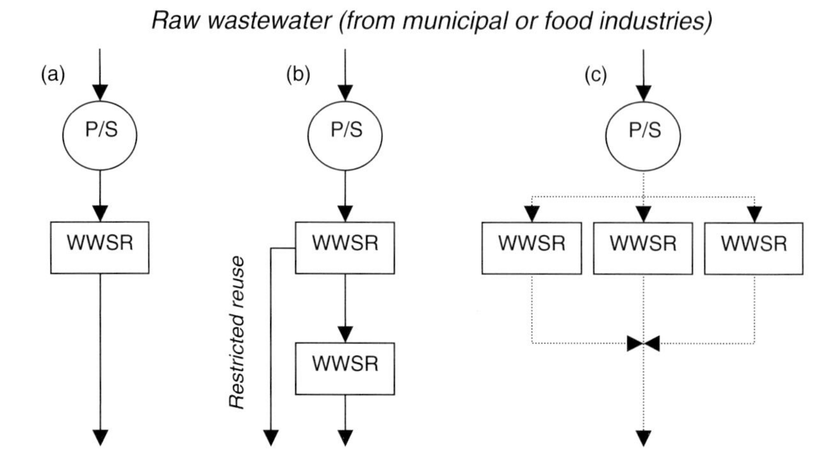

Figure 1.6 Different operation schemes for WWSRs (adapted from Cirelli, 2003) – P/S: primary and/or secondary treatments; WWSR: wastewater storage reservoir.

can remove a lot of contaminants, including heavy metals and other trace pollutants; however, the accumulation of persistent contaminants in the aquifer and the long-term effects on the quality of reclaimed water have not yet been completely assessed (Lin et al., 2004).

1.4 Framework for the selection of the optimal treatment train

Different treatment trains can be adopted for wastewater reclamation and use, the optimal solution finally depending on case-specific conditions. Parameters such as capital and running costs, existing treatments, flexibility, and availability of space are often critical in the decision process. However, some basic and established treatment trains can be considered and evaluated. As secondary treatments generally provide the same removal efficiency for the most critical contaminants, tertiary and quaternary treatments are key. Table 1.9 summarizes the potential for contaminant removal in the single treatment processes: the overall efficiency of a general treatment train can be theoretically estimated by a statistical combination of the single treatment efficiency presented in the table. The most common treatment reclamation trains are presented and discussed in terms of applicability, efficiency, diffusion, with particular reference to processes following conventional secondary treatments.

Conventional secondary treatments followed by disinfection

This is probably the most simple and basic treatment train for wastewater reclamation and use, and is also often used in developing countries. The main objective is to reduce the pathogens in the reclaimed water. If disinfection is obtained using oxidants, then a further reduction in the organic matter can be achieved. However, other contaminants such as

Table 1.9 Wastewater reclamation technologies and potential for contaminant removal (adapted and integrated from Culp, Wesner, Culp, 1979, in Tchobanoglous et al., 2003)

	Unit process or operation																				
Constituent	**Primary treatment**	**Activated sludge**	**Nitrification**	**Denitrification**	**Trickling filter**	**RBC**	**Coagulation-flocculation-sedimentation**	**Filtration after A/S**	**Carbon adsorption**	**Ammonia stripping**	**Selective ion exchange**	**Breakpoint chlorination**	**Reverse osmosis**	**Overland flow**	**Irrigation**	**Infiltration-percolation**	**Chlorination**	**Ozone**	**Ultrafiltration/microfiltration after A/S**	**Nanofiltration**	**MBR**
BOD	x	+	+	o	+	+	+	x	+		x		+	+	+	+		o	x	+	+
COD	x	+	+	o	+		+	x	x	o	x		+	+	+	+		+	x	+	+
TSS	+	+	+	o	+	+	+	+	+		+		+	+	+	+			+	+	+
NH_3-N	o	+	+	x		+	o	x	x	+	+	+	+	+	+	+			x	+	+
NO_3-N				+				x	o					x					x	+	
Phosphorus	o	x	+	+			+	+	+				+	+	+	+			+		x
Alkalinity		x					x	+								x			+		x
Oil and grease	+	+	+				x		x					+	+	+			x	+	+
Total coliform		+	+		o		+		+			+		+	+	+	+	+	+	+	+
TDS													+							x	
Arsenic	x	x	x				x	+	o										+	x	x
Barium		x	o				x	o											o		x
Boron							+				+		+								
Cadmium	x	+	+		o	x	+	x	o							o			x	x	+
Chromium	x	+	+		o	+	+	x	x										x	x	+
Copper	x	+	+		+	+	+	o	x							+			o	x	+

(Continued)

Table 1.9 (*Continued*)

	Unit process or operation																				
Constituent	**Primary treatment**	**Activated sludge**	**Nitrification**	**Denitrification**	**Trickling filter**	**RBC**	**Coagulation-flocculation-sedimentation**	**Filtration after A/S**	**Carbon adsorption**	**Ammonia stripping**	**Selective ion exchange**	**Breakpoint chlorination**	**Reverse osmosis**	**Overland flow**	**Irrigation**	**Infiltration-percolation**	**Chlorination**	**Ozone**	**Ultrafiltration/microfiltration after A/S**	**Nanofiltration**	**MBR**
Fluoride							x		o							o					
Iron	x	+	+		x	+	+	+	+										+	+	+
Lead	+	+	+		x	+	+	o	x							x			o	x	+
Manganese	o	x	x		o		x	+	x				+						+	x	x
Mercury	o	o	o		o	+	o	x	o										x	x	o
Selenium	o	o	o				o	+	o										+	x	o
Silver	+	+	+		x		+		x											x	+
Zinc	x	x	+		+	+	+		+							+				x	x
Color	o	x	x		o		+	x	+				+	+	+	+		+	x	+	x
Foaming agents	x	+	+		+		x		+				+	+	+	+		o	x	+	+
Turbidity	x	+	+	o	x		+	+	+				+	+	+	+			+	+	+
TOC	x	+	+	o	x		+	x	+	o	o		+	+	+	+		+	x	+	+
Stable trace organics							o		+			o	+					x	o	x	x

o = 25% removal of influent concentration, x = 25–50% removal of influent concentration, + = > 50% removal of influent concentration. Blank denotes no data, inconclusive results, or an increase.

A/S, activated sludge; BOD, biological oxygen demand; COD, chemical oxygen demand; MBR, membrane bioreactor; RBC, rotating biological contactors; TDS, total dissolved solids; TOC, total organic carbon; TSS total suspended solids.

suspended solids, metals, stable trace organic contaminants, etc., are not affected by the treatment; therefore, reclaimed water is only suitable for irrigation under restricted conditions. Disinfection is the basis for almost all the other treatment trains.

Conventional secondary treatments followed by coagulation, flocculation, clarification, media filtration, and disinfection

When all the treatment processes are implemented, the treatment train is also called "complete". If the clarification stage is omitted, then this is called "direct filtration"; whereas in "contact filtration" schemes, filtration directly follows coagulation and skips the flocculation tank. From a financial point of view, selection between complete treatment, direct filtration, and contact filtration is usually carried out on the basis of the turbidity level of the secondary effluent (Tchobanoglous et al., 2003). For good-quality secondary effluents, the most widespread treatment is contact filtration, for its simplicity, small footprint, reasonable costs, and flexibility. The treatment train provides reclaimed water, the quality of which is suitable for almost all conventional reuse applications. However, it requires chemicals, produces chemical sludge, and needs experienced management; therefore, it may not be suitable for applications in developing countries.

Conventional secondary treatments followed by advanced treatments

Advanced treatments (activated carbon adsorption, selective ion exchange, advanced oxidation processes, etc.) are usually combined with conventional tertiary treatments (coagulation-flocculation, filtration, disinfection, etc.). These treatment trains are particularly suitable for high-quality reuse applications such as industrial, household, etc., and they are not usually used in agriculture. Due to their associated costs and complexity, advanced treatments are not usually appropriate for use in developing countries.

Conventional secondary treatments followed by membranes

Membrane technology has developed in recent years and its application for agricultural reuse is becoming more and more common, and is likely to increase further in the short term, especially in developed countries. High-porosity membranes (usually following secondary treatments) can significantly remove suspended solids and the majority of pathogens, but without removing nutrients; thus, they are particularly indicated for agricultural reuse. Low-porosity membranes (usually following high-porosity membranes) are suitable when the control of salts in reclaimed water is a matter of concern, for example in agriculture. Their associated costs are also reasonable for agricultural reuse when the other main alternative for irrigation is the desalination of seawater.

Conventional secondary treatments followed by CWs

As a natural system, CWs are particularly suitable for small plants (although larger scale use exists, especially in North America) and for developing countries. Its simplicity of operation and the absence of required chemicals are the most attractive aspects. Possible drawbacks include the need for large areas and the costs of possible waterproofing. Some basic contaminants, such as suspended solids and pathogens may also not be adequately removed, or may even be introduced by the process itself; thus, this treatment train is usually more preferable for agricultural reuse and nature conservation than for high-quality applications.

Conventional secondary treatments followed by ponds, accumulation, solid removal, and disinfection

Ponds followed by accumulation basins allow the best reclamation in terms of the quantity of treated water. Their storage volumes act as buffers between the quantities of reclaimed water needed during irrigation and the amount of reclaimed water treated throughout the year. However, these treatments can lead to reduced removal efficiency for some compounds, as, and even more, than, in CW; therefore, they are usually followed by treatments to remove suspended solids (e.g. DAF systems) and disinfection stages. Thus, this treatment train is used mainly for agricultural purposes.

References

Baylis, C. L. and Petitt, S. B. (1997) The significance of coliforms and *Escherichia coli* and the Enterobacteriaceae in raw and processed foods. In: *Coliforms and* E. coli: *problem or solution?* (eds D. Kay and C. Fricker), pp. 49–53. The Royal Society of Chemistry, Cambridge, UK.

Beccari, M., Passino, R., Ramadori, R. and Vismara, R. (1993) *Rimozione di azoto e fosforo dai liquami* (Nitrogen and phosphorus removal from wastewater). Ulrico Hoepli Editore, Milano Italy.

Bixio D., Thoeye, C., de Koning, J., Joksimovic, D., Savic, D., Wintgens, T. and Melin, T. (2006) Wastewater reuse in Europe. *Desalination* **187**, 89–101.

Bouwer, H. (2000) Groundwater problems caused by irrigation with sewage effluent. *Journal of Environmental Health* **63** (3), 17–20.

Brodsky, M. H. (1997) Abolition of the term "fecal coliforms" seconded. *ASM News* **63**, 345–346.

Caretti, C. and Lubello, C. (2003) Wastewater disinfection with PAA and UV combined treatment: a pilot plant study, *Water Research* **37**, 2365–2371.

Causserand, C., Aimar, P. and Roques, C. (2003) Evaluation de la rétention bactérienne par des membranes (Evaluation of bacteria retention of membranes) *Récents Progrès en Génie del Procédes* **89**, 395–402.

CDEP – The Connecticut Department of Environmental Protection 2004. Connecticut Stormwater Quality Manual. Available on-line at http://dep.state.ct.us (last accessed January 2008).

Chen, H.H., Yeh, H.H. and Shiau, S. (2005) The membrane application on the wastewater reclamation and reuse from the effluent of industrial WWTP in northern Taiwan. *Desalination* **185**, 227–239.

Cirelli, G.L. (2003) *I trattamenti naturali delle acque reflue urbane – fitodepurazione, lagunaggio, accumulo in serbatoi* (Natural treatments of urban wastewater - constructed wetlands, pond treatments, reservoir storage). Esselibri, Napoli, Italy.

Cirelli, G.L., Consoli, S., and Di Grande, V. (2008) Long-term storage of reclaimed water: the case studies in Sicily (Italy). *Desalination* **218**, 62–73.

Culp, Wesner, Culp (1979) *Water reuse and recycling*. Office of Water Research and Technology, U.S. Department of the Interior, Washington DC, USA.

Del Pino, M.P. and Durham, B. (1999) Wastewater reuse through dual-membrane processes: opportunities for sustainable water resources. *Desalination* **124**, 271.

Eckenfelder, W.W. (2000) *Industrial water pollution control*. 3rd ed. McGraw-Hill, New York, USA.

EPA (1992) *Guidelines for water reuse*. EPA/625/R-92/004.

EPA (1999) *Sequencing Batch Reactors. Wastewater Technology Fact Sheet*. EPA 832-F-99-073, USA.

EPA (2000) *Trickling Filters. Wastewater Technology Fact Sheet*. EPA 832-F-00-014, USA.

EPA (2000a) *Chemical Precipitation. Wastewater Technology Fact Sheet*. EPA 832-F-00-018, USA.

EPA (2000b) *Constructed wetlands treatment of municipal wastewaters*. EPA/625/R-99/010, USA.

Feigin, A., Ravina, I. and Shalhevet, J. (1991) *Irrigation with treated sewage effluent*. Springer Verlag, Berlin, Heidelberg, New York, USA.

Fine, P., Halperin, R. and Hadas, E. (2006) Economic considerations for wastewater upgrading alternatives: an Israeli test case. *Journal of Environmental Management* **78**, 163–169.

Förtser L., Teichardt, R., Metz, R., Hübner, W. (1988) *Anfall und Einsatz von kommunalem Abwasser sowie ausgewählter Produktionsabwässer der Land- und Nahrungsgüterwirtschaft für Bewässerungszwecke. Forschungsberichte für die Landwirtschaft und Nahrungsgüterwirtschaft der Akademie der Landwirtschaftswissenschaften der DDR* (Treatment and use of urban wastewater and wastewater production of selected agricultural and food industry for irrigation purposes. Research for agriculture and food industry of the GDR Academy of Agricultural Sciences). Institut für Landwirtschaftliche Information und Dokumentation (Institute for Agricultural Information and Documentation) Berlin, Germany.

Galil, N.I. and Levinsky, Y. (2007) Sustainable reclamation and reuse of industrial wastewater including membrane bioreactor technologies: case studies. *Desalination* **2002**, 411–417.

Gerhart, V.J., Kane, R. and Glenn, E.P. (2006) Recycling industrial saline wastewater for landscape irrigation in a desert urban area. *Journal of Arid Environments* **67** (3) 473–486.

Ghermandi A., Bixio, D. and Thoeye, C. (2007) The role of free water surface constructed wetlands as polishing step in municipal wastewater reclamation and reuse. *Science of the Total Environment* **380**, 247–258.

Giraldi, D. and Iannelli, R. (2007) ET in constructed wetlands with subsurface flow: a limitation for water reuse? *Proceedings of the International Conference of Multifunctions of Wetland Systems*, Legnaro (PA), Italy, 26–29 June.

Grady, C.P.L., Daigger, G.T. and Lim, H.C. (1999) *Biological wastewater treatment*. 2nd ed., Marcel Dekker, Inc., New York, USA.

Hatziconstantinou, G.J. and Andreadakis, A.D. (2003) Partial nitrification activated sludge operation as an appropriate nitrogen removal scheme for the Mediterranean region. *Water Supply* **3** (4), 153–160.

Iannelli, R. and Giraldi, D. (2004) Una proposta di dimensionamento del processo MBR applicato ai reflui civili (A proposal of dimensioning of the MBR process applied to civil wastewater). *Proceedings of the International Symposium of Sanitary Environmental Engineering*, Taormina (Italy), 23–26 June 2004.

Into, M., Jönsson, A.S. and Lengdén, G. (2004) Reuse of industrial wastewater following treatment with reverse osmosis. *Journal of Membrane Science* **242** (1-2), 21–25.

Kadlec, R.H. and Knight, R.L. (1996) *Treatment wetlands*. CRC Press LLC. Boca Raton, FL.

Kadlec, R.H., Knight, R.L., Vymazal, J., Brix, H., Cooper, P. and Haberl, R. (2000) Constructed wetlands for pollution control: processes, performance, design and operation. *IWA Specialist Group on Use of Macrophytes in Water Pollution Control, Scientific and Technical Report No. 8*, IWA Publishing, London, UK.

Kim S.L., Chen, J.P. and Ting, Y.P. (2002) Study on feed pretreatment for membrane filtration of secondary effluent. *Separation and Purification Technology* **29**, 171–179.

Kobayashi T., Ono, M., Shibata, M. and Fujii, N. (1998) Cut-off performance of *Escherichia coli* by charged and noncharged polyacrylonitrile ultrafiltration membranes. *Journal of Membrane Science* **140**, 1–11.

Koivunen, J. (2005) Inactivation of enteric microorganisms with chemical disinfectants, UV irradiation and combined chemical/UV treatments. *Water Research* **39**, 1519–1526.

de Koning, J., Bixio, D., Karabelas, A., Salgot, M. and Schäfer, A. (2008) Characterization and assessment of water treatment technologies for reuse. *Desalination* **218**, 92–104.

Kretschmer N., Ribbe, L. and Gaese, H. (2002) Wastewater reuse for agriculture. In: *Technology resource management & development, vol. 2 – Special Issue: Water Management* (eds Institute of Technology in the Tropics), Cologne, ISSN 1618-3312.

Lee, C., Schmidt, C., Yoon, J. and von Gunten, U. (2007) Oxidation of N-nitrosodimethylamine (NDMA) precursors with ozone and chlorine dioxide: kinetics and effect on NDMA formation potential. *Environmental Science and Technology* **41**, 2056–2063.

Lin, C., Shacahr, Y. and Banin, A. (2004) Heavy metal retention and partitioning in a large-scale soil-aquifer treatment (SAT) system used for wastewater reclamation. *Chemosphere* **57**, 1047–1058.

Malato, S., Blanco, J., Alarcon, D.C., Maldonado, M.I., Fernandez-Ibanez, P. and Gernjak, W. (2007) Photocatalytic decontamination and disinfection of water with solar collectors. *Catalysis Today* **122** (1-2) 137–149.

Masotti, L. and Verlicchi, P. (2005) *Depurazione delle acque di piccole comunità, tecniche naturali e tecniche impiantistiche* (Treatment of wastewater from small communities, natural techniques and plant-engineering techniques). Hoepli, Milano, Italy.

Mino, T., Van Loosdrecht, M.C.M. and Heijnen, J.J. (1998) Microbiology and biochemistry of the enhanced biological phosphate removal process. *Water Research* **32**, 3193–3207.

Mitch, W.A., Sharp, J.O., Trussell, R.R., Valentine, R.L., Alvarez-Cohen, L. and Sedlak, D.L. (2003) N-Nitrosodimethylamine (NDMA) as a drinking water contaminant: A review. *Environmental Engineering Science* **20**, 389–404.

Mossel, D.A.A. (1997) Request for opinions on abolishing the term "fecal coliforms". *ASM News* **63**, 175.
van Nieuwenhuijzen, A.F., Evenblij, H. and van der Graaf, J.H.J.M. (2000) Direct wastewater membrane filtration for advanced particle removal from raw wastewater. *Proceedings of the 9th Gothenburg* Symposium, 2–4 October 2000, Istanbul, Turkey.
Pastor, L., Marti, N., Bouzas, A. and Seco, A. (2008) Sewage sludge management for phosphorus recovery as struvite in EBPR wastewater treatment plants. *Bioresource Technology* **99**, 4817–4824.
Pesticides Safety Directorate (2007) http://www.pesticides.gov.uk/approvals.asp?id=2062 (last accessed 16 July 2008).
Qin, J.J., Wai, M.N., Oo, M.H., Kekre, K.A. and Seah, H. (2006) Feasibility study for reclamation of a secondary treated sewage effluent mainly from industrial sources using a dual membrane process. *Separation and Purification Technology* **50**, 380–387.
Ravazzini A.M., van Nieuwenhuijzen, A.F. and van der Graaf, J.H.M.J. (2004) Direct ultrafiltration of municipal wastewater: comparison between filtration of raw sewage and primary clarifier effluent. *Desalination* **178**, 51–62.
Reith, C. and Birkenhead, B. (1998) Membranes enabling the affordable and cost effective reuse of wastewater as an alternative water source. *Desalination* **17**, 203–209.
Rousseau, D.P.L., Lesage, E., Story, A., Vanrolleghem, P.A. and De Pauw, N. (2008) Constructed wetlands for water reclamation. *Desalination* **218**, 181–189.
Shinde, M.H., Kulkarni, S.S., Musale, D.A. and Joshi, S.G. (1999) Improvement of the water purification capability of poly(acrylonitrile) ultrafiltration membranes. *Journal of Membrane Science* **162**, 9–22.
Smolders, G.J.F., Vandermeij, J., Vanloosdrecht, M.C.M. and Heijnen, J.J. (1994) Model of the anaerobic metabolism of the biological phosphorus removal process: Stoichiometry and pH influence. *Biotechnology and Bioengineering* **43**, 461–470.
Tchobanoglous, G., Burton, F.L. and Stensel, H.D. (2003) *Wastewater engineering: treatment and reuse*. 4th ed. Metcalf & Eddy, Inc., McGraw-Hill, New York, USA.
Teodosiu, C.C., Kennedy, M.D., Van Straten, H.A. and Schippers, J.C. (1999) Evaluation of secondary refinery effluent treatment using ultrafiltration membranes. *Water Research* **33**, 2172.
Tree, J.A., Adams, M.R. and Lees, D.N. (1997) Virus inactivation during disinfection of wastewater by chlorination and UV irradiation and the efficacy of F+ bacteriophage as a viral indicator. *Water Science and Technology* **35** (11), 227–232.
Wang, Z., Fan, Z., Xie, L. and Wang, S. (2006) Study of integrated membrane systems for the treatment of wastewater from cooling towers. Desalination **191** (1-3), 117–124.
Weber, B. and Juanicó, M. (2004) Salt reduction in municipal sewage allocated for reuse: the outcome of a new policy in Israel. *Water Science and Technology* **50**, 17–22.
Wintgens T., Melin, T., Schafer, A., Khan, S., Muston, M., Bixio, D. and Thoeye, C. (2005) The role of membrane processes in municipal wastewater reclamation and reuse. *Desalination* **178**, 1–11.
Wisniewski C. (2007) Membrane bioreactor for water reuse. *Desalination* **203**, 5–19.

Chapter 2
Health considerations in the recycling of water and use of treated wastewater in agriculture and other non-potable purposes

Hillel Shuval

2.1 Introduction

This chapter will deal primarily with the health considerations and the control of pathogenic microorganisms in water recycling and wastewater use in agriculture, as this is the most widely practiced form of reuse on a global basis. Increasingly, water specialists, resource planners and economists see wastewater as an economic good and, as time goes on, there will be an increased motivation to divert recycled water from low-income agriculture to areas where the added value of water is greater, such as industrial, non-potable urban, and environmental uses, including public parks, street cleaning, fire-fighting, open spaces, nature reserves, forests, river reclamation, green belts, golf courses, and football fields. There are some specific health guidelines and standards for such uses, which will not be reviewed in this chapter.

Raw domestic wastewater normally carries the full spectrum of pathogenic microorganisms – the causative agents of bacterial, viral, and protozoan diseases endemic in a community and excreted by infected individuals. It may also carry a range of toxic inorganic and organic chemicals, although such contaminants are considerably more important in wastewater from industrial sources. Water recycling and the use of wastewater for agriculture, industry and non-potable urban, and environmental purposes can be a highly effective strategy to exploit a sustainable water resource in water-scarce areas. It also can contribute significantly to nutrient conservation and environmental protection. It is essential, however, to understand the health risks involved and to develop appropriate strategies for the management of those risks.

Sound risk assessment and characterization are the necessary starting points in a process that leads to effective control measures for agricultural and other urban and environmental reuse. Control measures include establishing and enforcing microbial guidelines for

Treated Wastewater in Agriculture, First Edition, edited by Guy J. Levy, Pinchas Fine and Asher Bar-Tal

effluent quality, regulation of the types of crops to be irrigated, minimizing the potential for crop contamination by irrigation and the treatment of the wastewater to an appropriate degree. The resulting management of potential health risks will help protect farmers and their families, populations exposed to green open spaces and recreational areas, and the retailers and consumers of crops from infection with pathogenic microorganisms in the wastewater stream. This chapter will deal with the above issues.

2.2 Rationale: why should society allow, regulate, and thus encourage exposure of the population to known health risks?

Fundamental ethical and societal questions concerning controlled exposure of a population to known health risks must be considered when discussing the way to regulate such risks. First we must ask: Is it socially and morally justifiable to allow and regulate exposure to known health risks, which in effect implies approval and encouragement to take such risks? In other words, we must ask the question: "Is this trip necessary?"

The classical public health policy position on this question is that when considering optional exposures to health risks such as in the case of planned addition of chemical additives to foods to preserve nutrient value or protect food from spoilage, use of insecticides to protect food crops from destruction by insects and the addition of chemicals, e.g. chlorine, as a disinfectant to drinking water to destroy pathogens and prevent disease transmission, there must be clear evidence of the social, health, and economic benefits. In the cases above, the general public health view has been in the past that the health benefits far outweigh the health risks. Without doubt the introduction of chlorine disinfection of drinking water has saved hundreds of millions of lives in the past 100 years and the negative health effects have been limited. This attitude is changing, however, with respect to chlorine and some food additives and insecticides. This same consideration must apply to the question of allowing and regulating wastewater recycling and reuse for food production and other purposes.

The World Health Organization (WHO) has clearly emphasized the social, human welfare, and environmental sustainability benefits and importance of wastewater use in agriculture in their recently published *Guidelines for the Safe Use of Wastewater, Excreta and Greywater in Agriculture and Aquaculture* (WHO, 2006, hereafter referred to as "the third edition of the guidelines"). Bos and colleagues (2007) stated this succinctly, emphasizing that the third edition of the guidelines fits into the overall international policy framework for poverty alleviation and sustainable development as outlined by the Millennium Development Goals of the United Nations (UN) as follows:

Goal 1: Eradicate extreme poverty and hunger, relevant issues include:

- Wastewater, excreta and greywater make up an important resource for intensive agricultural production by the urban and rural poor and thereby strengthen their livelihood opportunities.
- Agricultural produce cultivated through the use of waste adds importantly to the food security of poor rural and urban communities.
- Reduced downstream ecosystem degradation resulting from the use of wastewater, excreta and greywater makes livelihood systems of the poor more secure.

Goal 7: Ensure environmental sustainability, the issues of relevance include:

- The safe use of wastewater, excreta and greywater contributes to less pressure on freshwater resources and reduces health risks for downstream communities.
- Improved sanitation in support of safe excreta use reduces flows of human waste into waterways, helping to protect human and environmental health.
- Improved water management, including pollution control and water conservation, is a key factor in maintaining ecosystem integrity.
- Waste-fed peri-urban agriculture can contribute importantly to improving the livelihood of slum settlers.

In addition to the international framework, there are important national policy perspectives of the safe use of wastewater, excreta and greywater, including poverty reduction, food security, protection of public health, protection of the environment, integrated water resources management, and energy reliance.

The above provides a clear statement of the potential health, social and environmental benefits of wastewater use in agriculture, which justifies the establishment of guidelines and regulations for such use. However, these socially important goals must be achieved with a minimum of health risks.

2.3 Persistence of pathogenic microorganisms in water, soil, and on crops from wastewater-irrigation

Pathogenic microorganisms in the wastewater stream can be transmitted to healthy individuals and cause disease if improper regulation and control methods in waste-water-irrigation are practiced. In order for disease-causing microorganisms (pathogens) in the wastewater effluent from a community to infect a susceptible individual they must be able to survive in the environment (i.e. in water, air, soil, or food) for the period of time during which they remain infectious, and they must be ingested in sufficient doses.

Factors that affect the survival of pathogens in soil include antagonism from soil bacteria, moisture content, organic matter, pH, sunlight, and temperature. Excreted enteric pathogens such as bacteria, viruses, protozoa, and helminth eggs do not usually penetrate undamaged vegetables but can survive for long periods in the root zone, in protected leafy folds, in deep stem depressions, and in cracks or flaws in the skin. Recent studies indicate that there is a possibility for pathogens to penetrate some root vegetables such as onions (Blumenthal and Peasey, 2002).

Data from numerous field and laboratory studies have made it possible to estimate the persistence of certain enteric pathogens in water, air, wastewater and soil, and on crops. These survival periods in the environment are presented in summary graphic form in Figure 2.1. For example, it appears that *Campylobacter* may survive in soil or on crops for only a few days, whereas most bacterial and viral pathogens can survive from weeks to months. The highly resistant eggs of helminths (worms), such as *Trichuris*, *Taenia*, and *Ascaris*, can survive for 9–12 months, but their numbers are greatly reduced during exposure to hostile environmental conditions.

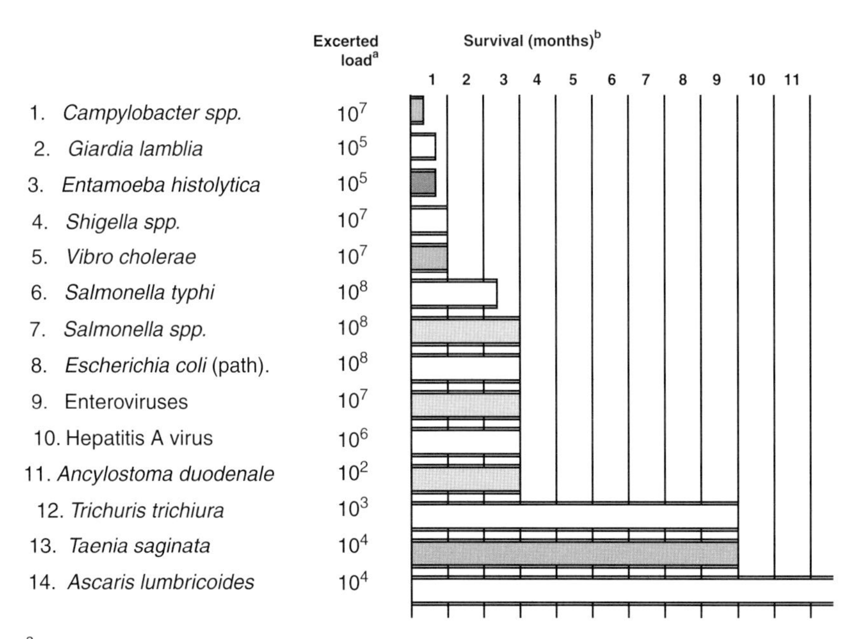

Figure 2.1 Survival times of enteric pathogens in water, wastewater, soil, and on crops (from Shuval et al., 1986, based partially on data from Feachem et al., 1983).

Our field studies in Israel have demonstrated that enteric bacteria and viruses can be dispersed for up to 730 m in aerosolized droplets generated by overhead (sprinkler) irrigation, but their concentration is greatly reduced by detrimental environmental factors such as sunlight and desiccation (Teltsch et al., 1980; Applebaum et al., 1984; Shuval et al., 1988, 1989a). Thus, whereas most excreted pathogens can survive in the environment long enough to be transported by wastewater to the fields and to the irrigated crops, their numbers are significantly reduced. The contaminated crops eventually reach the consumer, although by then the concentration of pathogens is greatly reduced, often below the level required for an infectious dose of the pathogen to be ingested by the consumer. Although the concentration of indicator organisms and pathogens can be significantly reduced by wastewater treatment processes, this is not the only phase in water recycling technology that achieves important reductions of the risk of exposure to pathogens. The rapid natural die-away of pathogens in the environment is an important factor in reducing the health risks associated with wastewater use and should be taken into account when estimating risk of infection. Studies have indicated that the mean rate of bacterial and viral reduction/inactivation in soil and on crops is about 0.5–2 logs/d (up to 99%/d) (WHO, 2006). The rate of pathogen inactivation may be faster under hot and sunny conditions. The third edition of the guidelines concludes that the mean minimal reduction of pathogens and enteric indicator organism under field conditions can be estimated conservatively at 2 logs or 99%. This is an important natural environmental risk reduction factor, which should not be overlooked.

2.4 Disease transmission by wastewater-irrigation

This section will provide an extensive review and evaluation of the research findings on disease transmission by wastewater-irrigation based on available scientific papers published in recognized journals and in numerous unpublished government reports, university theses, and private papers obtained during an intensive worldwide search carried out with the help of international and national agencies and individuals.

Over a thousand documents, some more than 100 years old, were examined in the course of our studies, but few offered concrete or reliable epidemiological evidence of health effects. Most of them based their conclusions on inference and extrapolation. Nonetheless, about 50 of these reports provided enough robust evidence based on sound epidemiological procedures to make a detailed analysis useful. Those studies are reviewed in detail in the United Nations Development Programme (UNDP)/World Bank report on which this chapter is partially based (Shuval et al., 1986). Our general conclusions of some of the more pertinent studies are presented below.

One of the goals of our research contributions to the UNDP/World Bank study (Shuval et al., 1986; Shuval, 1990), as described in this chapter, was to re-evaluate all the credible, scientifically valid and quantifiable epidemiological evidence of human health effects associated with wastewater-irrigation. Such evidence is needed to determine the validity of current and earlier regulations and to develop appropriate technical solutions for existing problems.

2.4.1 Illness associated with wastewater-irrigation of crops eaten raw

In areas of the world where the helminth infections caused by *Ascaris* and *Trichuris* are endemic in the population, and where raw, untreated wastewater is used to irrigate salad crops and/or other vegetables generally eaten uncooked, the consumption of such wastewater-irrigated salad and vegetable crops is likely to provide a significant transmission pathway, as demonstrated by Khalil (1931) in Egypt.

Similarly, our study in Jerusalem (Shuval et al., 1984) provided strong evidence that massive infections with both *Ascaris* and *Trichuris* may occur when salad and vegetable crops are irrigated with raw wastewater. These diseases almost totally disappeared from the community when raw wastewater-irrigation of vegetables was stopped (Fig. 2.2). Two studies from Darmstadt, Germany (Krey, 1949; Baumhogger, 1949) provide additional support for this conclusion.

These studies also indicate that regardless of the level of municipal sanitation and personal hygiene, irrigation of vegetables and salad crops with raw wastewater can serve as a major pathway for continuing and long-term exposure to *Ascaris* and *Trichuris* infections. Both of these infections are of a cumulative and chronic nature, so that repeated long-term reinfection may result in higher worm loads and increased negative health effects, particularly among children.

Cholera (pathogenic organism: *Vibrio cholerae*) can also be disseminated by vegetable and salad crops irrigated with raw wastewater. This possibility is of particular concern in non-endemic areas where sanitation levels are relatively high, and the common routes of cholera transmission, such as contaminated drinking water and poor personal hygiene, are closed. Under such conditions, the introduction of a few cholera carriers (or subclinical

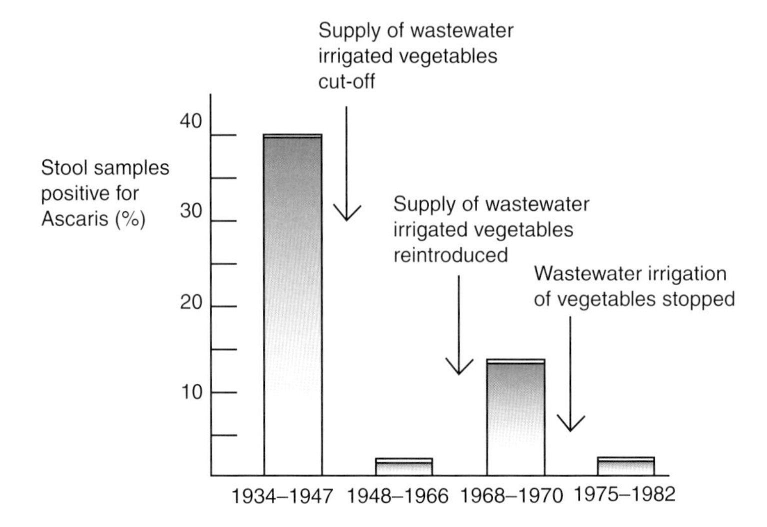

Figure 2.2 Relationship between *Ascaris*-positive stool samples in population of western Jerusalem and supply of vegetables and salad crops irrigated with raw wastewater in Jerusalem, 1935–1982 (from Shuval et al., 1986, based partially on data from Ben-Ari, 1962; Jjumba-Mukabu and Gunders, 1971; Shuval et al., 1984b).

cases) into a community could lead to massive infection of the wastewater stream and subsequent transmission of the disease to the consumers of the vegetable crops irrigated with the raw wastewater, as occurred in Jerusalem in 1970.

In Jerusalem the cycle of cholera transmission started with the first one or two imported cases from outside the city, most likely visitors from Jordan, which was experiencing a cholera epidemic at that time. Next, the wastewater flow to the irrigated vegetables carried the infectious organisms back to the residents (Fattal et al., 1986b). Similarly, our study from Santiago, Chile (Shuval, 1993) strongly suggests that typhoid fever and cholera can be transmitted by fresh salad crops irrigated with raw wastewater. Annually, there was a rapid rise in the number of typhoid fever cases in Santiago at the beginning of the irrigation season. The production of vegetables and salad crops was based on the irrigation of 16 000 hectares with, essentially raw, wastewater mixed with small flows of river water. The relatively high socioeconomic level, good water supply and good general sanitation in the city supports the hypothesis that irrigation with raw wastewater can become a major route for the transmission of such bacterial disease.

2.4.2 Cattle grazing on wastewater-irrigated pastures

Wastewater is often used to irrigate pastureland for cattle and sheep. What are the health risks associated with this practice? There is only limited epidemiological evidence to indicate that beef tapeworms (*Taenia saginata*) have been transmitted to populations consuming the meat of cattle grazing on wastewater-irrigated fields or fed crops from such fields. However, there is strong evidence from Melbourne, Australia (Penfold et al., 1937), and from Denmark (Jepson and Roth, 1949) that cattle grazing on fields freshly irrigated with raw wastewater or drinking from raw wastewater canals or ponds can become heavily infected with the parasite. This condition can become serious enough to require veterinary attention and may lead to economic loss. Irrigation of pastures with raw wastewater from communities where tapeworm infections prevail may provide a major pathway for the continuing cycle of transmission of the infection to animals and humans.

2.4.3 Exposure of wastewater farmers

Obviously the individuals most intensely exposed to pathogens at the farms where wastewater-irrigation is practiced are the farmers themselves. Some of the studies with the clearest epidemiological evidence relate to the health of such farm workers. Sewage farm workers exposed to raw wastewater in areas of India, where *Ancylostoma* (hookworm) and *Ascaris* infections are endemic, have much higher levels of infection than other agricultural workers (Krishnamoorthi et al., 1973). The risk of hookworm infection is particularly great in areas where farmers customarily work barefoot, as motile hookworm larva can penetrate the skin. Sewage farm workers in this study also suffered more from anemia (a symptom of severe hookworm infestation) than control farmers working in non-sewage irrigated fields. Thus, there is evidence that continuing occupational exposure to irrigation with raw wastewater can have a direct effect on human productivity and, as a result, on the economy.

Sewage farm workers are also more likely to become infected with cholera if the raw wastewater used for irrigation is from an urban area experiencing a cholera epidemic. This situation is particularly likely to arise in an area where cholera is not normally endemic and where the level of immunity among the sewage farm workers is low or non-existent. This proved to be the case in the 1970 cholera outbreak in Jerusalem (Fattal et al., 1986b).

Studies from industrialized countries have thus far produced only limited, and often conflicting, evidence of the incidence of bacterial and viral diseases among wastewater-irrigation workers exposed to partly or fully treated effluent, or among workers in wastewater treatment plants exposed directly to wastewater or wastewater aerosols (Camann and Moore, 1987). Most morbidity and serological studies have been unable to give a clear indication of the prevalence of viral diseases among such occupationally exposed groups. It is hypothesized that many sewage farmers or treatment plant workers have acquired relatively high levels of permanent immunity to most of the common enteric viruses endemic in their communities at a young age. Thus, by the time they are exposed occupationally, the number of susceptible workers is small and not statistically significant. Presumably this is also the case among infants and children in developing countries, because they are exposed to most endemic enteric viral diseases by the time they reach working age. Although this is not the case for some bacterial and protozoan pathogens, multiple routes of concurrent infection with these diseases may well mask any excess infection among wastewater-irrigation workers in developing countries. There is evidence, however, that aerosols heavily laden with pathogens, derived from raw or minimally treated wastewater spray irrigation or aeration tanks can cause disease in exposed workers or residents living in the immediate vicinity. Our study of the transmission of legionella bacteria among wastewater workers exposed to sprinkler irrigation revealed that they had a significantly higher level of legionella antibodies than non-irrigation workers in the same communities (Shuval et al., 1989c).

2.4.4 Exposure of residents in the vicinity of wastewater farms

A number of studies have evaluated the potential negative health effects that might result from living in the vicinity of farms where wastewater-irrigation, particularly sprinkler irrigation, is practiced. There is little evidence linking disease and/or infection among population groups living near wastewater treatment plants or wastewater-irrigation sites with pathogens contained in aerosolized wastewater.

Most studies have shown no demonstrable disease resulting from such aerosolized wastewater, which is caused by sprinkler irrigation and aeration processes. (Camann and Moore, 1987) Researchers agree, however, that most of the early studies have been inadequate. Our studies in Israel suggest that aerosols from sprinkler irrigation with poor microbial quality wastewater can, under certain circumstances, cause limited infections among infants living near wastewater-irrigated fields. The studies also concluded, however, that these were negligible and could be controlled by better treatment of the wastewater (Fattal et al., 1986a, 1987; Shuval et al., 1988, 1989b). Other studies have shown that young children who work or play in wastewater-irrigated fields can become infected in those cases where the microbial quality of the wastewater is very poor (WHO, 2006).

These findings support the conclusion that, in general, relatively high levels of immunity against most viruses endemic in the community block additional environmental transmission by wastewater-irrigation. Therefore, the additional health burden is not measurable. The primary route of transmission of such enteroviruses, even under good hygienic conditions, is through contact infection in the home at a relatively young age. Thus, as mentioned above, such person to person contact infection in the home with poor levels of hygiene is even more common in developing countries, so that using a town's wastewater for irrigation would not normally be expected to transmit viral disease in rural areas with high levels of immunity.

2.4.5 Epidemiological evidence of beneficial effects from wastewater treatment in reducing exposure to pathogens

When raw wastewater is used for irrigation there is no doubt that the wastewater stream carries very high concentrations of pathogens. Traditionally, conventional wastewater treatment plants have not been designed to significantly reduce the concentration of pathogenic microorganisms. However, such treatment can nonetheless provide a degree of removal up to about 85–99% reduction in coliform bacteria and pathogens. Some epidemiological studies have provided evidence that negative health effects can be reduced when wastewater is treated to partially remove pathogens. For example, Baumhogger (1949) reported that, in 1944, residents of Darmstadt who consumed salad crops and vegetables irrigated with raw wastewater experienced a massive infection of *Ascaris*; residents of Berlin, where biological treatment and sedimentation were applied to the wastewater prior to the irrigation of similar crops, did not. Another study on intestinal parasites was conducted on school children near Mexico City (Sanchez-Levya, 1976). The prevalence of intestinal parasites in children from villages that used wastewater-irrigation did not differ significantly from that in children from the control villages, which did not irrigate with wastewater. The lack of a significant difference between the two groups may have resulted from long-term storage of the wastewater in a large reservoir for weeks or months prior to its use for irrigation. It is assumed that sedimentation and pathogen die-away during long-term storage were effective in removing protozoa and the eggs of helminths, the pathogens of interest in this study. This study provides the first strong epidemiological evidence of the health protection provided by microbial reductions achieved in wastewater storage reservoirs. The degree of microbial reduction during the storage period in these reservoirs has been estimated at 99.9–99.99%, thus reducing the concentration of fecal coliforms (FC) from 10^7/100 ml to 10^3–10^4/100 ml.

Furthermore, the absence of negative health effects in population groups residing near wastewater treatment facilities or irrigation sites in Lubbock, Texas (Camann and Moore, 1987) and in Muskegon, Michigan (Clark et al., 1981) appears to be associated with the fact that well-treated effluents from areas of low endemicity were used for irrigation. Data from these field studies strongly suggest that pathogen reduction by conventional wastewater treatment, including long-term storage in wastewater reservoirs, can have a positive effect by providing protection for human health. In all the above studies, this positive protective effect was achieved despite the use of effluent that had not been disinfected and that contained a few thousand FC bacteria per 100 ml. These data suggest that appropriate conventional wastewater treatment resulting in effective reduction of FC to the level of a few thousand/100 ml, but not total removal, can provide a high level of health protection to exposed population groups.

2.4.6 *Critical review of epidemiological evidence of health effects of wastewater and excreta use in agriculture*

The most recent studies on this topic were done by Blumenthal and Peasey (2002), who carried out a detailed, authoritative update of the epidemiological evidence of health effects of wastewater use in agriculture for the London School of Hygiene and Tropical Medicine-(LSHTM). Their critical review of 42 selected studies of wastewater and excreta use in agriculture (including ten epidemiological studies of consumer risks from wastewater reuse, 25 studies of risks to farm workers and populations living near wastewater-irrigated fields, and seven studies of the effect of exposure to excreta or night-soil used as fertilizer) enabled the quantification of the risks of gastrointestinal infection from exposure to untreated wastewater, as well as an evaluation of the effects and benefits of wastewater treatment in reducing such risks. Their detailed study reconfirmed the main findings of earlier epidemiological evaluations, including our own, which provided the scientific basis for the second edition of the WHO *Guidelines for the Safe Use of Wastewater in Agriculture and Aquaculture* (1989) and which were used as a critical input in the preparation of the third edition (WHO, 2006). Their review of epidemiological studies of wastewater use shows that there are significant risks of gastrointestinal infection to consumers of crops, farm workers and their families and nearby populations exposed to untreated-raw wastewater or excreta that carry a heavy burden of pathogens. The studies showed that wastewater treatment prior to use can reduce those risks and provided an indication of the extent of wastewater treatment needed to protect exposed populations against risks from various wastewater-borne pathogens.

For unrestricted irrigation, including irrigation of vegetable and salad crops normally eaten uncooked they concluded that:

> When wastewater was only partially treated there was evidence that the risk of enteric infections (of bacterial and viral origin) was still significant when consumers ate some types of uncooked vegetables irrigated with water exceeding the 1989 WHO guideline values of 1000 FC/100 ml by a factor of 10. There is no evidence supporting any need to reduce the guideline value below the <1000 FC/100 ml in such circumstances. When sprinkler irrigation was used and the population was exposed to wastewater aerosols, there is an increased risk of infection when the quality of the wastewater was 10^6 thermotolerant coliforms (TC)/100 ml or essentially raw, but no increased risk of infection when the quality of the wastewater was 10^3–10^4 FC/100 ml.

Concerning the health risks to farm workers and their families, they concluded that "The data support the need for a fecal coliform guideline to protect farm workers, their children and nearby population from enteric viral and bacterial infections." They concluded that a guideline of 10^3 FC/100 ml would be safe for farmers and children under 15. They also concluded that there is a risk of *Ascaris* infection to farm workers and their families, particularly to children who are exposed to wastewater-irrigation. They concluded that ". . .the data suggest that the WHO nematode egg guideline of <one egg/liter is adequate if no children are exposed, but a guideline of <0.1 egg per liter would be safer if children are in contact with wastewater."

The recent extensive literature reviews by the WHO and their consultants (Blumenthal et al., 2000; Blumenthal and Peasey, 2002) have validated our initial findings. Table 2.1 presents a summary of the WHO findings (WHO, 2006).

2.4.7 Conclusions from the analysis of the epidemiological studies

It is possible to draw certain conclusions from the series of epidemiological studies on the health effects of wastewater use in agriculture. The studies from both developed and developing countries indicate that the following diseases are occasionally transmitted via raw or very poorly treated wastewater:

- The general public may develop ascariasis, trichuriasis, typhoid fever, or cholera by consuming salad or vegetable crops irrigated with raw wastewater, and possibly tapeworm by eating the meat of cattle grazed on wastewater-irrigated pasture. There may also be limited transmission of other enteric bacteria and protozoa.
- Wastewater-irrigation workers may develop ancylostomiasis (hookworm infection), ascariasis, possibly cholera, and, to a much lesser extent, infection caused by other enteric bacteria and viruses, if exposed to raw wastewater.
- Although there is no demonstrated risk to the general public residing in areas close to where wastewater is used in sprinkler irrigation, there may be minor transmission of enteric viruses to infants and children living in these areas, especially when the viruses are not endemic to the area and raw wastewater or very poor quality effluent is used.

Thus, the empirical evidence on disease transmission associated with raw wastewater-irrigation in developing countries strongly suggests that helminths are the principal

Table 2.1 Summary of health risks associated with the use of wastewater for irrigation (WHO, 2006)

Group exposed	Health threats		
	Helminth infection	**Bacteria/viruses**	**Protozoa**
Consumers	Significant risks of helminth infection for both adults and children with untreated wastewater	Cholera, typhoid and shigellosis outbreaks reported from use of untreated wastewater; seropositive responses for *Helicobacter pylori* (untreated); increase in non-specific diarrhea when water quality exceeds 10^4 thermotolerant coliforms/100 ml	Evidence of parasitic protozoa found on wastewater-irrigated vegetable surfaces but no direct evidence of infection
Farm workers and their families	Significant risks of helminth infection for both adults and children in contact with untreated wastewater; increased risk of hookworm infection to workers who do not wear shoes, risks for helminth infection remain, especially for children, even when wastewater is treated to <one helminth egg/l; adults are not at increased risk at this helminth concentration	Increased risk of diarrheal disease in young children with wastewater contact if water quality exceeds 10^4 thermotolerant coliforms/100 ml; elevated risk of *Salmonella* infection in children exposed to untreated wastewater; elevated seroresponse to *Norovirus* in adults exposed to partially treated wastewater	Risk of *Giardia intestinalis* infection reported to be insignificant for contact with both untreated and treated wastewater; another study in Pakistan estimated a threefold increase in risk of *Giardia* infection for farmers using raw wastewater compared to irrigation with fresh water. Increased risk of amoebiasis observed from contact with untreated wastewater
Nearby communities	Transmission of helminth infections not studied for sprinkler irrigation but same as above for flood or furrow irrigation with heavy contact	Sprinkler irrigation with poor water quality (10^{6-8} total coliforms/100 ml), and high aerosol exposure associated with increased rates of infection; use of partially treated water 10^{4-5} thermotolerant coliforms/100 ml or less in sprinkler irrigation is not associated with increased viral infection rates	No data for transmission of protozoan infections during sprinkler irrigation with wastewater

problem, with some limited transmission of bacterial and viral disease. In interpreting the above conclusions, one must remember that the vast majority of developing countries are in areas where helminthic and protozoan diseases such as hookworm infection, ascariasis, trichuriasis, and taeniasis are endemic. In some of these areas, cholera is endemic as well. It can be assumed that in most developing countries, in populations with low levels of personal and domestic hygiene, the children will become immune to the endemic enteric viral diseases when very young through contact infection in the home.

As pointed out earlier, these negative health effects were all detected in association with the use of raw or primarily treated wastewater of poor microbial quality and laden with pathogens. Therefore, wastewater treatment processes that effectively remove all, or most, of these pathogens, can and do reduce the negative health effects caused by the utilization of raw wastewater. Whereas helminths are very stable in the environment, bacteria and viruses rapidly decrease in numbers in the soil and on crops. Thus, the ideal treatment process prior to water recycling and wastewater use, should be particularly effective in removing helminth eggs, even if it is somewhat less efficient in removing bacteria and viruses. It is beyond the scope of this chapter to deal with water treatment processes appropriate for recycling and reuse.

In general, the above ranking of pathogens will not apply to the more developed countries or other areas in which helminth infections are not endemic. In those areas the negative health effects, if any, resulting from irrigation with raw or partly treated wastewater will probably be associated mainly with bacterial and protozoan diseases and, in a few cases, with viral diseases. Whatever the country or the conditions, however, the basic strategies for control are the same – the pathogen concentration in the wastewater stream must be reduced and/or the type of crops irrigated must be restricted.

Overall, our studies have demonstrated that the extent to which disease is transmitted by irrigation with partially or fully treated wastewater is much less than was widely believed to be the case by the public at large and public health officials in the past. It is possible to hypothesize that part of the early fears of serious public health risks of enteric disease transmission by wastewater-irrigation was based on psychological and cultural abhorrence and repulsion, as well as fears and uncertainties associated with human excreta and their associated odors. These fears were reinforced to some extent by early qualitative microbial studies, which showed that some pathogens of wastewater origin can survive in the wastewater stream, in the environment, in the soil, and on crops for days and even weeks. Little or no attention in those early evaluations was paid to the fact that the microbial concentrations were drastically reduced in the environment, very often to levels below that required for the minimal infectious dose or that their virulence was damaged or reduced by hostile environmental factors. Today the careful and critical evaluation of the mass of available epidemiological evidence researched by various groups of qualified experts in public health and epidemiology clearly shows that disease transmission by crops irrigated with partially treated wastewater carrying several thousand FC/100 ml is essentially non-existent.

Moreover, this study does not provide epidemiological support for the use of the much emulated California standard requiring a coliform count of 2/100 ml for effluent to be used in the irrigation of edible crops and even less support for the more recent and more stringent US Environmental Protection Agency (USEPA)/US Agency for International Development (USAID) (1992) recommended guideline of zero FC. No detrimental health effects were detected or reported when well-treated wastewater with much higher coliform counts was used.

2.5 Control of crops and irrigation methods to reduce health risks

Early in the development of wastewater-irrigation for agriculture, methods of reducing health risks by controlling the type of crops grown or the methods of irrigation have been proposed and in some cases used effectively. The risk of transmission of communicable disease agents to the general public by irrigation with raw or settled wastewater can be reduced by a number of agronomic techniques. Some of these restrict the types of crops grown, and others, through modification and/or control of irrigation techniques, prevent or limit the exposure of crops for human consumption to pathogens in the wastewater (e.g. Fine et al., 2006).

2.5.1 Regulating the type of crops

One of the earliest and still most widely practiced remedial measures is to restrict the type of crops irrigated with raw wastewater or with the effluent of primary sedimentation. As there is ample evidence that salad crops and other vegetables normally eaten uncooked are the primary vehicles for the transmission of disease associated with raw wastewater-irrigation, prohibiting the use of raw effluent to irrigate such crops can be an effective remedial public health measure. Although such regulations have been effective in countries with a tradition of civic discipline and an effective means for inspection and enforcement of pollution control laws, they will be of less value in situations where those preconditions are absent. In many arid and semiarid areas near major urban centers, where subsistence farmers irrigate with raw wastewater, the market demand for salad crops and fresh vegetables is very high. Thus, governmental regulations prohibiting farmers to grow such crops using wastewater to irrigate would be little more than a symbolic gesture. Even under the best of circumstances, it is difficult to enforce regulations that work counter to market pressures; to enforce regulations that prevent farmers from obtaining the maximum benefit from their efforts under conditions of limited land and water resources would be impossible.

2.5.2 Controlling irrigation methods

Basin irrigation of salad and vegetable crops usually results in direct contact of the crops with wastewater, thus introducing a high level of contamination. Sprinkler irrigation of salad crops also results in the deposit of wastewater spray on the crops and their contamination. The level of contamination may be somewhat less than basin irrigation. Many vegetables that grow on vines (e.g. tomatoes, cucumbers, squash) can be partially protected from wastewater contact if properly staked and/or grown hanging from wires that keep them off the ground. Some of these vegetables will, nevertheless, inevitably touch the ground.

Well-controlled ridge-and-furrow irrigation reduces the amount of direct contact and contamination. These methods cannot completely eliminate direct contact of the wastewater with leafy salad crops and root crops.

Drip irrigation causes much less contamination of the crops than any other irrigation method. In fact, our studies indicate that the use of drip irrigation tubes under polyethylene plastic surface sheeting used as mulch can vastly reduce or totally eliminate crop contamination. In our studies, we determined that the level of enteroviruses and bacterial indicator organisms on cucumbers grown under such protective drip irrigation systems was negligible (Sadovski et al., 1978). Drip irrigation is the most costly form of irrigation, but its hygienic advantages make it attractive as a safe method of wastewater-irrigation of sensitive vegetable and salad crops, even when the microbial quality of the effluent is not up to the strictest standards (see, e.g., Fine et al., 2006). In water-scarce areas drip irrigation is particularly attractive as it is the most water-efficient method of irrigation. Problems of clogging of tubes do arise but have been successfully controlled by use of pretreatment by sedimentation and/or filtration (see below). The third edition of the WHO guidelines suggests that with drip irrigation an effluent quality of 10 000 TC/100 ml provides a high degree of public health safety. Drip irrigation is also more efficient in water utilization and is beneficial to crop growth as the water dosage is more uniform and thus reduces stress from over- and undersaturation of the soil.

Fruit orchards do well with basin or ridge and furrow irrigation, but normal overhead sprinkler irrigation leads to direct contamination of the fruit. With low-level, low-pressure sprinkler irrigation, however, the main spray is below the level of the branches, and the fruit is less likely to be contaminated. In all cases, windfall picked from the ground will have been in contact with wastewater-contaminated soil.

Another possible control measure is to discontinue irrigation with wastewater at a specified period, such as 2 weeks, before harvesting the crop. This option is feasible for some crops, but the timing of a vegetable harvest is difficult to control. In addition, some types of vegetables are harvested daily, over long periods of time, from the same plot.

Some of the above irrigation control techniques can help reduce the danger of crop contamination, but they are feasible only in fairly advanced and organized agricultural economies. Health regulations dependent on any of the above procedures to protect certain high-risk crops from contamination must be backed up by legal sanctions and enforced by frequent inspections. If well-organized inspection and law enforcement systems are not present, as in some developing countries, the value of these options as a major remedial strategy may be limited. However, in the case of large, centrally operated sewage farms, managed by the government or large well-organized companies, such procedures can be of considerable value.

2.6 Development of health standards and guidelines for wastewater use

2.6.1 *The importance of health guidelines and standards for reuse*

One of the most important and widely practiced administrative methods for public health protection from the risks of uncontrolled wastewater-irrigation, particularly of vegetables and salad crops consumed uncooked, is the establishment of guidelines or legally binding standards for the microbial quality of wastewater used for irrigation. This section will review the scientific basis and historical and social forces that influenced the evolution of microbial standards and guidelines for safe wastewater use for agricultural purposes. This analysis will draw extensively on World Bank and WHO studies and reports, the goal of which was a cautious re-evaluation of the credible scientific evidence that could provide a sound basis for establishing feasible health guidelines for safe wastewater use (EAWAG, 1985; Shuval et al., 1986; WHO, 1989; Shuval, 1990; WHO, 2006).

The strict health regulations governing wastewater use that have been developed in the industrial countries over the past 70 years, such as those of the Department of Health of the State of California (Ongerth and Jopling, 1977), which require an effluent standard of 2 FC/100 ml for irrigation of crops eaten uncooked, and even the more recent USEPA/USAID (1992) recommended guideline values for unrestricted effluent use in agriculture of 0 FC/100 ml, have been based to some extent on early scientific data indicating that most enteric pathogens can be detected in wastewater and that they can survive for extended periods in wastewater-irrigated soil and crops (see Fig. 2.1). The early regulations were formulated before the full development of epidemiological evidence and risk analysis methods. As a result, policy makers, possibly out of caution and to some extent out of fear for litigation, used the cautious zero risk approach and introduced very strict regulations that they believed would protect public health (and safeguard themselves from criticism) against the potential risks thought to be associated with wastewater use. Most industrial countries were not concerned that these regulations were overly restrictive because the economic and social benefits of wastewater use were of only marginal interest.

2.6.2 The USEPA/USAID initiative for wastewater use health guidelines

In 1992, USEPA (USEPA/USAID, 1992), with the support of USAID, established its own rigorous recommended water quality guideline values for wastewater-irrigation of crops eaten raw: 0 FC/100 ml, a biochemical oxygen demand (BOD) of 10 mg/l, a turbidity of 2 nephelometric turbidity units (NTU) and a free chlorine residual of 1 mg/l. This quality of wastewater effluent can only be achieved in very costly high-tech, equipment-intensive wastewater treatment plants that require a high level of technological infrastructure for operation and maintenance, so that they can continuously meet such very rigorous standards. These guidelines were drafted by one of the leading American consulting engineering firms under contract to USAID. Such consulting engineering firms, often owned by wastewater treatment equipment manufacturers, naturally tend to favor such high-tech treatment processes. It is not unreasonable to assume that such high-tech wastewater treatment plants, that must be built to meet the requirement of these new US guidelines, are in the economic vested interests of the equipment manufacturers and their engineers. Again, these American guideline values are essentially as strict as those required for drinking water quality and reaffirm the conservative "no risk" or "fail safe" approach that has been taken by the early California standards. During the drafting stages, when the consulting engineers who authored the report of these new US guidelines were challenged by the author as the representative of the WHO as to why they choose such rigorous and unreasonably high standards when epidemiological and risk analysis research did not justify the need for such standards, the reply was "America can afford it and we prefer to play it safe". Some may hold different views to explain the motivation of the US authorities in insisting on such rigorous health guidelines for water reuse. One hypothesis is that the US authorities might have an interest in the need to establish such strict guidelines as a trade-barrier to keep certain agricultural products, irrigated with wastewater from being imported in competition with local farmers.

2.6.3 Impact of highly restrictive reuse guidelines and standards

Many of the current standards restrict the types of crops to be irrigated with conventional wastewater effluent to those not eaten raw. Regulations like those in California and Israel, requiring the effluent used for the irrigation of edible crops to have a bacterial standard approaching that of drinking water (2 FC/100 ml), or those currently recommended by the USEPA/USAID (1992) of 0 FC/100 ml, require high-tech treatment facilities, which are expensive to construct and to operate and which are usually not technically feasible or sustainable without highly skilled operators and a high-tech service infrastructure. This is particularly true for developing countries, but even applies to many developed countries. In reality, a standard of 0 or 2 FC/100 ml for irrigation is superior to the quality of drinking water for the majority of urban and rural poor in developing countries and many developed countries as well where FC concentrations are generally in the range of 10 or more/100 ml of drinking water.

In developed countries, where these crop restrictions can normally be enforced, vegetable and salad crops are not usually irrigated with wastewater. In the developing countries, many of which have adopted the same strict regulations, public health officials do not "officially" approve of the use of wastewater for irrigation of vegetable and salad crops eaten raw. However, when water is in short supply such crops are widely irrigated illegally with raw or poorly treated wastewater. This usually occurs in the vicinity of major cities, particularly in semiarid regions. It is estimated that currently some 50 countries throughout the world irrigate 50 million hectares of crops with essentially raw wastewater and produce 12% of the world's crops (Bos et al., 2007) with resulting health risks.

As the official effluent standards for vegetable irrigation are not within the obtainable range of common engineering practice and for economic considerations, new projects to improve the quality of effluent are not usually approved. With the authorities insisting on unattainable, expensive, and unjustifiable standards, farmers are practicing widespread uncontrolled and unsafe irrigation of salad crops with raw wastewater. Highly contaminated vegetables are supplied directly to the nearby urban markets, where such horticultural products can command high prices. This is a tragic case in which official insistence on the "strictest regulations" and the "very best" prevents cities and farmers from achieving the "feasible" and the "good". In other words: the perfect is the enemy of the good.

Some serous inconsistencies exist between the strict California or USEPA recommended guidelines and standards, which require edible crops to be irrigated with wastewater of drinking water quality, and the actual agricultural irrigation with normal surface water as practiced in the USA, Europe, and other industrialized countries with high levels of hygiene and public health. There are few, if any, microbiological limits on irrigation with surface water from rivers or lakes, which may be polluted with raw or treated wastewater. For example, USEPA's water quality criteria for unrestricted irrigation with surface water is 1000 FC/100 ml (USEPA, 1972). A United Nations Environment Programme (UNEP)/WHO world survey of river water quality (UNEP/WHO, 1996) has indicated that most rivers in Europe have mean FC counts of 1000–10 000/100 ml. And yet none of these industrialized countries have restrictions on the use of such river water for irrigation. A number of microbial guidelines have been developed for recreational waters considered acceptable for human contact and swimming. In Europe, guideline values vary from 100 coliforms/100 ml in Italy to 20 000 coliforms/100 ml in the former Yugoslavia. The European Union has in the past recommended a guideline value of 2000 FC/100 ml for recreational waters and only recently reduced them to 200/100 ml.

It is difficult to explain the logic of a 0 or 2 FC/100 ml standard for effluent irrigation when farmers all over the USA and Europe can legally irrigate any crops they choose with surface water from free-flowing rivers and lakes, which often have FC levels of over 1000/100 ml. There is some recent evidence of disease transmission by vegetable crops irrigated with polluted river water in the USA (C. Gerber, personal communication, 2007). However, so far the Americans have not set any legal irrigation requirements or limitations for the quality of natural, although often polluted, river water.

2.6.4 The World Bank/WHO initiative to re-evaluate wastewater use guidelines

Because of the above questions, and because of the fact that the strict coliform standards adopted by many countries were rarely enforced and often found to be unfeasible for economic reasons, and due to the lack of adequate technical infrastructure, in 1981, the World Bank and the WHO initiated an extensive multidisciplinary study on the health effects of wastewater-irrigation. The primary goal was to obtain an up-to-date scientific evaluation of the epidemiological and public health justification and validity of existing standards and guidelines, and to develop alternatives if this was deemed to be justified. These studies were carried out by three teams of epidemiologists, engineers, agronomists, and environmental specialists simultaneously and independently at three different environmental sciences and public health research centers – the London School of Hygiene and Tropical Medicine, the International Reference Centre on Waste Disposal (EAWAG), in Dübendorf near Zurich, and the Division of Environmental Sciences of the Hebrew University of Jerusalem, Israel. These groups, working independently, prepared reports summarizing their findings, analysis, and recommendations (Feachem et al., 1983; Blum and Feachem, 1985; Shuval et al., 1986).

In July 1985, the three independent teams and an additional group of environmental experts, including engineers and epidemiologists, met in Engelberg, Switzerland, under the auspices of UNDP, World Bank, WHO, UNEP and IRCWD (EAWAG). They reviewed the preliminary results of the scientific studies, which provided new epidemiological data and insights and from them formulated new proposed microbiological guidelines for treated wastewater use in agricultural irrigation (EAWAG, 1985). The group accepted the main findings and recommendations of our UNDP–World Bank study (Shuval et al., 1986) and concluded that: "Current guidelines and standards for human waste use are overly conservative and unduly restrict project development, thereby encouraging unregulated human waste use". This is one of the author's most important contributions in promoting safe water reuse.

The studies of all three groups confirmed that there was no epidemiological evidence of negative health effects from wastewater-irrigation of crops eaten uncooked with effluents having FC bacteria concentrations in the range of 1000/100 ml. At the meeting it was clear to all that the detailed epidemiological evidence supported the guideline recommendation of FC of 1000/100 ml for the irrigation of crops eaten uncooked. The group of scientists at Engelberg unanimously confirmed this recommendation, and the famous Engelberg Report (EAWAG, 1985) presents a historic breakthrough in the rational liberalization in health guidelines for water recycling and wastewater use in agriculture.

The guideline values recommended in the Engelberg Report were later accepted and approved by a WHO Meeting of Experts (WHO, 1989). Before the WHO Executive Board approved these guideline recommendations as the official WHO position, the draft report was sent out for review and approval by a panel of some 100 epidemiologists, public health officials, and environmental engineers. Thus, these guideline values for the microbial quality of effluent used for wastewater-irrigation of edible crops carry the stamp of approval of the highest international authority on public health and environmental matters. Other important international technical assistance agencies joined the WHO in supporting the new recommended guidelines for wastewater-irrigation including the World Bank, the Food and Agricultural Organization of the United Nations (FAO), the UNEP, and the UNDP. Meanwhile, a number of governments of both developing and developed countries have adopted these 1989 WHO recommended guidelines, including France, Portugal, and Mexico.

There were a number of innovations in these recommended guidelines. As the possibility of transmitting helminth infections to farmers by wastewater-irrigation of even non-edible crops was identified as a principal public health problem, a new, stricter approach to the use of raw wastewater was developed. The new WHO guidelines recommend effective water treatment in all cases to remove helminth eggs to a level of one or fewer helminth eggs per liter. The main innovation of the second edition of the WHO guidelines was: for crops eaten uncooked, an effluent must contain one or fewer helminth eggs per liter, with a geometric mean of FC not exceeding 1000/100 ml. This is a much more liberal coliform standard than the early California requirement of 2 total coliforms/100 ml.

An attractive feature of the second edition of the WHO guidelines (1989) was that effluent guideline values could be readily achieved with low cost, robust waste stabilization pond systems, and wastewater storage and treatment reservoirs that are particularly suited to developing countries in those areas where land and sun are freely available. In conjunction with alternating wastewater storage reservoirs, even higher degrees of treatment with an added safety factor can be achieved (Juanico and Shelef 1994). Most studies indicate that the critical design parameter to achieve the WHO microbial guidelines in the effluent is a long detention period of up to 25–30 d in maturation ponds or reservoirs.

2.6.5 Evaluating the WHO's 2006 health guidelines for wastewater reuse in agriculture

As mentioned, the 1989 second edition of the WHO guidelines recommended among other things that the quality of effluent for unrestricted irrigation of salad and vegetable crops normally eaten uncooked could be 1000 FC/100 ml. These guidelines have been very influential, and many UN agencies as well as a number of countries have adopted or adapted them for their wastewater use practices. However, some countries such as the USA (USEPA/USAID, 1992), Australia (2006), and Israel (Sviva, 2005) have recommended much stricter health guidelines for unrestricted agricultural use of wastewater in the range of 0–10 FC/100 ml.

After several years of intensive research, study and consultations with world experts on public health, epidemiology, agronomics and environmental sciences, and engineering, as well as other UN agencies, WHO published, in 2006, a third edition of the *Guidelines for the Safe Use of Wastewater, Excreta and Greywater in Agriculture and Aquaculture*, in four volumes. This third edition, which supersedes the previous editions, has been drafted by a panel of 35 experts. The third edition is recognized as representing the position of the UN system on issues of wastewater, excreta and greywater use, and health by "UN-Water", the coordinating body of the 24 United Nations agencies and programs concerned with water issues.

Since the publication of the second edition of the WHO guidelines in 1989, the development of Quantitative Microbial Risk Analysis (QMRA) methodology has enabled increasingly sophisticated and reliable analysis of health risks associated with wastewater use in agriculture. Quantitative Microbial Risk Analysis can estimate risks from a variety of different exposures and/or pathogens that would be difficult to measure through conventional epidemiological investigations due to the high cost and necessity of studying large populations. Shuval and colleagues (1997) were the first to develop and use QMRA to perform a comparative risk analysis of the USEPA/USAID (1992) wastewater reuse guidelines, Israeli recommended reuse guidelines, and the WHO guidelines of 1989. They based their study on laboratory research and a modified version of the Haas health risk analysis mathematical model (Haas et al., 1993). They found that when lettuce or other similar salad crops were irrigated with treated wastewater meeting the WHO recommended guidelines of 1000 FC/100 ml, the risk of mild enteric disease was insignificant. They estimated that the additional cost of treatment to avoid one case of enteric disease if the USA or Israel wastewater reuse guidelines were required could be roughly $500 000/case of mild disease prevented. They questioned whether such a standard and the high level of treatment required to achieve it was cost-effective and justified from a public health or economic point of view.

More recently, early studies by Shuval and colleagues (1997) have been extended by Professor Duncan Mara and colleagues for the 2006 WHO report to provide further information as a basis to evaluate the infection risks associated with the exposure to crops irrigated by wastewater, as well as exposure of irrigation workers. A combination of standard QMRA techniques (Haas et al., 1993) and 10 000-trial Monte Carlo simulations (Sleigh and Mara, 2003) was used. The risk estimates were determined using the β-Poisson dose–response model for bacterial and viral infections and the exponential dose–response model for protozoan infections.

In the 1990s, the World Bank and WHO developed a new approach to measuring disease burdens, allowing cost-effectiveness analysis of different health interventions for different diseases. This new and innovative approach assumes that the most appropriate metric for expressing the burden of a disease is the disability adjusted life year (DALY) (Murray and Lopez 1996; Box 2.1).

Box 2.1 Disability adjusted life years (DALYs) (WHO, 2006)

Disability adjusted life years are a measure of the health of a population or burden of disease due to a specific disease or risk factor. Disability adjusted life years attempt to measure the time lost because of disability or death from a disease compared with a long life free of disability in the absence of the disease. Disability adjusted life years are calculated by adding the years of life lost to premature death to the years lived with a disability. Years of life lost are calculated from age-specific mortality rates and the standard life expectancies of a given population. Years lived with a disability are calculated from the number of cases multiplied by the average duration of the disease and a severity factor ranging from 1 (death) to 0 (perfect health) based on the disease (e.g. watery diarrhea has a severity factor ranging from 0.09 to 0.12, depending on the age group) (Murray & Lopez, 1996). Disability adjusted life years are an important tool for comparing health outcomes, because they account not only for acute health effects but also for delayed and chronic effects – including morbidity and mortality (Bartram et al., 2001). When risk is described in DALYs, different health outcomes (e.g. cancer versus giardiasis) can be compared and risk management decisions can be prioritized.

The WHO has adopted, in the third edition of the *Guidelines for Drinking-water Quality*, a tolerable burden of waterborne disease from consuming drinking water of $\leq 10^{-6}$ DALY lost/person/yr (Bartram et al., 2001). This value corresponds to a tolerable excess lifetime risk of fatal cancer of 10^{-5} per person (i.e. an individual has a one in 100 000 lifetime chance of developing fatal cancer) from consuming drinking water containing a carcinogen at its guideline value concentration in drinking water. This level of disease burden can be compared with a mild but more frequent illness, such as self-limiting diarrhea caused by a microbial pathogen from wastewater-irrigation.

The estimated disease burden associated with mild diarrhea at an annual disease risk of one in 1000 (10^{-3}) (approximately one in ten lifetime risk) is also about 1×10^{-6} DALY (1 μDALY)/person/yr. Such a high level of health protection is required for drinking water, as it is expected to be "safe" by those who drink it. As food crops irrigated with treated wastewater, especially those eaten uncooked, are also expected to be as safe as drinking water by those who eat them, the third edition of the WHO guidelines requires the same high health protection level of $\leq 10^{-6}$ DALY lost/person/yr for wastewater use in agriculture.

2.6.6 WHO guidelines for monitoring and validation

For monitoring and validation purposes to assure safe wastewater use, the third edition of the WHO guidelines is based on analysis by QMRA methods and the criterion that a very high level of health protection of $\leq 10^{-6}$ DALY lost/person/yr for wastewater use in agriculture be assured. They are based as well on authoritative research studies on indicator organism and pathogen removal by various treatment processes, irrigation practices, and environmental factors. They assume that the level of TC (TC used in the WHO third edition are essentially equivalent to the FC used in early editions) in raw wastewater are in the range of 10^7/100 ml. The WHO report concludes:

> The Monte Carlo–QMRA results for unrestricted irrigation, based on the exposure scenario of lettuce consumption together with the relevant epidemiological evidence show that, in order to achieve $\leq 10^{-6}$ DALY/person/yr a total pathogen reduction of 6 log units for the consumption of leaf crops (lettuce) and 7 log units for the consumption of root crops (onions) is required".

In the third edition of the WHO guidelines, a pathogen reduction of 6–7 log units is used as the performance target for unrestricted irrigation to achieve the tolerable additional disease burden of $\leq 10^{-6}$ DALY/person/yr.

However, the WHO guidelines point out that:

> (A) 6–7 log unit pathogen reduction may be achieved by the application of appropriate health protection measures, each of which has its own associated log unit reduction or range of reductions. A combination of these measures is used, such that, for all combinations, the sum of the individual log unit reductions for each health protection measure adopted is equal to the required overall reduction of 6–7 log units.

To be on the conservative side, WHO assumes that the 7 log reduction of pathogens (or 99.99999%), or an equivalent reduction of TC bacteria, is required to assure the safe consumption of wastewater-irrigated lettuce and root crops. However, they do not assume that all of the pathogen/TC removal must be achieved solely through wastewater treatment processes.

They assume that the degree of post-wastewater treatment pathogen reduction and/or removal resulting from irrigation and exposure to sun, soil, and hostile environmental factors is some 0.5–2 logs/d or at least a reasonable minimum estimate for mean wastewater-irrigation conditions of 2 logs (99% reduction). They also assume a minimum pathogen/TC removal of 1 log (90%) by simple home rinsing and washing of wastewater-irrigated salad crops and vegetables. If detergents and mild disinfectants are used, the pathogen removal might be as high as 99%. They also assume an additional post-treatment 99% bacterial reduction when drip irrigation is used, as research has shown that there is little exposure of crops to wastewater pathogen when proper drip irrigation is practiced.

Thus, the third edition of the WHO guidelines points out that if these additional environmental factors, resulting in significant levels of pathogen and indicator organism inactivation and/or removal, are taken into account, the degree of TC removal required to monitor and validate the efficacy of the wastewater treatment process need not be more than 4 logs or 99.99%. As an example of one possible scenario the WHO report assumes an initial TC concentration in the raw wastewater of 10^7/100 ml, for which a 7 log unit pathogen reduction may be achieved by the application of appropriate health protection measures, each of which has its own associated log unit reduction or range of reductions. One of the examples given indicates that a minimum 2 log reduction is assumed for pathogen die-away under field conditions (higher in warm, sunny climates) and a 1 log

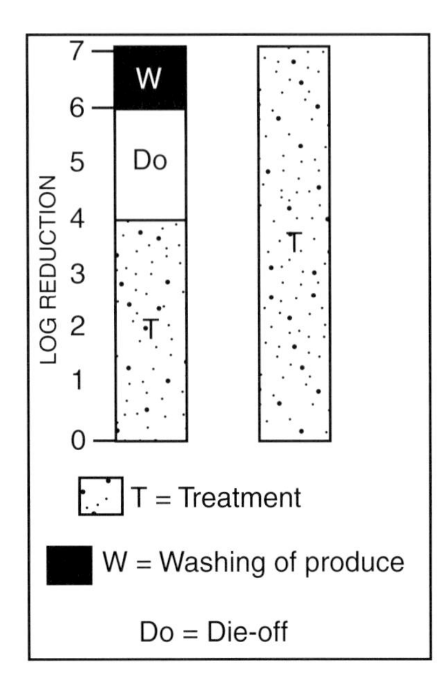

Figure 2.3 Graphic presentation of an example of the WHO monitoring and validation guidelines. In order to achieve a 7 log reduction in thermotolerant coliforms as recommended by the WHO for irrigation of crops eaten uncooked, two options are presented graphically.

reduction minimum is assumed for pathogen removal by home rinsing/washing. Based on the above assured reductions, then the wastewater treatment would require only a 4 log removal (99.99%) by the wastewater treatment process, or in other words an effluent quality of about 1000 FC/100 ml. An example of the application of the WHO guidelines for monitoring and validation is presented graphically in Figure 2.3.

In the option shown in the right column of Figure 2.3, chosen by some countries such as the USA, Israel, and Australia, the total 7 log (99.99999%) reduction in TC is achieved by wastewater treatment (T) only. This can be achieved technically by expensive full secondary and high-tech tertiary wastewater treatment. In the option shown in the left column, recommended by the WHO, it is assumed that there will always be at least a 2 log reduction (99%) of TC resulting from die-off (DO) after treatment as a result of exposure of pathogens to detrimental field conditions including ultraviolet radiation, exposure to heat, desiccation, bacterial competition, hostile soil environment such as pH, and natural die-off under generally unfavorable field conditions. An additional 1 log reduction (90%) will normally result from simple home kitchen washing of produce (W) before consumption. Thus, in order to achieve the total of 7 log reduction, the additional removal required by wastewater treatment would be a 4 log reduction (99.99%), which can be achieved either by low-cost, land-extensive treatment methods such as stabilization ponds in those areas where land and sun are available, or by relatively lower cost conventional wastewater treatment methods such as chemically assisted primary treatment and some newer combined methods including final membrane filtration.

The report states: "This is similar to the recommended required effluent quality of 1000 fecal coliforms (FC)/100 ml in the second edition of these guidelines (WHO, 1989)". Other scenarios assuming higher microbial removal by irrigation techniques and/or environmental factors resulting in lower levels of effluent requirements, such as 10 000 TC/100 ml are also presented, such as with the use of drip irrigation.

The section on unrestricted irrigation with wastewater concludes as follows in respect to those countries and authorities who have chosen the option of basing all of the risk reduction and pathogen/TC removal solely on advanced wastewater treatment technology:

> This option does not take into account pathogen reduction due to (a) natural die-off between final irrigation and consumption, and (b) specific food and/or cooking and overall health protection is therefore greater than even 10^{-6} DALY lost/person/yr. The very high costs and operational complexities of the wastewater treatment processes required by this option will generally preclude its applicant in many countries. Even in countries where this option is affordable it should be subject to robust cost-effectiveness analysis.

I have made my own simple economic analysis of the extra costs involved in meeting the USEPA/USAID, and the similar Australian and suggested Israeli, guidelines. All of these would require high-tech tertiary wastewater treatment to meet the standard of 0–10 TC/100 ml and their other strict requirements for low levels of BOD, total suspended solids, nitrogen, phosphorus, etc. Assuming that the WHO 2006 recommended guidelines for wastewater reuse for unrestricted irrigation of about 1000 TC/100 ml can be achieved at the costs associated with normal conventional secondary treatment, or even less if stabilization ponds can be used, then the minimal additional costs of tertiary treatment have been estimated as US$0.18/M^3 by Haruvi for conditions in Israel (Haruvi et al., 2004). Assuming a water application rate for crop irrigation with treated wastewater of between 5000 and 10 000 M^3/ha, then the additional cost of meeting those very rigorous standards can be estimated at some US$900–1800/ha. The WHO QMRA risk analysis studies indicate that little or no measurable additional health benefit would be gained from the significant extra expense associated with that extra degree of treatment. It is hard to find a public health, agronomic or economic rational justification for such additional expenditures. I doubt if agriculture can afford that extra expense in most cases.

Israeli authorities have recently recommended changing the approach toward wastewater effluent treatment from "crop-specific + barriers to pathogen infection" to upgrading almost all the effluent to a level ensuring $<$10 TC/100 ml. The first approach was formulated by the Israeli health authorities that considered irrigation with secondary effluent to be a complementary treatment. The additional cost involved was estimated at $67 million/yr or $356/ha/yr according to crop selection and area allocation (Fine et al., 2006). This cost difference also stems from the fact that most effluent is used for the irrigation of non-edible crops, for which partial or no pathogen removal is appropriate. These costs do not internalize the additional cost for the fertilizer value of the wastewater (Chapter 4).

The meeting of WHO Experts in Geneva that drafted the WHO new guidelines concluded in a resolution approved by consensus that:

> While each country can and should select the combination of risk reduction elements that suit its epidemiological, social and economic needs, however, the in-depth risk analysis studies carried out by the group, provide a sound epidemiological basis for concluding that the options presented (with the lower levels of wastewater treatment and microbial effluent quality) provide a high degree of health risk reduction and health protection which should meet the needs of most countries in a reasonable cost effective manner.

The group has concluded that these new risk assessment studies validated the 1989 WHO guideline recommendation of 1000 FC/100 ml for unrestricted irrigation of most vegetable and salad crops eaten uncooked.

2.7 Conclusions

Thus, after well over more than a century, health guidelines for wastewater reuse have gone through a complete cycle, from no regulation or control in the nineteenth century, to unreasonably and irrationally strict standards in the earlier and then latter part of the twentieth century. This was based to some extent at first on fear of the unknown and esthetic considerations, and, in my view, later to some degree partially on vested interests of the wastewater treatment equipment industry, to what now appears to be a more scientifically sound basis with a less restrictive approach as recommended by the third edition of the WHO guidelines and the other UN agencies befitting the more rational scientific and cost-effective approach of the twenty-first century. It is hoped that this new approach will encourage the development of controlled wastewater reuse for the benefit of mankind, while providing an appropriate level of health protection. The third edition of the WHO guidelines should be carefully studied and evaluated by the community of water scientists, engineers, public health officials, economists, and policy makers as to their possible implications for various national wastewater use guidelines.

Dedication

This chapter is dedicated to the memory of Professor Badri Fattal, my lifelong friend and colleague, who participated in most of the research studies reported on here.

References

Australia. (2006) National Guidelines for Water Recycling: Managing Health and Environmental Risks (Phase 1) 2006. Natural Resource Management Ministerial Council, Environment Protection and Heritage Council, Australian Health Ministers' Conference. 415 pp. www.ephc.gov.au/sites/default/files/WQ_AGWR_GL__Managing_Health_Environmental_Risks_Phase1_Final_200611.pdf (last accessed Sept. 2009).

Applebaum, J., Guttman-Bass, N., Lugten, M., Teltsch, B., Fattal, B. and Shuval, H.I. (1984) Dispersion of aerosolized enteric viruses and bacteria by sprinkler irrigation with wastewater. In: *Enteric Viruses in Water* (ed. J. L. Melnik) pp. 193–201. Monogr Virol: 15., Karger, Basel.

Bartram, J., Fewtrell, L. and Stenstrom, T.A. (2001) Harmonised assessment of risk and management for water-related infectious disease: an overview. In: *Water Quality: guidelines, standards and health: Assessment of risk and risk management for water-related infectious disease* (eds L. Fewtrelland J. Bartram) pp. 1–16. World Health Organization (WHO). IWA Publishing, London, UK.

Baumhogger, W. (1949) Ascariasis in Darmstadt and Hessen as seen by a wastewater engineer. *Zeitschrift fur Hygiene und Infektionskrankheiten* **129**, 488–506 (in German).

Ben-Ari, J. (1962) The incidence of *Ascaris limbriocoides* and *Trichuris trichuria* in Jerusalem during the period of 1934-1960. *American Journal of Tropical Medicine and Hygiene* **11**, 336–368.

Blum, D. and Feachem, R.G. (1985) *Health Aspect of Night Soil and Sludge Use in Agriculture and Aquaculture: an Epidemiological Perspective*. International Reference Centre for Wastes Disposal, Dubendorf, Switzerland.

Blumenthal U.J., Duncan, M., Peasey, A., Ruiz-Palacios, G. and Stott, R. (2000) Guidelines for the microbiological quality of treated wastewater used in agriculture: recommendations for revising WHO guidelines. *Bulletin of the World Health Organization* **78** (9), 1104–1116.
Blumenthal, U.J. and Peasey, A. (2002) *Critical Review of Epidemiological Evidence of the Health Effects of Wastewater and Excreta Use in Agriculture*. London School of Hygiene and Tropical Medicine. December 2002. www.who.int/water_sanitation_health/wastewater/whocriticalrev.pdf (last accessed Sept. 2009).
Bos, R., Thomas, A. and Redwood, M. (2007) The WHO Guidelines: Learning form implementation in four settings. *Proceedings of the 6th Conference on Wastewater Reclamation and Re-Use for Sustainability*, 9-12 October 2007, Antwerp, Belgium. International Water Association-IWA.
Camann, D.E. and Moore, B.E. (1987) Viral infections based on clinical sampling at a spray irrigation site. In: *Implementing Water Reuse. Proceedings of Water Reuse Symposium IV* pp. 847–863. 2–7 August 1987, Denver, CO, USA.
Clark C.S., Bjornson, H.S., Holland, J.W., Elia, V.J. and Majeti. V.A. (1981) *Evaluation of the Health Risks Associated with the Treatment and Disposal of Municipal Wastewater and Sludge*. EPA-600/S-1 81-030. US Environmental Protection Agency, Cincinnati, OH, USA.
EAWAG (1985), Health aspects of wastewater and excreta use in agriculture and aquaculture. *Report of Review Meeting of Environmental Specialists and Epidemiologists*, Engelberg, Switzerland. Sponsored by the World Bank/WHO. International Reference Centre for Wastes Disposal (EAWAG) Dübendorf, Switzerland.
Fattal, B., Wax, Y., Davies, M. and Shuval, H.I. (1986a) Health risks associated with wastewater irrigation: An epidemiological study. *American Journal of Public Health* **76**, 977–979.
Fattal, B., Yekutiel, P. and Shuval, H.I. (1986b) Cholera outbreak in Jerusalem 1970 revisited: The case for transmission by wastewater irrigated vegetables. In: *Environmental Epidemiology: Epidemiological Investigation of Community Environmental Disease* (ed. J.R. S Goldsmith) pp. 49–59. CRC Press, Boca Raton, LA, USA.
Fattal, B., Margalith, M., Shuval, H.I., Wax. Y. and Morag, A. (1987) Viral antibodies in agricultural populations exposed to aerosols from wastewater irrigation during a viral disease outbreak. *American Journal of Epidemiology* **125**, 899–906.
Feachem, R.G., Bradley, D.J., Garelick, H. and Mara, D.D. (1983) *Sanitation and Disease: Health Aspects of Excreta and Wastewater Management*. John Wiley and Sons, Chichester.
Fine, P., Halperin, R. and Hadas, E. (2006) Economic considerations for wastewater upgrading alternatives: an Israeli test case. *Journal of Environmental Management* **78**, 163–169.
Haas C. N., Rose, J. B., Gerba, C. and Regli, S. (1993) Risk assessment of virus in drinking water. *Risk Analysis* **13**, 545–552.
Haruvi, N., Shalhevet, S. and Ravina, I. (2004) Financial and managerial analysis of irrigation with treated wastewater in Israel. *Journal of Financial Management and Analysis* **16**, 65–73.
Jepson, A. and Roth, H. (1949) Epizootiology of cyiticercus bovis-resistance of the eggs of Taenia saginata. In: *Report of the 14th International Veterinary Congress* Vol. 2. His Majesty's Stationery Office, London, UK.
Jjumba-Mukabu, O.R. and Gunders, E. (1971) Changing patterns of intestinal helminth infections in Jerusalem. *American Journal Tropical Medicine and Hygiene* **20**, 109–116.
Juanico, M. and Shelef, G. (1994) Design operation and performance of stabilisation reservoirs for wastewater irrigation in Israel. *Water Research* **28**, 175–186.
Khalil, M. (1931) The pail closet as an efficient means of controlling human helminth infections as observed in Tura prison, Egypt, with a discussion on the source of ascaris infection. *Annals of Tropical Medicine and Parasitology* **25**, 35–62.
Krey, W. (1949) The Darmstadt ascariasis epidemic and its control. *Zeitschrift fur Hygiene und Infektions krankheiten* **129**, 507–518 (in German).
Krishnamoorthi, R.P., Abdulappa, M.R. and Anwikar, A.R. (1973) Intestinal parasitic infections associated with sewage farm workers with special reference to helminths and protozoa. In: *Proceeding of Symposium on Environmental Pollution*. Central Public Health Engineering Research Institute, Nagpur, India.
Murray, C.J.L. and Lopez, A.D. (1996) The global burden of disease. In: *Harvard School of Public Health, World Health Organization, World Bank, Global Burden of Disease and Injury Series* Vol. I., pp. 1–98. Boston, USA.

Ongerth, H.J. and Jopling, W.F. (1977) Water reuse in California. In: *Water Renovation and Reuse* (ed. H. I. Shuval), pp. 219–256. Academic Press, New York, USA.
Penfold, W.J., Penfold, H.B. and Philips, M. (1937) Taenia saginata: its growth and propagation. *Journal of Helminthology*, **15**, 41–48.
Sadovski, A.Y., Fattal, B., Goldberg, D., Katzenelson, E. and Shuval, H.I. (1978) High levels of microbial contamination of vegetables irrigated with waste water by the drip method. *Applied Environmental Microbiology* **36**, 824–830.
Sanchez-Levya, R. (1976) *Use of Wastewater for Irrigation in District 03 and 88 and its Impacts on Human Health.* Master's thesis, School of Public Health, Mexico City (in Spanish).
Shuval, H.I., Yekutiel, P. and Fattal, B. (1984) Epidemiological evidence for helminth and cholera transmission by vegetables irrigated with wastewater: Jerusalem a case study. *Water Science and Technology* **17**, 433–442.
Shuval, H.I., Adin, A., Fattal, B., Rawitz, E. and Yekutiel, P. (1986) *Wastewater Irrigation in Developing Countries: Health Effects and Technical Solutions.* World Bank Technical Paper Number 51 World Bank, Washington, D.C, USA.
Shuval, H.I., Wax, Y., Yekutiel, P. and Fattal, B. (1988) Prospective epidemiological study of enteric disease transmission associated with sprinkler irrigation with wastewater: An Overview. In: *Implementing Water Reuse, Proceedings of Water Reuse Symposium IV* pp. 765–781 American Water Works Association, Denver, USA
Shuval, H.I., Guttman-Bass, N., Applebaum, J. and Fattal, B. (1989a) Aerosolized enteric bacteria and viruses generated by spray irrigation of wastewater. *Water Science and Technology* **21**, 131–135.
Shuval, H.I., Wax, Y., Yekutiel, P. and Fattal, B. (1989b) Transmission of disease associated with wastewater irrigation: A prospective epidemiological study. *American Journal of Public Health* **79**, 850–852.
Shuval, H.I., Fattal, B. and Bercovier, H. (1989c) Legionnaires disease and the water environment in Israel. *Water Science and Technology* **20**, 11–12.
Shuval, H.I. (1990) Wastewater irrigation in developing countries - health effects and technical solutions. *Summary of World Bank Technical Paper no. 5, UNDP-World Bank wastewater and sanitation discussion paper series DP no. 2*, Washington D.C., USA.
Shuval, H.I. (1993) Investigation of typhoid fever and cholera transmission by raw wastewater irrigation in Santiago, Chile, *Water Science and Technology* **27**, 167–174.
Shuval, H.I., Lampert, Y. and Fattal, B. (1997) Development of a risk assessment approach for evaluating wastewater reuse standards for agriculture. *Water Science and Technology* **25**, 15–20.
Sleigh, P.A. and Mara, D.D. (2003) *Monte Carlo Program for Estimating Disease Risks in Wastewater Reuse.* Water and Environmental Engineering Research Group, Tropical Public Health Engineering, University of Leeds, Leeds. www.efm.leeds.ac.uk./CIVE/MCARLO
Sviva (2005) Israel Ministry of Environmental Protection. www.sviva.gov.il/Enviroment/bin/en.jsp?enPage=BlankPage&enDisplay=view&enDispWhat=Object&enDispWho=Articals^l3576&enZone=vaadot_tkina (in Hebrew) (last accessed 2 July 2008).
Teltsch, B., Shuval, H.I. and Tadmor, J. (1980) Die-away kinetics of aerosolized bacteria from sprinkler irrigation of wastewater. *Applied Environmental Microbiology* **39**, 1191–1197.
UNEP/WHO. (1996) *Global Environmental Monitoring System – GEMS Report for 1996.* WHO, Geneva.
USEPA. (1972) *Water Quality Criteria-1972. A report of the committee on water quality criteria-National Academy of Sciences.* USEPA, Washington, D.C., USA.
USEPA/USAID. (1992) *Guidelines for Water Reuse.* United States Environmental Protection Agency, Washington (Wash. Tech. Report no. 81, September 1992).
World Health Organization. (1989) *Health Guidelines for the Use Of Wastewater in Agriculture and Aquaculture. Report of a WHO Scientific Group-WHO Technical Report Series 778.* Geneva.
World Health Organization. (2006) *Guidelines for the Safe Use of Wastewater, Excreta and Greywater - Volume 2: Wastewater Use in Agriculture*, WHO, Geneva. www.who.int/water_sanitation_health/wastewater/gsuweg2/en/index.html (last accessed Sept. 2009).

Chapter 3
Irrigation with recycled water: guidelines and regulations

Nikolaos V. Paranychianakis, Miquel Salgot and Andreas N. Angelakis

Terminology definitions

Criteria Standards, rules or tests on which a judgment or decision can be based. Sometimes used interchangeably with "standards", "rules", "requirements", or regulations.
Guidelines Recommended or suggested standards, criteria, rules, or procedures that are voluntary, advisory, and non-enforceable.
Dual distribution systems Two independent piping systems that are used to deliver potable and recycled water.
Standard Applies to any enforceable rule, principle, or measure established by a regulatory authority. Often synonymous numerical water quality limits.
Regulations Criteria, standards, rules, or requirements that have been legally adopted and are enforceable by government agencies.
Restricted irrigation Irrigation with treated wastewater (TWW), of which the quality imposes restrictions on assessment of the irrigated area or in the use of the produced crop to protect public health.
Unrestricted irrigation Irrigation with TWW, of which the quality does not impose any restrictions on assessment of the irrigated area or in the use of the produced crop.
Water recycling The use of wastewater that is captured and redirected back into the same water use scheme, such as in agriculture. However, the term water recycling is often used synonymously with water reclamation.
Water reuse The beneficial use of TWW, for example in agricultural irrigation and industrial cooling.
(Source: Asano et al., 2007)

Treated Wastewater in Agriculture, First Edition, edited by Guy J. Levy, Pinchas Fine and Asher Bar-Tal

3.1 Introduction

Rapid population growth, improvement in living standards, expansion of irrigated land, and global warming cause increases in water demands worldwide. As a consequence, freshwater supplies are stretched to their limits, especially in semiarid and arid regions. Additionally, overexploitation of groundwater supplies in coastal aquifers often results in sea intrusion and subsequent degradation of quality. Thus, in order to meet future water demands and to conserve existing freshwater supplies, measures are required that aim to improve water use efficiency and/or use of alternative water sources like treated wastewater, stormwater, and/or saline waters.

Water reuse for various purposes is increasingly practiced throughout the world, particularly in arid environments including the south and west of the USA, Australia, Middle East, Mediterranean Region, etc. (Angelakis et al., 1999; U.S. Environment Protection Agency (USEPA), 2004). Water recycling is meant to help close the anthropogenic water cycle and enable sustainable reuse of available water resources. When integrated to water resources management, it can be considered as an integral part of pollution control and water management strategies. It may also result in benefits to public health, the environment, and economic development. Recycled water may provide additional renewable, reliable amounts of water and contribute to the conservation of freshwater supplies. Agricultural and landscape irrigation are currently the major uses of recycled water worldwide. Reuse of treated wastewater (TWW) in agriculture presents important benefits, including supplement of nutrients to crops and provision of further treatment of the applied TWW. However, severe drawbacks include nutrient release to the environment and spreading of pathogens.

The need to protect public health and ensure environmental sustainability has led environmental authorities to develop water quality standards depending on the water reuse application. Different approaches have been adopted to ensure the protection of public health and the environment depending mainly on the social, cultural, and economic status of the different countries and states. Legislation is distinguished into two forms: regulations and guidelines. Regulations are legally adopted, enforceable, and mandatory; guidelines are advisory, voluntary, and non-enforceable, but can be incorporated in TWW reuse permits and via this route can become enforceable requirements. Some international organizations and national agencies prefer adoption of guidelines to provide flexibility in regulatory requirements depending on site-specific conditions, which can result in different requirements for similar uses. This is the case in the World Health Organization (WHO), and USEPA and Australia. Most developed countries have established conservatively low-risk standards based on a high technology and cost approach, such as those of California (State of California, 2000). In contrast, developing countries adopt low-cost technologies and less strict water quality standards mainly influenced by the 1989 WHO recommendations (WHO, 1989). Furthermore, risk assessment and management analysis methodologies are increasingly adopted to establish criteria for water reuse.

The objectives of this chapter are to provide a brief summary of the existing criteria for water recycling for irrigation around the world, as well as the current trends and developments in this field. Water quality issues and the appropriate treatment schemes in order to maximize the benefits of water reuse and eliminate potential ecological and health hazards are also discussed.

3.2 Historical development of water recycling and reuse regulations

The first historical evidence of wastewater used for irrigation goes back to the Middle and Late Bronze Ages (*c*.3000–1400 BC) in the Minoan Civilization (Angelakis and Spyridakis, 1996). In more recent history, the first large-scale water reuse projects were developed at the beginning of 1800s when "sewage farms", a technology developed in Germany, were expanded to England and to other European countries until the end of nineteenth century (Gerhard, 1909; Stanbridge, 1976). "Sewage farming" became a common practice to protect public health and to control water pollution. However, sewage farms were principally operated as disposal sites aiming to maximize the amount of wastewater applied per surface unit rather than to recycle it efficiently for crop irrigation.

The first projects of the intended water reuse were carried out in California at the beginning of twentieth century. The pressure on water resources and the benefits of reclaimed water use to crop yield due to the increased concentrations of nutrients stimulated water recycling for irrigation. The State of California, recognizing the benefits and the potential health risks, set in 1918 the first regulations of water reuse in agriculture (California State Board of Health, 1918). These regulations encouraged water reuse for irrigation of non-edible crops and for crops cooked before being eaten, while prohibiting the use of raw wastewater for crop irrigation. Since then, the "California Regulations" have been continuously updated to cover new applications and to meet the stringent requirements for public health and of the environment.

In 1978 Israel adopted the first water reuse regulations for irrigation, and in 1999 they were revised by the Israeli Ministry of Health. In 1973 the first guidelines for water reuse were published by the World Health Organization addressing health criteria and treatment processes for water reuse applications (WHO, 1973). In 1989 following a meeting in Switzerland and taking into account the recommendations of the Engelberg report (International Reference Centre for Waste Disposal (IRCWD), 1985), the first revision of the WHO guidelines was published (WHO, 1989). Following the publication of the "California Regulations" and the WHO recommendations, many countries and states developed criteria for water reuse that were mainly influenced by these two distinct philosophies. The most important developments in the evolution of water reuse criteria are summarized in Table 3.1. In 2006 Australia completely revised its national guidelines for use of reclaimed water, published in 2000, based on a risk assessment and management approach. The same year the second revision of the WHO guidelines appeared (WHO, 2006), which was also based on a quantitative risk assessment methodology.

Technological developments in wastewater treatment processes in the twentieth century led to the "Era of Wastewater Reclamation, Recycling and Reuse" (Asano and Levine, 1998). Currently, existing technologies can produce water quality for any intended use. Water recycling is increasingly considered as an integral component of water resources planning and management, especially in water-limited regions.

Currently, the contribution of water recycling is estimated as only a small proportion of total water use; however, it is expected to increase particularly in water-limited regions. A statistical analysis of potable water use in New York City showed that when daily temperatures exceed 25°C, water use increases by 11 l/inh.°C (roughly 2%) (Intergovernmental Panel on Climate Change (IPCC), 2007). Furthermore, every time more stringent standards are

Table 3.1 Historical evolution of the TWW reuse standards for irrigation (adapted from Salgot and Angelakis, 2001)

Year	Data and quality criteria
Before 1918	TWW reuse for irrigation was practiced in various regions but there were no relevant standards. Some preventive measures were applied
1918	California State Board of Public Health set up the "First regulations for use of sewage for irrigation purposes in California"
1973	The first guidelines for TWW reuse by WHO (100 FC/100 ml, 80% of samples)
1978	State of California TWW reuse regulations: 2.2 TC/100 ml (Title 22)
1978	Israel regulations: 12 FC/100 ml in 80% of samples: 2.2 FC/100 ml in 50% of samples
1983	World Bank Report (Shuval et al., 1986)
1983	State of Florida: No detectable *E. coli*/100 ml for crops consumed raw
1984	State of Arizona: Standards for virus (one virus/40 l) and *Giardia* (one cyst/40 l)
1985	Feachem Report (Feachem et al., 1983)
1985	Engelberg report (IRCWD, 1985)
1989	The revised guidelines by WHO for TWW reuse: 1000 FC/100 ml < one nematode egg/l
1990	State of Texas: 75 FC/100 ml
1991	Sanitary French recommendations influenced by WHO guidelines
1992	USEPA Guidelines for water reuse: No detectable FC/100 ml
1999	Revised Israel regulations: Unrestricted irrigation <1 FC/100 ml and a multi-barrier approach
2000	Australian guidelines
2000	Revised criteria of water recycling by the State of California (Title 22)
2004	The revised guidelines of USEPA for Water Reuse (indirect potable use)
2006	The second revision of WHO Guidelines for using Treated Wastewater in Agriculture: Risk analysis and management
2006	"Australian guidelines for water recycling: Managing health and environmental risks": Risk analysis and management

FC, fecal coliforms; IRCWD, International Reference Centre for Waste Disposal; TC, total coliforms; TWW, treated wastewater; USEPA, United States Environmental Protection Agency; WHO, World Health Organization.

imposed by environmental agencies for TWW discharge in surface water bodies, these encourage water reuse applications like agricultural irrigation, even in regions without limitations in water resources. Water reuse projects have increased worldwide in both number and scale in recent years. In the USA, 6.4 Mm^3/d of TWW are reused and this figure is estimated to increase by 15% every year (USEPA, 2004). However, large differences are reported between states; Florida, California, Texas and Arizona account for the majority of the water reuse in the USA, a fact stimulated by the lower water availability compared to other states. By the year 2000, 457 wastewater treatment plants (WWTPs) were operated in Florida with a capacity of 1542 Mm^3/yr (Geselbracht, 2003). In California, water reuse was estimated at 2.0 Mm^3/d for the year 2002 and is expected to double by 2010. Water recycling for irrigation in California amounts to 68% of the total TWW use (DWR, 2003). In Australia the amount of TWW is estimated from 9% to 14% of the volume of produced wastewater, even though it has been estimated that recycled water could potentially supply approximately 50% of the water needs of urban users and a significant proportion of the irrigation water demands in Australia. The agriculture industry was the largest user of recycled water in 2000–1, accounting for 423.3 Mm^3 (Dimitriadis, 2005). In Spain the volume of water reused was estimated to be 400 Mm^3/yr in 2005, with 86% used for irrigation. By 2012 the volumes of reused TWW are expected to reach 1200 Mm^3 (USEPA, 2004). In Cyprus the major portion of wastewater from large cities is collected, treated, and used for irrigation of agricultural crops, green areas, and landscapes (Papadopoulos, 1995). In Tunisia, recycled water accounted for 4.3% of available water resources in 1996, and it is estimated to reach 11% in 2030. In Israel, approximately 70% of the collected wastewater is treated and reused for irrigation. Approximately 32% of overall yearly water use (which is *c.*48% of domestic water use) is reused for irrigation in agriculture. The overall use of marginal water (sewage TWW, runoff and saline) in Israeli agriculture accounts to 53% of yearly water withdrawal (2006 data; Israel Water Authority, 2008). The volume of TWW compared to the irrigation water resources is estimated to be about 8% in Jordan, and 32% in Kuwait (USEPA, 2004). In Japan, water reuse was estimated to be 257.5 Mm^3/yr in 2003 (Japan Sewage Works Association, 2003), but it is mainly used for non-potable urban applications. Agricultural irrigation accounts for up to 13% of total TWW use (Ogoshi et al., 2001).

In most developing countries the situation is very complicated. In many countries including India, Pakistan, Chile, and Mexico, water reuse contributes significantly to agricultural production, but there are no accurate data available for the total volumes used. In addition, the use of raw or partially treated wastewater for irrigation and the unplanned reuse has been associated with severe health risks and environmental degradation (WHO, 2006; USEPA, 2004).

3.3 Water recycling in agriculture: quality issues

The protection of public health of both farmers and consumers, the prevention of environmental degradation, and the reduction of adverse effects on crop yield are the main issues that must be considered when irrigation of agricultural crops with TWW is practiced. Currently, most of the TWW irrigation standards focus on parameters related directly or indirectly with public health. Such parameters include the concentration of indicator organisms, biodegradable organic matter (biochemical oxygen demand, BOD), suspended solids (total suspended solids, TSS), turbidity, and residual chlorine. With regard to the other constituents like nutrients, salts, toxic organics, and trace elements that are related to environmental sustainability and/or plant yield, they are commonly not regulated. Guidelines for assessing chemical quality of irrigation water and the potential adverse effects on irrigated crops have been published (Ayers and Westcot, 1988; Rowe and Abdel-Magid, 1995) and they are also considered to cover water reuse.

3.3.1 Pathogens

The transmission of infectious diseases in highly exposed population groups such as farmers, workers, consumers and the public, or to grazing livestock is the major concern when use of TWW is planned for irrigation of agricultural crops or urban areas. A great diversity of pathogens, including bacteria, protozoa, cyanobacteria, helminths, and viruses are found in untreated wastewater, which are associated with a wide range of diseases, mainly gastrointestinal illness (Table 3.2). The number and type of pathogens vary substantially in raw wastewater among different regions and seasons depending on the prevailing climatic conditions and the health status of the population served. In the

Table 3.2 Population of common pathogens and indicators in raw wastewater, infectious doses (median) and illnesses

Organism	Number/l	Median infection dose (N50)	Illness
Bacteria			
Total coliforms (indicators)	10^7–10^9		
Fecal coliforms (indicators)	10^5–10^8		
E. coli (indicators)	10^5–10^{10}	10^6–10^{10}	Gastroenteritis, haemolytic uraemic syndrome
Enterococci (indicators)	10^6–10^7		
Clostridium perfringens	10^5–10^6		
Campylobacter jejuni	10–10^4	1000	Gastroenteritis, Guillain–Barré syndrome
Pseudomonas aeruginosa	10^3–10^6		Skin, eye, ear infections
Salmonella	1–10^5		Gastroenteritis, reactive arthritis
Shigella spp.	1–10^3	10–20	Dysentery
Vibrio cholerae	10^2–10^5	10^6–10^8	Cholera
Protozoa			
Cryptosporidium parvum	1–10^4	1–10	Gastroenteritis
Entamoeba histolytica	1–10^2	10–20	Amoebic dysentery
Giardia lamblia	10^2–10^5	<20	Gastroenteritis
Virus			
Enteric virus	10^5–10^6	1–10	Gastroenteritis, respiratory illness, nervous disorders, myocarditis
Rotavirus	10^2–10^5	<10	Gastroenteritis
Adenovirus	10–10^4	<5	Gastroenteritis, respiratory illness, eye infections
Norovirus	10–10^4		Gastroenteritis
Somatic coliphages (indicator)	10^6–10^9		
F-RNA coliphages (indicator)	10^5–10^7		
Helminths			
Ascaris lubricoides	1–10^3	1–10	Roundworm
Ancylostoma duodenale	1–10^3		
Trichuris trichiura	1–10^2		Whipworm

Adapted from Feachem et al. (1983); Geldreich (1990); NRC (1996); Yeates and Gerba, (1998); Bitton (1999); Crittenden et al. (2005); Asano et al. (2007).

following paragraphs a short description of the main groups of pathogenic organisms found in wastewater is performed.

Viruses are obligate intracellular parasites and they are characterized by a higher infectivity than other pathogens found in wastewater (Table 3.2). They are also more resistant to treatment processes. Untreated wastewater contains a wide range of viruses that are pathogenic to humans. Concentrations in excess of 10^4 viral particles/l of wastewater have been assessed. To date, routine procedures for the detection and quantification of virus in wastewater samples have not been developed. The currently applied methods to determine the presence of viruses in a water sample take about 14 days, and an additional 14 days are required to identify the viruses (USEPA, 2004). Enteroviruses are the most common type of pathogenic viruses in wastewater (Table 3.2) and consist of small, single-stranded RNA viruses including the poliovirus types 1 and 2, echovirus, enterovirus, and coxsackievirus hepatitis A. The enteroviruses are known to cause a wide range of diseases in humans including poliomyelitis, respiratory infections, acute gastroenteritis, aseptic meningitis, pericarditis, myocarditis and viral exanthema, conjunctivitis, and hepatitis. However, there have been no documented cases of viral disease resulting from water recycling in the USA (USEPA, 2004).

Bacteria are single-celled organisms characterized by small size (0.2–10 μm). They constitute the most common pathogens found in wastewater (Table 3.2). A wide range of pathogenic bacteria, including opportunistic organisms, occur in wastewater. Pathogenic bacteria of concern include species of the genera *Shigella*, *Salmonella*, *Legionella*, and *Camplylobacter*. Gastrointestinal infections are the most common diseases associated with bacteria including diarrhea, cholera, salmonellosis, and dysentery. Non-enteric diseases transmitted by bacterial pathogens include legionellosis, leptospirosis, and melioidosis, a pneumonia-like disease.

Protozoa are single-celled organisms that lack a cell wall, but possess a flexible covering called pellicle (Asano et al., 2007). A number of protozoan species have been identified in untreated wastewater. Protozoa produce (oo)cysts, which can survive for long intervals in wastewater, aquatic and terrestrial environments. The most common species include *Entamoeba histolytica*, *Giardia intestinalis*, and *Cryptosporidium parvum*. They are enteric pathogens and the infection occurs after the consumption of food or water contaminated with (oo)cysts. Major outbreaks of *Cryptosporidium parvum* have occurred in drinking water. These three parasites are of major concern to operators involved in the recycling of TWW. They have been associated with dysenteric diseases and infections of the liver, lungs, pericardium, skin, and brain. Both *C. parvum* and *G. intestinalis* can cause acute diarrhea. The cysts of all three parasites are resistant to desiccation, temperature, pH variations, and chlorination.

Helminths are common intestinal parasites, which are usually transmitted by the fecal–oral route. Some of these parasites require an intermediate host before becoming infectious to humans. Helminth parasites commonly detected in wastewaters include *Ascaris lumbricoides* (round worm), *Ancylostoma duodenale* (hook worm), and *Trichuris trichiura* (whip worm). The prevalence of *Ascaris* infection is influenced by population density, education standards, sanitation levels, degree of agricultural development, and cultural and dietary habits (Khuroo, 1996). *Ascaris lumbricoides* is endemic in regions of Asia, India, South America, and Africa. However, infections have also been reported in the developed world, although at much lower levels. The WHO lists intestinal nematodes as the greatest health risk associated with agricultural use of untreated wastewater (WHO, 1989).

Despite variations in the number and species of pathogenic organisms in raw wastewater, treatment processes and their reliability definitely determine the populations of pathogenic organisms in TWW and hence the health risks derived. Currently, existing epidemiological studies reveal that recycling water that has received secondary or a higher level of treatment does not pose severe health risks when used in agriculture. Disease outbreaks have been documented only in cases where raw or primary treated wastewater was used to irrigate crops eaten raw or grazing crops (Crook, 1998). Furthermore, when TWW is applied to the land, pathogens are subject to various abiotic and biotic constraints, which substantially reduce their population and survival period (Paranychianakis et al., 2006 and references therein). The contribution of these factors is a function of the prevailing climatic conditions and the applied management practices. In wet and cool climates and in crops that develop a dense canopy, pathogens may remain viable for long periods increasing the risks of infection (WHO, 2006 and references therein).

Monitoring of all the pathogenic microorganisms present in wastewater is actually impossible. The evaluation of microbiological quality of TWW is, therefore, based on monitoring of surrogate organisms called "indicator organisms". Indicator organisms should be characterized by specific properties as noted by the National Research Council (NRC) (NRC, 2004), but to date ideal indicator organisms have not been identified. Total or fecal coliforms (TC or FC) are the most commonly used indicators worldwide. *Escherichia coli* is also used as indicator in water reuse criteria by some environmental agencies. It is a member of fecal group bacteria and it is the most representative species of the occurrence of fecal contamination. Due to the fact that the coliform group shows poor correlation with the occurrence of other pathogenic organisms (virus, protozoa, helminths), some environmental agencies include monitoring requirements of additional pathogens including F-specific bacteriophage MS2, or poliovirus. Furthermore, concern also has been addressed lately for some emerging pathogens including *Giardia* spp., *Cryptosporidium* spp. and the *E. coli* O157:H7, which are not represented adequately by the currently used indicator organisms. For example, some monitoring of protozoa is required in California and Florida. Wastewater treatment plants in Florida with a capacity higher than 3.78×10^3 m^3/d must be sampled at least every two years for *Giardia* spp. and *Cryptosporidium* spp., whereas in smaller treatment facilities sampling is required every 5 years. California also requires sampling for *Giardia* spp. and *Cryptosporidium* spp. in non-restricted applications if tertiary treatment does not include sedimentation between chemical coagulation and filtration processes.

3.3.2 Biodegradable organic matter

Despite the beneficial effects of organic matter on soil fertility through its effects on soil moisture and nutrient retention, its presence favors pathogen survival and transport of both pathogens and pollutants downward. Organic matter provides food for pathogens and reduces the efficiency of disinfection processes. The BOD is used as a gross measurement of biodegradable organic matter. Limits of BOD set by environmental agencies to assess TWW quality depend on its intended use. Typical values of TWW BOD used for unrestricted irrigation and irrigation of food crops are usually lower than 5 mg/l, whereas values for restricted irrigation and irrigation of non-food crops are lower than 30 mg/l.

3.3.3 Total suspended solids

Total suspended solids have been associated with disinfection efficiency. Disinfectants like chlorine and ozone may react with organic matter and protect pathogenic organisms from disinfectants. Increased concentrations of TSS, as in TWW from stabilization ponds, may also cause problems in TWW distribution and application pipes and hydraulic facilities, and reduce soil permeability. Furthermore, a significant portion of toxic organics and trace elements are sorbed onto organic matter. Recommended limits of TSS in TWW typically range between 5 and 30 mg/l depending on the intended use. Turbidity is routinely used as a surrogate parameter of TSS.

3.3.4 Residual chlorine

The concentration of chlorine in TWW is associated with the disinfection efficiency as well as pathogen regrowth. Residual chlorine is toxic to aquatic organisms and must be removed from the TWW if it is planned to be discharged to surface water bodies. It may also react with organic substances found in TWW or water to form toxic byproducts. In addition, several crops are sensitive to concentrations higher than 1 mg/l and should be considered when TWW irrigation is practiced. In general, threshold values up to 1 mg/are recommended in most water reuse criteria. Concentration of residual chlorine of 0.5 mg/l or slightly greater in the distribution pipes is necessary to reduce odors and pathogen regrowth.

3.3.5 Toxic organic compounds

There have been concerns regarding the fate of a wide spectrum of persistent and toxic organic substances that are found in wastewater, including pesticides, hydrocarbons, dioxins, pharmaceuticals, personal care products, hormones, disinfection byproducts, etc. Several of these compounds are considered as carcinogenic, teratogenic, or mutagenic, or to interfere with human and animal hormonal systems. Typically, under the condition that industrial wastewaters are not discharged in sewage, these compounds occur at very low concentrations, which do not represent significant risks for public health. Nevertheless, the revised Australian guidelines (Natural Resource Management Ministerial Council and the Environment Protection and Heritage Council (NRMMC–EPHC, 2006) suggest that the risks from these compounds should be considered. Concern for these chemicals is addressed mainly during aquifer recharge if groundwater is used for potable supply or because of the adverse effects on aquatic organisms when TWW is discharged to surface waters. When TWW is applied to the land, organic substances are efficiently inactivated in the soil matrix, through sorption, biodegradation, or volatilization (Paranychianakis et al., 2006). With regard to their potential uptake and accumulation by irrigated crops, numerous studies suggest that, although highly variable depending on the properties of the compound, organic substances occur at levels too low to pose any risk for crop yield or human health. In addition, plants possess metabolic pathways capable of transforming or degrading various groups of organic substances (Davis et al., 2002). However, there are no known studies dealing with the additive and/or synergistic impacts of the mentioned chemicals on soil microorganisms and their effects on biological diversity. Irrigation with TWW has been associated with the deep percolation of organic pollutants found in TWW, as well as with pesticides that already exist in the soil (Downs et al., 1999). Traces of organic pollutants were also reported in the runoff from a TWW-irrigated site (Pedersen et al., 2003).

3.3.6 Nutrients

The nutrients of major concern in recycled water are nitrogen and phosphorus. In general they occur in recycled water at higher concentrations than those required for optimum crop growth. Increased concentrations of nutrients, and particularly of nitrogen, may significantly affect the yield of the irrigated crops through excessive vegetative growth in expanse of yield and lower accumulation of carbohydrates in fruits. In addition this may result in nutritional imbalances. Excessive application of nutrients to the land may result in leaching of nitrates to underground aquifer and runoff of both phosphorous and nitrogen to surface waters. Nutrient transport to water sources may degrade quality, reduce biodiversity, and result in health risks if these supplies are used for potable use. High concentrations of nitrates in supplies used for potable use have been associated with health risks including methhemoglobinemia, cancer, thyroid disease, and diabetes, and limits for nitrates have been established. Recent studies question the impact of nitrates on human health (L'hirondel and L'hirondel, 2001; Avery and L'hirondel, 2003), whereas Fewtrell (2004) states that nitrates may be only one of the factors causing methhemoglobinemia. Nutrient accumulation into surface water, due to runoff, induces eutrophic conditions, characterized by algal development in excess, oxygen depletion, and pH changes, causing severe impacts on aquatic life. Additionally, toxins produced by some species of cyanobacteria have been associated with cancer, gastroenteritis, and nervous system disorders (Ling, 2000; WHO, 1986). Routinely, limits for nutrients are not included in water reuse criteria. Exceptions are some criteria that set limits for nitrogen, when recycling is planned for indirect reuse and for environmental application like natural wetlands or stream augmentation.

3.3.7 Salinity

Although the concentration of total dissolved solids in TWW depends mainly on that of water supply, domestic use of water increases the concentration of inorganic dissolved solids in the range of 150–500 mg/l (Crook, 1998). The enhanced levels of salts often found in recycled water may result in severe growth inhibition yield reductions and toxicity development (Miyamoto, 2003; Paranychianakis et al., 2004). A number of guidelines have been developed to predict and manage the impacts of salts contained in irrigation water on plant performance. These guidelines were integrated and modified by Ayers and Westcot (1988) to expand their usefulness. In order to eliminate the impacts of salinity on the productivity of agricultural land, appropriate management practices must be applied, including the selection of suitable irrigation methods (surface or subsurface irrigation versus sprinklers), plant cultural practices, and selection of salt-tolerant genotypes.

3.3.8 Trace elements

Trace elements are of great importance because of their potential impact on human or animal health, crop performance, and the population and activity of soil microflora and microfauna. Particular concern is given to the occurrence and accumulation of cadmium, copper, and molybdenum as these elements can be harmful to human and animals at concentrations that do not pose a significant risk for crops. Overall, trace elements in municipal TWWs having received secondary treatment do not pose a risk to plant growth or for entering the food chain, providing that industrial wastewaters are not discharged into sewerage systems. During the past few years a significant decrease has been observed in the concentrations of trace elements found in wastewater in the developed world as a result of the stringent disposal policy for industrial TWWs. This tends to eliminate the risks resulting from heavy metals when TWW irrigation is practiced. In developing countries, however, TWW application to the land often has been found to cause accumulation of trace elements at levels higher than those suggested by environmental agencies or organizations (Mapanda et al., 2005; Rattan et al., 2005). Threshold values of heavy metals and other toxic elements to prevent reductions in agricultural production and to protect public health have been suggested by Ayers and Westcot (1988) and Rowe and Abdel-Magid (1995).

3.4 Treatment requirements and TWW quality monitoring

Health and environmental risks can be eliminated either by reducing pathogen populations and concentration of toxic chemicals to levels that represent acceptable risks, or by adopting practices that reduce human exposure. In uses of high human exposure, such as irrigation of open access green areas or irrigation of food crops, reclaimed wastewater must receive the highest treatment level to eliminate health risks. By contrast, when exposure to reclaimed water is low or incidental a lesser level of treatment can be accepted.

It is now widely recognized that the adoption of water quality parameters is not sufficient by itself to characterize and guarantee the quality of reclaimed water. The use of surrogate parameters to estimate the occurrence of pathogenic organisms and organic substances is not representative of all the pathogenic organisms or toxic substances. Total and FC, which are commonly adopted as indicator organisms, although providing reliable parameters for bacteria, are not representative indicators for protozoa and viruses. In addition, total organic carbon (TOC) used as a surrogate parameter of organic carbon is not sufficient for ensuring public health protection when indirect potable reuse is practiced. Findings from research studies and gathered experience from the operation and monitoring of WWTPs reveal that both treatment processes and water quality standards are required to ensure that risks to public health and the environment are eliminated. This has led environmental agencies to describe treatment levels or to specify certain treatment technologies in addition to quality limits in their water reuse criteria. It is argued that inclusion of treatment processes in water reuse regulations may result in economic unfeasibility and may stifle the development of innovative technologies (Asano et al., 2007). However, states with comprehensive regulations for water reuse allow the use of alternative treatment technologies after thorough monitoring to demonstrate that they meet the requirements set by environmental authorities (Crook, 1998). Furthermore, requirements for treatment reliability are included in some cases to prevent the use of TWW that received a lower level of treatment than expected due to process upset, power outage, or equipment failure (USEPA, 2004). Such requirements include warning alarms, standby power sources, additional treatment units, and emergency storage and disposal. Inclusion of treatment levels and/or specific treatment processes ensures that reduction of pathogens at particular thresholds occurs. Typical removal efficiencies of pathogens for different treatment processes are shown in Table 3.3. In addition, the monitoring of certain microbiological, biochemical, and physicochemical parameters aims to optimize operation of treatment technologies. Such a combination of treatment technologies and quality thresholds actually limits monitoring requirements without negatively affecting monitoring reliability. Inclusion of preventive on-site measures like localized irrigation methods, control of public access, crop selection, and adoption of buffer zones may

Table 3.3 Indicative removal indicator organisms and different groups of pathogenic organisms under different treatment technologies[a]

Treatment	*E. coli*	Bacteria[b]	Viruses[c]	*Giardia*	Helminths
Primary treatment	0–0.5	0–0.5	0–0.1	0.5–1.0	0–2.0
Secondary treatment	1.0–3.0	1.0–3.0	0.5–2.0	0.5–1.5	0–2.0
Dual media filtration with coagulation	0–1.0	0–1.0	0.5–3.0	1.0–3.0	2.0–3.0
Membrane filtration	3.5–>6.0	3.5–>6.0	2.5–>6.0	>6.0	>6.0
Reverse osmosis	>6.0	>6.0	>6.0	>6.0	>6.0
Lagoon storage	1.0–5.0	1.0–5.0	1.0–4.0	3.0–4.0	1.5–>3.0
Surface wetlands	1.5–2.5	1.0	—	0.5–1.5	0–2.0
Subsurface wetlands	0.5–3.0	1.0–3.0	—	1.5–2.0	—
Chlorination	2.0–6.0	2.0–6.0	1.0–3.0	0.5–1.5	0–1.0
Ozonation	2.0–6.0	2.0–6.0	3.0–6.0	—	-
Ultraviolet light	2.0–>4.0	2.0–>4.0	>1.0 [d] >3.0 [e]	>3.0	-

[a]Specific features of the process, including detention times, pore size, filter depths, disinfectant dose have a significant effect on removal of pathogens Sources: NWQMS (2006); WHO (1989); Rose *et al.* (1996, 2001); NRC (1998); Bitton (1999); USEPA (1999, 2003, 2004); Mara and Horan (2003).
[b]Bacterial pathogens including *Campylobacter*.
[c]Includes adenoviruses, rotavirus, and enterovirus.
[d]Adenovirus.
[e]Enterovirus, hepatitis.

substantially reduce exposure to pathogens and requirements for treatment and quality standards (Table 3.4).

Monitoring compliance points differ between countries or states. Treated wastewater sampling is often required at certain intervals and points. Reclaimed water monitoring is commonly applied at the WWTP, and only in some cases in the distribution system. Another viewpoint requires sampling at the point of use. The major argument for this position is the potential regrowth of pathogens between the WWTP and reuse site (Asano et al., 2007). However, the regrowth is attributed to non-pathogenic organisms due to the lack of host cells. Under certain conditions the quality of TWW may deteriorate due to entry of pathogens from external sources. Storage of TWW, often practiced at periods of low irrigation demand, may result in external contamination, thus stressing the need to set critical points for TWW monitoring in the distribution system (USEPA, 2004). For example, *G. lamblia*, *C. parvum*, and *E. coli* O157:H7, which infect both animals and humans, can enter surface water bodies and increase risks for human infection.

3.5 Water reuse criteria: irrigation of agricultural crops and landscapes

Because TWW has been mainly used for crop irrigation in the past, it is not surprising that water reuse criteria are principally oriented to this application. Water reuse criteria can be in the form of regulations or guidelines depending on whether these are legally enforceable or not, respectively. Criteria principally concentrate on public health protection and set limits for pathogenic microorganisms and parameters affecting the effectiveness of disinfection. With regard to agricultural use, concern focuses on the potential contamination of irrigated

Table 3.4 Exposure reductions provided by on-site preventive measures

Control measure	**Log reduction in pathogen exposure**[a]
Cooking or processing of produce (e.g. cereal, wine grapes)	5–6
Removal of skins from produce before consumption	2
Drip irrigation of crops	2
Drip irrigation of plants/shrubs	4
Drip irrigation of crops with limited to no ground contact (tomatoes, capsicums)	3
Drip irrigation of crops with no ground contact (horticultural crops, grapes)	5
Subsurface irrigation of above ground crops	4
Subsurface irrigation of plants/shrubs or grassed areas	5–6
Withholding periods – produce (decay rate)	0.5 log/day
Withholding periods for irrigation of parks/sports grounds (1–4 hrs)	1
Spray drift control (microsprinklers, anemometer systems, inward-throwing sprinklers, etc.)	1
No public access during irrigation	2
No public access during irrigation and limited contact after (non-grassed areas) (e.g. food crop irrigation)	3
Buffer zones (25–30 m)	1

[a]Based on virus inactivation. Enteric bacteria are probably inactivated at a similar rate. Protozoa will be inactivated if withholding periods involve desiccation. Sources: Asano et al. (1992), Tanaka et al. (1998), Haas et al. (1999), van Ginnekin and Oron (2000), Petterson et al. (2001), Mara and Horan (2003), WHO (2006), NRMMC-EPHC (2006)

crops and the transmission of infectious diseases to consumers and farmers. Regulations and guidelines commonly include water quality standards, treatment requirements in combination with preventive measures like irrigation method, intended use of crop (food or non-food crops, consumed raw or after processing), setback distances, and control of site access. In the following sections, an effort is made to briefly present the basic methodologies and the trends in the development of regulations/guidelines for agricultural and landscape irrigation. More detailed emphasis is given on the two basic philosophies, the "WHO Guidelines" and "California Regulations", which have been used as models for the development of criteria throughout the world. The revised USEPA guidelines and the National Australian Guidelines, as well as guidelines/regulations for other countries in which water reuse is a widespread practice, are discussed.

3.5.1 World Health Organization

In September 2006 the second revision of WHO guidelines for wastewater use in agriculture was released, first published in 1973 (WHO, 1973). The WHO guidelines encompass a different philosophy to that adopted by the developed world and have influenced many countries to set up reuse criteria.

The 1973 edition of the WHO guidelines set a limit of 100 FC/100 ml in water used for irrigation of crops eaten raw. In the first revision this limit was expanded to a geometric mean of 1000 FC/100 ml, and standards for helminths (<1 egg/l) were also included (WHO, 1989). This arose from epidemiological studies indicating that helminths were the major hazard associated with agricultural use of untreated or partially treated wastewater.

These guidelines were heavily criticized by experts in developed countries, and Blumenthal et al. (2000) stressed the need for more stringent limits for helminth eggs (<0.1 egg/l). It should be stated, however, that the WHO guidelines are mainly oriented to meet the needs of developing countries, and overly strict standards may not be applicable and may lead to less health protection because they may be viewed as unachievable under the prevailing socio-economic conditions. Nevertheless, it must be realized that in cases where raw wastewater is directly reused, the WHO guidelines are a major step forward.

The second revision (2006) of the WHO guidelines is substantially different to the previous editions. They are intended to be used as the basis for the development of international or national approaches to assess and manage public health risks arising from the use of wastewater. The primary aim of the guidelines is to maximize public health protection and the beneficial use of important resources, and they are better adapted to the situation of the developing world.

A quantitative microbial risk assessment (QMRA) approach is applied to assess and manage the hazards associated with pathogens and toxic chemicals based on the Stockholm Framework methodology (Fewtrell and Bartram, 2001). This methodology suggests the integrated management of hazards by setting tolerable risks and taking into account all the exposures that may lead to a particular disease. Helminths are recognized as the greatest risk in cases where raw wastewater is used for irrigation of food crops. In addition, QMRA analysis revealed that risks of rotavirus infection resulting from different exposures were always higher compared to those of *Campylobacter* and *Cryptosporidium* (Table 3.5).

The burden of a disease is expressed in the number of disability-adjusted life years (DALYs) lost as a result of that disease. Disability-adjusted life years were launched by the World Bank and backed by the WHO as a measure of the burden of disease (World

Table 3.5 Summary of quantitative microbial risk assessment results for rotavirus[a] infection risks for different exposures (WHO, 2006)

Exposure scenario	Water quality[b] (*E. coli*/100 ml wastewater or 100 g soil)	Median infection risks per person per year	Notes
Unrestricted irrigation (crop consumers)			
Lettuce	10^3–10^4	10^{-3}	100 g eaten raw by person every 2 days 10–15 ml of wastewater remaining on crop
Onion	10^3–10^4	5×10^{-2}	100 g eaten raw by person per week for 5 months 1–5 ml of wastewater remaining on crop
Restricted irrigation (farmers or other heavily exposed populations)			
Highly mechanized	10^5	10^{-3}	100 days exposure per year 1–10 mg soil consumed per exposure
Labor intensive	10^3–10^4	10^{-3}	150–300 days exposure per year 10–100 mg soil consumed per exposure

[a]Risks estimated for *Campylobacter* and *Cryptosporidium* are lower.
[b]Non-disinfected treated wastewaters.
Reproduced with permission from *WHO Guidelines for the Safe Use of Wastewater, Excreta and Greywater*, Vol. 3 Wastewater Use in Agriculture, 3rd ed. World Health Organization, Geneva, Switzerland.

Bank, 1993; WHO, 1994). The DALY is the only quantitative indicator of burden of disease that reflects the total amount of healthy life lost, to all causes, whether from premature mortality of from some degree of disability over a period of time (Anand and Johnson, 1995). The intended use of the DALY is to assist in: (i) setting health service priorities; (ii) identifying disadvantaged groups and targeting of health interventions; and (iii) providing a comparable measure of output for intervention, program and sector evaluation, and planning.

The health-based targets, which actually define the level of health protection accepted as tolerable for each risk factor, can be achieved by the combination of health protection measures applied at different components of the reuse scheme. The tolerable risk in the revised guidelines has been defined at 10^{-6} DALYs per person per year for a water-borne disease. A multi-barrier risk reduction methodology is applied, including irrigation practices and food preparation measures. The natural die-off of pathogens is also taken into consideration. It should be stated, however, that these measures, as well as chemotherapy and vaccinations, are only complementary and cannot replace wastewater treatment. Indeed all different hazard/risk management scenarios provided in the revised guidelines include wastewater treatment technology. The WHO health-based targets for TWW irrigation of food crops are summarized in Table 3.6. They suggest, based on QMRA analysis, that rotavirus reduction is required to achieve the defined tolerable risk level (10^{-6} DALYs) from different exposures. With regard to helminths, epidemiological studies reveal that infections do not occur in cases of wastewater concentrations lower than 1 egg/l. The revised WHO guidelines (WHO, 2006) aim to develop realistic water reuse standards because the methodology adopted is based on the actual

Table 3.6 Health based targets for treated wastewater use in agriculture (WHO, 2006)

Exposure scenario	Health-based target (DALY/person yr)	Log pathogen reduction needed	No helminth eggs/l
Unrestricted irrigation	10^{-6a}		
Lettuce		6	$<1^{b,c}$
Onion		7	$<1^{b,c}$
Restricted irrigation	10^{-6a}		
Highly mechanized		3	$<1^{b,c}$
Labor intensive		4	$<1^{b,c}$
Localized irrigation	10^{-6a}		
High-growing crops		2	No recommendation[d]
Low-growing crops		4	$<1^{b,c}$

[a]Rotavirus reduction. The health-based target can be achieved, for unrestricted and localized irrigation, by a 6–7 log unit pathogen reduction (obtained by a combination of wastewater treatment and other health protection measures; for restricted irrigation, it is achieved by a 2–3 log unit pathogen reduction).
[b]When children under 15 are exposed, additional health protection measures should be used (treatment to <0.1 egg/l, protective equipment such as gloves/boots or chemotherapy).
[c]An arithmetic mean should be determined throughout the irrigation season. The mean value of $<$ one egg/l should be obtained in at least 90% of samples to allow for the occasional high-value sample (i.e. with >10 eggs/l). With some wastewater treatment processes (stabilization ponds), the hydraulic retention time can be used as a surrogate to assure compliance with $<$ one egg/l.
[d]No crops to be picked up from the soil.
DALY, disability-adjusted life years.
Reproduced with permission from *WHO Guidelines for the Safe Use of Wastewater, Excreta and Greywater*, Vol. 3 Wastewater Use in Agriculture, 3rd ed. World Health Organization, Geneva, Switzerland.

assessment of risks. This methodology suggests low-cost treatment technologies in combination with preventive measurements, which is the most appropriate philosophy for developing countries.

3.5.2 USEPA guidelines

In 1992 the USEPA in cooperation with the U.S. Agency for International Development (USAID) published guidelines for water reuse covering various applications (USEPA/USAID, 1992). In 2004 the revised USEPA guidelines were released, which included updated information on national water use and reuse practices and projects in the USA and outside. Emphasis was on issues dealing with indirect potable reuse, taking into account recent findings from research studies and practices. Direct potable reuse was also covered. Finally, attention was given to treatment technologies and health issues, including those arising from emerging pollutants and pathogens.

The primary purpose of the USEPA guidelines is to present and summarize water reuse guidelines, with supporting information, for the benefit of utilities and regulatory agencies. In states where criteria do not exist or are under revision, these guidelines can assist in developing reuse programs and appropriate criteria. They include both treatment processes and TWW quality limits for any intended use of recycled water. They use FC as indicator organisms and set limits for BOD and turbidity. A certain level of disinfection is included for all uses to avoid any consequences from inadvertent contact, or accidental or incidental misuse of TWW in reuse systems. With regard to irrigation of agricultural crops, secondary treatment with disinfection is the lowest level of treatment recommended. Furthermore, additional quality limits and preventive measures are included to minimize health risks and impacts on agricultural production including nutrient and heavy metal limits in recycled water, maximum concentrations of residual chlorine, and setback distances. The recommended treatment processes and TWW quality characteristics, as well as additional requirements for urban and crop irrigation are summarized in Table 3.7.

3.5.3 California regulations (Title 22)

The Department of Health Services has the responsibility to establish standards for water reuse in the State of California. "California regulations" include treatment process requirements, TWW quality standards, and operational and treatment reliability requirements (State of California, 2000). Total coliforms are used as indicator organisms, which must be monitored daily, and compliance is based on a running 7-d median number. For high-risk uses, such as irrigation of crops eaten raw or open access areas, the 7-d median of TC must not exceed 2.2/100 ml daily and 23/100 ml in 30 days, whereas any sample must not exceed 240 TC/100 m. Continuously monitored, daily average turbidity must not exceed 2 nephelometric turbidity units (NTU) within a 24-hr period and 5 NTU more than 5% of the time, and should never exceed 10 NTU. The lowest level of treatment required for the irrigation is secondary treatment, which is applied for non-food crops, whereas no requirements for pathogens are included. Reliability requirements include standby power supplies and treatment processes, alarms, emergency storage, and disposal of recycled

Table 3.7 USEPA suggested guidelines for treated wastewater reuse for irrigation of agricultural crops and landscapes

Reuse type	Treatment level	Water quality requirements	Setback distances	Comments
All types of landscape irrigation (e.g. golf courses, parks, cemeteries)	Secondary[a] Filtration[b] Disinfection[c]	pH = 6–9 <10 mg/l BOD <2 NTU[d] n.d. FC/100 ml[e,f] 1 mg/l Cl_2 (minimal)[g]	15 m to potable water supply wells	At controlled-access sites with limited public contact, a lower level, secondary treatment and disinfection to achieve <14 FC/100 ml, may be appropriate Chemical addition prior to filtration may be necessary to meet water quality The reclaimed water should not contain measurable levels of viable pathogens A higher chlorine residual and/or a longer contact time may be required to assure that viruses and parasites are destroyed A Cl_2 residual of 0.5 mg/l or greater in distribution system is recommended to reduce odors, slime, and bacterial regrowth Provide treatment reliability
Restricted access area irrigation Sod farms, silviculture sites, and other areas where public access is prohibited, restricted or infrequent	Secondary[a] Disinfection[c]	pH = 6–9 <30 mg/l BOD <30 mg/l TSS 200 FC/100 ml[e,f] 1 mg/l Cl_2 (minimal)[g]	90 m to potable water supply wells 30 m to areas accessible to the public (if spray irrigation)	Consultant recommended agricultural limits for metals For spray irrigation, TSS<30 mg/l to avoid clogging of sprinkler heads Provide treatment reliability

(continued)

Table 3.7 (*Continued*)

Reuse type	Treatment level	Water quality requirements	Setback distances	Comments
Food crops not processed[h] Surface or spray irrigation of any food crop, (crops eaten raw)	Secondary[a] Filtration[b] Disinfection[c]	pH = 6–9 <10 mg/l BOD <2 NTU[d] n.d. FC/100 ml[e,f] 1 mg/l Cl_2 (minimal)[g]	15 m to potable water supply wells	Consultant recommended agricultural limits for metals Chemical addition prior to filtration may be necessary to meet water quality The reclaimed water should not contain measurable levels of viable pathogens A higher chlorine residual and/or a longer contact time may be required to assure that viruses and parasites are destroyed High nutrient levels may adversely affect some crops during certain growth stages Provide treatment reliability
Processed food crops[h] Surface irrigation of orchards and vineyards	Secondary[a] Disinfection[f]	pH = 6–9 <30 mg/l BOD <30 mg/l TSS 200 FC/100 ml[f,g] 1 mg/l Cl_2 (minimal)[h]	90 m to potable water supply wells 30 m to areas accessible to the public (if spray irrigation)	Consultant recommended agricultural limits for metals For spray irrigation, TSS<30 mg/l to avoid clogging of sprinkler heads High nutrient levels may adversely affect some crops during certain growth stages Provide treatment reliability
Non-food crops Pasture, fodder, fiber, and seed crops	Secondary[a] Disinfection[c]			Consultant recommended agricultural limits for metals For spray irrigation, TSS<30 mg/l to avoid clogging of sprinkler heads High nutrient levels may adversely affect some crops during certain growth stages

Milking animals should be prohibited from grazing for 15 days after irrigation ceases. A higher level of disinfection (<14 FC/100 ml) should be provided if this period is not adhered to

Provide treatment reliability

[a]Secondary treatment processes include activated sludge processes, trickling filters, rotating biological contractors, and may include stabilization pond systems. Secondary treatment should produce TWW in which both the BOD and TSS do not exceed 30 mg/l.

[b]Filtration means the passing of wastewater through natural undisturbed soils or filter media such as sand and/or anthracite, filter cloth, or the passing of wastewater through microfilters or other membrane processes.

[c]Disinfection may be accomplished by chlorination, ultraviolet radiation, ozonation, other chemical disinfectants, membrane processes, or other processes. The use of chlorine as defining the level of disinfection does not preclude the use of other disinfection processes as an acceptable means of providing disinfection for reclaimed water.

[d]The recommended turbidity limit should be met prior to disinfection. The average turbidity should be based on a 24-hr time period. The turbidity should not exceed 5 NTU at any time. If TSS is used in lieu of turbidity, the TSS should not exceed 5 mg/l.

[e]Coliform limits are median values determined from the bacteriological results of the last 7 d for which analyses have been completed. Either the membrane filter or fermentation-tube technique may be used.

[f]Some stabilization pond systems may be able to meet this coliform limit without disinfection.

[g]Total chlorine residual should be met after a minimum contact time of 30 min.

[h]Commercially processed food crops are those that, prior to sale to the public or others, have undergone chemical or physical processing sufficient to destroy pathogens.

BOD, biochemical oxygen demand; FC, fecal coliforms; NTU, nephelometric turbidity units; TSS, total suspended solids; USEPA, United States Environmental Protection Agency.

Montoring requirements pH: weekly, BOD: weekly, turbidity continuous, FC: daily, Cl_2: continuous.

water not treated to desired level. Furthermore, setback distances are determined to ensure the protection of public health. A detailed description of the regulations of the State of California dealing with irrigation of agricultural crops and landscapes is provided in Table 3.8.

3.5.4 Florida

Both treatment processes and TWW quality limits required for any specific use of recycled water are determined by the State of Florida. An important difference, compared to criteria from other states, is that identical TWW quality requirements are imposed for both unrestricted and restricted urban access reuse. However, additional levels of treatment including filtration and possibly coagulation prior to irrigation are required in sites with public access. Florida criteria require a TSS concentration of 5.0 mg/l (monthly average) prior to disinfection. Continuous on-line monitoring of turbidity is required as an indicator

Table 3.8 California water recycling criteria: treatment and quality requirements for non-potable uses of reclaimed water (State of California, 2000)

Type of reuse	Total coliform limits[a]	Treatment required
Irrigation of fodder, fiber, and seed crops, orchards[b] and vineyards,[b] processed food crops, non-food-bearing trees, ornamental nursery stock,[c] and sod farms[c]	None required	Secondary
Irrigation of pasture for milking animals, landscape areas,[d] ornamental nursery stock and sod farms where public access is not restricted	≤23/100 ml ≤240/100 ml in more than one sample in any 30-d period	Secondary Disinfection
Irrigation of food crops,[b] restricted recreational impoundments, fish hatcheries	≤2.2/100 ml ≤23/100 ml in more than one sample in any 30-d period	Secondary Disinfection
Irrigation of food crops[e] and open access landscape areas[f]	≤2.2/100 ml ≤23/100 ml in more than one sample in any 30-d period 240/100 ml (maximum)	Secondary Coagulation[g] Filtration[h] Disinfection

[a]Based on running 7-d median.
[b]No contact between reclaimed water and edible portion of crop.
[c]No irrigation for at least 14 d prior to harvesting, sale, or allowing public access.
[d]Cemeteries, freeway landscaping, restricted access golf courses, and other controlled access areas.
[e]Contact between reclaimed water and edible portion of crop; includes edible root crops.
[f]Parks, playgrounds, schoolyards, residential landscaping, unrestricted access golf courses, and other uncontrolled access irrigation areas.
[g]Not required if the turbidity of the influent to the filters is continuously measured, does not exceed 5 nephelometric turbidity units (NTU) for more than 15 min and never exceeds 10 NTU, and there is capability to automatically activate chemical addition or divert the wastewater if the filter influent turbidity exceeds 5 NTU for more than 15 min.
[h]The turbidity after filtration through filter media cannot exceed 2 NTU within any 24-hr period, 5 NTU more than 5% of the time within a 24-hr period, and 10 NTU at any time. The turbidity after filtration through a membrane process cannot exceed 0.2 NTU more than 5% of the time within any 24-hr period and 0.5 NTU at any time.

that TSS threshold (5.0 mg/l) is not exceeded. Actually, no threshold value is specified, but turbidity should range between 2 and 2.5 NTU. The annual average carbonaceous biological oxygen demand (CBOD) must not exceed 20 mg/l after secondary treatment with filtration and high-level disinfection. Seventy-five percent of the FC samples, taken over a 30-d period, are required to be below detectable levels, with no single sample to exceed 25/100 ml. Monitoring of *Giardia* spp. and *Cryptosporidium* spp. is required. The sampling frequency is based on the capacity of the treatment plant. In WWTPs with a capacity less than 44 l/s, sampling is required once every 5 years, whereas in WWTPs with an equal or greater capacity sampling is required every 2 years. Samples are to be taken after disinfection. Residual chlorine should be at least 1 mg/l after a minimum acceptable contact time of 15 min during peak flow hour. Direct contact (spray irrigation) of edible crops, which are not peeled, skinned, cooked, or thermally-processed before consumption, with TWW is not allowed. Irrigation methods with indirect contact (ridge and furrow, drip, subsurface irrigation) can be used independently of the intended use of the crops. For irrigation of non-food crops a lower level of treatment is required; typically secondary treatment with disinfection is recommended. Florida requires that the annual average TSS not exceed 20 mg/l except when subsurface irrigation is used, in which case the single sample TSS limit is 10 mg/l. The annual average of FC must not exceed 200/100 ml and the monthly geometric mean 200/100 ml. The concentration of FC also must not exceed 400/100 ml in more than 10% of the samples in a 30-d period and not exceed 800/100 ml in any single sample. Finally, setback distances from buildings, potable water wells and surface waters for all irrigation uses are also included.

3.5.5 Other states of America

Forty-one states have set guidelines/regulations in the USA. These principally deal with urban and/or agricultural irrigation. All states include criteria for irrigation of non-food crops and 21 for irrigation of food crops. Unrestricted and restricted urban irrigation are also widespread with 28 and 34 states, respectively, having adopted relevant criteria (USEPA, 2004). Typically, secondary treatment and disinfection is the minimum level of treatment required prior to unrestricted irrigation of urban areas. Additional levels of treatment are included in some cases including coagulation, oxidation, and filtration. The concentration of indicator organisms in general ranges from non-detectable levels to 20/100 ml. The control of public access in restricted urban irrigation results in lower public exposure and less stringent treatment requirements and TWW quality standards.

Reuse of TWW for irrigation of food crops is prohibited in some states, whereas other states allow irrigation only in cases where crops are processed and not eaten raw. Treated wastewater quality standards vary greatly between states and they are mainly determined by irrigation method and the intended use and the specific characteristics of the crop. In general a high level of treatment, similar to that for unrestricted urban irrigation, is required. The limits of indicator organisms vary from non-detectable levels to 200/100 ml. Irrigation of non-food crops with recycled water results in lower risks regarding human exposure to the water, and hence a lower treatment level and water quality limits are required. Also, some states do not require disinfection of the TWW under the condition that certain requirements are met with regard to the irrigation method and intended use of crops.

3.5.6 Canada

In Canada water reuse is practiced on a relatively small scale, involving mainly agricultural and landscape irrigation. No federal guidelines on water recycling have been developed, although guidelines have been set in some provinces including British Columbia and Alberta.

3.5.7 Australian guidelines

The national guidelines of 2000 entitled "Guidelines for Sewerage Systems, Use of Reclaimed Water" provided advice on TWW quality standards, treatment levels, safeguards, controls, and monitoring (National Health and Medical Research Council and Agriculture and Resource Management Council of Australia and New Zealand (NHMRC-ARMCANZ), 2000). Thermotolerant coliforms were recommended as indicators of reclaimed water quality, which should not exceed 10 colony forming units (cfu)/100 ml for high contact uses. These guidelines were recently revised to provide a national approach for water recycling, to correct the inconsistencies in guidelines among state and territories, and to cover additional non-conventional water sources like greywater or stormwater.

The revised guidelines use a generic risk management framework that can apply to any system recycling water including TWW, greywater, and stormwater. Similar methodologies have been adopted by the food industry, through application of the hazard analysis and critical control point (HACCP) system, and more recently, in Australian Drinking Water Guidelines (National Health and Medical Research Council and Natural Resource Management Ministerial Council (NHMRC-NRMMC), 2004). An important innovation of these guidelines is that they focus on both health and environmental hazards. With regard to health risks they focus on pathogenic organisms, but concern also exists for chemical pollutants and particularly from long-term exposure to low concentrations. Particular emphasis is also given on the environmental impacts that may arise from water reuse with chemical substances receiving the principal focus; microbial hazards, such as the transfer of antibiotic-resistant bacteria into the environment are also considered (NRMMC-EPHC, 2006).

The risk management approach incorporates the concept of identifying and producing recycled water of a quality that is "fit-for-purpose". A classification system for recycled water is not included in the revised guidelines and the major reason is that such systems can limit flexibility. For example, uses such as dual reticulation, unrestricted irrigation, and irrigation of crops eaten raw are usually grouped together and are defined as high-exposure uses. However, the use of a risk assessment approach reveals that the pathogen removal requirements are different among these applications.

Disability-adjusted life years are adopted to convert the likelihood of infection or illness into burdens of disease, and set a tolerable risk of e.g. 10^{-6} DALYs per person per year (pppy). The tolerable risk is then used to set health-based targets to ensure that the risk remains below the tolerable value. Reference pathogens for each group are used to set human health-based targets. Dose–response relationships obtained from investigations of outbreaks or experimental human-feeding studies are used to determine how exposure to a particular dose of a hazard is associated with the incidence or likelihood of illness.

Treatment processes, and exposure reduction, either by using preventive measures at the site of use or by restricting uses of the recycled water are combined to protect human and environmental health. Table 3.9 indicates how treatment processes can be used alone or in combination with preventive measures to meet health-based log reduction targets for various uses of recycled water for irrigation.

3.5.8 Europe

At European level no regulations, guidelines, or good management practices of water reuse exist. However, reference to reuse is made in article 12 of the European Wastewater Directive (91/271/EEC) (European Union (EU), 1991) stating: "Treated wastewater shall be reused whenever appropriate". In order to make this statement fact, common definitions of what is "appropriate" are needed. The EU Water Framework Directive (WFD) introductory booklet "Tap into it!" refers to water scarcity and the need to reuse water. Although the WFD does not include reuse in the body of the directive, it introduces quantitative dimensions to water management, which may stimulate the consideration of water reuse. The Integrated Pollution Prevention Control (IPPC) legislation encourages water reuse and is included in the legislation within the WFD.

The principles of the WFD should fuel the discussion on recycled water use, management and its economic analysis, which will give to the recycled water the prospective of an economic good. The WFD discusses freshwater, but as recycled water is a byproduct of potable water use, it can be analogously extended to cover recycled water. This entails applying to recycled water all the principles and articles concerning management of other water sources. Recycled water demand is expected to grow with time at rates higher than supply, with relevant projects being implemented. Thus, the WFD could evidently be extended to cover recycled water as a water resource (Tsagarakis, 2005).

Water recycling is a common practice in southern European countries including Spain, France, Italy, Portugal, Malta, Cyprus, and Greece, where a severe water imbalance occurs during the summer period. Several EU countries have been developing national or regional guidelines for water reuse, directed mainly at irrigation of agricultural crops.

In Spain, a new Decree (R.D. 1620/2007) on "reuse of reclaimed water: quality criteria"; appeared in December 2007. The Decree permits several uses of reclaimed water, while prohibiting others. Reclaimed water can be used for residential purposes (garden irrigation, discharging cisterns); services (green spaces, street cleaning etc.); agricultural irrigation; environmental uses; industrial uses; and recreational purposes without public access. By contrast, reclaimed water cannot be used for human consumption (direct reuse) unless in a catastrophic situation (emergency); the majority of uses in the agrofood industry and for cooling purposes; uses in hospital; filtering shellfish growth; bathing water and fountains and other ornamental uses. The health authorities maintain the final decision of whether or not to allow reuse for each case. Depending on the use, there are several tables detailing the limits of amount of components that can be found in

Table 3.9 Treatment processes and on-site controls for designated uses of recycled water from treated sewage (NRMMC-EPHC, 2006)

Log reduction targets (V, P, B)[a]	Indicative treatment process	Log reductions achievable by treatment (V, P, B)	On-site preventive measures	Exposure reduction[b]	Water quality objectives[c]
Municipal use — open spaces, sports grounds, golf courses, dust suppression, etc or unrestricted access and application					
5.0 3.5	Advanced treatment Secondary, coagulation, filtration and disinfection	5.0 3.5	No specific measures		On case-by-case basis depending on technologies Could include turbidity criteria for filtration, disinfectant Ct or dose (UV)[d]
4.0	Secondary, membrane filtration, UV light	4.0			*E. coli* <1/100 ml
Commercial food crops consumed raw or unprocessed					
6.0 5.0 5.0	Advanced treatment to achieve complete pathogen removal (e.g. secondary, filtration and disinfection)	6.0 5.0 5.0	None required, although pathogen reduction will occur between harvesting and sale The recycled water can be used for all crops including spray irrigation of salad crops	0.5 V, B	On case-by-case basis depending on technologies Could include turbidity criteria for filtration, disinfectant Ct or dose (UV) *E. coli* <1/100 ml
Commercial food crops					
6.0 5.0	Secondary treatment with >25 d lagoon detention and disinfection	3.0–4.0 2.0–4.0	*Consumers* Crops with limited or no ground contact eaten raw (e.g. tomatoes, capsicums) – drip irrigation and no harvest of wet or dropped produce		BOD[e] <20 mg/l*d SS[e] <30 mg/l*d

5.0		>6.0	Crops with ground contact with skins removed before consumption (e.g. watermelons) – if spray irrigation, minimum 2 d from final irrigation to harvest Pathogen reduction between harvesting and sale *Public in vicinity of irrigation area*[f] No access and drip or subsurface irrigation No access during irrigation and if spray irrigation, minimum 25–30 m buffer	Disinfectant residual (e.g. minimum chlorine residual) or UV dose *E. coli* <100 cfu/100 ml
Commercial food crops				
6.0	Secondary treatment	2.0–3.0	*Consumers*	BOD <20 mg/l*d
5.0	with disinfection	1.0	Above-ground crops with	SS <30 mg/l*d
		>6.0	subsurface irrigation	
5.0			Crops with no ground contact and skins removed before consumption (e.g. citrus, nuts) – no harvest of wet or dropped produce – if spray irrigation, minimum 2 d from final irrigation to harvest Pathogen reduction between harvesting and sale *Public in vicinity of irrigation area* No access and drip or subsurface irrigation	Disinfectant residual (e.g. minimum chlorine residual) or UV dose *E. coli* <100 cfu/100 ml

(*continued*)

Table 3.9 (*Continued*)

Log reduction targets (V, P, B)[a]	Indicative treatment process	Log reductions achievable by treatment (V, P, B)	On-site preventive measures	Exposure reduction[b]	Water quality objectives[c]
			No access during irrigation and if spray irrigation, minimum 25–30 m buffer distance between irrigation area and nearest public access point		
Non-food crops – trees, turf, woodlots, flowers					
5.0	Secondary treatment or primary treatment with lagoon detention	0.5–1.0	Public in vicinity of irrigation area		*E. coli* <10 000 cfu/100 ml
3.5		0.5–2.0	No access and drip irrigation	6.0	
4.0		1.0–3.0	No access during irrigation and, if spray irrigation, 25–30 m minimum buffer distance between irrigation area and nearest point of public access, and spray drift control or extended buffer distances to >50 m	5.0	

[a]Log reduction targets are minimum reductions required from raw sewage on 95th percentiles.

[b]Exposure reductions are those achievable by on-site measures.

[c]Water quality objectives represent medians for numbers of *E. coli* and means for other parameters.

[d]Aim is to demonstrate reliability of disinfection and ability to consistently achieve microbial quality.

[e]BOD and SS are an indication of secondary treatment effectiveness.

[f]Log reductions for public in the vicinity of commercial food crop irrigation areas should comply with total log reductions required for municipal use.

B, enteric bacteria; BOD, biochemical oxygen demand; cfu, colony forming units; Ct, disinfectant concentration × time; P, enteric protozoa; SS, suspended solids; UV, ultraviolet; V, enteric virus.

reclaimed water. The number of components to determine could be over 60 depending on the circumstances, and could include organic and inorganic compounds, apart from microorganisms. Groundwater recharge is also limited by a number of compounds not to be found in any case. The samples to be controlled are huge, and can vary from once every month to three times per week; which can make control unaffordable from an economic point of view. The control points are at the exit of the treatment plant and sometimes the point of delivery; not the point of use.

In Italy the use of wastewater for irrigation was regulated from 1977 until 2003, in the frame of the 1976 Water Protection Act (Annex 5, CITAI, 1977), being considered an extensive treatment process. The TWW quality standards were, from a hygienic point of view, quite stringent, especially if we consider that, in many cases, Italian surface waters generally used for irrigation display a consistently lower microbiological quality. No standards were set for toxic or bioaccumulative substances and a specific evaluation was required for TWW volume, which could be applied yearly, depending on soil and crops. The irrigated area (access had to be kept under control) had to be surrounded by a buffer strip of at least 80 m (without buildings or roads), regardless of the TWW quality and irrigation method. Some of the Regional Governments (e.g. Puglia and Sicilia), using the powers given by the 1976 Water Protection Act, prepared and issued their own criteria.

Finally, following the frame of Law Decree n. 152/99, a new legislative set of rules was promulgated in 2006 (Ministry Decree, D.M. no 152/06; 05/05/2006). The new criteria have, at least partially, accepted previous independent proposals including proposals on electric conductivity and boron. Boron in treated wastewaters may occur at high concentrations in some districts: a survey carried out on the TWWs of 10 biological plants showed an average yearly concentration of 0.76 mg/l. On the contrary, other parameters including nematode eggs, viruses, and protozoa are not taken into consideration. The proposed criteria seem to follow a quite restrictive approach, especially for some chemical compounds; in some cases TWW quality standards are the same as those of drinking water. This approach will inevitably lead to difficulties in promoting water reuse applications, when compliance with some very strict standards will ask for advanced treatment requirements, with all the related consequences on the economic viability of reuse schemes. Another restrictive factor is the number of parameters taken into account and their monitoring protocols. In fact the number of parameters to be monitored exceeds 50 items and the sampling frequency can be very high, depending on the regional provisions. Also, it must be considered that no distinction is established among various crops to be irrigated with reclaimed wastewater (restricted and unrestricted irrigation) and no attention is paid to the influence of different irrigation practices (subsurface irrigation versus spray irrigation) to eliminate health risks.

Treated wastewater standards for reuse in France (Circular n 51 of July 22, 1991, of the Ministry of Health) are based on the WHO guidelines (WHO, 1989). Additionally, requirements including irrigation method, timing, distance, and other preventive measures are applied to eliminate health risks. Currently, TWW guidelines are under revision; new microbiological indicators for unrestricted irrigation (i.e. *Salmonella*, *Taenia* eggs) are being introduced, as well as more stringent operational restrictions.

The water reuse criteria in Cyprus have been influenced by both philosophies; WHO and California. They are stricter than the WHO (1989) guidelines, although they allow the adoption of low-cost technologies under certain conditions. They include both water quality standards and appropriate treatment levels. The disinfection level required for unrestricted irrigation is 5 FC/100 ml (80% of the time and a maximum value of 100 FC/100 ml). For food irrigation FC must not exceed 200/100 ml (maximum 1000 FC/100 m), whereas for irrigation of fodder and industrial crops the corresponding FC values are 1000 and 3000 FC/100 ml, respectively.

In Malta the TWW from the Sant Antnin WWTP has been used to irrigate 600 ha of crops by furrow and spray irrigation since 1983 (EUREAU, 2005). The water quality is suitable for unrestricted irrigation and is used to produce potatoes, tomatoes, broad and runner beans, green pepper, cabbages, cauliflower, lettuce, strawberries, clover, etc. Three major WWTPs exist in the country and there is a plan for reclamation and reuse of all the TWW produced. At national level no regulations, guidelines, or good management practices of water reuse exist and the WHO (1989) guidelines are adopted.

In January 2006 the Portuguese guidelines on water reuse were published (Portuguese Standard NP 4434). These guidelines do not focus only on the microbiological quality of reclaimed water, but also attention is given to the chemical composition of the TWW, the impacts on the soil, agricultural crops, and groundwater. Treated wastewater quality parameters (chemical and microbiological) and treatment levels are defined to protect public health, environment, and crop production. The crops are separated into four categories according to intended use and limits for microbiological parameters (FC and helminth eggs) and treatment requirements are suggested (Table 3.10). With regard to the TWW chemical composition emphasis is given to pH, salinity, sodium adsorption ratio (SAR), and trace element content. Geomorphological aspects of the area are taken into consideration. Treated wastewater irrigation is not recommended for areas with high slope (>20%) or karstic areas, whereas the groundwater table should be at depth 1–4 m, depending on the irrigation method. Buffer zones are set from the irrigated sites to nearest dwellings, which depend on the irrigation method and the treatment level.

There is now an effort to harmonize the various approaches to wastewater reuse at European level (Angelakis et al., 2003), just as the Australians have combined their different state guidelines to develop national guidelines. Recently, an EU MED Working Group on Water Recycling and Reuse has been established by the EU Directorate General (DG) for the environment. The first task undertaken by this working group was the preparation of a brief report that focuses on issues important to EU water directors, supported by the evidence developed within the EU and internationally. This report has been completed and it has been adapted by the water directors.

3.5.9 Non-European Mediterranean countries

The water scarcity in many Mediterranean countries has stimulated water reuse and in many it constitutes a significant component in water resources management. Most of these countries have developed frameworks for water recycling in order to eliminate risks to public health and protect the environment.

Israel is a pioneer in the field of water reuse and was among the first countries to set up water reuse criteria. The Israeli regulations for TWW irrigation follow the "zero risk" approach. It is achieved either through advanced treatment of wastewater to meet FC numbers lower than 10 cfu/100 ml (unrestricted irrigation), or by a combination of

Table 3.10 TWW reuse guidelines in Portugal (Marecos do Monte, 2007)

Class	Type of crop	FC (MPN or cfu/ 100 ml)	Helminth eggs (egg/l)	Appropriate treatment	Notes
A	Vegetables to be eaten raw	100	1	Secondary, filtration and disinfection or tertiary, filtration and disinfection	UV disinfection (self-cleaning lamps) or O_3 preferable to chlorination
B	Public parks and gardens, sport lawns, forest with public easy access	200	1	Secondary, filtration and disinfection or tertiary, filtration and disinfection	UV disinfection (self- cleaning lamps) or O_3 preferable to chlorination. Irrigation must avoid contact with people
C	Vegetables to be cooked, forage crops, vineyards, orchards	10^3	1	Secondary, filtration and disinfection or tertiary, filtration and disinfection or waste stabilization ponds (system with $\geq$ three ponds and tR$\geq$25 d)	UV disinfection (self-cleaning lamps) or O_3 preferable to chlorination. Irrigation of vineyards and orchards must avoid contact with fruit. Fruit fallen on the soil should not be collected
D	Cereals (except rice), vegetables for industrial process, crops for textile industry, crops for oil extraction, forest and lawns located in places of difficult or controlled public access	10^4	1	Secondary, maturation ponds$\geq$ (tR$\geq$10 d) or secondary filtration and disinfection	UV disinfection (self-cleaning lamps) or O_3 preferable to chlorination. Irrigation must avoid contact with people

TWW, treated wastewater; UV, ultraviolet.
Reproduced with kind permission of Springer Science and Business Media.

appropriate treatment levels and on-farm preventive measures (also called barriers). Crops are allowed to irrigate at a treatment level lower then the highest (unrestricted), yet require a certain minimum TWW level. In this event, the use of barrier(s) to pathogen infection makes possible the use of effluent treated to a yet lower disinfection level. The choice of barriers and the number of barriers needed depend on the susceptibility of the crop to pathogen transfer, the efficacy of the barrier in blocking pathogens, and the quality of the TWW. The barriers taken into consideration are sand filtration, long retention times in ponds and reservoirs, TWW ratio limited to 10% of the irrigation water, TWW disinfection, contact of the edible part of the crop with the soil, soil mulching by plastic cover, surface or subsurface drip irrigation, crops with pill or shell, and vegetables consumed after cooking (Table 3.11; Fine et al., 2006). Some barriers are considered more effective (e.g., subsurface drip irrigation, aerial distance) and are thus considered equivalent to two

Table 3.11 The number and type of barriers to pathogen transfer that are recommended by the Israel Ministry of Health according to crop type and TWW quality[a]

Crop	Disinfection[b]	Number of barriers required				Value of a barrier (by type and practice)				
		Low input[c]	Low input + >10 d detention[d]	High input[e]	Unrestricted irrigation[f]	Aerial distance from drip-emitters	Plastic ground cover	Subsurface drip.	Inedible pill or shell	Cooking obligatory
Leafy vegetables, strawberry		Prohib	Prohib	Prohib	only					
Vegetables eaten-fresh grown above ground (pepper, tomato, cucumber, paprika, and zucchini)	+	Prohib	Prohib	2	0		1	2		
Cooked vegetables with rind(eggplant, pumpkin)		Prohib	Prohib	2	0		1	2	1	1
Cooked vegetables grown in the ground (potatoes)		Prohib	Prohib	2	0					1
Peanuts		Prohib	Prohib	2	0					
Eaten-fresh vegetables grown in the ground (carrot, onion, radish)	+	Prohib	Prohib	2	0				1	
Beans	+	Prohib	Prohib	2	0		1	2		1
Vegetables with rind (watermelon, melon, peas)	+	Prohib	Prohib	2	0		1	2	1	
Artichokes		3	2-3	2	0	2	1	2		1
Corn (edible)		3	2-3	2	0	2	1	2	1	

Citrus		3	2-3	2	0	2		2	1	
Citrus, irrigated with pulsators or under-leaf sprinklers		3	2-3	2	0	1			1	
Citrus with edible peel (Chinese orange)		3	2-3	2	0	2		2		
Nuts, almonds, pomegranate, pistachios		3	2-3	2	0	2		2	1	
Deciduous trees (apple, prune, plum, pear, peaches, apricot) and cherry		3	2-3	2	0	2		2		
Tropical fruits (mango, avocado, persimmon)		3	2-3	2	0	1	1	2	1	
Tropical fruit with cutting of the lowest leaves		3	2-3	2	0	2	1	2	1	
Grapes with high trellis		3	2-3	2	0	2		2		
Grapes with regular trellis		3	2-3	2	0	1	1	2		
Grape with no trellis	+	Prohib	Prohib	2	0		1	2		
Cactus fruits		3	2-3	2	0	2	1	2	1	
Dates		3	2-3	2	0	3	1	2		
Olives		3	2-3	2	0	1	1	2		1
Flowers		3	2-3	2	0	1	1	2	1	

[a]Adapted from Halperin (1999); Fine et al., copyright 2006, with permission of Elsevier.
[b]Disinfection: (+) sign denotes disinfection is an obligatory barrier
[c]Low input treatment: BOD $>$ 60 mg/L, TSS $>$ 90 mg/l; e.g., oxidation pond TWW with detention time $\leq$ 10 d.
[d]Medium input treatment from oxidation ponds TWW with detention time $>$10 d (BOD 20–60 mg/l, TSS 30–90 mg/l).
[e]High-quality TWW from MBTP or equivalent (BOD: 20 mg/l, TSS: 30 mg/l).
[f]TWW suitable for unrestricted irrigation: wastewater treatment by any of the following methods: deep sand filtration, prolonged ponding ($\geq$60 d) or dilution in reservoir to $<$10% of the water (FC $\leq$10/100 ml and turbidity $\leq$5 NTU or TSS$\leq$10 mg/l and 1 mg/l residual chlorine at the site of irrigation.
[g]BOD, biochemical oxygen demand; MBTP, mechanical-biological treatment plant; NTU, nephelometric turbidity units; TSS, total suspended solids; TWW, treated wastewater.

to three other less effective barriers (Table 3.11). The combination of barriers depends on the crop, on the quality of the TWW, and on the irrigation method. Four qualities of TWW are considered:

(i) Very high-quality TWW ($\leq$10 *E. coli*/100 ml and more than 1 mg/l of residual chlorine), which does not require any barrier for crop irrigation.
(ii) High-quality TWWs (BOD$\leq$20 mg/l and SS$\leq$30 mg/l), which require two barriers.
(iii) Medium-quality TWWs (BOD and SS must not exceed 60 and 90 mg/l, respectively), which require three barriers. They cannot be used for irrigation of vegetables.
(iv) TWW from oxidation ponds (>15 d of retention time). Irrigation of fruits requires two barriers. Lower retention times require three barriers for fruit tree irrigation.

In Tunisia wastewater reuse in agriculture is regulated by the 1975 Water Code (law No. 75-16 of March 31, 1975), by the 1989 Decree No. 89-1047 (July 28, 1989), by the Tunisian criteria for the use of TWW in agriculture (NT 106-003 of May 18, 1989), by the list of crops that can be irrigated with TWW (Decision of the Minister of Agriculture of June 21, 1994), and by the list of requirements for agricultural wastewater reuse projects (Decision of September 28, 1995). They prohibit the irrigation of vegetables that might be consumed raw. Therefore, most of the recycled water is used to irrigate vineyards, citrus and other trees (olives, peaches, pears, apples, pomegranates, etc.), fodder crops (alfalfa, sorghum, etc), industrial crops (cotton, tobacco, sugarbeet, etc), cereals, and golf courses (Tunis, Hammamet, Sousse, and Monastir). Some hotel gardens in Jerba and Zarzis are also irrigated with recycled water. These criteria are relevant to WHO guidelines. The 1989 Decree stipulates that the use of recycled water must be authorized by the Minister of Agriculture, in agreement with the Minister of Environment and Land Use Planning, and the Minister of Public Health. It sets out the precautions required to protect the health of farmers and consumers, and the environment. Monitoring the physical-chemical and biological quality of recycled water and of the irrigated crops is planned: analyses of a set of physical-chemical parameters once a month, of trace elements once every 6 months, and of helminth eggs every 2 weeks on 24-hr composite samples, etc. In areas where sprinklers are used, buffer areas must be created. Direct grazing is prohibited on fields irrigated with wastewater.

Morocco has not developed standards for water reuse and the WHO guidelines are commonly applied in water reuse projects. In many cases, the use of raw or inadequately TWW has been associated with water-borne disease outbreaks. Improvement in water recycling practices is recognized as essential.

3.6 Conclusions

Because water shortages will continue to increase in the future, more sustainable solutions need to be developed for managing water resources. In addition, due to water scarcity, several regions of the world are very vulnerable to the impacts of climate change. Water resource planning is best served through a "twin track" approach to balancing water availability and demand management with sustainability. The approach of increasing water availability by developing and using non-conventional water resources such drainage water, breached water, and mainly TWW, could be considered practically, particularly in the regions under water stress. These water sources slow down the hydrological cycle and locally increase water availability, particularly in suburban and rural areas in which both landscape and agriculture irrigation should be promoted. As water resources are local and their availability varies to a great degree, any one-size-fits-all solution should be avoided. Furthermore, reduced and/or degraded water supplies induce restrictions on water uses and allocation policies among different user sectors. In water-limited regions, the competition for scarce water resources among users will inevitably reduce the supplies of freshwater available for crop irrigation. Thus, agriculture will increasingly be forced to utilize non-conventional water resources, such as recycled water, to meet its growing demands. Thus, recycled water could be a significant alternative water resource for irrigation (both agricultural and landscape) use particularly in dry arid and semiarid areas.

Currently, many countries consider TWW reuse as an integral part of pollution control and water management strategies as this can effectively contribute to closing the anthropogenic water cycle and enabling sustainable reuse of available water resources. Compared to other potential uses, like groundwater recharge and (in)direct potable uses, water recycling for crop and landscape irrigation constitutes a well accepted and practiced application throughout the world.

Water quality criteria constitute an indispensable part of TWW reuse in order to protect human health and the environment. However, important differences are observed between countries, states, or even within the same country. Arguments for the adoption of less strict regulations are often addressed, particularly in developing countries, based on the lack of documentation for adverse health effects rather than on the certainty that hazards and the potential risks resulting from TWW reuse are very low or non-existent. In fact, criteria development in most cases is based on semi-empirical and empirical criteria, rather than on the interpretation of available scientific data. Therefore, doubts arise with regard to the consistency of the criteria, and, as a consequence, environmental agencies are prompted towards more conservative regulations. In addition, the impacts on public health and the environment of the trace elements of natural and anthropogenic chemicals found in wastewater TWWs are largely unknown, and thus complicate both regulation setting and treatment requirements. The updated guidelines of WHO and Australia provide a scientific basis for criteria development based on QMRA methodology and could constitute a basis to set criteria worldwide and eliminate this variability.

Although TWW reuse for irrigation has not been shown to present unreasonable risks, information is needed in several areas to assist regulatory agencies to develop or improve their regulations or guidelines including:

(i) better indicator organisms to assess pathogenic populations in recycled water;
(ii) real time on-line monitoring of the population of indicator organisms and pathogens;
(iii) advances in risk assessment methodology to estimate risks to public health and the environment arising from the occurrence of pathogenic microorganisms and trace elements to make it a more useful tool;
(iv) determination of treatment effectiveness and reliability to achieve the designed removal of potentially hazardous pathogens and chemicals.

References

Anand, S. and Jonson, K. (1995) *Disability adjusted life year: A critical review. Harvard center for population and development studies working papers series (95. 06)*. Harvard, Boston.
Angelakis, A.N. and Spyridakis, S.V. (1996) The status of water resources in Minoan times - A preliminary study. In: *Diachronic Climatic Impacts on Water Resources with Emphasis on Mediterranean Region* (eds A. Angelakis and A. Issar), pp. 161–191. Springer-Verlag, Heidelberg, Germany.
Angelakis, A.N., Marecos do Monte, M.H., Bontoux, L. and Asano, T. (1999) The status of wastewater reuse practice in the Mediterranean basin. *Water Research* **33**, 2201–2217.
Angelakis, A.N., Bontoux, L., and Lazarova, V. (2003) Main Challenges and Prospectives for Water Recycling and Reuse in EU Countries. *Water Science and Technology: Water Supply* **3**(4), 59–68.
Asano, T., Leong, L.Y.C. and Rigby, M.G. (1992) Evaluation of the California wastewater reclamation criteria using enteric virus monitoring data. *Water Science and Technology* **26**, 1513–1524.
Asano, T. and Levine, A. (1998) Wastewater Reclamation, Recycling and Reuse: An Introduction. In: *Wastewater Reclamation and Reuse* (ed. T. Asano). Technomic Publishers, Lancaster, PA.
Asano, T., Burton, F.L., Leverenz, H.L., Tsuchihasi, R. and Tchobanoglous, G. (2007) *Water Reuse: Issues, Technologies and Applications*. Metcalf and Eddy, McGraw Hill, New York, USA.
Ayers, R.S. and Westcot, D.W. (1988) Irrigation water quality. In: *Irrigation with Reclaimed Municipal Wastewater - A Guidance Manual* (eds G.S. Pettygrove and T. Asano), pp. 3.1–3.36. Lewis Publishers, Chelsea, USA.
Avery, A. and L'hirondel, J.L. (2003) Nitrate and Methaemoglobinaemia. *Environmental Health Perspectives* **111**, 142–143.
Bitton, G. (1999) *Wastewater Microbiology*. 2nd ed. Wiley–Liss Inc, New York.
Blumenthal, U., Mara, D.D., Peasey, A., Ruiz-Palacios, G. and Stott, R. (2000) Guidelines for the microbiological quality of treated wastewater used in agriculture: recommendations for revising WHO guidelines. *Bulletin of the World Health Organization* **78** (9), 1104–1116.
Crittenden, J.C., Trussell, R. R, Hand, D.W., Howe, K.J. and Tchobanoglous, G. (2005) *Water Treatment: Principles and Design, 2nd edition*. John Wiley & Sons, Inc., Hoboken, NJ.
Crook, J. (1998) Water reclamation and reuse criteria. In: *Wastewater Reclamation and Reuse* (ed. T. Asano), pp. 627–703. Technomic Publishers, Lancaster, USA.
California State Board of Health (1918) *Regulations Governing Use of Sewage for Irrigation Purposes*. California State Board of Health, Sacramento, CA.
Davis, L.C., Castro-Diaz, S., Zhang, Q., and Ericson, L.E. (2002) Benefits of vegetation for soils with organic contaminants. *Critical Reviews in Plant Sciences* **21**, 457–512.
Dimitriadis, S. (2005) Issues encountered in advancing Australia's water recycling schemes. *Parliamentary Library, Information analysis and advice for the Parliament. Research Brief no* 2, 16 August 2005 (www.aph.gov.au/library).
Downs, T.J., Cifuentes-Garcia, E. and Suffet, I.M. (1999) Risk screening for exposure to groundwater pollution in a wastewater irrigation district of the Mexico City region. *Environmental Health Perspectives* **107**, 553–561.
DWR (2003) *Water recycling 2030: Recommendations of California's Recycled Water Task Force*. State of California Department of Water Resources, Sacramento, CA.
EU (1991) Council Directive of May 21, 1991 concerning Urban Wastewaters Treatment (91/271/EEC). *Official Journal of the European Communities*, L135/40, May 30.
EUREAU (2005) *Wastewater Recycling and Reuse in EUREAU Countries*, Draft Report. EUREAU, Brussels, Belgium.
Haas, C.N., Rose, J.B., and Gerba, C.P., (eds) (1999) *Quantitative Microbial Risk Assessment*. John Wiley and Sons Inc., New York, USA.
Fewtrell, L. and Bartram, J. (eds) (2001) *Water Quality: Guidelines, Standards and Health-Assessment of Risk and Risk Management for Water-Related Infectious Disease*. IWA Publishing on behalf of World Health Organization, London.
Fewtrell, L. (2004) Drinking-water nitrate and methaemoglobinaemia, Global burden of disease: A discussion, *Environmental Health Perspectives* **112** (14), 1371–1374.
Fine, P., Halperin, R. and Hadas, E. (2006) Economic considerations for wastewater upgrading alternatives: An Israeli test case. *Journal of Environmental Management* **78**, 163–169.
Gerhard, W.P. (1909) *Sanitation and Sanitary Engineering*. Self-published, New York.

Geldreich, E.E. (1990) Microbiological quality of source waters for water supply. In: *Drinking Water Microbiology: Progress and Recent Developments* (G.A. McFeters, ed.) Springer-Verlag, New York.

Geselbracht, J. (2003) Water reclamation and reuse. *Water Environmental Research* **75** (6), 1–63.

van Ginneken, M. and Oron, G. (2000) Risk assessment of consuming agricultural products irrigated with reclaimed wastewater: an exposure model. *Water Resources Research* **36**, 2691–2699.

Halperin, R. (1999) Regulations for water quality in Israel and internationally. *Health in the Field* **7**, 3–6 [in Hebrew].

IPCC (2007) *Climate Change 2007: The Physical Science Basis. Contribution of Working Group I to the Fourth Assessment Report of the Intergovernmental Panel on Climate Change* (Solomon et al., eds). Cambridge University Press, Cambridge, UK and New York.

IRCWD (1985) Health aspects of wastewater and excreta use in agriculture and aquaculture: The Engleberg Report. *International Reference Center For Waste Disposal News* **23**, 11–18.

Israel Water Authority (2008) www.water.gov.il/NR/rdonlyres/644ECF7F-7329-420D-AE22-2BCFBD6CC4BC/0/Matarot19962006.pdf. In Hebrew (last accessed October 2008).

Japan Sewage Works Association (2003) *Sewage Works in Japan*. Tokyo, Japan.

Khuroo, M.S. (1996) Ascariasis. *Gastroenterology Clinics of North America* **25**, 553–577.

L'hirondel, J. and L'hirondel, J.L. (eds) (2001) *Nitrate and Man: Harmless or Beneficial*. CABI, Wallingford, UK.

Ling, B. (2000) Health impairments arising from drinking water polluted with domestic sewage and excreta in China. In: *Water Sanitation & Health* (eds I. Chorus,et al.) IWA Publishing, London.

Mapanda, F., Mangwayana, E.N., Nyamangara, J. and Giller, K.E. (2005) The effect of long term irrigation using wastewater on heavy metal contents of soils under vegetables in Harare, Zimbabwe. *Agriculture Ecosystems & Environment* **107**, 151–165.

Mara, D. and Horan, N. (2003) *Handbook of Water and Wastewater Microbiology*. Academic Press, London.

Marecos Do Monte, M.H. (2007) Guidelines for good practice of water reuse for irrigation: Portuguese Standard NP 4434. In: *Wastewater Reuse: Risk Assessment, Decision-Making and Environmental Security* (ed. M. K. Zaidi), pp. 253–265. Springer, Amsterdam.

Miyamoto, S. (2003) Managing Salt Problems in Landscape. Use of Reclaimed Water in the Southwest. In: *Proceedings of the Reuse Symposium*. San Diego, CA.

NHMRC-ARMCANZ (National Health and Medical Research Council and Agriculture and Resource Management Council of Australia and New Zealand) (2000) *Guidelines for Sewerage Systems, Use of Reclaimed Water. National Water Quality Management Strategy*. NHMRC and ARMCANZ, Canberra.

NHMRC-NRMMC (National Health and Medical Research Council and Natural Resource Management Ministerial Council) (2004) *Australian Drinking Water Guidelines, Australian Government, Canberra*. National Health and Medical Research Council and Natural Resource Management Ministerial Council.

NRC (1996) *Use of Reclaimed Water and Sludge in Food Crop Production*. National Research Council, National Academy Press, Washington DC.

NRC (1998) *Issues in Potable Reuse*. National Research Council, National Academy Press, Washington DC.

NRC (2004) *Indicators for Waterborne Pathogens*. National Research Council, National Academy Press, Washington, DC.

NRMMC-EPHC (2006) *Australian Guidelines for Water Recycling: Managing Health and Environmental Risks*. Natural Resource Management Ministerial Council and the Environment Protection and Heritage Council.

Ogoshi, M., Suzuki, Y. and Asano, T. (2001) Water reuse in Japan. *Water Science and Technology* **43**(10), 17–23.

Papadopoulos, I. (1995) Present and perspective use of wastewater for irrigation in the Mediterranean basin. In: *Proceedings of the 2nd International Symposium on Wastewater Reclamation and Reuse* (eds A.N. Angelakis et al.), pp. 735–746. IAWQ. Iraklio, Greece, 17–20 October.

Paranychianakis, N.V., Aggelides, S. and Angelakis, A.N. (2004) Influence of rootstock, irrigation, and recycled water on growth and yield of Soultanina grapevines. *Agricultural Water Management* **69**, 13–27.

Paranychianakis, N.V., Angelakis, A.N., Leverenz, H. and Tchobanoglous, G. (2006) Treating wastewater through land treatments systems: A review of treatment mechanisms and plant functions. *Critical Reviews in Environmental Science Technology* **36**, 187–259.

Pedersen, J.A., Yeager, M.A. and Suffet, I.H. (2003) Xenobiotic organic compounds in runoff from fields irrigated with treated wastewater. *Journal of Agricultural Food Chemistry* **51**, 1360–1372.
Petterson, S.R., Ashbolt, N.J. and Sharma, A. (2001) Microbial risks from wastewater irrigation of salad crops: a screening-level risk assessment. *Water Environmental Research* **72**, 667–672.
Rattan, R.K., Datta, S.P., Chhonkar, P.K., Suribabu, K. and Singh, A.K. (2005) Long-term impact of irrigation with sewage TWWs on heavy metal content in soils, crops and groundwater - A case study. *Agricultural Ecosystem and Environment* **109**, 310–322.
Rose, J.B., Dickson, L.J., Farrah, S.R. and Carnahan, R.P. (1996) Removal of pathogenic and indicator microorganisms by a full-scale water reclamation facility. *Water Research* **30**, 2785–2797.
Rose, J.B., Huffman, D.E., Riley, K., Farrah, S.R., Lukasik J.O. and Hamann, C.L. (2001) Reduction of enteric microorganisms at the Upper Occoquan Sewage Authority water reclamation plant. *Water Environmental Research* **73**, 711–720.
Rowe, D.R. and Abdel-Magid, I.M. (1995) *Handbook of Wastewater Reclamation and Reuse*. CRC Press, Inc. Boca Raton, USA.
Salgot M. and Angelakis, A.N. (2001) Guidelines and regulations on wastewater reuse. In: *Decentralized Sanitation and Reuse, Integrated Environmental Technology Series*, pp. 446–466. IWA Publishing, London, UK.
Stanbridge, H.H. (1976) History of Sewage Treatment in Britain, Part 5 Land Treatment. The Institute of Water Pollution Control, Maidstone, Kent, England.
State of California (2000) *Code of Regulations, Title 22 Division 4, Chapter 3. Water Recycling Criteria.* Sections 60301 et seq., Dec. Berkeley, CA.
Tanaka, H., Asano, T., Schroeder, E.D. and Tchobanoglous, G. (1998) Estimating the safety of wastewater reclamation and reuse using enteric virus monitoring data. *Water Environmental Research* **70**, 39–51.
Tsagarakis, K.P. (2005) Recycled water valuation as a corollary of the 2000/60/EC water framework directive. *Agricultural Water Management* **72**, 1–14.
USEPA/USAID (1992) *Guidelines for Water Reuse*. Environmental Protection Agency, Office of Wastewater Enforcement and Compliance (Technical Report no. EPA/625/R-92/004), Washington, DC.
USEPA (1999) *Disinfection Profiling and Benchmarking Guidance Manual United States Environmental Protection Agency*. 815-R-99-013, EPA, Washington DC.
USEPA (2003) *Draft Ultraviolet Disinfection Guidance Manual.* EPA 815-D-03-007, EPA, Washington DC.
USEPA (2004) *Guidelines for Water Reuse*. EPA 625/R-04/108, EPA, Washington DC.
World Bank (1993) *World development report: Investing in health.*
WHO (1973) Reuse of TWWs: Methods of Wastewater Treatment and Health Safeguards. *Report of a WHO Meeting of Experts, Technical Report Series No. 17.* World Health Organization, Geneva, Switzerland.
WHO (1986) *Environmental Health Criteria 37: Aquatic (Marine and Freshwater) Biotoxins.* World Health Organization, Geneva, Switzerland.
WHO (1989) Health Guidelines for the Use of Wastewater in Agriculture and Aquaculture. *Report of a WHO Scientific Group, Technical Report Series 778.* World Health Organization, Geneva, Switzerland.
WHO (1994) *Global comparative assessment in the health sector. Disease burden, expenditures and intervention packages* (eds C.J.L. Murray and A. D. Lopez). World Health Organization, Geneva, Switzerland.
WHO (2006) *WHO Guidelines for the Safe Use of Wastewater, Excreta and Greywater, Vol. 3 Wastewater Use in Agriculture*, 3rd ed. World Health Organization, Geneva, Switzerland.
Yeates, M.V.,and Gerba, C.P., (1998) Microbial considerations in wastewater reclamation and reuse. In: *Wastewater Reclamation and Reuse* (ed. T. Asano), pp. 437–487. Technomic Publishers, Lancaster, USA.

Chapter 4
Economic aspects of irrigation with treated wastewater

Efrat Hadas and Yoav Kislev

4.1 Introduction

Water is a renewable resource. Precipitations enrich the reservoirs and freshwater can be withdrawn from rivers, lakes, or aquifers. Provision is sustainable only if withdrawals are constrained to safe yields, the supply is finite. Practically unlimited quantities of freshwater can be provided by seawater desalination, but at a cost in energy and damage to the environment. Water is actually consumed in agriculture in the sense that most of it evaporates either directly from the surface of the land or through the plants. Urban water is different; a great part of this water is not consumed but rather it is used as a carrier of dirt and refuse. Once treated, the sewage – now as recycled or reclaimed wastewater – can be reused. Although wastewater may be purified to potable quality, most of the treated effluent is directed to irrigation.

Some of the ingredients in wastewater may be useful for plant growth and replace fertilizers; others may be harmful to human health, the soil, or plants. Sewage treatment, aimed generally at removing harmful components, is expensive. This raises the issue of pricing and cost recovery: who should cover the cost, the producer of the sewage or the final user of the treated wastewater? In this chapter, we review the cost of treatment and discuss pricing.

Water, sewage, treatment plants, reclaimed wastewater and its application are all part of the environment in which we live. Great efforts were invested in recent years in economic assessment of the environmental impact of wastewater treatment and use. Much of the work in this area was motivated by the particular "external" nature of the impact – more often than not wastewater and the ways it is treated or reused affect the health and wellbeing of people not directly involved and markets do not exist to evaluate these health and environmental impacts. Economic studies of the environmental effects are, therefore, a critical part (but never the only part) of a sound policy formation process. Consequently, environmental impact assessment is an important component of the economics of the

Treated Wastewater in Agriculture, First Edition, edited by Guy J. Levy, Pinchas Fine and Asher Bar-Tal

wastewater sector. However, focusing on agriculture, we shall not discuss assessment of environmental impact in this chapter.

The chapter is built as an introduction to be read by economists, agriculturalists, engineers, and others who are interested but not yet experts in the economics of wastewater use in agriculture. Therefore, it focuses on basic information and relatively simple economic analysis. Further reading is found in the other chapters of this volume and in the vast economic literature on pricing and cost allocation.

4.2 Wastewater in agriculture

Urban and industrial wastewater accompanies almost all human settlements and its use in agriculture and other outlets is prevalent around the world. Farmers in poor countries often use untreated sewage, whereas developed countries carefully regulate the treatment and application of reused wastewater. Orderly statistics on the reuse of wastewater are not available, but by one count (Bixio and Wingens, 2006) there were, in 2003, more than 3300 reclamation plants around the world and the numbers are growing rapidly.

4.3 Wastewater and the regulation of its reuse

The composition of raw sewage varies greatly between communities depending on local attributes and the sectors using the water and producing the sewage. Households add salts and organic refuse; industrial enterprises contribute chemicals of diversified nature and composition. The following are broad groupings of sewage constituents (Hussain et al., 2002):

- organic matter;
- nutrients (nitrogen, phosphorus, potassium);
- inorganic matter (dissolved minerals);
- toxic chemicals;
- pathogens.

The regulation of reused wastewater mostly aims at preventing health hazards, both to irrigators and food consumers. Accordingly, different standards are set for restricted irrigation (of crops that are not intended for direct human consumption) and unrestricted irrigation. As a rule, rich countries set higher standards than developing countries and these differences are also reflected in the U.S. requirements as against the recommendations of the WHO (the UN World Health Organization). The American regulations have been criticized as being too high and less stringent requirement for poorer countries are justified on grounds that poor municipalities and farmers cannot and actually will not obey tough regulations.

Although microbial aspects have always predominated wastewater regulations, chemical guidelines are also found. Mostly their aim is to protect plants, the soil, and water reservoirs to which wastewater and their residues leak. In some places irrigation with reused wastewater is confined to areas from which leakage cannot reach aquifers or surface water reservoirs.

In addition to its application in agriculture, reclaimed wastewater is used for watering golf courses, gardens, and forests. Sometimes it may be found as coolant in industries. When not utilized in these ways, effluent, treated or untreated, may be disposed, or find its way, into rivers and oceans. Quite often, national and international regulations and conventions require higher standards for wastewater disposed in nature than for the effluent applied in agricultural irrigation.

4.4 Treatment

In principle, sewage can be treated and cleansed to high-standard potable water; and this is probably done, albeit indirectly, in several European countries. Treatment for irrigation can also be performed using many and varied technologies. The basic method, and in most cases the first stage if more advanced methods are implemented, is wastewater stabilization land ponds. They are relatively inexpensive, and, if maintained appropriately, very efficient. Other methods include wetland polishing; soil aquifer treatment; disinfection by chlorination, ozone, and UV application; microfiltration and reverse osmosis. The latter are membrane systems, considered expensive and rarely encountered. Supposedly they should be applied only where the recycled wastewater is intended for immediate drinking. Their advantage is, however, wider than this: membrane treatment removes salts and most other contaminants; irrigation with membrane treated wastewater will not damage soils and hurt plants; the treated wastewater can be freely used to recharge aquifers and other reservoirs – saving land, construction cost, water loss in open storage facilities, and separate distribution networks.

Conventionally, wastewater treatment is divided into different levels in accordance with accepted definitions (Halperin, 1999; Fine et al., 2006; Westcot, 1997) as follows:

- Preliminary treatment: removal of solids and other large materials from the row sewage.
- Primary treatment: removal of settleable organic and inorganic solids by sedimentation and removal of floating material by skimming. A half to two-thirds of the organic matter in the sewage is removed at this stage. Effluent from sedimentation plants is referred to as primary effluent.
- Secondary treatment: follows primary treatment and its aim is to remove further biodegradable dissolved and colloidal organic matter. Most of the removal is done at this stage by microorganisms and the process is usually enhanced by aeration and supply of oxygen. Following the treatment, the microorganisms are separated from the fluids by sedimentation and the sludge is removed for further treatment and disposal.
- Tertiary (advanced) treatment: further removes organic and inorganic components from the treated sewage. In most cases it also reduces significantly the nitrogen contents of the wastewater. It is used where specific health or environmental constraints are raised. In most places, edible products cannot be grown on wastewater unless it passed tertiary treatment.
- Quaternary treatment: mostly membrane treatment of effluent suitable for unrestricted household use.

Table 4.1 Investment in treatment plants

	Plant, million US dollars			Per CM, US dollars		
Capacity, CM/D	500	5000	50 000	500	5000	50000
Primary	0.266	1.888	4.888	0.53	0.38	0.09
Secondary	0.555	3.888	15.800	1.11	0.78	0.31
Tertiary	1.000	5.244	21.888	2.00	1.04	0.44
Tertiary + desalination	1.400	6.466	32.185	2.80	1.29	0.67

[a]CM cubic meter; CM/D CM per d.
[b]Calculated from data for 150 treatment plants in Israel in 2006.
[c]Exchange rate NIS 4.5 per US dollar.
Source: David Alkan, private correspondence, 2006.

4.5 Cost

As indicated, primary and secondary treatment is relatively inexpensive; further stages add markedly to the cost of reclaimed wastewater. Still, even the early stages are not free, they are capital-intensive, and may constitute a heavy burden on municipalities or farmers. Tables 4.1 and 4.2 present investment, capital cost, and operating costs for treatment of sewage by capacity of the facility.

The data in Table 4.1 were gathered from information supplied by planners of new plants. The data reveal economies of scale; capital cost per cubic meter (CM) of capacity in a large plant is less than half of the cost in a small plant. Operating costs are reported in Table 4.2. Their level reflects the specific treatment usually employed in the plants. They exhibit only modest economies of scale. Extensive plants are built in relatively small communities.

4.6 Replacement of fertilizers

As indicated above, sewage and recycled water usually contain nutrients of use to plants. Thus, they can replace fertilizers. Table 4.3 reports several examples. As the entries indicate, recycled water, and in some cases sludge, may replace fertilizer application. In several of the cases the replacement is significant, more than 50% of the requirement. In other cases, the replacement is modest. Also reported in the table are some cases in which irrigation with reused wastewater will result in too much of the nutrient being applied to

Table 4.2 Annual capital and operating cost in treatment plants, US dollars per CM

	Capital			Operating			Total		
Capacity, CM/D	500	5000	50 000	500	5000	50 000	500	5000	50 000
Primary	0.12	0.08	0.02	0.11	0.06	0.03	0.23	0.14	0.05
Secondary	0.21	0.17	0.07	0.17	0.13	0.09	0.38	0.30	0.16
Tertiary	0.41	0.23	0.10	0.28	0.22	0.15	0.69	0.45	0.25
Tertiary + desalination	0.86	0.28	0.15	0.54	0.40	0.29	1.40	0.68	0.44

Per unit capital cost was calculated for 25 years life expectancy at 7% interest.
Source: Table 4.1 and its source.

Table 4.3 Nutrients required and potential provision by recycled wastewater (kg per hectare and percent)

Crop	Nitrogen N		Phosphorus $P_2 0_5$		Potassium K_2O		Treatment in calculation
	kg/hectare	% from recycl.	kg/hectare	% from recycl.	kg/hectare	% from recycl.	
Citrus	285.3	55	61.7	21	306.7	90	Mechanical-biological treatment
Fruit trees	183.0	88	38.0	35	250.1	112	Mechanical-biological treatment
Dryland fruit trees	150.0	140	0.0		250.0	64	Sludge A
Open space vegetables	304.2	12	175.9	13	400.5	68	Treatment for all uses
Greenhouse vegetables	1159.0	5	748.3	3	1731.0	24	Treatment for all uses
Open space flowers	1596.4	4	655.6	3	1649.7	33	Mechanical-biological and sludge A
Flowers in a protected structure	208.2	19	140.7	17	224.0	127	Treatment for all uses
Irrigated field crops	185.2	35	73.2	7	161.4	70	Sludge A
Dryland field crops	77.0	351	73.0	85	0.0		Sludge A

Column 8 reports the sewage treatment assumed for each row.
Source: Hadas and Fine (2009).

Table 4.4 Typical cost of fertilizers and total cost of production (US dollars/hectare)

Crop	Nitrogen	Phosphorus	Potassium	Total fertilizers	Total cost
Citrus	304	76	224	604	7055
Fruit trees	162	38	140	340	17 916
Dryland fruit trees	96	0	111	197	3107
Open space vegetables	324	204	300	828	17 741
Greenhouse vegetables	1633	1222	1729	4584	122 542
Open space flowers	422	122	144	688	9731
Flowers in a protected structure	2649	1256	1882	5787	148 534
Irrigated field crops	131	64	78	273	2573
Dryland field crops	18	38	0	56	718

Exchange rate NIS 4.5 per 1 US dollar.
Source: Hadas and Fine (2009).

the plants. As indicated, overdoses may be harmful and their occurrence may require dilution of the water with freshwater or purposive removal of the nutrient from the effluent of the treatment plants.

"To be on the safe side" farmers often apply more than the recommended amounts of fertilizers and do not reduce application even when irrigating with recycled wastewater. This practice, often detrimental to soil and water resources, can be explained by the small share of fertilizers in the cost of production (Table 4.4).

4.7 Cost allocation and prices

The treatment of sewage is expensive; the product, effluent, is used in agriculture. Who should cover the cost, the producers of the sewage or the user of the effluent? The conventional wisdom is often that "the polluter pays". But this is not always so. Cost is allocated by prices, if farmers pay for the reused wastewater they share in covering its cost. We have, therefore, to examine prices. We shall distinguish between two cases: in the first, reclaimed wastewater is allocated by prices; in the second case, allocation is not based on prices.

As explained above, recycled water carries contaminations that are added to the soil and the groundwater. In many cases, the most important contaminations are salts as they are not removed in regular treatment of the sewage, not even in tertiary treatment. (Table salt, sodium chloride, is most abundant. Sodium is harmful to soil structure but, as chloride concentration is easier to measure, salinity is often expressed as ppm (part per million) chlorides. A more comprehensive measure of salinity is the electrical conductivity of the water commonly expressed in deciSiemens/meter.) Therefore, we take salts as representing all forms of contamination. Recycled wastewater is, however, not the only source of salts, freshwater also contains salt and therefore irrigation, whether with fresh or with reused water, adds salts to the soil. The salts added are drained to water sources and may have to be removed when their accumulation reaches harmful concentrations. The cost of removal will eventually become part and parcel of the cost of irrigation in general, and, in particular, the use of wastewater in agriculture. Moreover, the cost of reused water cannot be analyzed in isolation; the costs of freshwater and reused water have to be considered together. We begin with price allocation.

4.7.1 The role of prices

Prices provide, in the water economy and in general, three different functions:

(i) Prices convey information. Where prices are equal to marginal costs, the cost to the users of water, who pay the prices, is equal to the cost to the national economy. Water is used only if its contribution is economically justified.
(ii) Prices clear the market; in the sense that with appropriate prices all the available quantity of water, fresh and reused, is taken and shortages do not develop.
(iii) Prices cover cost. The revenue collected by the providers of water of any quality cover the cost of provision.

As we shall see, not always are all three functions fulfilled simultaneously.

4.7.2 The model economy

The principles of pricing are simple and well known; their application in the water economy is, however, generally not straightforward and depends on local and technological circumstances. Price determination is, therefore, presented here within the framework of a model economy. The model is based on a highly simplified description of the Coastal region in Israel (for a more detailed treatment, see Goldfarb and Kislev, 2007). In the model, water is supplied to the region from an outside source, Lake Kinneret (Sea of Galilee), and is used in urban communities and in agriculture. A constant share of urban water is collected as sewage, treated, and provided for reuse in agriculture. Irrigation adds salt to groundwater and it is removed by desalination. Prices determined in the model allocate the cost of freshwater and the reclaimed wastewater, including the cost of salt removal, between farmers and urban users.

Again for simplicity, the model envisages a water economy in a steady state: constant quantities of water are provided annually to the urban sector; agriculture receives year in and year out constant quantities of fresh and reclaimed water; and, also, the same quantities of salts as added yearly to the aquifer are removed by desalination. Consequently, the concentration of salt in the coastal groundwater is kept constant; it does not accumulate.

Formally, prices are determined in a cost minimization model. Agriculture is provided with the quantity X_A of freshwater (measured in CM per year) or its wastewater equivalent. Households (the urban sector) are provided annually with X_U. Freshwater from Lake Kinneret, M_{KH}, is supplied to households and agriculture. The ratio of sewage to water in households is r. Agriculture uses M_A CM of freshwater annually and R CM reclaimed wastewater. One CM of wastewater is equivalent, in its contribution to production, to γ CM of freshwater ($0 \leq \gamma \leq 1$). The system of equations describing these requirements is

$$\begin{aligned} M_A + M_U &= M_{KH} \\ M_A + \gamma R &= X_A \\ M_U &= X_U \\ R &= rM_U \end{aligned} \tag{4.1}$$

Irrigation water carries salt and it is deposited on the surface of the land and drained to the aquifer. Salt concentration, measured in grams of chlorides per CM (mg/l), is μ_K in Kinneret water, μ_R in wastewater, and μ_S in the water of the coastal aquifer. The quantity of desalinated coastal water is M_D CM/yr. For simplicity, we assume that the desalinated water (to remove salts) is returned to the aquifer and not consumed directly. We also disregard water lost in the concentrate. In the steady state all salts reaching the aquifer are removed by desalination

$$\mu_K M_A + \mu_R R = \mu_S M_D \tag{4.2}$$

The cost items in the model are conveyance from Lake Kinneret to the Coastal region, C_{KH} dollars/CM, sewage treatment, C_A dollars/CM (the index A stands for treatment to the level required for use in agriculture), and desalination C_D dollars/CM. We disregard the cost of intra-urban distribution and sewage removal, and extraction levies that are imposed in Israel on the withdrawal of water from reservoirs. The mathematical model is presented in Appendix 4.1.

4.7.3 Prices

The prices fulfilling the first function mentioned above (information) are equal to marginal cost and they will be, by Appendix 4.1,

$$\begin{aligned} P_A &= C_{KH} + C_D \frac{\mu_K}{\mu_S} \\ P_U &= C_{KH}(1-\gamma r) + rC_A + rC_D \frac{\mu_R - \gamma \mu_K}{\mu_S} \\ P_R &= \gamma P_A \end{aligned} \tag{4.3}$$

In equation 4.3, P_A, P_U, and P_R are, respectively, the price of water in agriculture, urban use, and the price farmers pay for reclaimed wastewater. The last price reflects the contribution of the recycled water in agriculture, relatively to the contribution of freshwater.

By these prices, farmers are charged for the transfer of water from Lake Kinneret plus the cost of the removal of salt this water adds to the aquifer. Urban dwellers are similarly charged for water transfer and salt removal, and, in addition, for sewage treatment. However, the recycled water, after being treated, is supplied to agriculture and the urban sector is credited in the model for its contribution in agriculture. In other words, the model envisages the urban sector as selling the treated effluent to agriculture at a price equal to its marginal contribution in that sector (not necessarily equal to treatment cost).

In many places, urban dwellers are charged separately for water and for sewage services. But even if these two items appear in different rows on the water bill, or in two different bills altogether, they are a single cost: a household cannot use water without incurring the cost of sewage removal.

The prices of the equation 4.3 can be calculated under the following reasonable assumptions:

$$\begin{aligned} &C_{KH} = 0.30 \quad C_D = 0.44 \quad C_A = 0.67 \\ &r = 0.60 \quad \gamma = 0.80 \\ &\mu_K = 250 \quad \mu_R = 350 \quad \mu_S = 400 \end{aligned} \tag{4.4}$$

The assumptions in equation 4.4 incorporate the estimate that urban use adds 100 g of chlorides/CM of effluent and that, in the steady state, chlorides concentration in the coastal water will be 400 g/CM (at present average concentration is 200 g/CM, but rising). In the calculation of prices we also incorporated the (reasonable) assumption that half the quantity of salts annually reaching the coastal water is removed by means other than desalination – drainage to the sea, fresh and reclaimed water "exported" to other regions. Accordingly, the prices, in dollars/CM, are:

$$P_A = 0.43 \quad P_U = 0.60 \quad P_R = 0.34$$

4.7.4 Cost recovery

We turn now to the third of the functions mentioned above – cost recovery. The first line in the following equation is the total revenue collected for fresh and reclaimed water supplied to urban dwellers and to agriculture; the last line is the total cost of provision, including conveyance, treatment, and desalination. The development of the equation relies on equations 4.1–4.3 above.

$$\begin{aligned} &M_A + R\gamma P_A + M_U \\ &\quad = (M_A + \gamma R)\left(C_{KH} + C_D\frac{\mu_K}{\mu_S}\right) + M_U\left(C_{KH}(1-\gamma r) + rC_A + rC_D\frac{\mu_R - \gamma\mu_K}{\mu_S}\right) \\ &\quad = M_A C_{KH} + \gamma R C_{KH} + M_U C_{KH} - \gamma R C_{KH} + R C_A + \frac{C_D}{\mu_S}(\mu_K M_A \\ &\quad + \gamma\mu_K R + R\mu_R - \gamma\mu_{KR}) \\ &\quad = C_{KH} M_{KH} + C_A R + C_D M_D \end{aligned} \tag{4.5-4.7}$$

Using equation 4.7, charging marginal cost prices for freshwater and wastewater allows the suppliers to cover all their costs. It should be added that cost is recovered with marginal prices for the particular cost structure assumed in this analysis (linear cost functions). With a different cost structure, charging marginal prices may not lead to cost recovery; this possibility will not be elaborated here.

4.8 Further considerations and alternatives

Up to this point the analysis has been done under the assumption that agriculture uses both fresh and recycled water. The farmers pay for the recycled water according to its (marginal) contribution to production. In such a situation, the price of freshwater in the urban sector may be lower than in agriculture – if γP_A is higher than the cost of treatment and salt removal. As urban dwellers do not consume all the water they get – mostly they use water for the removal of refuse – they can be seen, in analogy to car renting companies, as purchasing water, using it for a while, cleaning the sewage, and selling the treated effluent "second hand". The proceeds from the sale lower the cost of usage.

Although in principle the price of freshwater may be lower in town than on the farm, it will not always be so. If the marginal contribution of water in agriculture is relatively low, the price farmers will pay for recycled water will cover only part of the cost of sewage treatment. Then the cost of water provided to urban dwellers (including sewage treatment) will be higher than the price of freshwater in agriculture.

A situation may even arise when farmers will be paid for taking the recycled water. In a place with ample supply of freshwater, its cost and price to farmers will be low, and they will not be interested in taking a lower productivity effluent. The town may then "bribe" agriculture to help it get rid of the pollution it created.

Where urban communities are located close to oceans or rivers, they may dispose of the treated effluent in the nature. This option may not be open to inland communities. Where reclaimed water can be delivered to nature at a relatively low cost (although it is seldom the case), it will not be economically justified to allocate it to agriculture unless the contribution of the recycled water in agriculture is higher than the cost of provision. The existence of a non-agricultural option is sometimes termed "the zero alternative": the proceeds from turning the recycled water to agriculture must be comparatively high for the urban communities not to prefer the zero solution.

4.8.1 Will the polluter pay?

As we have seen, the recycled water may be "sold" to farmers, and, in this way, free the cities from the cost of sewage treatment. At the same time, it should not be forgotten that the fundamental approach of the preceding analysis is that the responsibility for the treatment of the sewage and its disposal lies with the users of the water: the urban dwellers and industry. If the treated effluent cannot be transferred for a price, the urban sector must cover all the cost of its disposal. The polluter pays in the sense of being responsible.

4.8.2 Scarcity rents and extraction levies

Water in most places is a scarce resource; its use is limited. Being scarce, it has economic value – seawater is not scarce and does not have, in general, economic value. Economic analysis justifies payment of a scarcity rent by water users. For example, in Israel, water providers pay an extraction levy for water withdrawn from aquifers or from surface sources, including Lake Kinneret (trying not to burden the argument, we disregarded the levy in the foregoing analysis). As water belongs to the public, the levy is paid as a tax to the treasury to be used for social benefit.

Where farmers take all the available recycled water, it is also scarce – they are willing to pay for additional quantities. This scarcity does not, however, justify the imposition of a scarcity levy on recycled water. The basic scarcity is that of freshwater at the source, and it is there that the levy should be charged. Once paid, the water "belongs" to the urban sector and any payment for it should go to whomever does the treatment and recycling. Imposing a tax on the recycled water will introduce a wedge between the value of the treated effluent in agriculture and the cost of its collection and treatment in the city and lower the incentive of the urban sector to recycle its water.

4.9 Agreements

The provision of recycled water for a price is appropriate where the number of users is large. But sewage treatment and reuse is often a local affair: a town disposes of the reclaimed water and an agricultural cooperative accepts it to be distributed to its members. In such a case, the town and the cooperative may enter negotiations on the conditions of the transfer. The agreement the parties reach will be bounded: the town will not transfer the treated wastewater to the farmers if the associated cost is higher than that of the (zero) alternative. Likewise, the farmers will not take the wastewater unless its contribution to their fields is higher than its cost to them. But between these two limits, there is room for negotiations. Where will they settle? One possibility is that they will follow, even if unknowingly, the bargaining solution proposed by Nash (1950). The solution is a framework for the analysis of cost allocation under simplifying and reasonable assumptions (not detailed here).

4.9.1 Nash bargaining solution

The town operates a sewage treatment plant and it disposes of its effluent into the ocean at a cost of C_S dollars/CM. A suggestion is raised to redirect the effluent to neighboring agriculture. The cost of treating and provision of the recycled water to agriculture is C_A dollars/CM and it is lower than the cost of ocean disposal, $C_S > C_A$. The recycled water will be given to the farmers at a price P_R lower than the price they pay for freshwater (disregarding, for simplicity, quality differences) $P_A > P_R$. The utility the town draws from the shift of the treated effluent from ocean disposal to agriculture is

$$U = R(C_S - C_A + P_R) \tag{4.8}$$

The utility of the farmers

$$A = R(P_A - P_R) \tag{4.9}$$

Using the Nash Solution, the transfer price to be set in the bargaining process will maximize the product of the utilities of the bargaining parties. The maximization problem is: choose a transfer price P_R to maximize N in the following equation

$$N = UA \tag{4.10}$$

By the first order condition, the price is

$$P_R = \frac{P_A - (C_S - C_A)}{2} \tag{4.11}$$

With this transfer price, the parties to the agreement – agriculture and the urban sector – will derive equal utilities

$$U = A = R\left(\frac{P_A + (C_S - C_A)}{2}\right) \tag{4.12}$$

The nominator in equation 4.12 is the utility, measured in dollars CM of recycled water: the saving of freshwater in agriculture, P_A, plus the saving in cost of treatment in town, $C_S - C_A$, and this sum is divided equally between the parties. Notice that according to equation 4.12, the agreed price, P_R, may be negative: the agreement reached will determine that the town pays the farmer for taking the effluent of its treatment plant. As indicated above, such a solution can be expected if the contribution of water in agriculture is comparatively low and the cost of disposing of the treated effluent in the ocean or river is high.

4.9.2 Remarks

Game theory is the discipline dealing with mutual relations, economic or otherwise. Substantial efforts were devoted in game theory to the question of cost allocation. The Nash Solution presented here is probably the simplest of those problems and solutions. In reality, allocation problems may be more complex and difficult; for example, the sewage may be treated to several quality levels, or the agreement may have more than two bargaining parties. In such cases the solutions that game theory offers may be more complex but the fundamental principle is the same: a cost allocation that all parties can accept (for an application, see Dinar et al., 1986).

The Nash Solution can be viewed from two different angles. Accepting the solution as the finding of a positive analysis, the solution points to the agreement the parties are expected to reach once they realize that it will be for the benefit of both if they cooperate in directing the reclaimed wastewater to agriculture and in sharing the cost. From a normative point of view, the solution can serve the government or a mediator to suggest or force a transfer price – where the parties do not reach an agreement on their own.

The difficulty is that an outside agent, trying to set a transfer price, has to know the utility of the parties sharing the effluent.

Also, notice that the transfer price reached or dictated by the Nash Solution need not be a "market clearing" price. The Nash Solution price applies where an agency or a cooperative allocates the recycled water to its users. Where farmers are free to take as much of the effluent as they see fit at the price P_R of equation 4.12, the quantity demanded may exceed or fall short of the available supply.

A problem that often haunts parties to agreements is of potential opportunistic behavior. Once the agreement has been reached and the appropriate infrastructure installed, the farmers can threaten the town that they will not take the reclaimed water unless the transfer price is lowered. Similarly, the town – if it has an alternative – may threaten the farmers that the treated effluent will not be provided unless the price is increased. These potential threats are often real obstacles to reaching and implementing an agreement.

4.10 The role of the government

Governments provide public goods. In our case, first and foremost are regulations pertaining to treatment and reuse of wastewater. The regulations and their enforcement are public goods – as we have seen, wastewater and its disposal may have severe negative external effects. The health of farmers, their families, food consumers, and water using households cannot be left "to the market". There does not exist a market to regulate these external effects.

Further government involvement may also be justified sometimes. Enforcement is not costless, a city may dispose of its raw sewage into a nearby river at no cost and farmers may use slightly treated wastewater to irrigate sensitive crops. The government may find that it is more efficient and cost-effective to subsidized treatment and recycling than to engage endlessly in futile monitoring and enforcement. Economists who generally oppose subsidies and government economic intervention should be careful when criticizing efforts to achieve environmental public goals.

4.11 Concluding comments

As world population grows and standards of living increase, more water is used and more food is consumed. Irrigated agriculture will have to give up, in many places, freshwater to be diverted to human utilization. Replacement, at least partly, will come in the form of recycled wastewater. The treatment of wastewater is expensive and choice of technique and degree of purification will have to be made carefully. With cost comes the question of its recovery, hence the prices and levies charged in the urban sector and agriculture. These issues can be solved once it is recognized that treated wastewater is both a burden and an input – in agriculture and possibly in nature. Water is often carried over long distances, sewage is in general more local in nature. In this chapter, we demonstrated the basic approach to pricing and cost allocation, both where treated wastewater is provided to a large number of farmers and where a relatively small community comes to an agreement with surrendering agriculturalists. Needless to say, the economic models formulated were founded on a simplifying assumption; reality is more complex and application always requires careful analysis and planning. It is not superfluous, however, to re-emphasize the role of the government as regulator and arbitrator. Because of its strong external effects and complex economic issues, wastewater cannot be left to free market rule. Efficient and health preserving management of the sector can be expected only where the government is responsible and is able to convince the various stakeholders of the need to follow its guidelines.

References

Bixio, D. and Wingens, T. (2006) *Water reuse system, management manual.* AQUAREC, Directorate-General for Research Global change and ecosystems, European Commission.
Dinar, A., Yaron, D. and Kannai, Y. (1986) Sharing regional cooperative gains from reusing effluent for irrigation. *Water Resource Research* **22**, 331–338.
Fine, P., Halperin, R. and Hadas, E. (2006) Economic considerations for wastewater upgrading alternatives: An Israeli test case. *Journal of Environmental Management* **78**, 163–167.
Goldfarb, O., and Kislev, Y. (2007) *Cost and pricing in a regime of sustainable water resources.* http://departments.agri.huji.ac.il/economics/yoav-home.html.
Hadas, E. and Fine, P. (2009) Options to Reduce Greenhouse Gases Emission during Wastewater Treatment for Agricultural Reuse. *Handassat Mayim (Water Engineering)* **61**, 22–28 (Hebrew).
Halperin, R. (1999) Regulations for water quality in Israel and internationally. *Health in the Field* **7**, 3–6 (in Hebrew).
Hussain, I., Raschid, L., Hanjra, M.A., Marikar, F. and van der Hoek, W. (2002) *Wastewater use in agriculture, review of impacts and methodological issues in valuing impacts, Working paper 37.* International Water Management Institute, Colombo, Sri Lanka.
Nash, J. (1950) The Bargaining Problem. *Econometrica* **18**, 155–162.
Westcot, D.W. (1997) *Quality control of wastewater for irrigated crop production. Water reports – 10.* Food and Agriculture Organization of the United Nations, Rome (1020–1203).

Appendix 4.1 The mathematical model of pricing in the steady state

The model is formulated under the assumption that all the requirements in equations 4.1 and 4.2 in the text are realized and all the variables are positive (no zeroes). The Lagrange function is

$$\begin{aligned} L = {} & C_{KH}M_{KH} + C_A R + C_D M_D + \lambda_A(X_A - M_A - \gamma R) + \lambda_U(X_U - M_U) \\ & + \lambda_M(M_A + M_U - M_{KH}) + \lambda_D(\mu_K M_A + \mu_R R - \mu_S M_D) + \lambda_R(rM_U - R) \end{aligned} \quad \text{(A.1)}$$

First order conditions

$$(i) \quad \frac{\partial L}{\partial M_{KH}} = C_{KH} - \lambda_M = 0$$

$$(ii) \quad \frac{\partial L}{\partial M_D} = C_D - \lambda_D \mu_S = 0$$

$$(iii) \quad \frac{\partial L}{\partial M_U} = -\lambda_U + \lambda_M + \lambda_R r = 0$$

$$(iv) \quad \frac{\partial L}{\partial M_A} = -\lambda_A + \lambda_M + \lambda_D \mu_K = 0$$

$$(v) \quad \frac{\partial L}{\partial R} = C_A - \gamma\lambda_A + \lambda_D \mu_R - \lambda_R = 0$$

$$(vi) \quad \frac{\partial L}{\partial \lambda_A} = X_A - M_A - R = 0$$

$$(vii) \quad \frac{\partial L}{\partial \lambda_U} = X_U - M_U = 0$$

$$(viii) \quad \frac{\partial L}{\partial \lambda_M} = M_A + M_U - M_{KH} = 0$$

$$(ix) \quad \frac{\partial L}{\partial \lambda_D} = \mu_K M_A + \mu_R R - \mu_S M_D = 0$$

$$(x) \quad \frac{\partial L}{\partial \lambda_R} = rM_U - R = 0 \tag{A.2}$$

Using conditions *(i)*, *(ii)*, *(iv)*

$$\begin{aligned} \lambda_M &= C_{KH} \\ \lambda_D &= C_D \frac{1}{\mu_S} \\ \lambda_A &= C_{KH} + C_D \frac{\mu_K}{\mu_S} \end{aligned} \tag{A.3}$$

And incorporating conditions (*iv*), (*v*)

$$\begin{aligned} \lambda_R &= C_A - \gamma C_{KH} + C_D \frac{\mu_R - \gamma \mu_K}{\mu_S} \\ \lambda_U &= C_{KH}(1 - \gamma r) + rC_A + rC_D \frac{\mu_R - \gamma \mu_K}{\mu_S} \end{aligned} \tag{A.4}$$

The shadow prices reflect marginal cost of provision to the sectors. Prices of urban water, freshwater in agriculture, and effluent are therefore

$$\begin{aligned} P_A &= \lambda_A = C_{KH} + C_D \frac{\mu_K}{\mu_S} \\ P_U &= \lambda_U = C_{KH}(1 - \gamma r) + rC_A + rC_D \frac{\mu_R - \gamma \mu_K}{\mu_S} \\ P_R &= \gamma P_A \end{aligned} \tag{A.5}$$

The last price, $P_R = \gamma P_A$, is not derived from the first order conditions; it reflects the quality ratio of effluent to freshwater. The prices of equation A.5 are presented in equation 4.3 in the text of the chapter.

The shadow price of effluent, λ_R, reports their cost to the national economy: by how much would the cost of the national product increase with an additional MC of effluent – when freshwater supply was not expanded?

$$\lambda_R = C_A - \gamma C_{KH} + C_D \frac{\mu_R - \gamma \mu_K}{\mu_S} = C_A + C_D \frac{\mu_R}{\mu_S} - \gamma P_A \tag{A.6}$$

The components of the shadow price are the cost of sewage treatment plus removal of salt in the effluent minus the contribution of the effluent to production in agriculture. In the prices suggested in equation A.5, the cost of effluent is allocated to the sectors, households and agriculture.

Part II

Impacts on the Soil Environment and Crops

Chapter 5

Major minerals

5.1 Nitrogen in treated wastewater-irrigation

Asher Bar-Tal

5.1.1 Introduction

Nitrogen (N) is one of the major essential nutrient elements for all living organisms. It is one of the major elements in amino acids, proteins, nucleic acids, alkaloids, purine bases, and enzymes. It is also one of the main elements in the chlorophyll in plants and algae. Consequently, large quantities of nitrogen are required by plants and especially by agricultural crops. Typical deficiency symptoms are very slow-growing, weak, and stunted plants; leaves light green to yellow in color, beginning with the older leaves; plants will mature early, and dry weight and fruit yield will be reduced.

Raw sewage water contains high concentrations of organic and mineral N, the range depending on the water source. The N content in raw municipal sewage water is, wide, depending on the standard of living from 20 to 100 mg/l (Feigin et al. 1991). The concentration of N in the treated wastewater (TWW) decreases with the progress in the treatment from primary to secondary and to further tertiary and specific treatment for removing N from the effluent as described in Chapter 2. Consequently, total N in the TWW used for irrigation is in the range of 5–60 mg/l (Feigin et al. 1991). The average concentration of total N (organic and mineral) in TWW used for irrigation of citrus in Israel is about 31 mg/l and the concentration of ammonium is 23 mg/l (Anonymous, 2006). Nitrogen concentration as nitrate in TWW is usually below 1–2 mg/l. For some crops, N added to soils through TWW irrigation can be similar to or even exceed the amount commonly applied via fertilization with freshwater (Feigin et al. 1991). Consequently, TWW may be considered as a source of N for crops, which is a beneficial characteristic for TWW use in agriculture (Adeli and Varco, 2001; Feigin et al. 1981; Feigin et al. 1984; Haruvy et al. 1999; Oron et al. 1982). If TWW irrigation is well managed, it can be beneficial to crop growth and significantly reduce the amounts of applied fertilizers. As the

Treated Wastewater in Agriculture, First Edition, edited by Guy J. Levy, Pinchas Fine and Asher Bar-Tal

concentration of N in the TWW decrease to low values below 10 mg/l the amount applied insignificant even when a high irrigation head is used. For example, if the annual water head is in the range of 500–1000 mm and the total N concentration is 5 mg/l, the annual applied total N through the water is 25–50 kg/ha, a relatively small fraction of the range of 200–500 kg/ha fertilizer N applied to various irrigated crops. Therefore, the focus of this chapter is secondary and tertiary TWW that contains N concentration above 10 mg/l. The two major concerns in using TWW as a source of N are: (i) the availability to crops of mineral and organic N in the effluent that is applied to the soil; and (ii) environmental pollution hazards from nitrogen due to irrigation with TWW. The main processes that influence N availability and its environmental pollution hazards are the chemical transformations of the added inorganic and organic N in the soil.

5.1.2 Nitrogen transformations

Nitrogen gas (N_2) is the main component of the atmosphere (about 78%); however, most plants are not able to utilize this source of N directly, but are dependent on the chemical transformation of N to two ions, NH_4^+ and NO_3^-, that are taken up from the soil solution by the plant root system (Marschner, 1995). The paradox in N nutrition of plants is that N is taken from the soil rather than air, although the total N content of soils is much lower than that of air, it ranges from <0.02% in uncultivated mineral soils to >2.5% in peats; the surface layer of most cultivated soils contains between 0.06 and 0.5% N (Bremner and Mulvaney, 1982); in top soils in the USA a range of 0.039–3.2% has been reported (Bronson, 2008).

Soil N is composed of an organic fraction that consists of up to 90% of the total soil N (Stevenson, 1982). Only the mineral fraction of the soil N, ammonium and nitrate, are available for plants (Marschner, 1995). Although it has been shown in the last decade that plants can absorb small organic molecules like amino acids, the main form of available N is ammonium and nitrate (Jones et al., 2005). A fraction of the organic soil N, in the range of 10–40%, can be mineralized (Stanford and Smith, 1972), and become available to plants following mineralization and transformation to the available ionic forms. It should be noted that these transformations involve a wide range of oxidation levels of the N atom from -3 for NH_3 (or NH_4^+) to+ 5 for NO_3^- (Stevenson, 1982). A schematic presentation of the N cycle in the environment is shown (Fig. 5.1.1). Most of these transformations in the soil involve microorganisms. The relevant chemical processes that affect the fate of N applied with TWW are presented and discussed below.

5.1.2.1 Mineralization and immobilization

Nitrogen mineralization is the production of inorganic N from organic N, whereas nitrogen immobilization is the incorporation of inorganic N into organic forms. These reactions are mainly biologically mediated processes, although they may include also chemical production and consumption of N. Net N mineralization is the sum of the two opposing processes of N mineralization and N immobilization, measured by increasing inorganic N concentration. Net immobilization is defined as the sum of these two opposing processes when the inorganic N concentration decreases. Assimilation is the uptake of inorganic N by microorganisms and incorporation into biomass. The terms assimilation and immobilization are often used synonymously, although they are not identical because

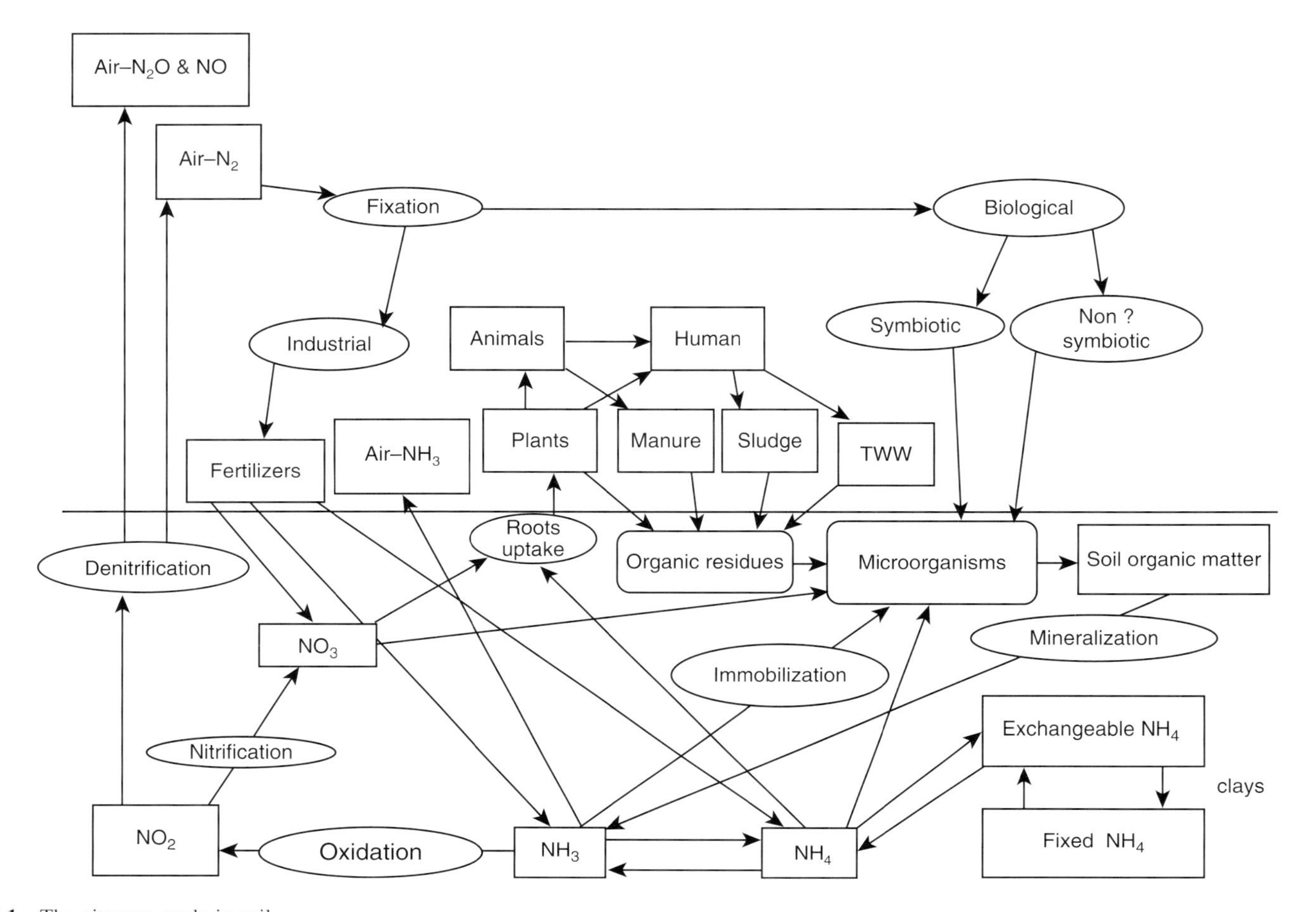

Figure 5.1.1 The nitrogen cycle in soil.

microorganisms also assimilate small organic molecules like amino acids (Myrold and Bottomley, 2008). Microbial assimilation of N (immobilization) is determined by the C flow and the C/N ratio of the heterotrophic decomposers. Mineralization and immobilization always accompany each other operating in reverse directions. Under N limiting conditions, immobilization takes over and soil microorganisms assimilate the inorganic N species (NH_4^+, NH_3, NO_3^- and NO_2^-) into organic constituents of their cells and tissues, thus making them unavailable to plants. However, immobilized N is eventually remineralized due to the turnover of the organic material and the decomposer community (Shindo and Nishio, 2005). Hence, the heterotrophic biomass is in a dynamic, regenerating state. The continuous transfer of mineralized N into organic products of synthesis and of immobilized N back into inorganic decay products can be defined as mineralization-immobilization turnover (MIT). The net effects of this heterotrophic activity are easy to measure, but the differentiation between the processes is only possible using labeled N (Myrold and Bottomley, 2008). However, the amount of mineral N resulting from the mineralization and immobilization processes corresponds to the net mineralization, which equals the extent to which mineralization exceeds immobilization.

Rates of decomposition of organic materials in soil determine the amount of carbon (C) that is mineralized and released as CO_2 versus the amount of C that is retained in various forms in the soil. Decomposition rates greatly influence the amount of N that is mineralized and becomes available for plant uptake or susceptible for leaching versus that which is retained in soil organic matter (SOM) or lost via gaseous emission. In the last 25 years of research it has been found that the dissolved organic matter (DOM) plays a critical role in the processes of organic matter decomposition because its rate of decomposition is higher than the solid fraction of SOM (Myrold and Bottomley, 2008). In TWW most of the organic matter is in the DOM form, hence the organic matter in TWW is expected to decompose faster than solid applied organic residues. The rate of mineral N release from the organic component of applied sewage effluent to the soil is an important factor in synchronizing nutrient release and plant uptake processes.

Factors that affect decomposition rates of organic residues have been identified by several authors. Generally, decomposition and nutrient release patterns are influenced by climatic factors (i.e. temperature and humidity), edaphic factors (i.e. soil moisture, aeration, temperature, microbial biomass, and nutrient status) (Swift et al., 1979) and residue quality factors (Swift et al., 1979; Heal et al., 1997). Work by Fu et al. (1987) and Xu et al. (2006) have shown that soil pH affects mineralization of residues.

MacDonald et al. (1995) investigated temperature effects on the kinetics of microbial respiration and net N mineralization. They found that cumulative respired C and mineralized N increase with temperature. Lomander et al. (1998) reported $-0.83°C$ as the minimum temperature for decomposition activity. They showed that the response of Q_{10} (the factor of decomposition increase with increase of 10°C) increased temperature response with decreasing temperature ($Q_{10} = 2.2$ at 25°C and 12.7 at 0.3°C). Production of respired C and mineralized N closely fit first order kinetic models. Contrary to common assumptions, rate constants estimated from first order models were not consistently related to temperature, but to apparent pool sizes of C and N, which were highly temperature-dependent. Andersen and Jensen (2001) and Nicolardot et al. (1994) discovered that decomposition of recalcitrant substances, such as structural carbohydrates was more limited at low temperatures than that of the more degradable substances. Zak et al. (1999) reported significant temperature-matric potential interactions, based on the greatest decline of substrate pools for microbial respiration and net N mineralization between −0.01 and −0.3 MPa at 25°C. They concluded that high rates of microbial activity at warm soil temperature are limited by the diffusion of substrate to metabolically active cells.

Oxygen levels also affect decomposition. The process of immobilization is enhanced under aerobic conditions and impaired under less aerobic conditions. When large amounts of organic materials are added, conditions leading to anaerobiosis may develop, which affect the ability of soil microorganisms to assimilate C and N and seriously impair N immobilization (Vinten et al., 2002).

Nitrogen concentration in tissues ranging from 18 to 22 g/kg is the critical value for the transition from net immobilization to net mineralization (Müller et al., 1988). Nitrogen is required for microbial growth and proliferation, and theoretically the optimum C/N ratio of the decomposable substrate is 25 (Heal et al., 1997). Hence, application of organic matter having a C:N ratio greater than 25 will result in N immobilization, whereas at lower C:N ratios mineralization will be favored. However, Vigil and Kissel (1991) showed that the break-even point between net N immobilization and mineralization was at a C:N ratio of 40 when net mineralization of residues after extended incubation (medium- to long-term experiments) was considered. This critical C:N ratio is, however, higher than values often reported, and therefore may not be useful for estimating short-term mineralization or mineralization kinetics but points to the role and size of more recalcitrant biochemical fractions influencing later stages of decomposition. In the last 25 years it has been shown that the prediction of mineralization or immobilization can be improved by determining organic matter quality and the C/N ratio in its different components (Myrold and Bottomley, 2008). The typical ratios of organic C to organic N in TWW are around 5 or lower. Therefore, it may be expected that irrigation with TWW will result in net mineralization rather than N immobilization.

The percentage of organic N in TWW was estimated by Feigin et al. (1991) to range between 10% and 30%. In the National Wastewater Survey in Israel (Anonymous, 2006) it was found that the average fraction of organic N in TWW is 35% of the total N, with an increasing proportion of organic N in effluents of lower quality (e.g. higher biochemical oxygen demand, BOD). There is limited knowledge regarding the dynamics of organic N (from effluent) in soils and on the influence of factors such as BOD, Eh+ pH, bicarbonate and total C on N-transformations in TWW irrigated soils. Most of the studies dealing with effluent N have concentrated on the transformations of mineral N (Feigin et al. 1981; Magesan et al. 1998). It is difficult to separate soil transformations of NH_4^+ and organic N originating in effluents, particularly when these are also masked by transformations of soil N.

Zhou et al. (2003) investigated the mineralization of organic N from TWW after removing the mineral N fractions. They found in suspension-based experiments performed with a microbially active calcareous clay soil that nitrification started after about a 1–4 d lag (higher lag associated with higher BOD), and the total mineral N reached plateau values after about 9–14 d. The time estimated for completion of ammonification of the organic N in well-mixed and aerated suspensions was 3–6 d. The following rate coefficients of ammonification of the effluent-originated organic N found in a sandy loam, loess and calcareous clay soils were 0.3 wk, 0.4 wk and 1.1 wk, respectively. The relatively fast movement of the effluent-originated organic N in soil and its mineralization characteristics indicate that this fraction significantly affects short (d) and middle (wk) range transformations of N in effluent-irrigated soils.

Stenger et al. (2001) investigated net carbon and nitrogen mineralization in intact soil cores following surface application of dairy farm effluent (DFE) onto three grassland top soils from two major dairying regions in New Zealand. Nitrogen mineralization was faster in the coarser soils than in the finer soil. After a 183 d incubation period, net N mineralization in the two sandy loams was equivalent to 38% and 44% (the more porous soil) of the organic DFE-N soil, whereas N immobilization was observed in the silty clay soil. Their results demonstrate the influence of soil texture on mineralization of organic N contained in DFE. Although just 38–44% of the organic N of the DFE was mineralized in the 183 d incubation period, it should be noted that the total N concentration in this effluent was 450 mg/l, of which 80% was organic, whereas total N concentration in TWW is between 10 and 100 times lower and the organic fraction is much lower.

Barkle et al. (2001) investigated the microbial turnover of DFE irrigated onto soils and the microbial availability of the soluble DFE fraction compared with that of a glucose plus ammonium solution. Net carbon mineralization in the standard DFE treatment was finished within 13 d and amounted to $29.7 \pm 2.4\%$ of the carbon applied, whereas in the high DFE treatment it continued until the end of the measurements (Day 50) and reached $48.4 \pm 0.5\%$. Dairy farm effluent application at the standard rate had only a minor effect on soil microbial biomass, whereas the high DFE application supported a higher microbial biomass over a longer period. Nitrogen immobilization persisted in the standard DFE treatment throughout the experiment, whereas the high DFE treatment shifted to the net mineralization phase by Day 113. Approximately 60% of the amended C was mineralized within 13 d in both the soluble DFE and glucose treatments, indicating similar microbial availability. Microbial growth, however, occurred only in the glucose treatment.

5.1.2.2 Ammonium and ammonia reactions

Ammonium being a monovalent cation is partitioned between the soil solution and the exchange complex of the clay minerals. The cation exchange capacity (CEC) of the soil governs the concentration of NH_4^+ in the soil solution. Consequently, the CEC of the soil influences both the formation of NH_3 and the mobility of NH_4^+ in soils. The source of the CEC and the exchange reactions of cations in soils and of NH_4^+ specifically have been described in numerous publications and are out of the scope of this chapter. A brief description of the governing forces will be given as they relate to specific features of irrigation with TWW. The CEC of clay minerals is mainly a constant negative charge formed by isomorphic substitution in the clay structure. Therefore, ions with positive charge are adsorbed to clay particle surfaces. The attraction of the cations to the clay surfaces increases with dilution of the soil solution. Adsorption strength to the soil surface is generally determined by the ionic charge and the degree of hydration of the ion. Therefore, the strength of adsorption of the major cations to soil surfaces is in the following order: $Al^{+3} > Ca^{+2} > Mg^{+2} > K^+ = NH_4^+ > Na^+$. Consequently, the adsorption of ammonium increases as the fraction of the monovalent cations Na^+ and K^+ increases and decreases as the fraction of the tri- and divalent cations Al^{+3}, Ca^{+2} and Mg^{+2} increases. However, the ionic strength of the soil solution has a strong impact on the adsorption preference of these ions. The preference for the tri- and divalent cations over the monovalent cations increases as the ionic strength decreases (soil solution is diluted). Consequently, drying will enhance the adsorption of NH_4^+, and therefore the concentration of NH_4^+ in the soil solution will not increase linearly as a function of the water content. Irrigation with TWW with higher salt concentration and sodium adsorption ratio (SAR) than freshwater will enhance the adsorption of NH_4^+ to the soil surfaces. The preference towards exchangeable tri- and divalent cations also becomes higher as the surface charge density increases. Therefore, the preference of Ca^{+2} over Na^+ is stronger on montmorillonite than kaolinite. Thus, clay type also influences the partitioning of NH_4^+ between the solution and the solid phases.

Cation exchange can be described by several equations based on thermodynamic considerations. Although the Gapon equation is not included among these equations, it is the most practical and widely used equation for quantifying the exchange of Na^+ with Ca^{+2} and Mg^{+2}, because it has a relative constant exchange coefficient for a reasonable range of SAR values. It was also successfully adopted by Izaurralde (1985) (in Kissel et al. 2008) to describe the adsorption of NH_4^+ to soil over a relatively wide range of pH values from 6.5 to 8.7, with NH_4^+ concentration values in the range of 1.6–10.8 mmol/l.

$$[NH_4{}^+{}_{Soil}]/[Ca_{Soil}] = K_g[NH_4{}^+{}_{aq}]/[Ca_{aq}^{+2}]^{0.5} \tag{5.1.1}$$

A practical ratio was developed by Schofield (1947) and adopted by Kissel et al. (2008) for NH_4^+:

$$[NH_4{}^+{}_{aq}]/[Ca_{aq}^{+2}]^{0.5} = K_r \tag{5.1.2}$$

Kissel et al. (2008) used this ratio with published soil solutions of Ca^{+2} and NH_4^+ data to illustrate that with a 10-fold drop in water content at the soil surface, such as from field capacity to air dryness at the soil surface, the concentration of Ca^{+2} would increase 16.7-fold, whereas the NH_4^+ will increase only four times.

The mineral composition of the soil solution and the presence of high soluble organic matter have been shown to influence the exchange of NH_4^+ with Na^+ and the sodicity of the soil (Levy et al. 1986). They found that a preference on the exchange complex for NH_4^+ and K^+ ions was an order of magnitude higher than for Na^+, and therefore the sodicity of the soils decreased as the concentrations of NH_4^+ and K^+ ions were higher. They also found that the ionic activity product of $CaCO_3$, which precipitated during the treatment with TWW, was higher than that of the untreated soils. According to Levy et al. (1986) this difference was attributed to the inhibiting effect of the TWW organic matter. The higher solubility of $CaCO_3$ affected the sodicity level of the soil and the NH_4^+ partitioning between the solution and the exchange complex. The strong impact of dissolved organic matter in solution on NH_4^+ partitioning between the solution and the exchange complex was also demonstrated by Fernando et al. (2005). They found that the sorption rate of NH_4^+ in soils equilibrated with liquid swine waste was slower than in inorganic solution, whereas the sorption capacity was higher. Chung and Zasoski (1994) reported that organic matter addition to soil increased selectivity for NH_4^+ over K^+ but decreased the selectivity for NH_4^+ over Ca^{2+}. Thus, applied organic matter can have contrasting effects with regard to the preference of exchangeable NH_4^+ with other cations. Further research is required to help fully understand the effects of TWW soluble organic matter on NH_4^+ adsorption in various soils.

Phillips and Sheehan (2005) investigated the adsorption of NH_4^+ in soils that contain both permanent and variable charge surfaces. They showed that the adsorption affinity of NH_4^+ increases in most of the soils that contain a high fraction of variable charge surfaces. The mechanism for this increased affinity was due to an increase in negative charge by displacing Al^{+3} from the complex. However, in some soils the effect of elevating pH was to increase Ca^{+2} and Mg^{+2} and therefore there was no change or even a reduction was observed in the adsorption of NH_4^+ with higher pH values. Chung and Zasoski (1994) observed that increasing pH increased the selectivity for NH_4^+ over Ca^{+2}, whereas pH effects on NH_4^+-K^+ were insignificant. As the common pH of waste water is neutral or above, the application of waste water would increase the capacity of most soils with variable charge to adsorb NH_4^+, reducing the risk of NH_4^+ leaching.

Three layer or 2:1 type clay minerals in soil affect NH_4^+ and potassium concentration in soil solution by an additional mechanism called fixation (Kissel et al. 2008). It is well accepted that fixed NH_4^+ and K+ ions are present as unhydrated forms between the negatively charged layers of the clays. In some clay minerals the layers collapse and the trapped NH_4^+ and K+ ions are not available for exchange. The 2:1 type clay minerals with isomorphic substitution within the tetrahedral layers, such as vermiculite, have greater fixation capacities than those with isomorphic substitution within the octahedral layers, like montmorilonite. Consequently, vermiculite has the highest capacity for fixed NH_4^+ and K+ ions, followed by illite, whereas fixation in montmorilonite occurs only after drying. The cation exchange sites including the "fixed" sites are probably heterogeneous in relation to their energy of adsorption. However, for simplicity, it is widely accepted to divide soil NH_4^+ into the following groups: soil solution, exchangeable, intermediate, and fixed. Intermediate NH_4^+, which falls between the exchangeable and fixed forms occupies interlayer sites of the clay plates that are in transition to collapse. Fixation of new added NH_4^+ from irrigation with TWW containing NH_4^+ probably occurs at these sites.

The environmental factors that control NH_4^+ fixation in addition to clay mineralogy are: NH_4^+ concentration, wetting and drying cycles, freezing and thawing, the solution ion composition, base saturation, pH and organic matter. Many studies have shown that the amount of NH_4^+ fixation increases with increasing NH_4^+ concentrations in solution, whereas the fraction of fixed NH_4^+ of the total decreases (Kissel et al., 2008). As NH_4^+ is the dominant form of N in secondary TWW, it is expected that irrigation with TWW will increase the fixed NH_4^+ in soil following irrigation with such water. This mechanism may partly explain the reduced recovery of N under irrigation with TWW.

Many investigations have shown that drying wet soil increased NH_4^+ fixation and that several cycles of drying and wetting enhanced the fixation of NH_4^+ (Kissel et al., 2008). As soil in an irrigated field undergoes cycles of wetting and drying, it is expected that under irrigation with TWW there is a greater potential for NH_4^+ fixation in soils with the appropriate clay mineralogy. The freezing and thawing cycle has been shown to enhance fixation, but these conditions generally do not occur when soils are irrigated with TWW.

Ion composition in soil and in irrigation water may affect NH_4^+ release and fixation processes in soils directly through competition on the fixation sites or indirectly through displacement of exchangeable NH_4^+. Although the effect of K^+ on NH_4^+ has been studied intensively, the effect is not predictable and it is dependent on the timing of application of

K^+ before, simultaneously with, or after NH_4^+ application. As the concentrations of K^+ and NH_4^+ in TWW are higher than in freshwater, the most relevant case for TWW irrigation is simultaneous addition. Most studies have shown that K^+ application before NH_4^+ blocked NH_4^+ fixation for a short time, but no effect was obtained when both ions were applied simultaneously (Nonmik, 1957). The strong impact of the pre-saturating cation on NH_4^+ fixation in soils or release from clay in soils has been shown by several authors. For example, in data presented by Kissel et al (2008) (from Harada and Katsuna (1954)), much higher NH_4^+ fixation in soil was obtained in soil saturated with Na^+, Ca^{2+}, and Ba^{2+} than with K^+, Al^{3+}, and H^+. The effect of K^+ was probably a specific effect on fixation, whereas the strong effects of Al^{3+} were through the strong adsorption of the three valence Al on the exchangeable sites and H^+ was probably replaced by Al. Feigenbaum et al. (1994) showed that the release of recently clay-fixed NH_4^+ was considerably higher when the clay was presaturated with Ca than when it was saturated with K. Both works dealt with presaturation with cations; however, there is no evidence of a significant effect of the simultaneous application of different cations similar to irrigation with TWW.

Increasing the pH of the solution above 5.5 and increasing the base cation concentration have been found to increase NH_4^+ fixation in soils. This effect probably stems from the displacement of exchangeable K^+ by Al^{3+} at low pH values. Consequently, the typical increase in pH by TWW irrigation may enhance fixation of the applied NH_4^+.

Organic matter has been shown to enhance the release of fixed NH_4^+ (Nommik and Vahtras, 1982) and to prevent or reduce the fixation of newly added NH_4^+ (Hinmnman, 1966; Porter and Stewart, 1970). In contrast to these findings, Fernando et al. (2005) found higher NH_4^+ fixation in soils treated with liquid swine waste than in the same soils treated with inorganic solutions with similar ionic composition. They attributed this higher sorption capacity to new binding sites added to the soils by DOC sorption to the clay fraction. These new sites behave as variable charge sites and the charge increases with the higher pH in the liquid swine waste. Porter and Stewart (1970) studied the effect of different soluble organic matter components on NH_4^+ fixation and found that NH_2 groups reduced NH_4^+ fixation in soils relative to COOH functional groups. As TWW contains much higher soluble organic matter than freshwater, these findings could be very relevant when irrigating with TWW. Further research is required to better understand the possible effects of soluble organic matter in TWW on NH_4^+ fixation in soils.

Ammonia loss to atmosphere by volatilization is an important factor affecting the efficiency of ammonium and urea fertilizers. The main forms of N in TWW, ammonium, and organic N, decompose in soils to form molecules similar to urea. Therefore, when considering the nutritional benefits in applying TWW to crops one must not disregard the risk of ammonia loss. The governing reactions and the environmental conditions that affect ammonia volatilization from inorganic fertilizers and organic sources applied to soil have been reviewed by Nelson (1982), Fenn and Hosner (1985), Harrison and Webb (2001), Kissel et al. (2008) and Francis et al. (2008). A brief description and specific considerations for TWW are given here. Ammonia evolution is a function of its concentration in solution, which is governed by the ammonium concentration and pH of the solution, and is described by the following chemical equation:

$$NH_4^+{}_{(aq)} \rightarrow NH_{3(aq)} + H^+_{(aq)} \tag{5.1.3}$$

The thermodynamic equilibrium equation is:

$$K_a(T) = \langle[NH_{3(aq)}]\rangle\langle[H^+_{(aq)}]\rangle/\langle[NH_4^+{}_{(aq)}]\rangle \quad (5.1.4)$$

where T is temperature in Kelvin units. From reaction 5.1.3 and equation 5.1.4 it is clear that volatilization of ammonium increases as the pH increases (the concentration of protons decreases). The value of the pK at 25°C is 9.25 and there is a strong effect of temperature according to the following equation:

$$K_a(T) = 10.0^{(-2728.3/T\ -0.094219)} \quad (5.1.5)$$

Consequently, temperature has a very strong impact on the partitioning between ammonium and ammonia. Kissel et al. (2008) calculated (using equations 5.1.4 and 5.1.5) that at a pH 8.5 and at a temperature range of 10–40°C, the fraction of ammonia is 5.5% and 33%, respectively. As the temperature and pH increase the fraction of ammonia increases, enhancing the toxicity of NH_3 to germinating seeds and young seedlings (Bennett and Adams, 1970) and potential loss of nitrogen by ammonia volatilization.

In calcerous soils ammonium fertilizers react with calcite as follows:

$$X(NH_4)_2Y + DCaCO_3 \rightarrow D(NH_4)_2CO_3 + Ca_nY_x \quad (5.1.6)$$

$$(NH_4)_2CO_3 + H_2O \rightarrow 2NH_4{}^+OH^- + CO_2 \quad (5.1.7)$$

where Y represents the anion of the NH_4 salt, and D, X, and n are dependent on the valences of the anions and cations. The amount and rate of $(NH_4)_2CO_3$ formed depends on the solubility of Ca_nY_x and its rate of formation. The more and faster this salt becomes insoluble, the more $(NH_4)_2CO_3$ is formed (equation 5.1.6), resulting in a higher pH (equation 5.1.7) and higher volatilization of ammonia (equations 5.1.3 and 5.1.7). Consequently, ammonia volatilization following ammonium fertilization is higher in calcerous soils than in neutral or acidic soils.

Temperature affects NH_3 volatilization through its influence on the equilibrium constants. As temperature increases, the concentration of $NH_{3(aq)}$ and the potential loss of the NH_3 gas from the soil increases proportionally. Consequently, the rate of NH_3 volatilization follows a diurnal cycle that closely matches that of the solar radiation (Hoff et al. 1981). Hengnirun et al. (1999) assumed a first order reaction for NH_3 volatilization and proposed a relationship between its rate and the temperature of the soil system:

$$\log(R_{volat}) = \log(R_{volat}(T_{opt})) + (T - T_{opt})\log(\theta_{volat}) \quad (5.1.8)$$

where R_{volat} is the first order reaction rate constant corrected for temperature T, $R_{volat}(T_{opt})$ is the first order rate constant as determined under optimum temperature conditions T_{opt}, and θ_{volat} is the temperature correction coefficient, which according to Steenhuis et al. (1976) is approximately 1.08. Hengnirun et al. (1999) developed a model VOLAT, in which a first order rate constant integrates the effects of temperature, CEC and air flow rate. This model assumes instantaneous equilibrium between NH_4^+ and NH_3 and relates $NH_3(aq)$ to $NH_{3(g)}$ by means of Henry's law.

The concentration of the cation NH_4^+ is governed by CEC and the composition of cations in the soil as described above. Consequently, a high CEC decreases potential NH_3 volatilization, whereas cations that displace exchangeable NH_4^+ enhance NH_3 volatilization (Avnimelech and Laher, 1977; Fenn et al. 1982a,b; Fine et al. 1989). Organic matter may also exhibit a CEC that may affect NH_4^+ concentration and NH_3 volatilization. Carbonate precipitation subsequent to the release of Ca^{2+} and Mg^{2+} by exchange with NH_4^+ can also mitigate a rise in pH (Fenn and Hosner, 1985). Ammonia volatilization from soils is dependent on soil texture; it increases as the soil texture is lighter; and decreases as the clay fraction increases due to ammonia adsorption to the soil solid surface and because of ammonium adsorption by the cation exchange complex. Reductions in ammonia losses from 90% of applied N in sandy soil to 50% in clay soil with a CEC of 58 meq/100 g soil have been reported by Fenn and Kissel (1973). Temperature influences the ratio of ammonium/ammonia in solution and on the volatilization of ammonia. Ammonia loss from ammonium fertilizers is strongly enhanced with increasing temperature (Fenn and Kissel, 1974).

Temperature also has an indirect effect on ammonia volatilization through its influence on the mineralization of organic N to ammonium. The rate of mineralization increases linearly in response to temperature up to 35°C. Ammonia volatilization decreases as the placement depth of inorganic ammonium fertilizers and urea increases. High emission of greenhouse gases and ammonia are a major environmental problem due to the common methods used in applying animal manures and slurries in agricultural fields (Hansen et al., 2003; Rodhe et al., 2006). Nitrogen loss as the gaseous form ammonia may account for a large fraction of the mineralized N from soil amended with organic waste (OW) (Ryan and Keeny, 1975; Fine et al. 1989). Fine et al. (1989) found that up to 87% of the ammonified sludge N was lost as gaseous species, mostly as ammonia. The amount of N lost as NH_3 may increase as OW amendment rates increase (Ryan and Keeny, 1975; Fine et al. 1989). Management and irrigation method influence the potential of ammonia volatilization. Ammonia emission from soil-applied manure is not only governed by the ammonium content and pH of the manure, but also by the application technique, weather conditions, and soil properties (Huisman et al. 1997). Injection of animal slurries to a relatively shallow depth of 2–5 cm has been found as an efficient method to reduce ammonium emission to a very low or undetectable level, whereas large quantities of up to 40% of the total added ammonical nitrogen from surface application were released to the air (Hansen et al., 2003; Rodhe et al., 2006). Sommer et al. (2006) found that the major factor that influenced ammonia volatilization from soils applied with different livestock slurries was the infiltration depth of the slurry. The ammonium applied with TWW by irrigation percolates into the soil, therefore it is expected that it is less exposed to volatilization loss than mineral and organic fertilizer applied to the soil surface.

There is a lack of information on N loss from fields irrigated by TWW. From the background information on ammonia loss from applied ammonium and urea fertilizers, we may conclude that irrigation with TWW with a high ammonium content may result in ammonia volatilization, especially in alkaline soils. Smith et al. (1996) measured a considerable amount of ammonia volatilization following irrigation with TWW, but much less than from an equivalent application of fertilizer N in urea form (Fig. 5.1.2). The ammonium concentration in the TWW, pH at the soil surface, and temperature are factors that directly influence the equilibrium between NH_4^+ and NH_3 in the soil solution

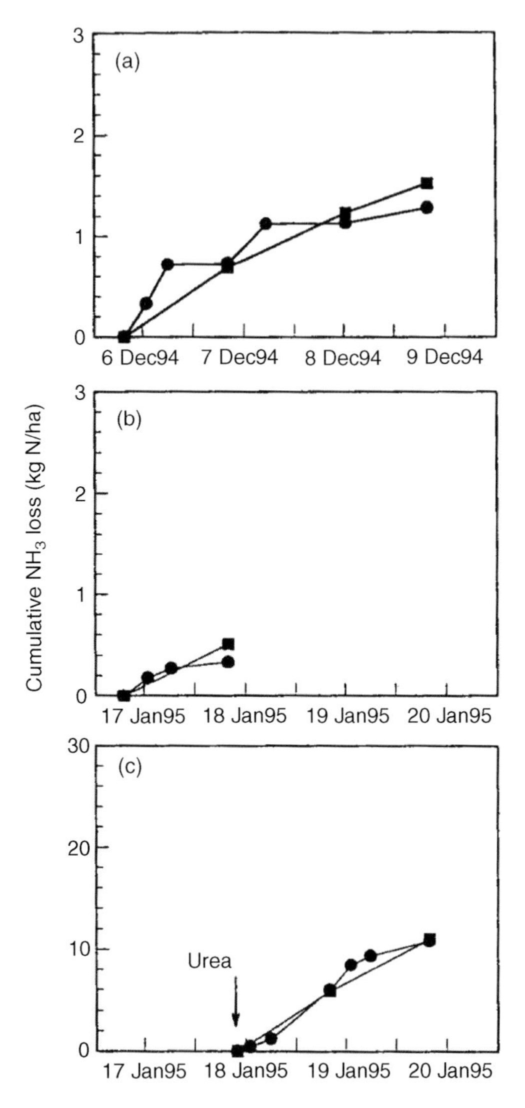

Figure 5.1.2 Cumulative ammonia evolution measured with the samplers developed by Leuning et al. (1985) (●) and the samplers and procedures described by Schjoerring et al. (1992) (■). (a) Effluent applied in January, and (c) urea applied in January. Reproduced with permission from the *Australian Journal of Soil Research* **34**(5), 789–802 (C. Smith, J.R. Freney and W.J. Bond). Copyright CSIRO 1996. Published by CSIRO PUBLISHING, Collingwood, Victoria Australia.

and the quantity of NH_3 that is being volatilized (Smith et al., 1996). The same authors reported that the loss of NH_3 from the soil after TWW application under maximum evaporative conditions, high temperature, and wind speed, reached 24% of the TWW applied ammonium in 2 d. A strong linear relationship between ammonia emission and water evaporation following irrigation with TWW was presented by these authors (Fig. 5.1.3). They noted that the volatilization of ammonia can be reduced by frequent application of small quantities of TWW. Irrigation methods that increase the contact of water drops with the air accelerate volatilization, whereas drip irrigation decreases it.

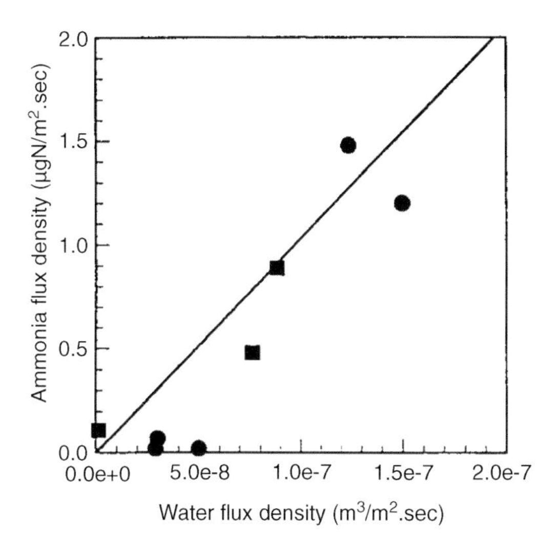

Figure 5.1.3 Relationship between measured NH_3 flux density and the water flux density following the application of effluent. Reproduced with permission from the *Australian Journal of Soil Research* **34**(5), 789–802 (C. Smith, J.R. Freney and W.J. Bond). Copyright CSIRO 1996. Published by CSIRO PUBLISHING, Collingwood, Victoria Australia.

Consequently, it is expected that the order of potential N losses through ammonia volatilization is sprinklers > surface drip irrigation > subsurface drip irrigation. The time required for mineralization of the organic fraction of N to produce ammonium also reduces the potential rate of N loss through ammonia volatilization. In conclusion, N loss through ammonia volatilization from TWW irrigation is usually low and can be minimized by proper management.

5.1.2.3 Nitrification

Nitrification is the process of oxidation of ammonium to nitrate. It is a two-step microbial process, consisting of the conversion of ammonium (NH_4^+) to nitrite (NO_2^-), and its subsequent conversion to nitrate (NO_3^-). This process is mediated by a group of aerobic chemlitho-autotrophic bacteria that gain energy from oxidation of N and fix inorganic C (CO_2) (Norton, 2008). Although some low levels of heterotrophic nitrification do occur, rates are low and quantities of nitrate produced are relatively small compared to the chemoautotrophs. Different groups of bacteria are responsible for the two steps involved in nitrification. The first step of the ammonium oxidation is carried out by ammonia-oxidizing bacteria (AOB) and includes species from five genera, the most common being *Nitrosomonas*; whereas the second step is by nitrite-oxidizing bacteria (NOB) such as *Nitrobacter* or *Nitrospira*. The ammonium and nitrite ions are oxidized according to the following stoichiometry:

$$NH_3 + O_2 + 2H^+ + 2e^- \rightarrow NH_2OH + H_2O \rightarrow NO_2^- + 5H^+ + 4e^- \quad (5.1.9)$$

$$NO_2^- + H_2O \rightarrow NO_3^- + 2H^+ + 2e^- \quad (5.1.10)$$

High levels of NH_4^+ can inhibit the nitrification processes (Focht and Verstraete, 1977; Niemiera and Wright, 1987b; Lang and Elliott, 1991; Russell et al., 2002) and it has been shown that nitrite (NO_2) oxidizing bacteria are more sensitive to high NH_4^+ levels than the NH_4^+ oxidizers (Nakos and Wolcott, 1972). In addition to the specific effect of NH_4^+, high solute concentrations, in themselves, might inhibit nitrification, and the degree of inhibition was found to be related to both osmotic pressure and the specific anion that accompanied the NH_4^+ ions (Darrah et al., 1985, 1987). The osmotic pressure threshold for inhibition was found to be 3.3 atm; however, chloride inhibited nitrification to a greater extent than its contribution to the osmotic pressure would indicate (Darrah et al., 1987). The specific inhibitory effect of chloride must be considered carefully, because progressive chloride accumulation in the root zone as a result of high uptake of water and low frequency of irrigation is common under semiarid conditions.

Other biochemical pathways for nitrification have been suggested by several researchers, as well as the chemical oxidation of NO_2 to NO_3. Photochemical oxidation of NH_3 to NO_2 (aq) is negligible in soils. On the other hand, chemical oxidation of NO_2 to NO_3 could have some significance in soils (Haynes, 1986), e.g. non-microbial conversion of NO_2 to NO_3 may help explain the lack of NO_2 accumulation in soils. The species diversity of the autotrophic nitrifier organisms in soil is relatively narrow, hence the nitrification process is much more sensitive to external environmental conditions than most other N transformations, which are carried out by a more diverse group of microorganisms. All of the nitrifiers are obligate aerobes, i.e. require atmospheric O_2 and nitrification is especially sensitive to soil moisture content and does not occur in waterlogged soils. Soil pH, temperature and heavy metals such as Cd, Cu, Zn also influence the nitrification process (Ishikawa et al., 2003; Sierra et al., 2001; Sierra, 2006). Gilmour (1984) described the effects of temperature, water content, and pH on the kinetics of nitrification and nitrification inhibition. He found the kinetics of nitrification to be of zero-order, whereas the effect of pH was found to have a linear impact on the rate coefficient.

Nitrification increases with rising temperature, within moderate limits (5–35°C), and the relationships between temperature and nitrification rates are usually described with the Arrhenius equation (Focht and Verstraete, 1977; Niemiera and Wright, 1987a; Rodrigo et al., 1997; Russell et al., 2002). Nitrite-oxidizing bacteria were found to be more sensitive to low temperature than the NH_4^+ oxidizers, therefore NO_2, the intermediate product of the nitrification process, may be accumulated at low temperatures (Russell et al., 2002). Nitrite is well known to be toxic to plants; therefore, NH_4^+ fertilization in a cold climate should be treated with caution. However, under hot conditions the general trend of increasing nitrification rates with elevating temperature changes direction, and above 35°C nitrification rates drop.

The pH of the medium has been reported to be an important influencing factor in nitrification processes, and the tendency for nitrification rates to decrease under acid conditions is generally recognized (Schmidt, 1982; Niemiera and Wright, 1987b; Lang and Elliott, 1991; Olness, 1999; Ste-Marie and Paré, 1999; Russell et al., 2002; Kyveryga et al., 2004). However, the shape and magnitude of the effect of pH depends on medium properties, water content, and N sources (Stevens et al., 1998; Ste-Marie and Paré, 1999; Russell et al., 2002). The optimal pH for the growth and metabolism of autotrophic nitrifiers is normally within the narrow range of between 7 and 8, but nitrification occurs at much lower pH values of 6.5 (Burton and Prosser, 2001) and nitrification has been confirmed in soils with pH values from 3 (DeBoer and Kowalchuk, 2001). It is well accepted that the nitrifying bacteria exist as isolated colonies rather than being homogeneously distributed through the soil volume (Darrah et al., 1987; Strong et al., 1997). Consequently, the heterogeneity within low-buffered soil in pH and in other chemical features driven by nitrification processes may be large. A high fraction of ammonium relative to nitrate in acid forest soil following irrigation with TWW was reported by Smith et al. (1996), but Phillips (2002a) reported a rapid nitrification rate of ammonium from piggery wastewater in different soil samples from two sites in Australia where the pH value ranged from 4.4 to 6.4. He showed that adsorption governed ammonium concentration in the soil solution for a short period of 3 d after the wastewater application, whereas nitrification was the major factor later on.

The nitrogen source may affect the rhizosphere pH via three mechanisms (Marschner, 1995; Bar-Yosef, 1999): (i) displacement of H^+/OH^- adsorbed on the soil particles; (ii) nitrification/denitrification reactions; and (iii) release or uptake of H^+ by roots in response to NH_4 or NO_3 uptake, respectively. Mechanisms (i) and (ii) are not associated with any plant activity, and affect the whole volume of the irrigated soil, whereas mechanism (iii) is directly related to the uptake of nutritional elements, and may be very effective because it affects a limited volume in the immediate vicinity of the roots (Moorby et al., 1984; Gahoonia and Nielsen, 1992; Marschner and Römheld, 1996; Taylor and Bloom, 1998; Bloom et al., 2003). The extent of the pH changes caused by the three mechanisms described above depends on soil properties, irrigated volume, plant activity, and all the environmental factors that affect nitrification rate, as detailed above.

The most common form of mineral N in effluents is NH_4^+. However, rapid adsorption on the surface sites of the solid phases, and nitrification reactions diminish the NH_4^+ content in soils, and hence, nitrate is the main mineral N found in neutral to calcareous soils of fields irrigated with TWW (Feigin et al. 1978; 1984; Anonymous, 2006; and numerous others), whereas ammonium is the predominant form in acid soils irrigated with TWW (Livesley et al. 2007). Master et al. (2004) reported that irrigation with secondary TWW stimulates nitrification. They related these results to the long-term exposure of the nitrifier population to TWW irrigation. In their research the average gross nitrification rate estimates were 11.3 and 15.8 mg N/kg/soil d for freshwater- and TWW-irrigated bulk soils, respectively. Average gross mineralization rate estimates were about 3 mg N/kg/soil d for the two water types. However, high levels of nitrite in soil solution following application of TWW to soils have been reported (Master et al., 2003, 2004). In a short-term laboratory study, Master et al. (2003) found that application of TWW to a heavy textured soil (Grumosol) inhibited nitrification by 15–25% and induced nitrite accumulation in soils incubated at field-capacity moisture content (Fig. 5.1.4). Possible reasons for these findings are the higher sensitivity of nitrite (NO_2) oxidizing bacteria than the NH_4^+ oxidizers (Nakos and Wolcott, 1972) to high NH_4^+ levels, the combined effect of the elevated NH_4^+ concentration and high pH (Burns et al., 1996), and the effect of the increase in osmotic pressure and chloride concentrations (Darrah et al., 1985, 1987). Other possible factors for the accumulation of nitrite in soil irrigated with TWW are the presence of low molecular weight dissolved organic matter that delays the oxidation of nitrite (Stueven et al., 1992); oxygen stress created by the consumption of oxygen for

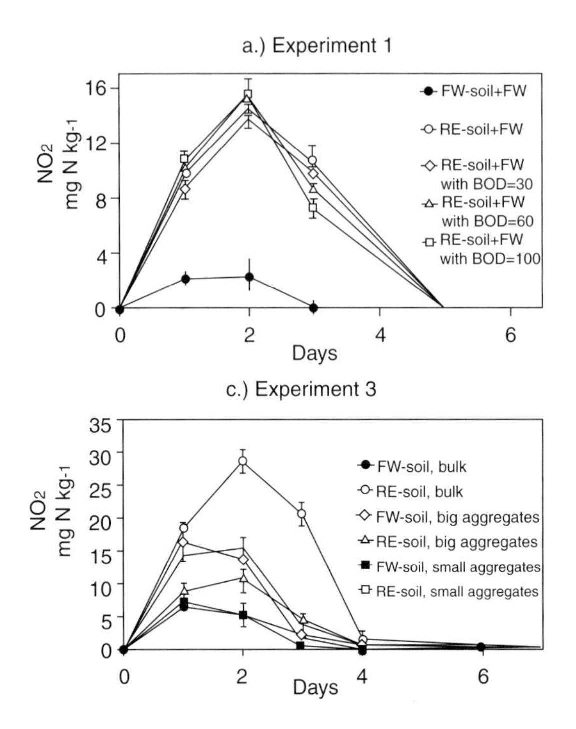

Figure 5.1.4 The effect of TWW application on accumulation of Nitrite and Nitrate in soil. Adapted from Master et al. 2004. *Journal of Environmental Quality* **33**, 852–860, with permission.

decomposition of the TWW organic matter. An additional reason could be the weakening of the soil nitrifying microbial population due to the prolonged exposure to O_2-consuming conditions. In a relatively long-term incubation study of 10–14 d Master et al. (2004) obtained one order of magnitude higher nitrite levels in TWW-irrigated soil (up to 30 mg N/kg soil) than in freshwater irrigated soil (Fig. 5.1.4). Higher levels of NO_2^- were observed at a moisture content of 60% than at 70% and 40% w/w. Nitrite levels were also higher when TWW was applied to a relatively dry Grumosol (20% w/w) than at subsequent applications of TWW to soil at 40% w/w. In this research, isotopic labeling indicated that the majority of NO_2^- was formed via nitrification.

5.1.2.4 Denitrification

Denitrification is the transformation of oxidized nitrogen into gaseous N compounds (Firestone, 1982), starting from nitrate to N_2 through the following pathway:

$$NO^-_{3\,(aq)} \rightarrow NO^-_{2\,(aq))} \rightarrow NO_{(g))} \rightarrow N_2O_{(g))} \rightarrow N_{2(g)}$$

The first step in the denitrification pathway is:

$$NO^-_{3\ (aq)} \rightarrow NO^-_{2\ (aq)} + 0.5O_{2(g)} \tag{5.1.11}$$

The second step produces the unstable gas NO:

$$NO^{-}_{2\,(aq)} \rightarrow NO_{(g)} + 0.5O_{2(g)} \tag{5.1.12}$$

The third step produces molecular nitrogen:

$$NO_{(g)} \rightarrow 0.5N_{2(g)} + 0.5O_{2(g)} \tag{5.1.13}$$

Many types of bacteria are known to be responsible for denitrification, including *Pseudomonas*, *Bacillus*, *Micrococcus*, and *Achromoonbacter*. Denitrification occurs under anoxic conditions as can be expected from equations 5.1.11–5.1.13. The availability of organic carbon as electron donors to microorganisms capable of reducing nitrogen oxides is an important factor in denitrification (Firestone, 1982). Burford and Bremner (1975) reported that 1 μg of C was involved in the production of 1.17 μg N_2O or 0.99 μg N as N_2. The conditions that enhance denitrification are: wet soil and waterlogging, application of organic matter, and a high nitrate concentration (Coyne, 2008). Coyne (2008) describes denitrification as "a cascade of events that begins when O_2 cannot quickly diffuse through soil or is removed from the environment by respiration. Denitrifiers need C to grow, and in the absence of O_2 they need NO_3 for growth on C". Denitrification increases with elevating pH and temperature (Coyne, 2008). As irrigation with TWW tends to increase soil pH, it may enhance denitrification. Irrigation with TWW is common in arid and warm climates, conditions that enhance denitrification.

Very high denitrification rates occur in soil treated with sewage sludge. A high organic matter content in soil irrigated with effluent, together with low oxygen levels due to the decomposition of the organic matter, may enhance denitrification under irrigation with effluent.

Feigin et al. (1981) added ^{15}N-enriched NH_4^+ to effluents, which were used to study ammonium transformations in the soil–plant system in a pot greenhouse experiment. From the greater amount of unaccounted tagged N in TWW than fresh irrigated pots, they concluded that TWW enhanced denitrification (Table 5.1.1). Dag et al. (1984) obtained

Table 5.1.1 Recovery of N added as fertilizer, effluent and mineral-solution-N to the soil plant system (Feigin et al. 1981, JEQ 10:284-287)

N-fertilizer	Water type	N recovery	Soil-N		Tagged-N
Added		Plant-N	Organic	total	
mg/pot			%		
60	Demineralized	55.4	32.2	34.1	89.5
120	Demineralized	69.3	23.3	24.3	93.6
60	Mineral solution	57.3	34.4	36.4	93.7
120	Mineral solution	63.0	21.0	22.2	85.2
60	Sewage effluent	62.7	25.2	27.7	90.4
120	Sewage effluent	65.1	20.2	22.0	87.0
0	Mineral solution	65.7	16.0	17.1	82.8
0	Sewage effluent	60.9	14.0	15.4	76.3

Reproduced with permission.

higher losses of N from TWW irrigated cotton plots (40–55%) than those fertilized and irrigated with freshwater (10–40%). They related these to enhanced denitrification, ammonia volatilization, and nitrate leaching in TWW-irrigated plots.

In the USA, federal guidance documents for design of forested land treatment systems indicate the expected range for denitrification to be up to 25% of applied N, and most forest land treatment systems are designed using values from 15% to 20% of applied N. However, few measurements of denitrification following long-term wastewater applications at forested land treatment sites exist. In a study on N denitrification from spray irrigation of forested land in the Georgia Piedmont that has been operating for more than 13 yrs, Meding et al. (2001) found that only 2.4% of the TWW-applied N was lost through denitrification. Barton et al. (1999) reported that the annual denitrification rate in forest soil irrigated with TWW was 2.4 kg N/ha/yr, only slightly higher than the unirrigated soil (1.7 kg N/ha/yr). They concluded that denitrification was limited by excessive aeration in the free-draining soils. Singleton et al. (2001) investigated the effects of adding dairy shed effluent (DSE) to pasture, on the amounts and forms of N leached from large undisturbed soil monolith lysimeters of a Gley Soil, under water table at three depths: 25 (high), 50 (medium), or 75 (low) cm below the soil surface. The smaller amount of NO_3-N leached from the high-water table treatment was attributed to enhanced denitrification. Smith and Bond (1999) used stable N isotopes to determine the fate of TWW-derived N in an irrigated tree plantation used as an alternative to the direct discharge of sewerage effluent into inland rivers. Based on the high recovery of the labeled N in the soil and the trees, they concluded that there was little loss of N by denitrification, even though the soil was sufficiently wet for leaching. Monnett et al. (1995) studied in a column laboratory experiment the effect of irrigation head and the frequency of spray irrigation of filtered sewage water containing a high concentration of N (32.9, 7.5 and 11.5 mg/l of NO_3^-, NH_4^+, and organic N, respectively) on the loss of N by denitrification. As expected, increasing the effluent application rate from 1.25 and 2.5 cm/wk elevated the average gaseous losses of N from 5.3% and 26.2% of the applied N, respectively. Reducing the rate of application from three times to one time per day decreased the gaseous losses of N. This effect of irrigation frequency stemmed from the strong impact of water content in the soil pores of the top soil layer on denitrification. It is well documented that the N_2O-N fluxes increase steeply following irrigation or raining and then decrease steeply. This behavior indicates that denitrification occurs when the soil pores of the top soil layers are filled with water leading to anaerobic conditions. The data of Monett et al. (1995) provides information on the total gaseous losses of N by denitrification, but no conclusion on the air pollution caused by the N_2O emission because they used the acetylene block technique, which blocks further reduction of the N_2O. Tozer et al. (2005) used stable N isotopes to determine the fate of TWW-derived N within a land-based municipal TWW irrigation scheme of forest and wetland for 11 yrs. Denitrification including that occurring within the wetland accounted for 3% of the applied N. Barton et al. (2000) investigated the factors limiting the denitrifying population in a forested soil, by studying the individual and combined effects of soil aeration, water content, nitrate, and carbon on denitrification enzyme activity (DEA). The size of the soil denitrifying population in the soil appeared to be limited by soil aeration, and limiting oxygen availability increased the denitrifying population above that observed in the field. Furthermore, TWW irrigation altered the short-term response of denitrifiers to anaerobic soil conditions. Under low oxygen conditions, denitrifiers in the TWW-irrigated soils produced enzymes sooner and at a greater rate than soils without a history of TWW irrigation. However, the size of the denitrifying population cannot be expected to be large in free-draining, coarsely textured soils even when provided with additional nitrogen and water inputs. Irrigation with TWW containing relatively high organic matter levels may enhance denitrification and high emissions of N_2O. In a laboratory study, Master et al. (2003) found that the amount of N_2 and N_2O losses from soil irrigated with low-quality TWW were negligible. However, when the soils were water saturated with concentrated TWW, the N losses by denitrification reached a high value of 3.1% of total applied N. Bhandral et al. (2007) reported that the emissions rate of N_2O from different farm wastewaters (treated farm dairy effluent (TFDE), untreated farm dairy effluent (UFDE), treated piggery farm effluent (TPFE) and treated meat effluent (TME)) was affected by the type of the effluent with TPFE emitting the highest amount (0.585 kg N_2O-N/ha or 2.17% of the total added effluent-N) during autumn application and TME emitting the highest amount (0.286 kg N/ha or 0.84% of the total effluent-N added) during winter application. The differences in the N_2O emissions among the effluents could be attributed to the differences in their C:N ratios. However, the return to preapplication N_2O emissions rates within 2 wks of autumn effluent application and 3 wks of winter effluent application indicates that the effect of effluent application on flux is short-lived (Bhandral et al., 2007).

Kremen et al. (2005) developed a model for reactive, multi-species diffusion to describe N transformations in spherical soil aggregates, emphasizing the effects of irrigation with TWW. Aggregate size and soil respiration rate were identified as the most significant parameters governing the existence and extent of the anaerobic volume in aggregates. At the aggregate level, significant differences between apparent and gross rates of N-transformations were predicted (e.g. NH_4^+ oxidation and N_2 formation), resulting from diffusive constraints due to aggregate size. With increasing anaerobic volume, the effective nitrification rate determined at the aggregates level decreases until its contribution to nitrification is negligible. It was found that the nitrification process was predominantly limited to aggregates <0.25 cm. Assuming that nitrification is the main source for NO_3 formation, denitrification efficiency is predicted to peak in medium-sized aggregates, where aerobic and anaerobic conditions coexist, supporting coupled nitrification and nitrification processes. In TWW-irrigated soils, the predicted NO_2 formation rate in small aggregates is enhanced when compared to freshwater-irrigated soils. The difference vanishes with increasing aggregate size as anaerobic NO_2 consumption exceeds aerobic NO_2 formation due to the coupling of nitrification and denitrification. The above research indicates that the effects of irrigation with TWW on soil physical properties like aggregate size and soil pore distribution have a strong impact on N transformations, specifically nitrite accumulation and denitrification in soil irrigated with TWW. It also demonstrates the importance of the "hotspot" phenomenon for denitrification in TWW-irrigated soils.

The quantities of applied N losses as gasses from TWW-irrigated soils are relatively small. The quality of the TWW, as well as the management and method of irrigation, is expected to influence N losses. Low-quality TWW with a high organic matter content is expected to enhance denitrification. The denitrification rate increases by increasing the rate and frequency of effluent irrigation. Higher denitrification is anticipated in heavy and poorly drained soils than in coarse textured, well drained soils. The environmental pollution caused by greenhouse gas emissions due to denitrification (N_2O) is a potential problem in using TWW for irrigation.

5.1.3 TWW-N availability to plants

The availability of nitrogen from TWW to plants is an important agricultural, environmental, and economical issue. If the N in TWW is available to plants, then secondary treated domestic sewage water can serve as a N source. From an agricultural point of view, the available N in TWW has to be considered in the fertilization budget. From an economic point of view, the replacement of N fertilizer by the TWW-N has to be considered in calculating the benefits of TWW containing N. From an environmental point of view, the availability of organic and mineral N in TWW may reduce the threat of N pollution of water resources by irrigation with TWW, therefore the required standard of N content in TWW has to be determined according to its availability to plants.

Prior to crop uptake, the organically bound N in TWW applied to soil must be transformed and mineralized into plant-available mineral N. Feigin et al. (1981) studied the fate of mineral N when applied as fertilizer with freshwater or as TWW containing N to corn plants grown in a soil in a pot experiment for 43 d. They found that total recovery of the tagged N by corn plants ranged between 55% and 69% of the total quantity supplied, 14% to 34% recovered in the soil organic matter, and 6% to 24% were lost, probably due to denitrification (Table 5.1.1). The highest N loss was from the effluent treatment, probably due to the high carbon supply.

No differences were found between the effects of N supplied via irrigation with swine waste lagoon and inorganic commercial fertilizer N on forage grass yield and N utilization (Adeli and Varco, 2001; Adeli et al., 2005). In a 3-yr field experiment, Bieloray et al. (1984) reported that yields and biomass of cotton irrigated with secondary TWW without N fertilization was similar or higher than under freshwater irrigation with optimum N fertilization (5 N levels were examined with the freshwater). In this experiment it was shown that the uptake of N by plants irrigated with effluent was similar to the optimal fertilization with freshwater irrigation (Feigin et al. 1984). Rhodes grass was shown to very efficiently utilize N from applied TWW. Irrigating with an effluent containing 18 mg/l inorganic N resulted in 105.2% N recovery as plant uptake (calculated as the ratio of N plant uptake to applied N) when no fertilizer was applied (Vaisman et al., 1982). To maximize yield of the Rhodes grass, 250 kg N/ha was supplied resulting in 74.8% recovery and saving 45% of the recommended fertilizer amendment. Positive yield response and larger N uptake of trickle irrigated sorghum was recorded with TWW containing 30 mg N/l over freshwater (Papadopolus and Stylianou, 1988). Bar-Tal et al. (2004) conducted a corn field study for 3 yrs on a fine-clayey Vertisol soil in Israel. The overall annual quantity of mineral N applied by fertigation or effluent irrigation was 100–145 kg/ha yielding 20–22 t/ha, regardless of water type, whereas N concentration in plant organs was similar and adequate in the two treatments. High grapefruit yields with no reduction in leaf N content were obtained in a 4-yr experiment where N fertilization was adjusted to the concentration of inorganic N in the irrigating TWW, (0.32 mg/l N-NH_4 and 5.6 mg/l N-NO_3) (Maurer et al., 1995). Grapevines grown for 3 yrs in sand in 30 l pots and irrigated with TWW containing low N concentrations (4.2–6.8 mg/l N-NH_4 and 4.1–5.3 mg/l N-NO_3) had a lower leaf-N content than under

Table 5.1.2 Yield and N status responses of various crops to irrigation with TWW of different quality and N level and forms

Nitrogen				Crop	Yield	N content	Reference
TN	**Organic**	**NH_4^+**	**NO_3^-**				
mg/l							
412	67	347	8	Bermudagrass and sorghum	P	P	Adeli and Varco, 2001
341	52	280	9	Bermudagrass	P	P	Adeli et al., 2003
426	60	363		Bermudagrass and sorghum	P	P	Adeli et al., 2005
35	5	30		Grapefruit	P	P	Bar-Tal et al., 2004
35	5	30		Cotton and Corn	P	P	Bar-Tal et al., 2007[a]
15	6.7	5.3	3.0	Pasture	P	P	Barton et al., 2005
30		25	<5	Rose	P	P	Bernstein et al., 2006
35–70	6–8	28-59	1.4–2.8	Cotton	P	P	Bieloray et al., 1984 and Feigin et al., 1984
48, 32				Bermudagrass	P	P	Da Fonseca et al., 2007
35–70	6–8	28–59	1.4–2.8	Cotton	P	P	Dag et al., 1984
35–70	6–8	28–59	1.4–2.8	Cotton, Rhodes	P	P	Feigin et al., 1978
70	21	49		Corn	P	P	Feigin et al., 1981
45	7	35	3	Corn	P	P	Fine et al., 1994[b]
149	40	109		Turntip (Rape)	P	P	Jacobs and Ward, 2007
69	20	49		Cotton and Corn	P	P	Mandal et al., 2008
		0.32	5.6	Grapefruit	NR	NR	Maurer et al., 1995
10.2			3.6	Grape	P	N	Neilsen et al., 1989a
10.2			3.6	Apple	P	P	Neilsen et al., 1989b
10.2			3.6	Vegetables	P	NR[c]	Neilsen et al., 1989b
25–30				Rhodes	P	P	Papadopolus and Stylianou, 1988
		5.6	4.7	Grape	N	N	Paranychianakis et al., 2005
11.3	4.3	2.5	4.5	Eucalyptus	P	NR	Snow et al., 1999
		15–30	0	Rhodes	P	P	Vaisman et al., 1982
6		1	5	Citrus	P	NR	Zekri and Koo, 1994

[a]Bar-Tal et al. 2007. Unpublished data.
[b]Fine et al., 1994. Unpublished data.
[c]In majority of the examined vegetable crops no effect of the N-TWW on leaf N content was found whereas in several vegetable crops positive response was recorded.

freshwater irrigation with half Hogland solution (approximately 112 mg/l N); the TWW treatment reached deficient levels early in the third year (Paranychianakis et al., 2005). These results do not indicate that the availability of TWW-N is lower than that of N in freshwater, but that additional N fertilization has to be applied when the TWW-N is below the recommended fertilization value.

Barton et al. (2005) conducted a study to determine the effect of TWW application on plant uptake in four contrasting soils. The soils, from which pasture is regularly cut and removed, varied in their ability to assimilate nutrients from secondary treated domestic effluent under high hydraulic loadings, in comparison with unirrigated, fertilized pasture. The effluent-irrigated treatment received between 746 and 815 kg N/ha over 2 yrs of irrigation, and unirrigated treatments received 200 kg N/ha of dissolved inorganic fertilizer over the same period. Applying effluent significantly increased plant uptake of N in all soil types. For the effluent-irrigated soils, plant N uptake ranged from 186 to 437 kg N/ha/yr. In a study in Brazil, utilization of secondary-treated sewage effluent (STSE) as an alternative source of water and nitrogen (N) for bermudagrass pasture saved 32.2–81.0% of the recommended N rate without a loss of grass dry matter and concentration of protein yield (Da Fonseca et al., 2007). Jacobs and Ward (2007) reported a linear increase in total nitrogen uptake by a summer turnip (*Brassica rapa* L.) crop over three consecutive summer periods as a function of the applied dose of secondary-pond dairy effluent that contain a high level of organic nitrogen. Bar-Tal et al. (2004) reported that fruit yield and N status of indicator leaves of trees irrigated by TWW containing 25–35 mg/l was not different from those irrigated with freshwater and similar inorganic N content. Bernstein et al. (2006) found no differences in N concentration in diagnostic leaves of rose plants grown in soilless culture and irrigated with either TWW or with freshwater plus inorganic fertilizer N.

There are very few direct measurements of the amount of N taken up by plants from TWW, for example Feigin et al. (1981). However, there is a lot of indirect evidence from field experiments indicating that the response and N uptake of various crops to the nitrogen in TWW was not different from applied inorganic fertilizer N. A summary of the above reviewed literature is presented in Table 5.1.2. From 24 published works, only in one experiment yield and leaf N content were reduced by TWW irrigation (Paranychiaakis et al. 2005), one report showed reduction of leaf N content (Neilsen et al. 1989a), and in four reports no effect of the TWW on leaf N could be measured (Maurer et al., 1995; Neilsen et al., 1989c; Snow et al., 1999; Zekri and Koo, 1994). In the remaining reviewed works, positive yield response and increase in leaf N or total N uptake was obtained. It should be noted that in all the studies that failed to show a positive contribution of TWW-N to crop yield and plant N status the N concentration in the effluent was relatively low, less than 12 mg/l. Differences in N efficiency from TWW relative to inorganic fertilizer N might stem from higher N losses in gaseous forms (by ammonia emission and denitrification) and higher N leaching, which may result from poor coupling of application time with crop uptake. Nitrogen losses in gaseous form are usually low, therefore nitrogen in TWW should be regarded as a source of nitrogen replacing fertilizer inorganic N. Whereas fertilizer N application is totally controlled by the grower who can adjust the supply to a known consumption curve as a function of time, the supply by TWW only partially follows a N consumption curve when N concentration is kept constant in the TWW and the N consumption is proportional to water uptake.

5.1.4 TWW-N leaching and environmental pollution

Irrigation with TWW increases the threat to groundwater quality from uncontrolled leaching of various constituents, including soluble organic matter, salts and nitrates (Amiel et al., 1990; Ronen and Magaritz, 1985; Fine et al., 2002), and may result in significant economical damage (Haruvy et al. 2000). The major components of N in TWW are ammonium and dissolved organic N. In most soils ammonium transport and leaching in soil are slower than that of nitrate due to the exchangeable cation complex of the soil particles. Consequently, N leaching under irrigation with TWW should be slower than under fertigation with nitrate. However, in an acid soil, ammonium and organic N from effluent irrigation were found as the main forms of excess N transported below the root zone, whereas nitrate leaching was relatively small under a normal irrigation dose (Livesley et al., 2007). Under conditions of rapid nitrification, the main inorganic ion form of N in soils irrigated by effluent is nitrate; and therefore the transport and leaching of nitrate in soils under irrigation with TWW might be rapid. Phillips (2002b) reported very large leaching of N following application of piggery wastewater from two sources containing high N concentrations (159 and 713 mg/l, of which 80% was ammonium and 20% organic N) to large undisturbed soil cores. Most (~90%) of the N in the soil solution and the leachate was nitrate indicating fast nitrification, whereas the concentration of ammonium in water and 2MKCl extracts were very low. Therefore, the capacity of the two soils to retain N and to prevent downward transport following rainfall was found to be very limited. Although the dissolved organic nitrogen (DON) can be adsorbed by soil colloids (Kaiser and Zech, 2000), it has been found to be more mobile in soils than ammonium (Zhou et al., 2003). In a lysimeter study with dairy shed effluent (DSE), Singleton et al. (2001) found that most of the leached N (80–90%) was in organic N form. They also found that raising the load of N applied by irrigation from 511 kg N/ha/yr in the first year to 1518 kg N/ha/yr in the second years, resulted in an increase of the total amount of leached N from 33.3 to 131.4 kg N/ha/yr, respectively, but the fraction of the leached N remained similar 7% to 9%. In this study the total amount of leached N and the fraction of the organic form of N decreased as the water table depth increased from 25 to 75 cm. In a similar study, when irrigating pasture soil cores, Barton et al. (2005) found that 69–87% of the N leached from secondary TWW treatments were in organic form. Thus, irrigation with TWW containing high DON could be a threat to the environment especially under poor drainage conditions.

Optimization of irrigation, fertilization, and cropping management can reduce the transport of pollutants (e.g. nitrate) below the root zone (Hu, 1997; Hu and Pigram, 2001; Letey and Pratt, 1984; Vaisman et al., 1981). When considering secondary effluent as an N source, the amount and timing of nutrient uptake by the crop have to be considered. Feigin et al. (1978) were able to reduce the quantity of nitrate leached below the root zone by applying better management controls for secondary effluent irrigated cotton. Minimizing the leaching of TWW-N in an irrigated Rhodes grass field was achieved by accounting for the TWW-N in the fertilization application and by optimizing the irrigation head (Vaisman et al., 1981). Similar results were obtained for sorghum by adjusting the fertilization levels by taking into account the TWW-N (Papadopoulos and Stylianou, 1988). Similarly, Fine et al. (2002) virtually eliminated nitrate leaching under effluent irrigated and fertigated Eucalyptus. Adeli et al. (2003) reported that for late-season irrigation the apparent N recovery was 59% less, with a delay in irrigation from 1 September to 1 October. Proper fertilization and adequate plant growth can also retard downward leaching of salts by enhancing water uptake by the crop. Feigin and Halevy (1994) demonstrated this effect in a study of a soil profile to a depth of 12 m in a 30-yr permanent fertilization plot experiment. Feigin et al. (1984) managed to accommodate the N, P, and K requirements of freshwater and secondary effluent irrigated cotton. They also managed to maintain most of the soil nitrate within the upper 0.6 m layer of the soil throughout the irrigation season. However, residual nitrate not taken up by the crop was leached later on, during the rainy season. Jemison and Fox (1994) found that total nitrate mass in the 1.2-m soil profile following corn harvest was useful to predict the annual

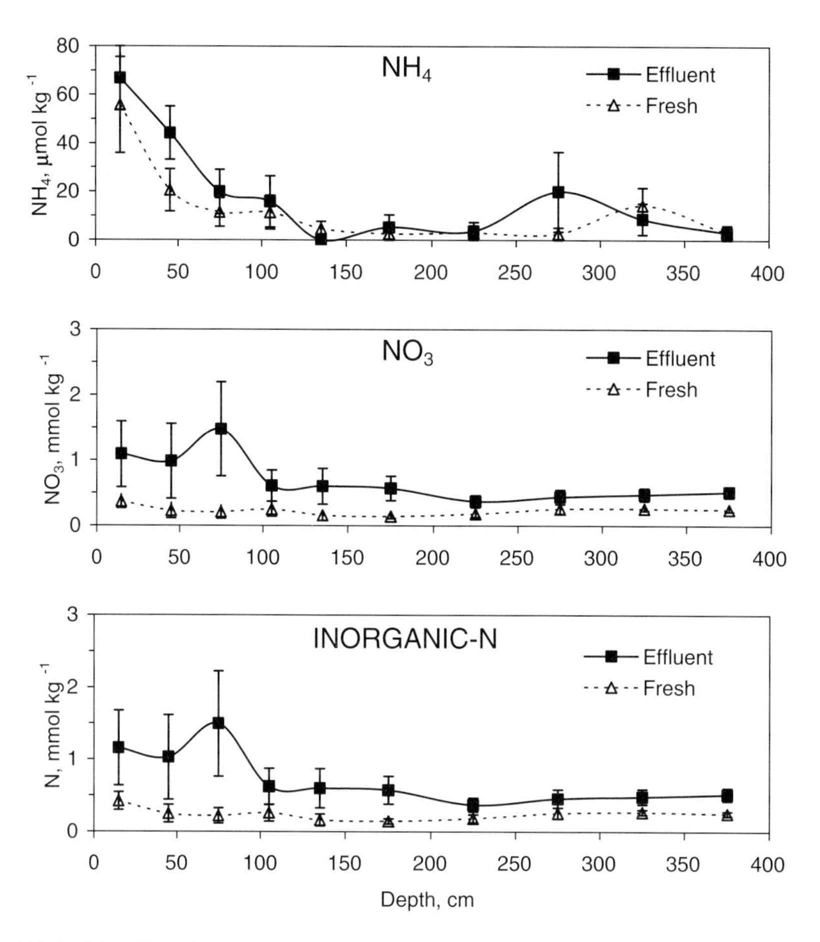

Figure 5.1.5 The effect of 7 years of irrigation of grapefruit orchard with TWW or freshwater on the distribution of NH_4 and NO_3 in the soil profile, October 2003. (Bar-Tal et al. 2004, unpublished data.)

amount of nitrate leached over winter. Downward movement of N and higher inorganic N concentrations below the root zone were recorded in several long-term field experiments conducted in Israel to examine the possibility of minimizing N leaching from TWW irrigation of citrus and field crops (Fig. 5.1.5 and Fig. 5.1.6, respectively).

A long-term survey of the effects of irrigation with TWW in comparison to freshwater in various regions and soil types in citrus and Avocado orchards has been conducted in Israel (Anonymous, 2006). In this survey it was found that the mean annual load of inorganic N in plots irrigated with TWW was higher by 150–190 kg/ha than in freshwater irrigated plots in the period of 1999–2001. The reason for this huge difference stemmed from the fact that growers tended to ignore the N contribution of the TWW based on the assumption that TWW-N is unavailable to the trees. Later on, from 2002 to 2005, there was a gradual reduction in this excess N load from 100 to 50 kg/ha due to two factors: (i) the efforts of the Extension Service to distribute fertilization recommendations that are adjusted for the composition of the TWW; and (ii) the increase in quality of TWW including reduction in organic and inorganic N concentration by construction of new purification plants that meet

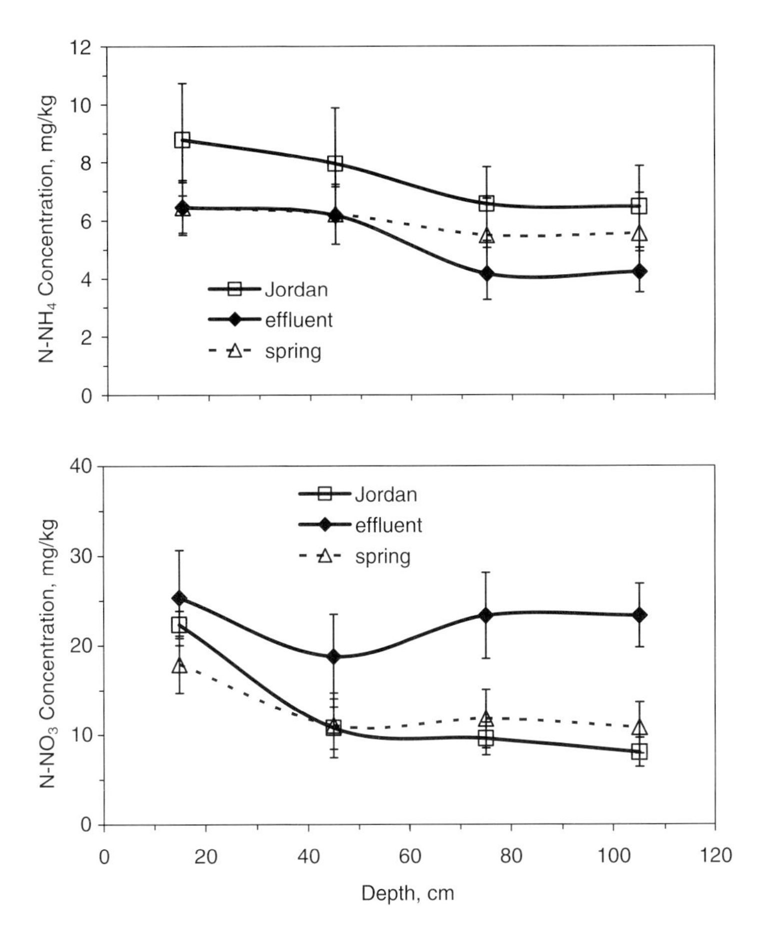

Figure 5.1.6 The effect of 6 years of irrigation of field crops, corn and cotton, with TWW or freshwater on the distribution of NH_4 and NO_3 in the soil profile, May 2007. (Bar-Tal et al. 2007. Unpublished data.)

new standards adopted by the Ministry of Environmental Protection. It is not surprising that irrigation with TWW was found as one of the factors included in a statistical model for the difference in inorganic N (mainly nitrate) in the top soil layers (0–30 and 30–60 cm) at the end of the irrigation season. In most plots, these layers were leached during the following winter by rainfall below the root zone; the excess N load in the TWW irrigated plots probably contributed to groundwater pollution. In heavy textured soils (clay fraction $> 40\%$) the effect of TWW irrigation on increasing the mean inorganic N in the whole measured profile of 0–120 cm was significant at the end of the irrigation season and at the end of the rainy season, whereas no significant effect was found in lighter textured soils. The difference in inorganic N concentration between TWW and freshwater-irrigated plots decreased from the first to the last years of the survey, in general agreement with the decrease in surplus N load in the TWW plots. In such a survey there are many factors that contribute to differences between plots, therefore the comparison is very complex and the known factors may explain only part of the differences. However, it is clear that overfertilization in TWW irrigation plots increased the risk of groundwater pollution by nitrate.

Runoff and horizontal transport of excess N from TWW might pollute aboveground water resources. In a study of a land-based municipal TWW irrigation scheme in which 4.68 t N/ha was applied over 11 yrs (426 kg N/ha/yr) to wetland and forest systems, the transport of effluent N via the stream was estimated as 263 t (29%) of the applied N (Tozer et al., 2005).

Irrigation with TWW containing high levels of N in organic and inorganic forms may be considered as a resource for N, replacing the expensive and energy expending inorganic fertilizer. However, the risks of environmental pollution by excess organic and inorganic N as a result of irrigation with TWW have to be considered. The location of the irrigated fields in relation to above and below groundwater, soil types, crops, climate, and rainfall, as well as management strategies, will determine the proper N concentration in TWW.

5.1.5 Modeling TWW-N fate

The fate of N originating from TWW irrigation in soil is governed by the numerous processes and factors described above. Therefore, it is very difficult to predict N fate in agricultural fields irrigated by TWW. Models are powerful tools to predict processes and identify critical factors controlling such systems. Numerous models have been developed to simulate N transformations and fate in the plant–soil systems.

Simulation models are valuable tools in the analysis of agricultural systems, particularly for assessing the impact of climatic variability and long-term consequences of alternative management strategies (Probert et al., 1998). Models are complimentary to experimentation, which is invariably constrained by the prevailing seasonal conditions, limitations in the treatments imposed, and the duration of experimentation. Thus, models have traditionally been used to extrapolate results from agricultural experiments to other situations such as different soils of differing fertility, other seasons, and different management schemes such as crop sequence, tillage, and residue management practices (Probert et al., 1995; 1998). They also permit scenario analysis of alternative management strategies in terms of production and economics, risks, and consequences for the resource and the environment (Probert et al., 1995).

The principles of modeling the nitrogen cycle in soils and agriculture have been recently described by Cabrera et al. (2008). More detailed reviews of different C and N models can be found in other reviews (Molina and Smith, 1998; Ma and Shaffer, 2001; McGechan and Wu, 2001). Models that predict carbon and nitrogen mineralization are classified into two groups, depending on the mode of formulation and how they represent the decomposition process. Models are either empirical (fitted to experimental data) or mechanistic (process-based models). Empirical models are statistically derived factor-based equations and the most widely used rate equations to predict decomposition and accumulation of N, are the single (Stanford and Smith, 1972) (equation 1.3.2) and double exponential (Molina et al., 1980) kinetic models. They do not include microbial processes explicitly (Stanford and Smith, 1972) or treat them very simply (Richter et al., 1982).

Many computer process-based models have been developed to simulate the effect of OM application under various conditions on the fate of N. Rodrigo et al. (1997) compared the ability of nine different models to predict the influence of moisture and temperature on C and N transformations in soils. They found that several of these models predicted fairly well the trends of N forms and fate without any additional calibration processes. These studies enriched the understanding of N fate in complex soil–plant systems; they shed light on the most important factors controlling nitrogen uptake by plants, leaching, and emission as gas. Unlike first order kinetic models, which are based on mathematical regression models, that have been used to describe net N and C mineralization, such as the equation of Stanford and Smith (1972), process-oriented dynamic models can be modified to account for individual processes that may limit decomposition or transformation rates thereby giving a better prediction of N turnover in soils.

Vogeler et al. (2006) used a deterministic model for assessing the risk of groundwater contamination by nitrate from land-based sludge disposal. The processes of water and nutrient transport were modeled using a mechanistic scheme based on Richards' equation for water transport and the convection–dispersion equation (CDE) for nutrient transport. These equations were both linked to a sink term for plant uptake. The model simulated well the transport of water and movement of bromide in the four different lysimeters. The agreement between measured and simulated nitrate leaching was also reasonable considering the simplified model. Uptake of N by trees reduced the quantity of N available for leaching. This model could aid in the development of sustainable management of land-based sewage sludge disposal in terms of nitrate leaching.

Although numerous studies examined the use of models for simulation of the N process in agricultural fields amended with N fertilizer or various organic residues including sewage sludge and composts (Rodrigo et al. 1997), there are just a few publications on the use of models to simulate N processes in fields irrigated with TWW. Snow et al. (1999) used the APSIM model to simulate the fate of N from TWW applied to a eucalyptus plantation. Large quantities of N leaching below the root zone were measured and predicted. Kremen et al. (2005) demonstrated the importance of aggregate size for nitrification-denitrification processes in simulating N fate in lysimeters irrigated with TWW. They also showed by using a modeling approach that modification of soil physical properties due to irrigation with TWW had a strong impact on N fate in the soil. The results of this work indicate that a comprehensive model that takes into account various physical and chemical processes affected by TWW are required for realistic predictions of N fate in TWW-irrigated fields. Moreover, TWW effects on the microorganism population and their activity should be taken into account. All the above mentioned models use a one-dimensional approach that assumes a homogeneous field. However, the importance of spatial variability in the field for simulation of water and solutes under field and regional scales has been demonstrated in numerous publications (Chapter 14). Therefore, there is a need for developing models and evaluation of such models for simulation of N fate in agricultural soils irrigated with TWW.

5.1.6 Conclusions

The typical ratios of organic C to organic N in TWW are around 5 or lower. Therefore, it may be expected that irrigation with TWW will result in net mineralization rather than N immobilization. Consequently, the net result of irrigation with TWW containing N in inorganic and organic forms is mineralization of the organic form.

When the main mineral form of N in TWW is ammonium in well-aerated neutral and basic soils, it is oxidized to nitrite and nitrate shortly after irrigation, with nitrate becoming the dominant inorganic N in soil solution of such soils. As temperature increases (up to optimum in the range of 30–40°C), the process is more rapid. There is a risk of nitrite accumulation following irrigation with low-quality TWWs that contain high organic matter and under excess irrigation in poorly drained soils. In acid soils with a pH below 5.0–5.5 and under low temperatures, the oxidation of ammonium slows down and it becomes the dominant inorganic N form in soil solution of fields irrigated with TWW containing organic and ammonium N forms.

Nitrogen loss through ammonia volatilization from TWW irrigation is usually low and it can be minimized by proper management. Potential losses by ammonia volatilization are higher in high pH and coarse-textured soils. High temperature and strong wind enhance the potential losses. Irrigation methods that increase the contact of water drops with the air accelerate N volatilization, whereas drip irrigation decreases it. Consequently, it is expected that the order of potential N losses through ammonia volatilization is sprinklers>surface drip irrigation>subsurface drip irrigation. The volatilization of ammonia can be reduced by frequent application of small quantities of TWW. The time required for mineralization of the organic fraction of N to produce ammonium also reduces the potential rate of N loss through ammonia volatilization.

The amount of applied N losses as gasses from TWW irrigated soils is relatively small. The quality of the TWW, as well as management and irrigation methods is expected to influence N losses. Low-quality TWW with a high organic matter content is expected to enhance denitrification. High rate and frequency of irrigation enhances gaseous losses by denitrification. Anaerobic conditions following a high dose of water typical for a flooded system will probably enhance denitrification. Sprinklers and drip irrigation systems are expected to minimize denitrification per irrigation event, but the higher the frequency of irrigation events the higher will be the denitrification. Higher denitrification is anticipated in heavy, poorly drained soils than in well-drained coarse-textured soils. The emission of the greenhouse gas N_2O, formed during denitrification, is an environmental concern in TWW-irrigated soils.

Some studies demonstrated the high availability of nitrogen in TWW for plants, especially the inorganic forms. There is a lot of indirect evidence from field experiments indicating that the response and N uptake of various crops to N in TWW was not different from applied inorganic fertilizer N. In some studies a difference in N recovery between N from TWW and from inorganic fertilizers was observed. These differences probably stem from N evolution as gas and the different N distribution through the irrigation season. Synchronization of N supply through TWW has to be maximized with maximum growth rate and N consumption in order to minimize potential offsite movement of residual soil N.

Irrigation with TWW may cause soil and groundwater contamination. Irrigation with TWW increases the risk of nitrate leaching below the root zone in neutral to alkaline soils or ammonium and organic N in acid soils, probably due to the higher content of total N (including the organic N) than inorganic fertilization, the continuous N load until the end of the irrigation season and the enhanced downward movement of the organic component. A lack of grower awareness of the need for adjusting N fertilization by taking into consideration the composition of TWW, increases the load of nitrogen in fields irrigated with TWW and the potential of environmental pollution. Moreover, N downward leaching during the rainy season occurs even under a deficit supply of N. Careful monitoring of the soil moisture profile may reduce the rate of downward leaching. Irrigation of summer crops with effluent or with freshwater combined with fertilizers will increase the potential pollution of groundwater with nitrate. Further studies on means to reduce potential groundwater pollution, like the use of a catch crop during winter time are required. Management methods have to be applied to minimize runoff near aboveground freshwater resources. It is recommended to restrict or avoid irrigation with TWW above groundwater in coarse-textured soils and to apply TWW just above deep groundwater. Reducing the N content of TWW to the minimal agronomic value is required, especially in sensitive environments like shallow groundwater and near aboveground freshwater reservoirs.

Adjustment of fertilization recommendations to TWW and distribution of this information to growers are essential tools for reducing the risks of excess N application and environmental pollution. If growers lack this information or are not convinced of the availability of the TWW-N to crops they might apply excess N with high risk of environmental pollution.

There is a need for developing models and evaluation of such models for simulation of N fate in agricultural soils irrigated with TWW. These models have to include the effects of the water quality on soil physical properties and aggregate formations that influence the biological N transformations, as well as solution and gas transport in soil, and field scale variability in soil properties that has strong impact on transport processes.

References

Adeli, A. and Varco, J.J. (2001) Swine lagoon effluent as a source of nitrogen and phosphorus for summer forage grasses. *Agronomy Journal* **93**, 1174–1181.

Adeli, A., Varco, J.J. and Rowe, D.E. (2003) Swine effluent irrigation rate and timing effects on Bermudagrass growth, nitrogen and phosphorus utilization, and residual soil nitrogen. *Journal of Environmental Quality* **32**, 681–686.

Adeli, A., Varco, J.J., Sistani, K.R. and Rowe, D.E. (2005) Effects of swine lagoon effluent relative to commercial fertilizer applications on warm-season forage nutritive value. *Agronomy Journal* **97**, 408–417.

Amiel, A.J., Magaritz, M., Ronen, D. and Lindstrand, O. (1990) Dissolved organic carbon in the unsaturated zone under land irrigated by wastewater effluent. *Water Pollution Control Federation Bulletin* **62**, 861–866.

Andersen, M.K. and Jensen, L.S. (2001) Low soil temperature effects on short-term gross N mineralisation-immobilisation turnover after incorporation of a green manure. *Soil Biology and Biochemistry* **33**, 511–521.

Anonymous (2006) *National Wastewater Effluent Irrigation Survey 2003–2005*. Ministry of Agriculture and Rural Development, State of Israel, November, 2006.

Avnimelech, Y. and Laher, M. (1977) Ammonia volatilization from soils - equilibrium considerations. *Soil Science Society of America Journal* **41**, 1080–1084.

Barkle, G.F., Stenger, R., Sparling, G.P., and Painter, D.J. (2001) Immobilisation and mineralisation of carbon and nitrogen from dairy farm effluent during laboratory soil incubations. *Australian Journal of Soil Research* **39**, 1407–1417.

Bar-Tal, A., Fine, P., Yofe, M., Keinan, M., Soriano, S., Markovitz, T., Erner, Y., Tagari, E., Artzi, B., Assouline, S., Hayimovitz, A., Rosner, M., Steinhardt, R. and Dasberg, S. (2004) Long term effects of effluent irrigation management on the fruit yield and soil in grapefruit orchid in Ramat Hacovesh. *Hanotea* **57**, 524–528 (Hebrew)

Barton, L., McLay, C.D.A., Schipper, L.A. and Smith, C.T. (1999) Denitrification rates in a wastewater-irrigated forest soil in New Zealand. *Journal of Environmental Quality* **28**, 2008–2014.

Barton, L., Schipper, L.A., Smith, C.T. and McLay, C.D.A. (2000) Denitrification enzyme activity is limited by soil aeration in a wastewater-irrigated forest soil. *Biology and Fertility of Soils* **32**, 385–389.

Barton, L., Schipper, L.A., Barkle, G.F., McLeod, M., Speir, T.W., Taylor, M.D., McGill, A.C., van Schaik, A.P., Fitzgerald, N.B. and Pandey, S.P. (2005) Land application of domestic effluent onto four soil types: plant uptake and nutrient leaching. *Journal of Environmental Quality* **34**, 635–643.
Bar-Yosef, B. (1999) Advances in fertigation. *Advances in Agronomy*, **65**, 1–77.
Bennett, A.C. and Adams, F. (1970) Concentration of $NH_3(aq)$ required for incipient NH_3 toxicity to seedlings. *Proceedings of the Soil Science Society of America* **34**, 259–263.
Bernstein N., Bar-Tal, A., Friedman, H., Snir, P., Chazan, A. and Ioffe, M. (2006) Application of treated wastewater for cultivation of roses (Rosa hybrida) in soil-less culture. *Scientia Horticulturae* **108**, 185–193.
Bhandral, R., Bolan, N.S., Saggar, S. and Hedley, M.J. (2007) Nitrogen transformation and nitrous oxide emissions from various types of farm effluents. *Nutrient Cycling in Agroecosystems* **79**, 193–208.
Bieloray, H., Vaisman, I. and Feigin, A. (1984) Drip irrigation of cotton with treated municipal effluents. I. Yield response. *Journal of Environmental Quality* **13**, 231–234.
Bloom, A.J., Meyerhoff, P.A., Taylor, A.R. and Rost, T.L. (2003) Root development and adsorption of ammonium and nitrate from the rhizosphere. *Journal of Plant Growth Regulation* **21**, 416–431.
Bremner, J.M. and Mulvaney, C.S. (1982) Nitrogen – Total. In: *Methods of Soil Analysis, part 2: Chemical and Microbiological Properties* (eds A.L. Page, R. H. Miller and D. R. Keeney), pp. 595–624. ASA, CSSA and SSSA, Madison, WI.
Bronson, K.F. (2008) Forms of inorganic nitrogen in soil. Chapter 2. In: *Nitrogen in Agricultural Systems* (eds J.S. Schepers, W.R. Raun, R.F. Follet, R.H. Fox and G.W. Randall), pp. 31–55. Agronomy Monograph 49, ASA, CSSA and SSSA, Madison, WI.
Burford, J.R. and Bremner, J.M. (1975) Relationships between the denitrification capacities of soils and total, water soluble and readily decomposable soil organic matter. *Soil Biology and Biochemistry* **7**, 389–394.
Burns, L.C., Stevens, R.J. and Laughlin, R.J. (1996) Production of nitrite in soil by simultaneous nitrification and denitrification. *Soil Biology and Biochemistry* **28**, 609–616.
Burton, S.A.Q., and Prosser, J.I. (2001) Autotrophic ammonia oxidation at low pH through urea hydrolysis. *Applied Environmental Microbiology* **67**, 2952–2957.
Cabrera, M., Molina, J.A.E. and Vigil, M. (2008) Modeling the nitrogen cycle. In: *Nitrogen in Agricultural Systems* (eds J.S. Schepers, W.R. Raun, R.F. Follet, R.H. Fox and G.W. Randall), pp. 695–730. Agronomy Monograph 49, ASA, CSSA and SSSA, Madison, WI.
Chung, J.B. and Zasoski, R.J. (1994) Ammonium-potassium and ammonium-calcium exchange equilibria in bulk and rhizosphere soil. *Soil Science Society American Journal* **58**, 1368–1375.
Coyne, M.S. (2008) Biological denitrification. In: *Nitrogen in Agricultural Systems* (eds J.S. Schepers, W.R. Raun, R.F. Follet, R.H. Fox and G.W. Randall), pp. 201–253. Agronomy Monograph 49, ASA, CSSA and SSSA, Madison, WI.
Da Fonseca, A.F., Melfi, A.J., Monterio, F.A., Montes, C.R., de Almeida, V.V. and Herpin, U. (2007) Treated sewage effluent as a source of water and nitrogen for Tiffon 85 bermudagrass. *Agricultural Water Management* **87**, 328–336.
Dag, J., Feigin, A., Giskin, M., Golan, S., Jerome, G., Davidov, S. and Keinan, M. (1984) *Response of cotton to effluent irrigation in a grumusol soil in the Yizre'el Valley; 1976–1982 experiments. 1. Cotton yield.* Agric. Res. Org. Bet Dagan, Israel. Pamphl 228:47 pp. (Hebrew with English summary)
Darrah, P.R., Nye, P.H. and White, R.E. (1985) Modelling growth responses of soil nitrifiers to additions of ammonium sulphate and ammonium chloride. *Plant Soil*, **86**, 425–439.
Darrah, P.R., Nye, P.H. and White, R.E. (1987) The effect of high solute concentrations on nitrification rates in soil. *Plant Soil* **97**, 37–45.
DeBoer, W. and Kowalchuk, G.A. (2001) Nitrification in acid soils: Microorganisms and mechanisms. *Soil Biology and Biochemistry* **33**, 853–866.
Feigenbaum, S., Hadas, A., Sofer, M. and Molina, J.A.E. (1994) Clay-fixed labeled ammonium as a source of available nitrogen. *Soil Science Society of America Journal* **58**, 980–985.
Feigin, A., Bielorai, H., Dag, Y., Kipnis, T. and Giskin, M. (1978) The nitrogen factor in the management of effluent-irrigated soils. *Soil Science* **125**, 248–254.
Feigin, A., Feigenbaum, S. and Limoni, H. (1981) Utilization efficiency of nitrogen from sewage effluent and fertilizer applied to corn plants growing in a clay soil. *Journal of Environmental Quality* **10**, 284–287.
Feigin, A. and Halevy, J. (1994) Irrigation-Fertilization-Cropping management for maximum economic return and minimum pollution of ground water. In: *Contamination of groundwaters* (eds Adriano,

D.C., Iskandar, A.K. and Murarka, I.P.). Advances in Environmental Science. Northwood Science Reviews. UK.

Feigin, A., Ravina, I. and Shalhevet, J. (1991) Irrigation with Treated Sewage Effluent. Management for environmental Protection. *Advances Series in Agricultural Sciences 17.* Pub. Pub. Springer-Verlag. Berlin, Heidelberg, New York.

Feigin, A., Vaisman, I. and Bielorai, H. (1984) Drip irrigation of cotton with treated municipal effluent: II. Nutrients availability in soil. *Journal of Environmental Quality* **13**, 234–238.

Fenn, L.B. and Kissel, D.E. (1973) Ammonia volatilization from surface application of ammonium compounds on calcerous soils. I. General theory. *Soil Science Society of America Journal* **37**, 855–859.

Fenn, L.B. and Kissel, D.E. (1974) Ammonia volatilization from surface application of ammonium compounds on calcerous soils. II. Effects of temperature and rate of ammonium nitrogen application. *Soil Science Society of America Journal* **38**, 606–610.

Fenn, L.B. and Hosner, L.R. (1985) Ammonia volatilization from ammonium or ammonium-forming nitrogen fertilizers. *Advances in Soil Science* **1**, 123–169.

Fenn, L.B., Matocha, J.E. and Wu, E. (1982a) Soil cation-exchange capacity effects on ammonia loss from surface-applied urea in the presence of soluble calcium. *Soil Science Society of America Journal* **46**, 78–81.

Fenn, L.B., Matocha, J.E. and Wu, E. (1982b) Substitution of ammonium and potassium for added calcium in reduction of ammonia loss from surface-applied urea. *Soil Science Society of America Journal* **46**, 771–776.

Fernando, W.A.R.N., Xia, K. and Rice, C.W. (2005) Sorption and desorption of ammonium from liquid swine waste in soils. *Soil Science Society of America Journal* **69**, 1057–1065.

Fine, P., Mingelgrin, U. and Feigin, A. (1989) Incubation studies of the fate of organic nitrogen in soils amended with activated sludge. *Soil Science Society of America Journal* **53**, 444–450.

Fine, P., Hass, A., Prost, R. and Atzmon N. (2002) Organic carbon leaching from effluent irrigated lysimeters as affected by residence time. *Soil Science Society of America Journal* **66**, 1531–1539.

Firestone, M.K. (1982) Biological denitrification. In: *Nitrogen in Agricultural Soils. Agronomy Monograph. 22* (ed. F.J. Stevenson) pp. 289–326. ASA, CSSA and SSSA, Madison, WI.

Focht, D.D. and Verstraete, W. (1977) Biochemical ecology of nitrification and denitrification. *Advances in Microbiol Ecology* **1**, 135–214.

Francis, D.D., Vigil, M.F. and Mosier, A.R. (2008) Gaseous losses of nitrogen other than through denitrification. Chapter 8. In: *Nitrogen in Agricultural Systems* (eds J.S. Schepers, W.R. Raun, R.F. Follet, R.H. Fox and G.W. Randall), pp. 255–279. *Agronomy Monograph 49*, ASA, CSSA and SSSA, Madison, WI.

Fu, M.H., Xu, X.C. and Tabatabai, M.A. (1987) Effect of pH on nitrogen mineralization in crop residue treated soils. *Biology and Fertility of Soils* **5**, 115–119.

Gahoonia, T.S. and Nielsen, N.E. (1992) Control of pH at the soil-root interface. *Plant Soil* **140**, 49–54.

Gilmour, J.T. (1984) The effects of soil properties on nitrification and nitrification inhibition. *Soil Science Society of America Journal* **48**, 1262–1266.

Hansen, M.N., Sommer, S.G. and Madsen, N.P. (2003) Reduction of ammonia emission by shallow slurry injection: Injection efficiency and additional energy demand. *Journal of Environmental Quality* **32**, 1099–1104.

Harrison, R. and Webb, J. (2001) A review of the effect of N fertilizer type on gaseous emissions. *Advances in Agronomy* **73**, 65–108.

Haruvy, N., Offer, R., Hadas, A. and Ravina, I. (1999) Wastewater Irrigation-Economic concerns regarding beneficiary and hazardous effects of nutrients. *Water Resources Management* **13**, 303–314.

Haruvy, N., Hadas, A., Ravina, I. and Shalhevet, S. (2000). Cost assessment of averting groundwater pollution. *Water Resources Management* **13**, 303–314.

Haynes, R.J. (1986) Effects of soil acidification and subsequent leaching on levels of extractable nutrients in a soil. *Plant and Soil* **95**, 327–336.

Heal, O.W., Anderson, J.M. and Swift, M.J. (1997) Plant litter quality and decomposition: A historical overview. In: *Driven by Nature: Plant Litter Quality and Decomposition* (eds G. Cadisch, and K. E. Giller), pp. 3–30. CAB Int., Wallingford, England.

Hengnirun, S., Barrington, S., Prasher, S.O. and Lyew, D. (1999) Development and verification of a model simulating ammonia volatilization from soil and manure. *Journal of Environmental Quality* **28**, 108–114.

Hinmnman, W.C. (1966) Ammonium fixation in relation to exchangeable K and organic matter content of two Saskatchewan soils. *Canadian Journal of Soil Science* **46**, 223–225.

Hoff, J.D., Nelson, D.W. and Sutton, A.L. (1981) Ammonia volatilization from liquid swine manure applied to cropland. *Journal of Environmental Quality* **10**, 90–95.

Hu, X.D. (1997) Sustainability of effluent irrigation schemes: Measurable definition. *Journal of Environmental Engineering-ASCE* **123**, 928–932.

Hu, X.D. and Pigram, J. (2001) Computer-aided design of effluent irrigation. *Environmental Modelling Software* **16**, 47–52.

Huisman J.F.M., Hol, J.M.G. and Bussink, D.W. (1997) Reduction of ammonia emission by new slurry application techniques on grassland. In: *Gaseous nitrogen emission from grasslands* (eds S.C. Jarvis and B.F. Pain), pp. 281–285. CABI Publ., London, UK.

Ishikawa, T., Subabarao, G.V. Ito, O. and Okada, K. (2003) Suppression of nitrification and nitrous oxide emission by the tropical grass *Brachiaria humidicola. Plant and Soil* **255**, 413–419.

Jacobs, J.L. and Ward, G.N. (2007) Effect of second-pond dairy effluent on turnip dry matter yield, nutritive characteristics, and mineral content. *Australian Journal of Agricultural Research* **58**, 884–892.

Jemison, J.M. and Fox, R.H. (1994) Nitrate leaching from nitrogen-fertilized and manured corn measured with zero-tension pan lysimeters. *Journal of Environmental Quality* **23**, 337–343.

Jones, D.L., Healey, J.R., Willet, V.B., Farrar, J.F. and Hodge, A. (2005) Dissolved organic nitrogen uptake by plants – an important N uptake pathway? *Soil Biology and Biochemistry* **37**, 413–423.

Kaiser, K. and Zech, W. (2000) Sorption of dissolved organic nitrogen by acid subsoil horizons and individual mineral phases. *European Journal of Soil Science* **51**, 403–411.

Kissel, D.E., Cabrera, M.L. and Paraasivam, S. (2008) Ammonium, ammonia and urea reactions in soils. Chapter 4. In: *Nitrogen in Agricultural Systems* (eds J.S. Schepers, W.R. Raun, R.F. Follet, R.H. Fox and G.W. Randall), pp. 101–155. Agronomy Monograph 49, ASA, CSSA and SSSA, Madison, WI.

Kremen, A., Bear, J., Shavit, U. and Shaviv, A. (2005) Model demonstrating the potential for coupled nitrification denitrification in soil aggregates. *Environmental Science and Technology* **39**, 4180–4188.

Kyveryga, P.M., Blackmer, A.M., Ellsworth, J.H. and Isla, R. (2004) Soil pH effects on nitrification of fall-applied anhydrous ammonia. *Soil Science Society of America Journal* **68**, 545–551.

Lang, H.J. and Elliott, G.C. (1991) Influence of ammonium: nitrate ratio and nitrogen concentration on nitrification activity in soilless potting media. *Journal of the American Society for Horticultural Science* **116**, 642–645.

Letey, J. and Pratt, D.F. (1984) Agricultural pollutants and ground water quality. In: *Behaviour of pollutants in unsaturated zones* (eds Yaron B., Dagan, G. and Goldshmid, Y.), pp. 211–222. Springer-Verlag, Berlin.

Levy, R., Fine, P. and Feigin, A. (1986) Sodicity levels of soils equilibrated with wastewater. *Soil Science Society of America Journal* **50**, 35–39.

Livesley, S.J., Adams, M.A. and Grierson, P.F. (2007) Soil water nitrate and ammonium dynamics under a sewage effluent irrigated Eucalyptus plantation. *Journal of Environmental Quality* **36**, 1883–1894.

Lomander, A., Kätterer, T. and Andrén, O. (1998) Modeling the effects of temperature and moisture on CO_2 evolution from top- and subsoil using a multi-compartment approach. *Soil Biology and Biochemistry* **30**, 2023–2030.

Ma, L. and Shaffer, M.J. (2001) A review of carbon and nitrogen processes in nine U.S. soil nitrogen dynamic models. In: *Modeling Carbon and Nitrogen Dynamics for Soil Management* (eds M.J. Shaffer et al.) pp. 55–102. Lewis publishers, Boca Raton, FL. USA.

MacDonald, N.W., Zak, D.R. and Pregitzer, K.S. (1995) Temperature effects on kinetics of microbial respiration and net nitrogen and sulfur mineralization. *Soil Science Society of America Journal* **59**, 233–240.

Magesan G.N., McLay, C.D.A. and Lal, V.V. (1998) Nitrate leaching from a free-draining volcanic soil irrigated with municipal sewage effluent in New Zealand. *Agricultural Ecosystems and the Environment* **70**, 181–187.

Mandal, U.K., Warrington, D.N., Bhardwaj, A.K., Bar-Tal, A., Kautzky, L., Minz, D. and Levy, G.J. (2008) Evaluating impact of irrigation water quality on a calcareous clay soil using principal component analysis. *Geoderma* **144**, 189–197.

Marschner, H. (1995) *Mineral Nutrition of Higher Plants*, pp 889. Academic Press, London, GB.

Marschner, H. and Römheld, V. (1996) Root-induced changes in the availability of micronutrients in the rhizosphere. In: *Plant Roots: the Hidden Half*, 2nd edition (eds Waisel, Y., Eshel, A., Kafkafi, U.), pp. 557–579. Marcel Dekker, New York.

Master, Y., Laughlin, R.J., Shavit, U., Stevens, R.J. and Shaviv, A. (2003) Gaseous nitrogen emissions and mineral nitrogen transformations as affected by reclaimed effluent application. *Journal of Environmental Quality* **32**, 1204–1211.

Master, Y., Laughlin, R.J., Stevens, R.J. and Shaviv, A. (2004) Nitrite formation and nitrous oxide emissions as affected by reclaimed effluent application. *Journal of Environmental Quality* **33**, 852–860.

Maurer, M.A., Davies, F.S. and Graetz, D.A. (1995) Reclaimed waste-water irrigation and fertilization of mature Redblush grapefruit trees on Spodosols in Florida. *Journal of the American Society for Horticultural Science* **120**, 394–402.

McGechan, M.B. and Wu, L. (2001) A review of carbon and nitrogen processes in European soil nitrogen dynamic models. In: *Modeling Carbon and Nitrogen Dynamics for Soil Management* (eds M.J. Shaffer et al.) pp. 103–171. Lewis publishers, Boca Raton, FL. USA.

Meding, S.M., Morris, L.A., Hoover, C.M., Mutter, W.L. and Cabrera, M.L. (2001) Denitrification at a long-term forested land treatment system in the piedmont of Georgia. *Journal of Environmental Quality* **30**, 1411–1420.

Molina, J.A.E., Clapp, C.E., and Larson, W.E. (1980) Potentially mineralizable nitrogen in soil: The simple exponential model does not apply for the first 12 weeks of incubation. *Soil Science Society of America Journal* **44**, 442–443.

Molina, J.A.E. and Smith, P. (1998) Modeling carbon and nitrogen processes in soil. *Advances in Agronomy* **62**, 253–297.

Monnett, G.T., Reneau, Jr., R.B. and Hagedorn, C. (1995) Effects of domestic wastewater spray irrigation on denitrification rates. *Journal of Environmental Quality* **24**, 940–946.

Moorby, H., Nye, P.H. and White, R.E. (1984) The influence of nitrate solution on H^+ efflux by young rape plants (*Brassica napus* cv. Emerald). *Plant Soil* **84**, 403–415.

Müller, M.M., Sundman, V., Sininvaara, O. and Merilainen, A. (1988) Effect of chemical composition on the release of nitrogen from agricultural plant materials decomposing under field conditions. *Biology and Fertility of Soils* **6**, 78–83.

Myrold, D.D. and Bottomley, P.J. (2008) Nitrogen Mineralization and Immobilization. In: *Nitrogen in Agricultural Systems* (eds J.S. Schepers, W.R. Raun, R.F. Follet, R.H. Fox and G.W. Randall) pp. 157–172. Agronomy Monograph 49, ASA, CSSA and SSSA, Madison, WI.

Nakos, G.G. and Wolcott, A.R. (1972) Bacteriostatic effect of ammonium to Nitrobacter agilis in mixed culture with *Nitrosomonas europaea*. *Plant Soil* **36**, 521–527.

Neilsen, G.H., Stevenson, D.S. and Fitzpatrick, J.J. (1989a) The effect of municipal waste-water irrigation and rate of fertilization on petiole composition, yield and quality of Okanagan Riesling grapes. *Canadian Journal of Plant Science* **69**, 1285–1294.

Neilsen, G.H., Stevenson, D.S., Fitzpatrick, J.J. and Brownlee, C.H. (1989b) Nutrition and yield of young apple-trees irrigation with municipal waste-water. *Journal of the American Society for Horticultural Science* **114**, 377–383.

Neilsen, G.H., Stevenson, D.S., Fitzpatrick, J.J. and Brownlee, C.H. (1989c) Yield and plant nutrient content of vegetables trickle-irrigation with municipal wastewater. *Horticultural Science* **24**, 249–252.

Nelson, D.W. (1982) Gaseous losses of nitrogen other than through denitrification. In: *Nitrogen in Agricultural Soils* (ed. F.J. Stevenson), pp. 327–364. Agronomy Monograph. 22. ASA, CSSA and SSSA, Madison, WI.

Nicolardot, B., Fauvet, G. and Cheneby, D. (1994) Carbon and nitrogen cycling through soil microbial biomass at various temperatures. *Soil Biology and Biochemistry* **26**, 253–261.

Niemiera, A. and Wright, R.D. (1987a) Influence of temperature on nitrification in a pine bark medium. *Horticultural Science* **22**, 615–616.

Niemiera, A. and Wright, R.D. (1987b) Influence of NH_4-N application rate on nitrification in a pine bark medium. *Horticultural Science* **22**, 616–618.

Nonmik, H. (1957) Fixation and defixation of ammonium in soils. *Acta Agriculture Scandinavica* **9**, 395–436.

Nommik, H. and Vahtras, K. (1982) Retention and fixation of ammonium and ammonia in soils. In: *Nitrogen in Agricultural Soils* (ed. F.J. Stevenson), pp. 123–171. Agronomy Monograph. 22. ASA, CSSA and SSSA, Madison, WI.

Norton, J.M. (2008) Nitrification in agricultural soils. In: *Nitrogen in Agricultural Systems* (eds J.S. Schepers, W.R. Raun, R.F. Follet, R.H. Fox and G.W. Randall), pp. 173–199. Agronomy Monograph 49, ASA, CSSA and SSSA, Madison, WI.

Olness, A. (1999) A description of the general effect of pH on formation of nitrate in soils. *Journal of Plant Nutrition and Soil Science* **162**, 549–556.

Oron, G., Ben-Asher, J. and deMalach, Y. (1982) Effluent in trickle irrigation of cotton in arid zones. *Journal of Irrigation and Drainage Engineering* **108**, 115–126.

Papadopolus, I. and Stylianou, Y. (1988) Trickle irrigation of Sudax with treated sewage effluent. *Plant and Soil* **110**, 145–148.

Paranychianakis, N.V., Nikoolantonakis, M., Spanakis, Y. and Angelakis, A.N. (2005) The effect of recycled water on the nutrient status of Soultanina grapevines grafted on different rootstocks. *Agricultural Water Management.* Vol p.

Phillips, I.R. (2002a) Phosphorus sorption and nitrogen transformations in soils treated with piggery wastewater. *Australian Journal of Soil Research* **40**, 335–349.

Phillips, I.R. (2002b) Nutrient leaching in large undisturbed soil cores following surface applications of piggery wastewater. *Australian Journal of Soil Research* **40**, 515–532.

Phillips, I.R. and Sheehan, K.J. (2005) Importance of surface charge characteristics when selecting soils for wastewater re-use. *Australian Journal of Soil Research* **43**, 915–927.

Porter, L.K. and Stewart, B.A. (1970) Organic interactions in the fixation of ammonium in soils and clay minerals. *Soil Science* **100**, 229–233.

Probert, M.E., Dimes, J.P., Keating, B.A., Dalal, R.C. and Strong, W.M. (1998) APSIMs Water and nitrogen Modules and simulation of the dynamics of water and nitrogen in fallow systems. *Agricultural Systems* **56**, 1–28.

Probert, M.E., Keating, B.A., Thompson, J.P. and Parton, W.J. (1995) Modelling water, nitrogen and crop yield for a long-term fallow management experiment. *Australian Journal of Experimental Agriculture* **35**, 941–950.

Richter, J., Nuske, A., Habenicht, W. and Bauer, J. (1982) Optimized N-Mineralization parameters of loess from incubation experiments. *Plant & Soil* **68**, 379–88.

Rodhe, L., Pell, M. and Yamulki, S. (2006) Nitrous oxide, methane and ammonia emissions following slurry spreading on grassland. *Soil Use Management* **22**, 229–237.

Rodrigo, A., Recous, S., Neel, C. and Mary, B. (1997) Modelling temperature and moisture effects on C-N transformations in soils: Comparison of nine models. *Ecological Modelling* **102**, 325–339.

Russell, C.A., Fillery, I.R.P., Bootsma, N. and McInns, K.J. (2002) Effect of temperature and nitrogen source on nitrification in a sandy soil. *Communications in Soil Science and Plant Analysis* **33**, 1975–1989.

Ronen, D. and Magaritz, M. (1985) High concentration of solutes at the water table of a deep aquifer under sewage irrigated land. *Journal of Hydrology* **30**, 311–323.

Ryan, J.A. and Keeny, D.R. (1975) Ammonia volatilization from surface-applied wastewater sludge. *Journal of the Water Pollution Control Federation* **47**, 386.

Schmidt, E.L. (1982) Nitrification in soil. In: *Nitrogen in Agricultural Soils* (ed. F.J. Stevension), pp. 253–288. Agronomy Monograph 22. ASA, CSSA and SSSA, Madison, WI.

Shindo, H., and Nishio, T. (2005). Immobilization and remineralization of N following addition of wheat straw into soil: determination of gross N transformation rates by ^{15}N-ammonium isotope dilution technique. *Soil Biology and Biochemistry* **37**, 425–432.

Sierra J. (2006) A hot-spot approach applied to nitrification in tropical acid soils. *Soil Biology and Biochemistry* **38**, 644–652.

Sierra, J., Fontaine, S. and Desfontaines, L. (2001) Factors controlling N mineralization, nitrification, and nitrogen losses in an Oxisol amended with sewage sludge. *Australian Journal of Soil Research* **39**, 519–534.

Singleton, P.L., Mclay, C.D.A. and Barkle, G.F. (2001) Nitrogen leaching from soil lysimeters irrigated with dairy shed effluent and having managed drainage. *Australian Journal of Soil Research* **39**, 385–396.

Smith, C.J., Freney, J.R. and Bond, W.J. (1996) Ammonia volatilization from soil irrigated with urban sewage effluent. *Australian Journal of Soil Research* **34**, 789–802.

Smith, C.J. and Bond, W.J. (1999) Losses of nitrogen from effluent-irrigated plantation. *Australian Journal of Soil Research* **37**, 371–389.

Snow, V.O., Smith, C.J., Polglase, P.J. and Probert, M.E. (1999) Nitrogen dynamics in a eucalyptus plantation irrigated with sewage effluent or bore water. *Australian Journal of Soil Research* **37**, 527–544.

Sommer, S.G., Jensen, L.S., Clausen, S.B. and Sogaard, H.T. (2006) Ammonia volatilization from surface-applied livestock slurry as affected by slurry composition and slurry infiltration depth. *Journal of Agricultural Science* **144**, 229–235.

Stanford, G., and Smith, S.J. (1972) Nitrogen mineralization potentials of soils. *Soil Science Society of America Journal* **36**, 465–472.

Steenhuis, T.S., Bubenzer, G.D. and Converse, J.C. (1976) *Ammonia Volatilization of Winter Spread Manure*. ASAE paper no. 76-4514. ASAE, St. Joseph, MI.

Ste-Marie, C. and Pare, D. (1999) Soil pH and N availability effects on net nitrification in the forest floors of a range of boreal forest stands. *Soil Biology and Biochemistry* **31**, 1579–1589.

Stenger, R., Barkle, G.F. and Burgess, C.P. (2001) Mineralisation and immobilisation of carbon and nitrogen from dairy farm effluent (DFE) and glucose plus ammonium chloride solution in three grassland topsoils. *Soil Biology and Biochemistry* **33**, 1037–1048.

Stevens, R.J., Laughlin, R.J. and Malone, J.P. (1998) Soil pH affects the processes reducing nitrate to nitrous oxide and di-nitrogen. *Soil Biology and Biochemistry* **30**, 1119–1126.

Stevenson, F.J. (1982) Organic forms of soil nitrogen. In: *Nitrogen in Agricultural Soils* (ed. F.J. Stevenson). Nitrogen in Agricultural Soils. Agronomy Monograph. 22. ASA, CSSA and SSSA, Madison, WI.

Strong, D.T., Sale, P.W.G. and Heylar, K.G. (1997) Initial soil pH affects the pH at which nitrification ceases due to self-induced acidification of microbial microsites. *Australian Journal of Soil Research* **35**, 565–570.

Stueven, R., Vollmer, M. and Bock, E. (1992) The impact of organic matter on nitric oxide formation by *Nitrosomonas europeae*. *Archives in Microbiology* **158**, 439–443.

Swift, M.J., Heal, O.W., and Anderson, J.M. (1979) *Decomposition in Terrestrial Ecosystems. Studies in Ecology 5*. University of California Press Berkeley, *CA*.

Taylor, R.A. and Bloom, A.J. (1998) Ammonium, nitrate, and proton fluxes along the maize root. *Plant Cell Environment* **21**, 1255–1263.

Tozer, W.C., Wilkins, K.J., Wang, H., Van de Heuvel, M., Charleson, T. and Silvester, W.B. (2005) Using ^{15}N to determine a budget for effluent-derived nitrogen applied to forest. *Isotopes in Environmental and Health Studies* **41**, 13–30.

Vaisman, I., Shalhevet, J., Kipnis, T. and Feigin, A. (1981) Reducing ground-water pollution from municipal waste-water irrigation of Rhodes-grass grown on sand dunes. *Journal of Environmental Quality* **10**, 434–439.

Vaisman, I., Shalhevet, J., Kipnis, T. and Feigin, A. (1982) Water regime and nitrogen fertilization for Rhodes grass irrigated with municipal waste-water on sand dune soil. *Journal of Environmental Quality* **11**, 230–232.

Vigil, M.F. and Kissel, D.E. (1991) Equations for estimating the amount of nitrogen mineralized from crop residues. *Soil Society of America Journal* **55**, 757–761.

Vinten, A.J.A., Whitmore, A.P., Bloem, J., Howard, R. and Wright, F. (2002) Factors affecting N immobilsation/mineralisation kinetics for cellulose-, glucose- and straw-amended sandy soils. *Biology and Fertility of Soils* **36**, 190–199.

Vogeler, I., Green, S.R., Mills, T. and Clothier, B.E. (2006) Modelling nitrate and bromide leaching from sewage sludge. *Soil and Tillage Research* **89**, 177–184.

Xu, J. M., Tang, C., and Chen, Z.L. (2006) The role of plant residues in pH change of acid soils differing in initial pH. *Soil Biology and Biochemistry* **38**, 709–719.

Zak, D.R., Holmes, W.E., MacDonald, N.W., and Pregitzer, K.S. (1999). Soil temperature, matric potential, and the kinetics of microbial respiration and nitrogen mineralization. *Soil Society of America Journal* **63**, 575–584.

Zekri, M. and Koo, R.C.J. (1994) Treated municipal wastewater for citrus irrigation. *Journal of Plant Nutrition* **17**, 693–708.

Zhou, J., Green, M. and Shaviv, A. (2003) Mineralization of organic N originating in treated effluent used for irrigation. *Nutrient Cycling in Agroecosystems* **67**, 205–213.

5.2 Phosphorus

Bnayahu Bar-Yosef

5.2.1 Introduction

Irrigation with reclaimed municipal wastewater (effluents) may affect phosphorus (P) status in soil differently from base fertilization with the same amount of P, or adding inorganic P (P_i) via the water. The reasons for the difference are: (i) P in effluents is applied continuously at nearly constant concentration along the entire irrigation period, whereas in conventional fertigation P is supplied according to plant demand and P availability level in soil; (ii) effluents include both P_i and organic P (P_o), which differ in their chemical and mobility properties and utilization by plants; (iii) effluents include macro organic carbonaceous molecules (e.g. humic acid (HA) and fulvic acid (FA)) that might sorb P, complex Ca^{2+} and Fe^{3+}, and thus affect P mobility in soil. Effluent irrigation differs from preseeding soil amendment by biosolids as well. Under effluent irrigation P adsorption by soil is in equilibrium with the effluent solution P concentration. In biosolid amendment the added P is mixed in the plowed soil layer, thus the initial sorption and concentration in soil solution are considerably higher than in effluent irrigation. When irrigating biosolid amended soils with freshwater, the P_i concentration in soil solution declines due to plant uptake, microbe consumption and leaching, and adsorbed P is released to re-equilibrate with the decreasing P_i solution concentration. Another difference between effluent irrigation and biosolid application is the greater impact of biosolid added organic matter and solutes in the top soil layer on water infiltration and decreased P runoff losses (Sharpley et al., 2001; Gburek et al., 2005).

The objectives of this chapter are to elucidate some of the processes that determine reclaimed wastewater P behavior and fate in soil, evaluate means to assess P leaching and runoff risks, and present case studies summarizing long-term soils and crop responses to effluent irrigation. The growing acceptance of effluents as a source of water substituting the dwindling freshwater resources in agriculture and landscaping (Wastewater Effluent Irrigation Taskforce, 2006; Parsons et al., 1995; California Water Resources Control Board, 2003; Pescod, 1992), give these objectives urgency, particularly in developing

Treated Wastewater in Agriculture, First Edition, edited by Guy J. Levy, Pinchas Fine and Asher Bar-Tal

guidelines for optimal reclaimed wastewater use, similar to guidelines for P application in animal manure (Shober and Sims, 2003).

Previous reviews on agricultural use of effluents (Bouwer and Chaney, 1974; Pettygrove and Asano, 1985; Feigin et al., 1991; Pescod, 1992; Stevens, 2006) are less applicable for assessing environmental risks associated with reclaimed wastewater P, because they did not consider the geochemistry and transport of all P species present and the potential contribution of organic P to plant uptake.

5.2.2 Phosphorus forms and species in effluents

The predominant P sources in sewage sludge are human excretion (estimated in UK as 1.5 g P/day/person) and detergents (0.4 g P/day/person) (White and Hammond, 2007). As daily per capita water use in UK is 334 l (Hinrichsen et al., 1998), the expected total P concentration in sludge is (1.9/0.334=) 5.7 mg P/l. This is within the range of total P usually reported for primary effluents (5–17 mg P/l, Feigin et al., 1991). About 70–90% of total P in sewage sludge is in P_i form and the rest as P_o (Kirkham, 1982). Using nuclear magnetic resonance (NMR) techniques revealed that the P_i fraction includes orthophosphate and a small quantity of pyrophosphate (Escudey et al., 2004). X-ray diffraction analysis of untreated sludge indicated the presence of dicalcium phosphate dihydrate (DCPD) crystals (Huang and Shenker, 2004). Other studies (reviewed by Hedley and McLaughlin, 2005) reported the presence of octa-calcium phosphate (OCP) too. The different solubility of these minerals may explain the considerable range of P_i concentrations in sewage solutions reported by Feigin et al. (1991).

The P_o species in fresh sewage sludge have not been sufficiently identified. In sewage activated sludge ~50% of P_o was reported to be phospholipids, 30% inositol hexaphosphate ($C_6H_6(H_2PO_4)_6$, IHP), and 20% humic compounds (Fine and Mingelgrin, 1996). These proportions differ from reported P_o components in soil: 50–70% as IHP, ~5% as phospholipids and phosphoglycerides, 1–2% as nucleotides, and the rest as unaccounted for recalcitrant species (Bar-Yosef, 2003; Turner et al., 2003). Secondary (biological) treatment of effluents does not usually affect the total P concentration in solution (Feigin et al., 1991). The concentration is reduced, however, by tertiary treatments, due to addition of alum or lime designed to coagulate suspended solids and biochemical oxygen demand (BOD). This addition may decrease total P concentration in solution to 0.2–1.0 mg P/l (Feigin et al., 1991; Pescod, 1992), depending on solution pH, dominant cation, and reaction time. Another approach that reduces P_i concentration in solution is sturvite ($MgNH_4PO_4{\cdot}6H_2O$) harvesting; the sturvite may then be used as P fertilizer (Web and Ho, 1992; Woods et al., 1999). Conditions inducing microbial P consumption exceeding metabolic requirements also result in decreased solution P_i concentration. Elliott et al. (2002) reported that under such soil conditions microbial biomass KCl soluble P exceeded P minerals KCl extractable P.

The activity coefficients (γ) of P species in effluents are smaller than those in freshwater due to higher solution ionic strength. Regular Cl^-, Na^+, $Ca^{2+}+Mg^{2+}$, and HCO_3^- concentrations in effluents are 60–80% higher than in low-quality freshwater characterizing arid zones, the SAR is ~2.3-fold higher and the pH is 0.2 units higher (Table 5.2.1). Besides their effect on γ, Cl^- and Na^+ may suppress yield of salt-sensitive crops, and Na^+ and HCO_3^- may adversely affect soil hydraulic conductivity (Bresler et al., 1982) and

Table 5.2.1 Comparison between ionic composition of secondary effluents and freshwater in Israel (mean of 95 locations and 6 yrs of water sampling[a])

Component	Units	Freshwater	Secondary effluents
Cl^-	mg/l	160	240
Na^+	mM	3.7	7.2
$Ca^{2+} + Mg^{2+}$	mM	3.6	3.4
SAR[b]	$(mmol/l)^{1/2}$	1.6	3.9
pH		7.6	7.8
HCO_3^-	mM	4.8	8
NO_3-N	mg/l	8	3
NH_4^+-N	mg/l	0	15
Total P	mg/l	0	5
Orthophosphate-P	mg/l	0	3.5
BOD[c]	mg/l	0	
TSS[c]	mg/l	0	25

[a]Recalculated from Effluent Survey Team Force (2006).
[b]The lower reported SAR (sodium adsorption ratio, $C_{Na}/(C_{Ca+Mg})^{1/2}$) in freshwater relative to the value calculated from the presented Na and Ca+ Mg data ($1.95\ mM^{1/2}$) stem from the fact that the comparison is based on some different water sources in the two tests.
[c]Since 2006 the permitted BOD (biological oxygen demand) and TSS (total suspended solids) in Israel is 10 and 10 mg/l.

soil aggregate stability (Shainberg and Levy, 1992). The difference between arid zone freshwater and effluents $Ca^{2+} + Mg^{2+}$ concentration is too small to be effective (Table 5.2.1).

5.2.3 Sorption, transport and transformations of effluent P in soil

5.2.3.1 *Inorganic P*

5.2.3.1.1 *Adsorption*

Adsorption reactions of P_i by soil and soil constituents have been thoroughly studied in the literature (for review see Bar-Yosef, 2003; Pierzynski et al., 2005). Best predictions of P_i adsorption (A, mg/kg) by clay minerals and metal oxides as a function of P_i concentration in solution (C_P, $mg\,L^{-1}$), pH, and ionic strength (I) have been obtained by multi-layer surface complexation-electrostatic models (Goldberg and Sposito, 1985; Dzombak, 1990). The drawbacks of these models are: (i) they require a large number of parameters that must be defined for each system; and (ii) their application to soils has not been established yet. In soils maintaining nearly constant pH and I under a wide range of P application levels, the Langmuir adsorption model (equation 5.2.1) has been established as a satisfactory predictor of P adsorption (Hedley and McLaughlin, 2005). The model has two parameters: maximum adsorption, b (mol/kg or mg/kg), and ion affinity to the surface, K_M (l/mol or l/mg P).

$$A = K_M\, b\, C_P/(1 + K_M\, C_P) \tag{5.2.1}$$

To account for simultaneous multi-anion adsorption (e.g. $H_2PO_4^-$ and HPO_4^{2-}) and pH effect, a phenomenological adsorption model derived from the Langmuir competitive adsorption model was developed (equation 5.1.2, Bar-Yosef et al., 1988).

$$A_i = \frac{T \cdot \sum_{i}^{n} K_i \cdot C_i}{1 + \sum_{j}^{m} K_j \cdot C_j} \tag{5.2.2}$$

Here, $T = T_o \cdot e^{[(RPH/pH)-1] \cdot G}$ (mol/l soil), RPH is a soil constant related to the pH at the inflection point of the surface titration curve, T_o and G are soil constants (mol/l soil and dimensionless, respectively), K is the aforementioned K_M for participating ions (mol/l), C is concentration in solution (mol/l), i = 1,. . .n are the adsorbing P species, and j = 1,. . .n are the P species plus OH^- and other competing anions. When hydroxyl adsorption is calculated, i = 1. The equation needs fewer parameters than multi-layer models and is differentiable with respect to both P and H^+, thus accounting for soil pH-dependent P buffering capacity, which is required in the P transport equation (Olsen and Kemper, 1968).

Several studies showed a decrease in P sorption in the presence of organic matter (OM), suggesting competition between adsorption of decomposed OM components and P on common adsorption sites (Guppy et al., 2005a). Competitive adsorption also has been shown to play a role in the presence of HA and FA (Guppy et al., 2005a, 2005b). Sibanda and Young (1986) found a decrease in P_i adsorption by goethite, gibbsite, and soil in the presence of HA (at an initial high concentration of 16 g/l HA). They attributed it to competitive adsorption with HA and noted that the effect of HA on P_i adsorption diminished with increasing pH, and vanished at pH ~ 7.6. Ge and Hendershot (2004) found that the maximum surface charge in a two-site pH-dependent P adsorption model they had developed was strongly affected by the organic C concentration in soil (%OC) and was linearly correlated with the log (%OC). Another plausible mechanism explaining the impact of organic compounds on P retention in soil is P complexation by carbonaceous macromolecules. Mobile, stable complexes enhance P_i losses from soil due to reduced P adsorption (Guppy et al., 2005b). Immobile, stable complexes reduce P_i losses and increase its sorption relative to un-amended soil. An example of a soil-attached (immobile) macromolecule is HA, which binds P_i in proportion to its Al content (Larsen et al., 1959; Heng, 1989) and is therefore expected to reduce P_i transport in soil. HA-P_i complexes have not been investigated yet with respect to chemical stability and pH dependency.

The aforementioned P adsorption models assume reversible adsorption–desorption processes. However, some studies showed smaller desorption than adsorption (for review see Bar-Yosef, 2003). Goldberg and Sposito (1985) suggested that adsorbed P_i becomes kinetically irreversible because a reversibly adsorbed monodentate P_i is converted into an irreversible bidentate form, but no satisfactory theory explaining such transformations was offered. It is also possible that the conventional method to determine desorption, namely, excessive suspension dilution, enhances apparent irreversibility by exposing new adsorption sites due to dissolution of clay broken edges. Due to uncertainties in quantifying irreversible desorption and possible impact of desorption conditions on P partitioning, irreversible adsorption–desorption is still unaccounted for in soil P transport models.

Phosphorus sorption involves two distinct phases: a fast surface reaction reaching equilibrium after a few minutes; and a slow reaction that may last for days (e.g. Barrow and Shaw, 1975; Luengo et al., 2006, Arai and Sparks, 2007). The fast reaction involves surface complexation, and the slow one an intra- and inter-particle diffusion controlled processes. The sorption reactions are usually determined by macroscopic methods, but recently they were also validated in goethite by attenuated total reflection infrared spectroscopy (Luengo et al., 2006). Two points should be emphasized, however: (i) the fast adsorption accounts for the majority of total P sorption; and (ii) the slow sorption is practically complete after 24–48 hrs (Arai and Sparks, 2007).

The partitioning of P between solid and liquid phases in soil is calculated by equation 5.1.3. The total mobile P in soil (Q_p, mg/kg) equals adsorbed P (A, defined in equation 5.2.1) plus P found in solution at given solution:solid ratio (θ, l/kg).

$$Q_P = A + \theta C_P \tag{5.2.3}$$

Determining the Langmuir constants for each soil is impractical. To circumvent this problem, the relationships between b and K_M and certain soil properties were investigated. The best correlations were obtained with soil clay content (Bar-Yosef and Akiri, 1978, Kuo et al., 1988) and examples of such functions in calcareous soils are presented in Figure 5.2.1. Similarly, Ge and Hendershot (2004) found useful correlations between maximum surface charge and soil cation exchange capacity and organic C content (not presented).

The determination of adsorbed P or C_p in each soil is also impractical; therefore, these were replaced by experimentally simple soil extraction terms. The most common sorbed P extractors are Olsen, Bray, and Mehlich solutions (Kuo, 1996) that recover a certain fraction of A, the value of which depends on soil conditions and properties, and are well correlated with the P fraction available to plants. The C_p is approximated by soil water extract at a given solution:soil ratio. To use equation 5.2.3 with extractable soil P data, the P recovery percentage ($REC = 100^*$ Extractable P/A) must be known. Bar-Yosef and Akiri (1978) investigated REC of $NaHCO_3$ (Olsen) extractable P in representative Israeli soils and found good correlation with % clay in soil (Fig. 5.2.2) over a range of 4–40% clay and P levels between 5% and 100% of maximum adsorption (b). Similar REC versus % clay relationships were obtained for Bray and Mehlich extractions in neutral and slightly acidic soils, respectively (Kuo, 1996).

5.2.3.1.2 Precipitation-dissolution

Crystallization and dissolution kinetics theories are beyond the scope of this chapter (see Stumm and Morgan, 1996, and Lasaga, 1998). A practical question is, however, what is the rate and order of P precipitation in supersaturated solutions of Ca^{2+} or Fe^{3+} and Al^{3+}. In the case of Ca^{2+}, the first minerals to precipitate are dicalcium phosphate dihydrate (DCPD, $CaHPO_4 \cdot 2H_2O$, $pK_{sp} = 6.6$) and amorphous calcium phosphate (ACP, $Ca_3(PO_4)_2 \cdot xH_2O$, $pK_{sp} = 24–25.5$). Dicalcium phosphate dihydrate may precipitate at $pH < 7$, and ACP at $pH > 7$. With time, ACP transforms into the thermodynamically more stable tricalcium phosphate (TCP, $Ca_3(PO_4)_2$ $pK_{sp} = 32.6$), octacalcium phosphate (OCP, $Ca_4H(PO_4)_3 \cdot 2.5H_2O$, $pK_{sp} = \sim 46$) and hydroxyapatite (HAP, $Ca_{10}(PO_4)_6(OH)_2$,

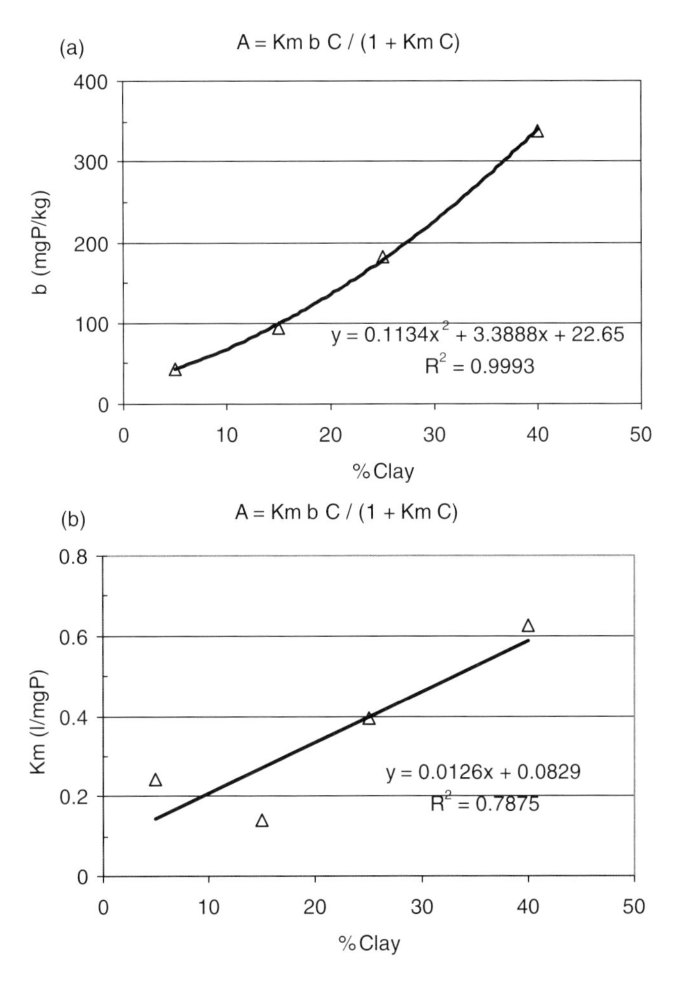

Figure 5.2.1 Langmuir coefficients b and K_m (equation 5.2.1) as a function of soil clay content. Data derived from Bar-Yosef and Akiri (1978).

$pK_{sp} = 49–57$) (Musvoto et al., 2000a). The crystal growth of the stable minerals is very slow, therefore, the equilibrium attainment is also slow, taking between a minimum of 1 month and a number of years (Musvoto et al., 2000b). Experimental data by Grossl and Inskeep (1991) reveal that crystallization rate of DCPD in an initially supersaturated solution maintained at constant pH of 5.7 varied between 17 μmol/l/s (0–5 min) and 0.22 μmol/l/s (5–120 min). The mean rate of HAP precipitation (initial P and Ca = 0.86 mM) was reported to be 0.007 μmol/l/s (Inskeep and Silvertooth, 1988). As the degree of supersaturation decreased, the precipitation rate declined. Due to the slow transition between precipitated P minerals, Musvoto et al. (2000a) used ACP when simulating Ca-P crystallization in wastewater, whereas Cho (1991) assumed that the short-term governing Ca-P mineral is DCPD.

The aforementioned precipitation rates can be compared with P_i adsorption kinetics. Taking as an example a soil suspension of 10 g/l and adsorption of 1 mmol P/kg soil reaching equilibrium in 1 min, gives an adsorption rate of (10 μmol P/l/min =)

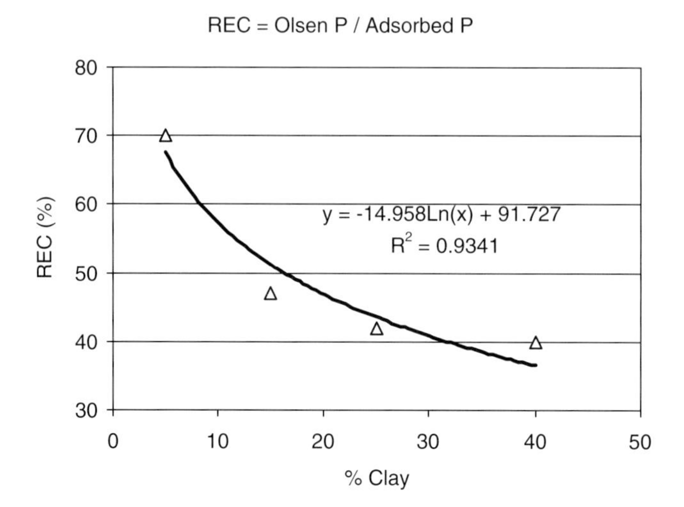

Figure 5.2.2 Recovery percentage of adsorbed P by $NaHCO_3$ 0.5 M (REC) as a function of soil clay content. The obtained function is appropriate for P adsorption in the range of 5–100% of maximum P adsorption at each % clay. Data derived from Bar-Yosef and Akiri (1978).

0.16 μmol/l/s. As under field conditions the soil:solution ratio is ~1000 g/l, the corresponding adsorption rate is ~16 μmol/l/s, much higher than the cited precipitation rates, except short-term DCPD crystallization. As soil organic matter reduces appreciably the precipitation rate of Ca-P minerals due to "crystal poisoning" (Inskeep and Silvertooth, 1988; Grossl and Inskeep, 1991), it seems safe to conclude that under soil conditions the fast disappearance of P_i from solution is mainly due to adsorption. The "poisoning effect" also implies that Ca-P crystallization rate under effluent irrigation is slower than under freshwater irrigation.

Fine and Mingelgrin (1996) reported, based on Ca-P solubility diagram analysis (Lindsay, 1979), that in two different Mediterranean soils, equilibrated for 100 d after amendment with domestic waste-activated sludge, the soil solution P concentration was undersaturated with respect to DCPD, supersaturated with respect to TCP, and most probably controlled by OCP. In the presence of high NH_4^+ concentration P_i may precipitate as struvite (pK_{sp} 12.6–13.15, Web and Ho, 1992). The minerals controlling solution P concentration in sludge-amended soils (Fine and Mingelgrin, 1996) are similar to minerals controlling P concentration in biologically treated effluents (see "P forms and speciation in effluents" above). This implies that quasi-equilibrium soil solution P concentration in sludge-amended soils should resemble the soil solution P concentration in effluent-irrigated soils. This pertains while P minerals determine P concentration in soil solution, but the similarity in concentration is not expected to change much when P adsorption becomes dominant.

In addition to the aforementioned thermodynamic P solubility considerations, P disappearance from soil solution was also expressed by semi-empirical rate equations, e.g. first and second order rate reactions, parabolic diffusion-controlled processes, power functions, and the Elovich equation. For a review of empirical P kinetic models in soils the reader is referred to Sanyal and De Datta (1991).

5.2.3.2 Organic P

5.2.3.2.1 Adsorption

Meager information is available on IHP adsorption by soils (Anderson et al., 1974). The pH-dependent adsorption of IHP and RNA by clay minerals was studied by Goring (1952) and Bartholomew (1950). Their data was recalculated by Bar-Yosef (2003) in the form of pH-specific Langmuir adsorption isotherms. The data indicate that at similar C_P ($mg\,L^{-1}$ P) and pH, adsorption of IHP-P ($mg\,kg^{-1}$) exceeds the adsorption of P_i by bentonite. Similar results were published by Celi et al. (2001) comparing pH-dependent IHP and P_i adsorption by goethite. The greater IHP adsorption is expected as its molecule includes 12 ionizable protons rendering it a much higher charge density than P_i. From a surface coverage point of view, each IHP molecule occupies an area equivalent to four adsorption sites of orthophosphate (Condron et al., 2005) leaving two of the six $H_2PO_4^-$ groups free.

The greater IHP retention is compatible with data showing that in competitive IHP–P_i adsorption on Fe-oxide strips (Sharpley et al., 1995) and soils (Evans, 1985), the affinity of IHP to the surface is stronger. Other studies of IHP adsorption by goethite (Ognalaga et al., 1994), illite, kaolinite and goethite (Celi et al., 1999), and calcite (Celi et al., 2000) indicate Langmuir-like adsorption within a certain range of solution P concentration, and a change in phase when a concentration threshold is exceeded.

The rate of IHP adsorption was studied in a short range-ordered Al precipitate at initial pH of 6.8. The rate was very fast during the first hour, and slowed down considerably between 1 and 24 hrs, similar to the behavior of a P_i control (Shang et al., 1990). A first order kinetic model was used to describe adsorption during both stages. As expected from molecule size considerations, the rate constant of Pi exceeded that of IHP. The activation energies of P_i and IHP adsorption were 48 and 89 KJ/mol P adsorbed, respectively. The adsorption kinetics of P_i and IHP on short-range iron precipitate (Shang et al., 1992) were similar to the kinetics in short-range Al precipitate. Adsorption of glycerophosphates was found to be much smaller than IHP (Evans, 1985, Rauschkolb et al., 1976), therefore their mobility in soil should exceed IHP.

Data in Figure 5.2.3 summarize the differences between IHP and P_i behavior in soil. At similar P concentration in equilibrium solution, the adsorption of IHP-P by a loess soil is approximately tenfold greater than P_i, and adsorption of both forms obeys a Langmuir adsorption isotherm. This explains two other observations from Figure 5.2.3: (i) negligible P_i adsorption in the presence of IHP (top right), stemming from the competitive adsorption mentioned above; and (ii) negligible IHP leaching in soil (bottom right). In addition, the presence of IHP had a negligible effect on the shape of the P_i breakthrough curve (compare inflow of P_i, bottom left, and inflow of P_i+ IHP, bottom right, Fig. 5.2.3).

5.2.3.2.2 Precipitation

The dissociation constants of IHP are 6.3 ($pK_{7,8}$) and 9.7 (pK_{9-12}) (Cosgrove, 1980), lower than the corresponding P_i constants of 7.2 ($pK_{H_2PO_4}$) and 12.3 (pK_{HPO_4}) (Lindsay, 1979). According to Celi et al. (2001), IHP can complex with Ca^{2+} by forming two soluble species: Ca_1-IHP and Ca_2-IHP. Reaction with a third Ca^{2+} ion would result in the precipitation of Ca_3-IHP. Experimentally, the precipitation (in the presence of initial IHP

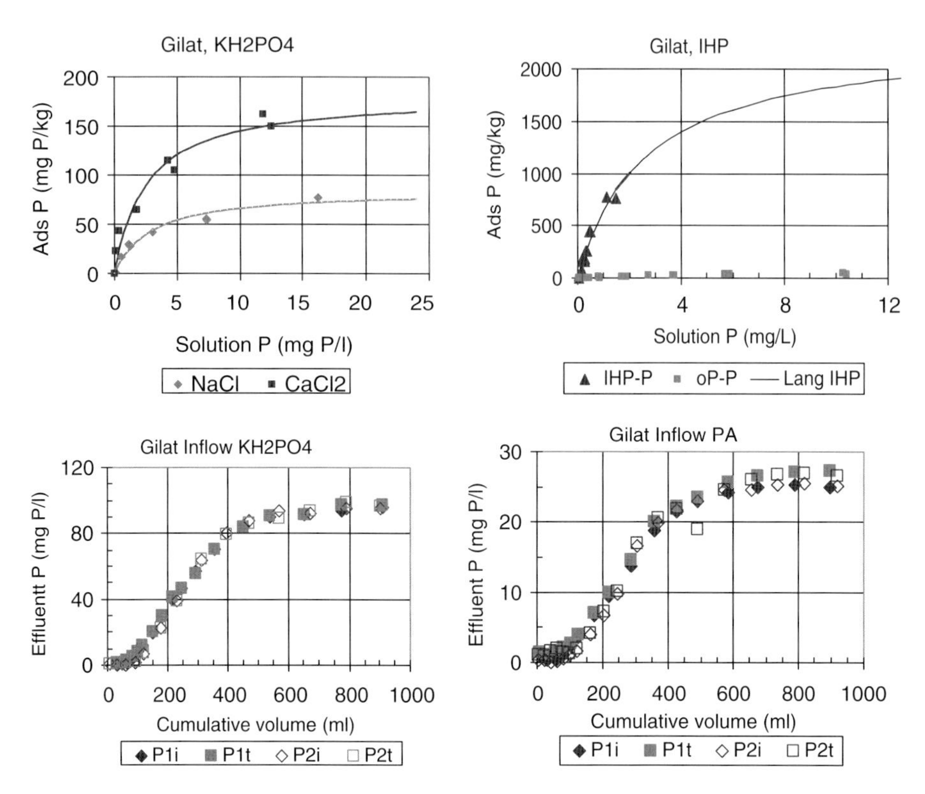

Figure 5.2.3 Orthophosphate (P_i) and IHP adsorption and transport in Gilat loess soil (Israel, 15% clay). (a) Adsorption isotherms of P_i at 10 mM NaCl (pH 7.7) and 5 mM $CaCl_2$ (pH 7.0). (b) Adsorption isotherm of IHP in the presence of P_i (IHP-P:P_i in added solution = 3:1) at 10 mM NaCl, pH 7.6. In both top panels the solution:soil ratio was 20:1, and the lines were fitted by the Langmuir adsorption model (equation 5.2.1). The parameters for P_i were derived from Fig. 5.2.1, 15% clay). (c) Breakthrough curves of P_i and total P (P_t). The inflow P_i concentration was 96 mg/l P; one pore volume = 50 ml. (d) As bottom left only for inflow solution of 72 mg/L IHP-P and 25 mg/L P_i. The two replicates of the column experiments are presented (1 and 2 in legend). Note that (i) the outflow (effluent) P concentration in the plateau equaled the inflow P_i concentration, and (ii) outflow P_i and Pt were almost identical. These observations strongly indicate that IHP transport in soil was negligible. Source: original data by the author.

and Ca^{2+} concentrations of 0.8 mmol/l P and 3.3 mmol/l Ca^{+2}) started at pH 5–6 and caused complete disappearance of IHP from solution at pH > 6. No solubility product constant for the IHP-Ca salt, or others, could be found in the literature.

5.2.3.2.3 Mineralization/Immobilization

Assessment of P_o mineralization is done by variety of methods, from rough experimental data to complex modeling of the process. Barber (1995) estimated that the rate of P_i production via mineralization equals the rate of OM decomposition multiplied by the P concentration in the OM. In a 0–40 cm soil layer, soil bulk density of 1.25 kg/l, 2% OM in soil, OM decomposition rate of 2% per year, and 0.5% P in the OM, the rate of P_i release is 10 kg P_i/ha/yr. The role of orthophosphatase enzymes in P_o mineralization seems to be less important than the role of organic C decomposition (for review see Condron et al., 2005).

To induce P_o mineralization of OM added to soil, it should contain at least 0.2% total P, otherwise net P immobilization will occur. Another empirical rule is that C:P ratio <100 induces mineralization, whereas C:P >300 leads to P immobilization (Iyamuremye and Dick, 1996).

The P isotopic dilution technique (e.g. Oehl et al., 2001) opened new avenues in the determination of P_o mineralization in soils. Using this technique, Oehl et al. (2004) determined mineralization rates of 1.4–2.5 mg P/kg soil/d in Typic Hapludalf soil (pH 6.4 ± 0.4, 14 g C/kg) amended with 330 mg P_i and 370 mg P_o/kg. Previously, P_o mineralization rates in several Northern American Mollisols were reported to be 0.9–4.2 mg P/kg/d (Lopez-Hernandez et al., 1998); in two acid organic soils they were found to be 0.68–1.36 mg P/kg/d (cited by Condron et al., 2005); in 25 U.S. soils amended with 20–400 mg P_O/kg soil and incubated at 40°C, they were 4–45 mg P/kg soil per 25 d (Tompson et al., 1954); in long-term mineralization trials under field conditions in the UK and North America, they were 0.5–42 kg P/ha/yr (cited by Condron et al., 2005). Assuming steady mineralization rate (zero order reaction) and using 1.0 mg P/kg soil = 2.5 kg P/ha (0–20 cm soil layer), the above 0.5–42 kg P/ha/yr are equivalent to 0.6×10^{-3} to 47.0×10^{-3} mg P/kg/d, which is 150–1000-fold slower than the abovementioned P_o mineralization rates obtained in short-term incubation experiments. The big difference stems from the fact that P_o mineralization is plausibly a first order, rather than zero order, reaction.

According to Tarafdar and Claassen (1988), plant P uptake from soil amended by IHP or P_i was comparable, and closely related to the activity of phosphatase in soil. The authors cited published P_o mineralization rate results characterizing diverse soil conditions: (i) mineralization under field conditions in the tropic 67–157 kg P/ha/yr; (ii) short-term incubation of acid organic soils 0.68–1.36 mg P/kg/d; (iii) 7 d incubation of bio-organic cropping soil 1.7 mg P/kg/d.

The considerable variability in P_o mineralization rate results could stem from different incubation periods, and/or different contribution of microbial mass to organic P released to soil via cell lysis. Cell lysis is stimulated by soil drying–wetting cycles, and these differed in the aforementioned studies.

Despite the predominance of IHP and orthophosphate diesters in organic P, few attempts have been made to estimate their individual mineralization rates. Hinedi et al. (1988), who used a ^{31}P NMR technique, found that in acid and alkaline soils amended with sludge and incubated for 1, 28, 70, and 140 d, the P_i diesters (mainly ATP and NAD) completely hydrolyzed after 28 d, whereas IHP (a monoester) was still present after 140 d. The sludge in this study was incorporated at a rate equivalent to 100 ton/ha to give an initial concentration of ~18 g P/kg. Mineralization rates of NAD and ATP also were studied by Frossard et al. (1996). The soil was clay-loam (pH 7, total P concentration 890 mg/kg), amended by 100 mg/kg NAD-P or ATP-P. Forty-six percent of the added NAD-P was mineralized in 0.7 d, but only part of this was found in solution due to resorption of P_i. The mineralization rate of ATP was ~15% slower than NAD. Suzumara and Kamatani (1995) studied mineralization rates of IHP in coastal sediments and found that the half life was 60 d. Under anaerobic conditions the mineralization rate was much higher and the entire quantity of IHP was mineralized in 40 d.

Organic P mineralization models can be divided into two groups: (i) simple (zero, first order and parabolic) models; and (ii) multi-component models that are based on a set of first order equations describing P_o mineralization in several organic P pools, and mass transfer among them (for review of these models see Bar-Yosef, 2003). The models differ in number of pools and consequently number of parameters needed for the solution. Grierson et al. (1998) compared several group (i) models and concluded that a zero order model best described P_o mineralization in moist sandy soils. Among group (ii) models the best known is the Erosion-Productivity Impact Calculator (EPIC) (Jones et al., 1984). Besides incorporating the set of first order mineralization reactions, this model includes an expression accounting for P uptake and release by microbial mass. This addition increased the flexibility of simulating net P_i release to soil, and indeed EPIC provides realistic P_o mineralization predictions under field conditions. Despite the disadvantage of numerous parameters, EPIC is one of the best predictors of P_o transformations in soil.

5.2.3.3 Transport

5.2.3.3.1 Leaching and movement in soil

Free P_i ions and ion pairs differ in their sorption and mobility in soil. In calcareous soils and pH between 5 and 9, the relevant P_i species are $H_2PO_4^-$, HPO_4^{2-}, $CaH_2PO_4^+$, $CaHPO^o$, and $CaPO_4^-$ (Schecher, 2001). In the absence of Ca-complexion agents, the mole fractions of the sum of $CaH_2PO_4^+$, $CaHPO^o$, and $CaPO_4^-$ at pH 5.5, 7.0, 8.0 and 9.0 are 0.25, 0.65, 0.83, and 0.93, respectively (total P and Ca concentrations = 1 mM and 10 mM, respectively). In the presence of 10 mM citrate, a moderate Ca complexion agent, the corresponding mole fractions are 0.06, 0.15, 0.27, and 0.49, implying a higher concentration of free P_i ions in solution. Carbonate has a much smaller effect on Ca-P ion pairing than citrate (Bar-Yosef, 1996). In acid soils Fe-P and Al-P ion pairs dominate. Citrate and oxalate form stable complexes with Fe^{3+} and Al^{3+} (Bar-Yosef, 1996), thereby increasing the mole fraction of free P_i species in solution. Gardner et al. (1983) suggested that citrate forms a soluble ferric-hydroxyphosphate polymer ($Fe|O|OH|H_2PO_4|Cit$), which moves P from the bulk soil solution to root surfaces. Stevenson and Cole (1999) suggested the existence of a P macromolecule, the phosphohumate complex (HA-Al-phosphate), which can explain experimental P retention by acid organic soils.

All aforementioned P species move in the soil profile by diffusion and convection, and both depend strongly on soil water content (θ, l/l). The governing equation describing change in P concentration in a given soil subvolume with time is equation 5.2.4:

$$\frac{\partial Q}{\partial t} = -\frac{\sum_{i=1}^{n} \partial F_i}{\partial z} - PR + M - U \tag{5.2.4}$$

Here Q is the sum of the n mobile P_i species concentration in the soil (mol/l soil, equation 5.2.5), F_i is the flux of the ith species (mmol/cm^2/s, equation 5.2.6), t is time, z is depth (cm), PR is orthophosphate precipitation and dissolution rate, M is P_o mineralization rate, and U is P_i uptake rate (all in mol/l soil).

$$Q = \sum_{1}^{n} (Ap_i + \theta \cdot C_i) \text{ (symbols defined in equations 5.2.1 to 5.2.3)} \tag{5.2.5}$$

$$F_i = -Dp_i \frac{\partial C_i}{\partial z} + q \cdot C_i \qquad (5.2.6)$$

In equation 5.2.6 Dp is the ion diffusion coefficient in soil solution (cm^2/s), which is a function of θ (Bar-Yosef, 2003), C_i is the concentration of the ith P species in soil solution (mol/l), $q = -K(h)\left(\frac{\partial h}{\partial z} + 1\right)$, where h is the water pressure head (cm) and $K(h)$ (cm/s) is the soil hydraulic conductivity function. The solution velocity v (cm/s) equals to $v = q/\theta$. As dissociation reactions are practically instantaneous, all the P_i species can be expressed in terms of $H_2PO_4^-$. Equations 5.2.4–5.2.6 must be solved for soluble P_o species too, only in this case U = 0. To solve equations 5.2.5 and 5.2.6, the spatial and temporal q, θ and pH must be known. The latter is a prerequisite for P speciation and pH-dependent adsorption calculations (equation 5.2.2). Consequently, at each time step the solution of equation 5.2.4 must be preceded by solving the Richards equation (equation 5.2.7) and prediction of pH(z,t).

$$\frac{\partial \theta}{\partial t} = -\frac{\partial q}{\partial z} - W_u \qquad (5.2.7)$$

The W_u term describes the spatial water uptake rate by plant roots (l/l soil/s), which depends on root length and h distributions in the soil. Further discussion of the simultaneous solution of equations 5.24–5.27 and their parameterization is outside the scope of this chapter. An example of a model incorporating equations 5.2.4–5.2.7 was published by Grant and Heaney (1997). A simpler approach to simulate P reactions and transport in soil was suggested by Schoumans and Groenendijk (2000). Their model requires fewer parameters, particularly in the simulation of water flow, but as in the case of the Grant and Heaney model it has not been sufficiently validated and evaluated. The partial confidence in P transport model predictions under field conditions emphasizes the need to have solid field data in order to estimate P leaching under diverse soil, treated wastewater composition, crop, and environmental conditions.

5.2.3.3.2 Transport in agricultural runoff

Phosphorus transport in runoff depends on two processes: (i) inducing runoff and erosion (e.g. Yu, 2003); and (ii) releasing P from soil to runoff solution (Knisel, 1980; Ahuja et al., 1982; Sharpley and Smith, 1989; Sharpley et al., 2001; Pote et al., 1999; Ebeling et al., 2002; Condron et al., 2005). Several models were developed to simulate process (i) under steady rainfall and sloping soils (e.g. Wallach and van Genuchten, 1990), but they did not account for P flow. Basin scale runoff, erosion and solute transport models (Hopstaken and Ruijgh, 1994; Cassel, 1998) also disregard P transport. Runoff and erosion processes were incorporated in a model that assesses P ecological risks (Hession et al., 1996) and in a model that considers economic implications of rising P concentration in runoff (Govindasamy et al., 1994). The risk assessment model allows simultaneous analysis of the impact of factors affecting runoff induction and P release to runoff, on P surface transport.

All current runoff models disregard biological activity and chemical reactions that affect soil hydraulic properties via sealing and clay dispersion mechanisms (Shainberg and Levy, 1992; Shainberg et al., 1997). The tedious parameterization of numerical runoff models and their inadequate treatment of chemical and biological reactions explain the fact that P transport in agricultural runoff is currently estimated by empirical rather than numerical models. For example, Sharpley and Smith (1989) assumed, based on empirical data, that soluble phosphorus concentration in runoff (C_{Pr}) can be predicted from extractable P concentration and the effective depth of the soil flow zone.

$$P_r = K_{pr} \cdot P_s \cdot E \cdot \varrho \cdot t^{\alpha} \cdot W_p^{\beta} / Vp \tag{5.2.8}$$

Here P_r is the average C_{Pr} in runoff per individual event (mg/l), P_s the extractable P content (in this case Bray, mg/kg) of surface soil (0–50 mm) before each runoff event, E the effective flow zone depth (mm), ϱ the soil bulk density (kg/l), t the duration of the runoff event (min), W_p the runoff water:soil (suspended sediment) ratio, V_p the total runoff volume per event (mm), and K_{pr}, α, and β soil constants. The range of values of these constants in the cited study were 0.05–0.54, 0.12–0.17, and 0.3–0.62, respectively. To separate between soluble and particulate P, runoff was filtered (0.45 μm) and soluble P_i determined in the filtrate. The particulate P concentration, which is also regarded as bioavailable, is determined by NaOH extraction. E is a function of runoff volume, rain intensity, and impact energy. As this function is usually unavailable, it was replaced by measured soil loss (kg/ha) (Sharpley and Smith, 1989):

$$ln\,(E) = i + 0.576\,ln(soil\ loss) \tag{5.2.9}$$

Here i relates to soil texture and ranges between –1.2 and –2.5. If soil loss is unknown, a constant E value of 0.3 mm is used. According to Heathwaite et al. (2005), silt loam soils support a higher runoff P concentration than finer-textured loams and clay loam for any given surface soil P concentration, which is compatible with the order of P retention by these soils. They also report that manure application influences the potential erosion and runoff of soils. When broadcasted, manure temporarily reduced erosion from tilled soils by providing a protective ground cover and by enhancing the formation of stable soil aggregates. The improved soil aggregation and aggregate stability may also explain the beneficial effect of long-term manure application on enhanced soil water-holding capacity and infiltration rate (Heathwaite et al., 2005).

5.2.3.3.3 Eutrophication threshold P concentrations in water

Excess phosphorus and other nutrients in the soil-root volume have a major effect on nutrient accumulation in surface and underground water bodies. The nutrients induce autotroph proliferation, especially algae and cyanobacteria, which are followed by increasing bacterial population, enhanced respiration rate, declined dissolved oxygen concentration, and subsequently loss of aquatic animals and release of toxic materials into the water. Phosphorus is the main growth-limiting factor in autotroph development, because, unlike N and C, it cannot be acquired from the atmosphere and must reach the water body by transport. Orthophosphate is the only P form that autotrophs can assimilate, but P_o must also be considered as it is hydrolyzed by extracellular enzymes.

Considerable progress has been made in understanding and predicting eutrophication processes (Thomann and Linker, 1998). Available steady-state models (e.g. Moustafa, 1998) can simulate algal biomass production as a function of total P input/outflow per unit lake area. Transient state models simulate algae growth as a function of P_i uptake, algae decomposition, and mineralization, P_i sorption by suspended particles and water reservoir sediments, and P diffusive transport in the system (Clement et al., 1998). More detailed eutrophication models account for organic P and N transformations as well (DUFLOW, 1992). The body of experimental and simulated results indicate that soluble P concentration between 0.01 and 0.02 mg P/l in surface water are critical values above which eutrophication is accelerated (Daniel et al., 1998).

In U.S. eutrophication-inflicted areas, P application to soils is strictly regulated. Allowed P addition is based on soil test and permitted (threshold) P concentration in soil, the latter assigned in relation to soil type and cover crop. The soil test approach is cheap, simple, and fast, but requires reliable calibration versus eutrophication (Maguire et al., 2005). Another approach regulating P application in US is the "P Index" (Sharpley et al., 2001). The method is based on estimating the impact of P source (level in soil and form and quantity added) and transport pathway (rain erosion and runoff; irrigation erosion) on annual P loss risks using a tabulated questionnaire (for more details see Heathwaite et al., 2005). The "P Index" method is purely empirical and can be adjusted for different states in the USA by choosing a state-specific weight factor for each variable. In the UK, the P loss risk is assessed by a decision support system called the "Nutrient Export Risk Matrix" (NERM) (MAFF, 2000). It encompasses a three-dimensional matrix. The first axis describes the current P level in soil, expected P application rate and forms, crop P consumption, and tillage and husbandry management, all integrated to give a single estimate of how much of the P is prone to direct mobilization by runoff, drainage or subsurface horizontal flow. The second axis incorporates soil properties and reflects the soil propensity under given conditions to lose P by runoff, erosion, and leaching. The third dimension (axis) describes flow connectivity and is used to assess the surface topography, cultivation tracks, and buffer strips. In its dynamic watershed scale version, the NERM utilizes a simplified watershed model calculating surface and subsurface flows in the watershed. The NERM estimates the variations in P export from land receiving different rates of biosolids and/or fertilizer P, in relation to soil and topography properties. More details on P loss models at the basin scale can be found in Heathwaite et al. (2005).

5.2.4 Treated wastewater irrigation

5.2.4.1 Phosphorus application versus consumption rate

Under all regulations of treated wastewater use for irrigation, P application via wastewater should not exceed P consumption by the crop. To comply, temporal P consumption by irrigated field and orchard crops must be known (Table 5.2.2). Phosphorus uptake by orchards was meagerly reported in the literature, therefore data on citrus only are presented. The data for field crops (Table 5.2.2) allow estimation of weekly permitted

Table 5.2.2 Daily phosphorus consumption (kg/ha/d) by field crops allowed to be irrigated by secondary effluents (drip irrigation) as a function of time after emergence or planting

Part 1: vegetables

Days after emergence or planting	Lettuce	Celery	Chinese cabbage	Broccoli	Sweetcorn	Carrot	Musk-melon
1–10	0.01	0.03	0.10	0.00	0.10	0.06	0.03
11–20	0.10	0.04	0.16	0.01	0.15	0.16	0.03
21–30	0.50	0.11	0.31	0.12	0.20	0.12	0.07
31–40	0.60	0.08	0.51	0.13	0.55	0.12	0.18
41–50	0.55	0.20	0.87	0.20	0.85	0.19	0.25
51–60	0.45	0.23	0.81	0.13	1.15	0.20	0.25
61–70	—	0.35	0.45	0.36	0.80	0.29	0.35
71–80	—	0.29	0.28	0.46	0.20	0.27	0.45
81–90	—	0.39	00.28	0.38	—	0.27	0.43
91–100	—	0.17	—	0.32	—	0.24	0.27
101–110	—	0.18	—	0.18	—	0.30	0.13
111–120	—	0.30	—	0.09	—	0.59	0.07
121–130	—	0.54	—	0.09	—	0.58	—
131–140	—	0.69	—	0.04	—	0.91	—
141–150	—	—	—	0.01	—	1.32	—
151–160	—	—	—	—	—	0.88	—
161–170	—	—	—	—	—	0.81	—
Total (kg P/ha)	22	36	29	26	40	73	25

Derived from Bar-Yosef (1999).

Part 2: Additional vegetables and cotton (kg P/ha/d) and citrus (kg P/ha/month)

Days after emergence or planting	Processing tomato	Fresh tomato	Bell pepper	Eggplant	Cotton	Citrus (Shamouti)[a] Month	
1–10	0.02	0.01	0.01	0.01	0.01	Jan	0
11–20	0.05	0.02	0.10	0.01	0.04	Feb	0
21–30	0.16	0.03	0.18	0.01	0.05	Mar	0.35
31–40	0.19	0.03	0.28	0.01	0.05	Apr	0.70
41–50	0.75	0.03	0.33	0.02	0.10	May	1.40
51–60	0.80	0.04	0.28	0.08	0.10	June	3.15
61–70	1.80	0.04	0.75	0.09	0.30	July	4.55
71–80	0.50	0.18	0.28	0.05	0.40	Aug	4.20
81–90	0.50	0.22	0.43	0.05	0.90	Sep	3.50
91–100	0.89	0.10	0.44	0.05	0.50	Oct	3.15
101–110	—	0.30	0.23	0.09	0.55	Nov	2.80
111–120	—	0.60	0.18	0.15	0.85	Dec	0.70
121–130	—	0.45	0.10	0.27	0.40	Total (kg/ha)	27.6
131–150	—	0.17	0.30	0.31	0.08		
151–180	—	—	—	0.38	—		
181–220	—	—	—	0.35	—		
Total (kg P/ha)	59	24	37	33	44		

[a]Based on evaluated monthly N consumption (Legaz and Primo-Millo, 1981, 1984, 1988; Kato, 1986) and the ratio %N/%P in plant = 14 (which is the ratio in leaves according to Bielorai et al., 1985).

P application rates by effluents. For example, under an irrigation dose of 5 mm/day and 3.5 mg P_i/l, the daily P supply (175 g/ha) is smaller than consumption during most of the growing season of all vegetable and field crops, but larger than the daily P consumption by citrus. This conclusion is valid for the seasonal P demand too (Table 5.2.2), assuming effluent application of 500 mm per season.

5.2.4.2 Phosphorus losses from wastewater-irrigated soils

5.2.4.2.1 General considerations

Runoff and leaching, the main pathways of P losses in soil, are controlled by different physical and chemical mechanisms (Hedley and McLaughlin, 2005). In the case of leaching, the P sorption capacity, soil hydraulic conductivity, amount and continuity of soil macropores, and organic P form and method of application (solution or solid) are the predominant factors. In the case of runoff, the P retention by soil, particulate P transport, and effects of effluent organic matter on soil hydraulic properties are the influential factors. Organic matter components increase water infiltration and macroporosity (Hedley and McLaughlin, 2005), thereby reducing runoff from soils. Of note, macroporosity effects on leaching cannot be expressed in shallow box runoff experiments, therefore laboratory-scale rain simulators tend to overestimate P runoff in biosolid and wastewater-treated soils. Kleinman et al. (2004) estimated that using shallow runoff boxes causes a ~threefold reduction in infiltration, 1.5-fold increase in runoff and ~fourfold increase in erosion in comparison with field runoff plots.

5.2.4.2.2 Laboratory-Scale runoff observations

Laboratory runoff experiments showed that runoff P_i concentration (RP, mg/l) can be predicted from Mehlich-3 extractable P concentration in soil (MP, mg/kg) according to equation 5.1.10 (Kleinman et al., 2004):

$$\log(RP + 1) = 0.009 + 0.0005\,MP\ (r^2 = 0.88,\ MP\ \text{range}\ 13\text{–}400\ \text{mg/kg P}) \quad (5.2.10)$$

The slope of the equation was valid for both acidic and alkaline soils, and it did not differ significantly from slopes obtained in field runoff experiments. Fang et al. (2002) studied bioavailable P in shallow box rain simulator runoff. They found that soluble reactive P in runoff of Minnesota River basin calcareous soils was significantly correlated with soil Mehlich-III, Olsen, water-extractable, and P saturation index P. The critical levels of soil Mehlich-III and Olsen P that induced stream eutrophication were 65–85 and 40–55 mg/kg P, respectively.

Penn et al. (2005) determined the effect of soil mineralogy on P_i sorption and runoff losses in rain simulator. The P retention by soil was strongly correlated with Al-bearing minerals, such as hydroxy-interlayered-vermiculite, gibbsite, and amorphous Al, and inversely related to P runoff losses. The P retention was found to be negatively correlated with the kaolinite content in soil.

5.2.4.2.3 Laboratory-Scale leaching observations

Mamo et al. (2005) assessed the effect of short- and long-term land application of potato-processing wastewater on P leaching from a sandy loam soil. The experiment was carried out in soil columns 0.4×0.16 m height and diameter, respectively, with initial P concentrations of 200 and <25 mg/kg P, representing long- and short-term exposure to wastewater irrigation. When inflow wastewater of low-P concentration (<1 mg/l) was added to the high-P soil, the soil acted as a P source, and the total phosphorus (TP) concentration in the leachate was 3.5 times higher than the input TP concentration. When 6 mg/l P wastewater was applied to the high-P soil, the soil acted as a P sink retarding the TP concentration in the leachate by 80%. This indicates that, depending on P levels of the soil and wastewater, reduction or increase in leachate P will occur below the surface soil. Mamo et al. (2005) found that P desorption was greater at 10°C than at 4°C, and Br tracer proved that no preferential flow in the soil columns that could affected P leaching occurred.

Elliott et al. (2002) studied P leaching in water-saturated sandy soil columns (0.15×0.4 m) differing in P sorption capacity. The soils were amended with two P levels (56 and 224 kg P/ha) added as triple superphosphate (TSP), chicken manure, or secondary sludge. The sludge treatment mimics effluent irrigation effects on P leaching because both secondary effluents and solution flowing through soil-sludge are in quasi-equilibrium with the same Ca-P minerals (see "precipitation and dissolution" above). After four leachings with KCl (~25% soil pore volume each), total P leached, expressed as percentage of the soil maximum P sorption capacity, was 22% in the high-P, TSP treatment, ~3% in the high P, manure chicken treatment, and <0.5% in the sludge treatment. The inorganic fraction of leached P in the sludge and manure treatments was ~65% ± 10%. Another observation was that leachate P concentration was roughly correlated with the KCl extractable P of the amending materials.

5.2.4.2.4 Field observations

In contrast with the soil column experiment described above, there is growing evidence that under field conditions P applied to soil in wastewater or biosolid is prone to deeper leaching than P added as mineral fertilizer and leached by freshwater. For example, Adriano et al. (1975) reported a P downward movement to a depth of 6.6 m in sandy soils after long-term spray application of food-processing wastewater. Even though no comparison with P_i was presented, the reported leaching depth exceeded the expected leaching of fertilizer P. Sommers et al. (1979) and King et al. (1990) reported a significant increase in total P concentration in the 0.30–0.60 m soil layer in a sandy loam soil after 12 yrs of irrigation with municipal wastewater, and P movement to depth of 0.75 m in loamy sand after 11 yrs of swine lagoon effluent application. In these studies, leaching of fertilizer P was not determined either, but its movement depth is expected to be considerably shallower.

Sims et al. (1998) found that in manure-amended soils P_o species moved deeper than P_i species, which may partly support the above observations. In P fertilized soils, however, the concentration of soluble P_o in drainage water was shown to be lower than P_i. For example, in Rothamsted arable soils the soluble P_o constituted <5% of the total P in tile drainage and in grassland ~15% (Condron et al., 2005). In soils irrigated with secondary effluents and receiving 330 kg P/ha in 2 yrs, the P_o fraction in drainage was ~45% (Barton et al., 2005). Using strong anion exchange resins, Espinosa et al. (1999) detected in filtered leachate of silt clay soil traces of IHP, organic polyphosphate, and sugar phosphates. Condron et al. (2005) stated that the fraction of dissolved P_o in runoff and leachate is usually smaller than in the particulate P (>0.45 μm) (~20:80%), indicating stronger P_o than P_i retention by the solution particles.

Woodard et al. (2007) studied effects of 3 yrs of dairy effluent irrigation on P movement in the upper 1.5 m soil layer of a deep, well-drained sand. Effluent doses providing 70, 110, and 165 kg P/ha/yr to forage crops resulted in time averaged leachate P concentration ≤0.1 mg/l below the rooting zone (1.5 m). Mehlich-1-extractable P, Al and Ca concentrations in the 25 cm topsoil increased with time, but no change in concentrations in the 25–122 cm soil layer were observed. Topsoil Ca increased with dairy effluent rate due to high Ca levels in the effluents (~305 mg/l) and supplemental irrigation water (~145 mg/l); Al concentration increased due to its presence in dairy effluent (~31 mg/l). The enhanced Ca and Al concentrations increased P retention in the topsoil. The effect of effluent quantity on P and Al concentrations in the topsoil was negligible due to large initial concentrations in soil. The P saturation ratio ((Mehlich P)/(oxalate-extractable Al+ Fe)) increased with time in the topsoil, but not in subsoil layers because oxalate-extractable Al concentration in the subsoil was very high, 119–229 mg/kg Al. In the Netherlands, soils with saturation index >25% were defined by Breeuwsma et al. (1995) as prone to groundwater P contamination, whereas in Florida sandy soils the threshold ratio was estimated as 30% (Nair et al., 2004).

Johnson et al. (2004) studied the fate of four annual P application rates added as dairy wastewater (0, 47, 94, 188 kg P/ha) or poultry litter (0, 147, 295, 590 kg P/ha) to loamy sand soil grassland over a 4-yr period. The P recovery by the forage, and Bray-1 extractable P (BEP) were monitored twice annually as a function of depths (down to 180 cm). Increasing annual P rates increased P removal by forage. At the highest P rate of each amendment, less than 13% of the applied P was found in harvested plants. Compared with controls, all annual P rates in the dairy wastewater had negligible effects on BEP concentrations at depths below 30 cm. In contrast, the highest annual P rate (590 kg/ha) in poultry litter increased BEP concentration in comparison with the control at depth intervals down to 120 cm (Fig. 5.2.4). Increases in BEP concentration below 30 cm were interpreted as indicating excessive P rates contributing to non-point pollution in subsurface drainage.

Comparing the 188 kg P/ha applied as dairy wastewater (230 mg P/l wastewater) and the 147 kg P/ha applied as poultry litter can indicate differences in P leaching between these two organic matter sources (Fig. 5.2.4). In dairy wastewater the extractable P concentration in the 0–30 cm soil layer was higher than in poultry litter (80 versus ~40 mg/kg, insignificant difference), whereas in deeper soil layers the concentrations were somewhat lower. The deeper poultry P movement could stem from a higher P_i/P_o ratio, but unfortunately this ratio was not reported in the paper. In both amendments and all P application rates P leached to a depth of 150 cm (Fig. 5.2.4), compatible with previous data.

Tarkalson and Mikkelsen (2004) studied effects of P source (P_i versus broiler litter) and application rate on runoff P losses. As P level increased, P concentration in runoff was consistently higher in soils amended by broiler litter than by P_i fertilizer. Penn and

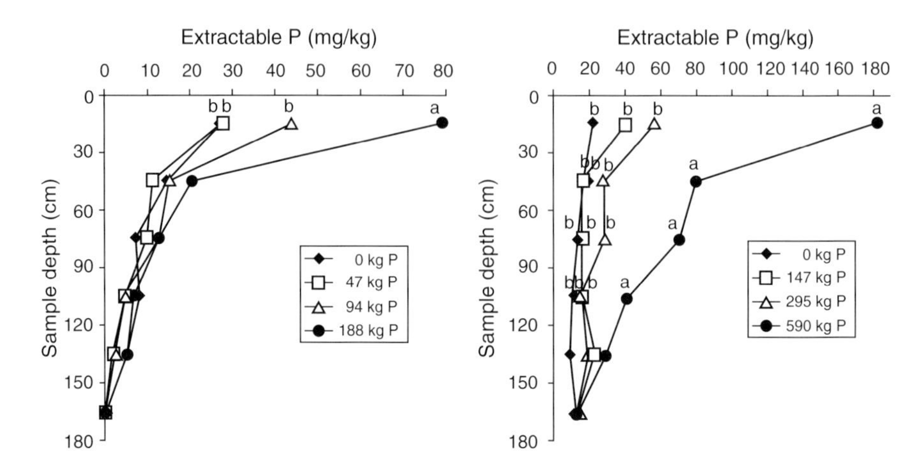

Figure 5.2.4 Effect of four levels of P added over 4 yrs as dairy wastewater (left hand side) or poultry litter (right hand side) to Florida loamy sand soil on Bray 1 extractable P concentration as a function of depth. Reproduced from Johnson et al. (2004), with permission.

Sims (2002) showed that biosolid application increased dissolved and total P concentration in runoff; the same effect is expected when secondary effluent irrigation takes place.

The presented field-scale data prove that assessing potential runoff and drainage P losses from biosolid or wastewater-treated soils requires integrative approach. The unique effect of some factors, e.g. P_i and P_o adsorption by different soils and P level in soil are known and were treated quantitatively in the review section. The impact of other factors, ostensibly organic matter soil amendment effect on P mobility is still poorly understood and needs more research. The possible role of organic matter in preferential P flow is discussed below.

5.2.4.2.5 Preferential P flow in effluent irrigated soils

Barton et al. (2005) studied P leaching in well and slowly draining soils irrigated with secondary effluents. The leaching depth in the slowly draining soil was greater. The mechanism explaining this result was that flow was locally forced to macropores (see later) and soil cracks, whereas in well-drained soil matrix flow dominated. Macropore flow decreased the P contact time with soil and the soil surface area available for adsorption, thus decreasing P depletion from solution relative to well-drained soil. Kleinman et al. (2003) used the preferential macropore flow mechanism to explain deep P transport (to rain tiles) in manured soil; they also reported that macropore clay films along the soil profile had a higher concentration of sorbed P in comparison with bulk soil clay films. Bergen Jensen et al. (2000) found rapid P leaching in water-saturated soil columns (50 cm diameter, 50 cm deep) packed with undisturbed structured clayey soil amended with cattle feces, and obtained a breakthrough P concentration of 12 mg/l P at about 0.15 pore volume of the soil column. The continuous macropores were attributed to earthworm burrows. Dissolved P_i and P_o accounted for 37% and 8% of the leached P; particulate P_i and P_o constituted 22% and 33% of the leached P, respectively. An alternative mechanism to explain preferential P flow in soil is pore coating by hydrophobic material found in effluents or biosolids (Dr Ellen Graber, Agricultural Research Organization, Israel, personal communication). The coating decreases the soil surface area available for reaction, and hence P adsorption, and the hydrophobicity is more effective in reducing water flow in micropores than in macropores. The reduced micropore flow induces greater water transport via macropores, and thus water movement in soil is accelerated. The overall effect is that in effluent- or biosolid-treated soils water flow is faster than in freshwater-irrigated soils, P adsorption is reduced, and P leaching is enhanced.

5.2.5 Case studies

Case studies showing the fate of P in soil under effluent versus freshwater irrigation should satisfy at least two prerequisites: (i) the duration of effluent application should be long enough to warrant measurable changes in P status in soil; and (ii) the studied soils should differ in physical and chemical properties to ensure wide-range P sorption and transport characteristics.

5.2.5.1 Long-term studies with a single soil

Wang et al. (2003) investigated 29 physical, chemical, and biological soil characteristics (one of them P accumulation in soil) obtained in a 2000 ha Bakersfield (CA) field after 70 yrs of irrigation with reclaimed wastewater. The characteristics were compared with those obtained in an adjacent field irrigated with freshwater and grown with the same crops. The wastewater contained 10–20, 5–15, and 5–10 mg/l of BOD, total nitrogen, and total phosphorus, respectively. On average, 1.4 m of water was applied annually. The soil pH was 7.4 and the clay content 14%. The parameters that were found to be most suitable for evaluating effluent irrigation effects on soil health were: total porosity (TPOR), pH, EC, and total Mg, P, and Zn concentration in soil. However, only TPOR and Mg were significantly different in the effluent- and freshwater-irrigated fields, the values in the freshwater treatment being higher. The reduced TPOR under effluent irrigation is compatible with lab-scale results reviewed earlier. The lack of significant differences in P, Zn, EC, and pH may stem from the fact that soils were sampled in the top 0–15 cm layer only, whereas main effects plausibly occurred deeper in the soil profile.

In another case study carried out in a semi-arid zone in Turkey, the macro- and micronutrient distributions in the profile of an alluvial soil (9% clay, 25% silt, 66% sand) were investigated. The site was irrigated for 30 yrs with secondary effluents (pH 8.3, EC = 1.7 dS/m, P = 3 mg/l) or freshwater (Angin et al., 2005). The wastewater reduced soil pH from 7.7 in freshwater to 7.35 (mean 0–90 cm), increased % organic matter in the 0–30 cm soil layer from 2% to 3.2%, and increased available P concentration in the 0–30, 30–60 and 60–90 cm soil layers from 25 to 42, 28 to 38 and 38 to 46 mg/kg, respectively, once again proving deeper P leaching under effluent than freshwater irrigation.

In the last case study Rusan et al. (2007) investigated nutrient accumulation and transport in soil after 0, 2, 5, and 10 yrs of effluent irrigation (pH 7.3, P 15.5 mg/l, NO_3-N 30 mg/l) in fine-textured calcareous soil (pH 7.8, CEC 32 cmol/kg). The pH profile in soil was unaffected by effluents even after 10 yrs of irrigation. The percent organic matter in the top 0–20 cm soil layer rose from 0.7% to 1.2%, and bicarbonate extractable P increased in the 0–20, 20–40 and 40–60 cm soil layers from 15 to 42, 6 to 34 and 5 to 32 mg P/kg, respectively, indicating again deep P leaching.

5.2.5.2 Long term wastewater irrigation survey

A long-term effluent irrigation survey was conducted in Israel between 1998 and 2006 (Tartchichky et al., 2002, 2004, 2006). It included fields of known history and 2–25 yrs of continuous irrigation with effluent or freshwater. The fields were 116 commercial plots differing in soil physical and chemical properties and climatic conditions (130–650 mm/yr rainfall and 1350–2000 mm/yr class A pan evaporation). The main crop was grapefruit and data pertinent to this crop only will be presented. Soils were sampled twice a year (end of winter (April) and end of irrigation period (September)) at six points between adjacent trees in the row to depths of 0–30, 30–60, 60–90 and 90–120 cm. The two nearest samples were combined so that each field plot comprised three replicates. The soil samples were analyzed for EC, pH, and Ca+ Mg concentrations in saturated paste extract; mineral N was extracted by K_2SO_4, and P by $NaHCO_3$ 0.5 M (Olsen-P). In the autumn of 2002 P_i was also determined in the soil-saturated paste (C_{ps}). In the first (1998) sampling date soils were analyzed for % clay, % organic matter, % carbonates, and soil cation exchange capacity (CEC). Citrus leaves were sampled in October, and analyzed for acid- and water-soluble nutrients. Irrigation water was sampled in each field four times a year. The effluent P concentration varied with time between 3.1 and 4.3 mg/l P_i (time average 4.1) and between 4.4 and 5.4 mg/l P_t (time average 5.2). Organic P was estimated as P_t-P_i. In the years 1998/9 the freshwater was enriched by P fertilizer to give 2 mg/l P_i; later on no P was added. The irrigation rate during the survey period was 700 mm/yr $\pm$ 15%.

5.2.5.2.1 Factors affecting soil solution P

When evaluating effluent versus freshwater effects on soil P, the soil solution concentration (C_{pl}, approximated by C_{ps}) is as important as Olsen P. The reason is that C_{pl} is a better indicator of P mobility in soil, and together with solution pH and Ca^{2+} concentration it allows assessment of which solid phase controls its concentration in solution. However, Olsen P is determined at much higher frequency than C_{ps}, therefore a functional relationship between the two might be helpful in evaluating P status in soil. To obtain a significant relationship the survey field plots had to be sorted by soil clay content (% clay). For each % clay class a Langmuir-like Olsen P versus C_{ps} isotherm could be fitted (Fig. 5.2.5). The obtained affinity term (K_{ol}) was very similar in all % clay groups except 1–10% where it was twofold smaller than the rest. The maximum sorption (B_{ol}) increased from 60 to 220 mg P/kg as % clay increased from 1–10 to 75% (Fig. 5.2.5), and was smaller at any % clay>10% than b in the Langmuir model (Fig. 5.2.1). The difference between B_{ol} and b is consistent with the P REC (= Olsen P/Adsorbed P) concept discussed above (Fig. 5.2.2).

The effect of effluents versus freshwater on C_{ps} was assessed by correlating them with several soil and water parameters and comparing the regressions. The tested parameters were Olsen-P, % clay, CEC, carbonate and OM concentration in soil (%), saturated paste ion concentrations (all in the top 0–30 cm soil layer), and irrigation water pH and ion concentrations. Stepwise regressions were computed for 77 effluent- and 34 freshwater-irrigated plots. Spring and autumn samples were treated as separate parameters. The regressions having highest R^2 are summarized in Table 5.2.3.

Under freshwater irrigation the main factor affecting C_{ps} was Olsen P (0.33 out of R^2 of 0.49; pH range 6.9–8.3), followed by % OM and saturated paste Ca concentration (C_{Cas}).

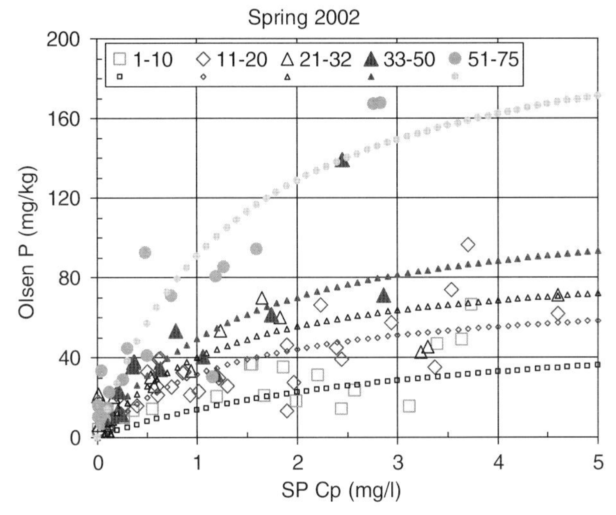

Parameter	% clay group (and number of fields)				
	1-10 (n=16)	11-20 (n=35)	21-32 (n=16)	33-50 (n=13)	51-75 (n=16)
K_{ol} ± s.e.	0.3 ± 0.05	0.74 ± 0.35	0.80 ± 0.65	0.70 ± 0.55	0.72 ± 0.54
B_{ol} ± s.e.	60 ± 28	74 ± 26	90 ± 21	120 ± 72	220 ± 122
F, Pr>F	63, <0.0001	65, <0.0001	32, <0.0001	38, <0.0001	109, <.0001

K_{ol} = -0.0004 * %clay^2 + 0.0298 * %clay + 0.257 (R^2 = 0.68)
B_{ol} = 52.738 * Exp(0.0218 * %clay) (R^2 = 0.99)
Note the equations % clay is given as fraction.

Figure 5.2.5 Relationship between Olsen ($NaHCO_3$ extractable) P (mg/kg P) and P_i concentration in saturated paste (SP C_p, mg/l) (0–30 cm soil layer) in spring of 2002. Eighty percent of the samples were effluent-irrigated and 20% freshwater-irrigated. Data points (large symbols) were sorted by % clay in soil (1–10, 11–20, 21–32, 33–50, 51–75%); each group was fitted by a Langmuir-like equation (Olsen P = K_{ol} B_{ol} Cp/(1+ K_{Ol} C_p), small symbol lines). The estimated parameters, their standard error (s.e.), and model F value and significance level (Pr > F) are summarized in the table above. The estimated K_{ol} and B_{ol} were correlated versus % clay (employing the mean % clay in each group); the regression are presented at the bottom of the table.

Table 5.2.3 Stepwise regression between P concentration in saturated paste extract (C_{ps}, mg/l P) and system variables (year 2002)[a]

1 C_{ps}	= **3.932**	− **0.0432**	* **pH^2**	+ **0.0218**	* **Olsen P**	− **0.0285**	* **% clay**	**R^2**
±sd	1.377	0.0243		0.0044		0.0068		0.425
F	8.2**	3(p = 0.08)		25***		18***		
Part. R^2			0.19		0.096		0.137	
2 C_{ps}	= **0.774**	− **0.0288**	* **C_{Ca}**	+ **0.0212**	* **Olsen P**	− **0.545**	* **%OM**	**R^2**
±sd	0.255	0.0172		0.0048		0.23		0.485
F	9.2**	3(p = 0.10)		20***		5.6*		
Part. R^2			0.048		0.331		0.107	

[a]pH and Olsen P (mg/kg soil) are spring samples; C_{Ca} (mg/l) is concentration in saturated paste extract in autumn.
*, **, *** P = 0.05, 0.01 and 0.001, respectively.
Estimated parameters are accompanied by standard deviations (±sd) and significance level (F and Pr > F). Case 1 accounts for effluent-irrigated plots (n = 77) and case 2 for freshwater-irrigated plots (n = 34). No other variable (see text) met the 0.15 significance level for entry into presented models.

The pH did not enter the freshwater regression. The Ca result indicates that Ca-P minerals played a role in determining C_{ps} concomitantly with adsorption. Under effluent irrigation the factor affecting most C_{ps} was pH^2 (0.19 out of 0.43) followed by % clay and Olsen P. Ca concentration and % OM did not enter the effluent irrigation regression; probably because the organic matter added via effluents masked their effect on C_{ps}. The fact that OM reduces the precipitation rate of Ca-P minerals was discussed in the "precipitation – dissolution" section above. Notably the R^2 of the effluent regression was lower than the freshwater regression, indicating higher variability. The extra variability could stem from the uneven number of years plots were treated by effluents, and from preferential water and P flow induced by effluents' hydrophobic substances (see "preferential P flow" above). The low overall R^2 of both regressions stemmed from unaccounted for climate, management, and culturally related variability that was identical in the two studied systems.

5.2.5.2.2 *P leaching profiles*

Under effluent drip irrigation, P accumulation in the top 0–120 cm soil layer exceeded the accumulation under freshwater drip irrigation, and the P movement in the soil profile was deeper (compare Fig. 5.2.6a and Fig. 5.2.6b). This trend was masked by large experimental variability discussed above. To reduce variability and enhance insight, one plot in each % clay group (the one receiving effluents for the longest period of time) was selected for repeating the Olsen P profiles comparison. In this case the concentration profiles were more consistent, and deeper P penetration could be observed as % clay decreased (Fig. 5.2.6c). In the top 0–30 cm soil layer, Olsen P concentration, under effluent irrigation increased with increasing % clay, even though irrigation water C_p was nearly identical in all fields. This stemmed from the fact that as % clay increased, P sorption in equilibrium with C_p also increased. Results in Figure 5.2.5 show that Olsen P concentrations in equilibrium with 3.5 mg P_i/l (effluent concentration in most years) in the 10, 20, 32, 50, and 70% clay classes were 33, 52, 65, 83, and 155 mg P/kg, respectively, which is comparable with the 0–30 cm data in Figure 5.2.6a, and Figure 5.2.6c. The soils in the 1–10 and 11–20% clay groups showed Olsen P concentrations >30 mg/kg at a depth of 120 cm, whereas, in % clay classes >50%, P did not move below the 30–60 cm soil layer (Fig. 5.2.6c).

To evaluate more closely the potential P leaching to underground water, deeper soil samples (to 240 cm) were taken in representative plots in the year 2002 (Fig. 5.2.7). In 1–10% clay soils that were irrigated with effluents applied through microjet sprinklers (JE10%, Fig. 5.2.7a), P reached a depth of 240 cm, and the concentration peak was found between 90 and 120 cm. In jet-irrigated freshwater and 1–10% clay soils (JF10%, Fig. 5.2.7a), P did not move below 1.2 m. In 11–20% clay soils, P penetrated to a depth of ~60 cm. Comparison between drip-irrigated effluents (DE10%, Fig. 5.2.6c,) and jet-irrigated effluents (JE10%, Fig. 5.2.7a) revealed that the Olsen P concentration profiles in 1–10% clay soils were similar. In 11–20% clay soils, P leached deeper under drip- than under jet-irrigation (DE20%, Fig. 5.2.6c versus JE20%, Fig. 5.2.7a). This result is expected in view of the smaller irrigated area under drip-irrigation, and the same P quantity per m^2 field in both cases.

The P concentration versus soil depth results can be approximated by a step function having a vertical constant concentration that changes to 0 at the P concentration front in

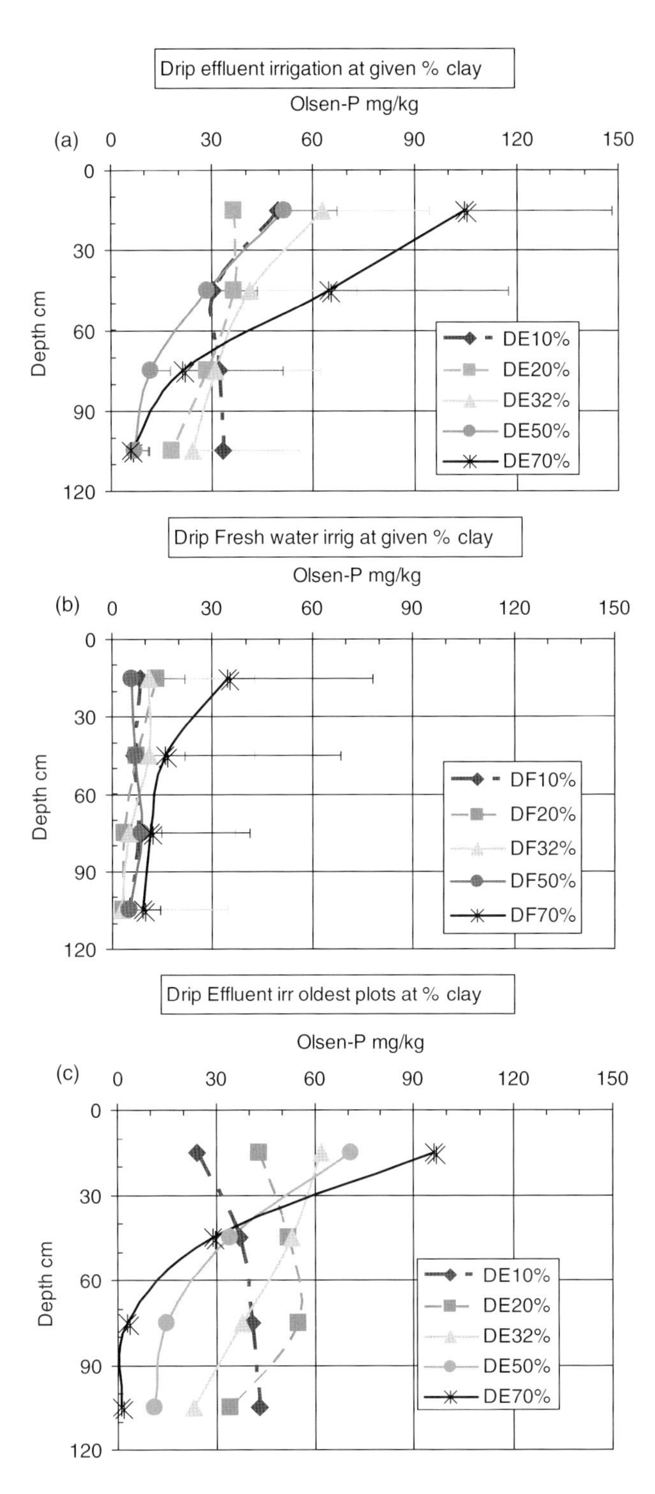

Figure 5.2.6 Olsen P concentration profiles under increasing soil clay contents (10–70%). (a) Drip effluent irrigation (DE); spring 2005 sampling. Number of plots per % clay group=5 ± 2; soils differed in effluent application period (8–23 years). Bars (one sided) show one standard deviation from the mean. (b) As for (a), only for freshwater (DF) instead of effluents. (c) As for (a), only for a single field plot in each % clay group that received effluents for the longest period of time (18–23 years). Recalculated from Tartchichky et al. (2002).

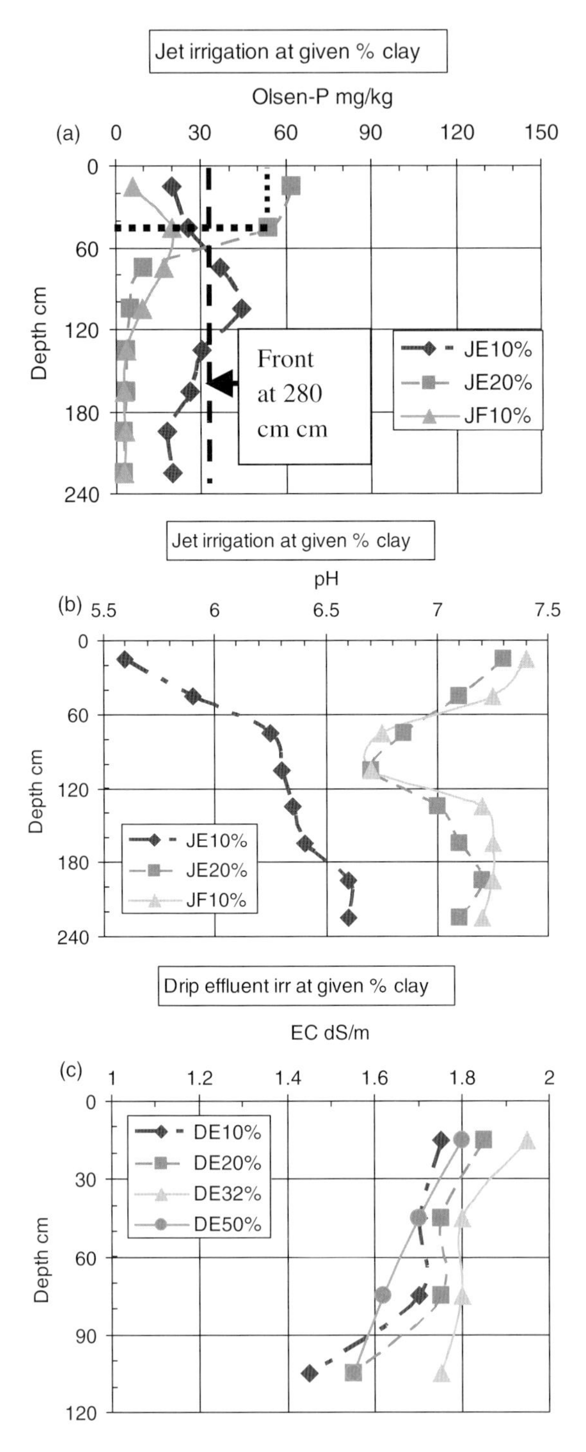

Figure 5.2.7 (a) Olsen P profile to depth of 240 cm in three treatments (no replicates): JE10% and JE20% (jet effluent irrigation in 1–10% and 10–20% clay soils), and JF10% (jet freshwater irrigation, 1–10% clay soil). Black dashed lines are the predicted step functions (see text) for treatments JE10% and JE20%. (b) pH profile in same soils. (c) EC profiles (to depth of 120 cm) under drip effluent-irrigation in soils differing in % clay (1–10, 11–20, 21–32, 33–50%). Each treatment consisted of 5 ± 2 soils. Recalculated from Tartchichky et al. (2002).

soil ("piston P movement"). The step function can be calculated when the irrigation solution has a constant P concentration (C_P, mg/l), and Langmuir adsorption (A, mg/kg P) at equilibrium with C_P is the only sink for P in the system:

$$AS\varrho h = VCp \tag{5.2.11}$$

Here, S is the irrigated soil area (m^2), ϱ is soil bulk density (kg/l), h is the depth where the P concentration front is located (m), and V is the solution volume (m^3) added over S. The A is calculated by equation 5.2.1 for appropriate Langmuir constants. As soils in the survey were not fitted with Langmuir constants, but were defined in terms of Langmuir-like Olsen P parameters (table in Fig. 5.2.5), the relationship A = Olsen P/REC was used to estimate A in equation 5.2.11. The value of REC can be approximated from the known % clay of the soil (Fig. 5.2.2). For example, let us estimate h after 20 yrs of effluent irrigation at a rate of 700 mm/yr (V = 14000 l/m^2), C_P = 3.5 mg/l, S = 0.33 m^2 (33% wetted area) in 10% clay soil and ϱ = 1.3 kg/l. Choosing the 10% clay K_{ol} (= 0.3 l/mg) and B_{ol} (= 60 mg/kg) gives Olsen P = 31 mg/kg. As REC for 1–10 % clay soils = 0.57 (Fig. 5.2.2), the estimated A = 54 mg/kg. Inserting A in equation 5.2.11 gives h = 2.1 m. Repeating the calculation for 20% and 50% clay soils yields h = 1.0 and 0.47 m, respectively.

Field plot JE10% (Fig. 5.2.6a, 10% clay) received effluents for 37 yrs since 1968; the afore-calculated Olsen P concentration for this soil was 31 mg/kg and the estimated REC = 0.57. For V = 700 × 37 (=25900 m^3), the value of h is 2.8 m. This step function was included in Figure 5.2.7a. For field plot JE20% (20% clay, 9 yrs of effluent application since 1996, Fig. 5.2.7a) the calculated Olsen P was 53 mg P/kg, REC = 0.47 and estimated h = 0.45 m. This step function is also included in Figure 5.2.7a. The two step functions give reasonable estimates of the depth of the Olsen P concentration front under the studied experimental conditions. Also, they reasonably estimate the depth-averaged Olsen P concentration in the profile. If total P concentration in effluents (5 mg/l) is used in the calculation, the predicted step function overestimates the experimental results, implying that organic P contribution to Olsen P transport is negligible. The body of results in this section indicates that P_i leaching under effluent irrigation could be explained in all survey soils by P adsorption theories without resorting to preferential P transport mechanisms.

5.2.5.2.3 Evaluating the role of IHP in survey results

About 30% of total P in survey effluents was P_o and it is estimated that 65% of it was IHP. According to data in Figure 5.2.3 the adsorption of IHP is ~tenfold greater than that of P_i, and its mobility in water-saturated soil is negligible relative to P_i. This means that the adsorption front of added IHP (in equilibrium with 1.5 mg/l IHP-P) in loess soil (~15% clay) is expected to be within the top 5 cm soil layer (h = 700* 10* 1.5/(300* 600* 1.3)).

Data in the "mineralization/immobilization" section indicate that the half life of IHP mineralization under aerobic conditions is 60–100 d (for first order reaction this is equivalent to a rate constant of 0.007 1/d). At this rate the contribution of IHP to P_i transport or uptake is minor unless the concentration of IHP in soil is very high. The contribution of total soil organic P (P_{os}) mineralization to P_i under field conditions was estimated in that section as ~10 kg P/ha/yr (out of ~300 kg P_{os}/ha), comprising ~3% mineralization per year (equivalent to first order reaction rate constant of ~0.0001 1/d).

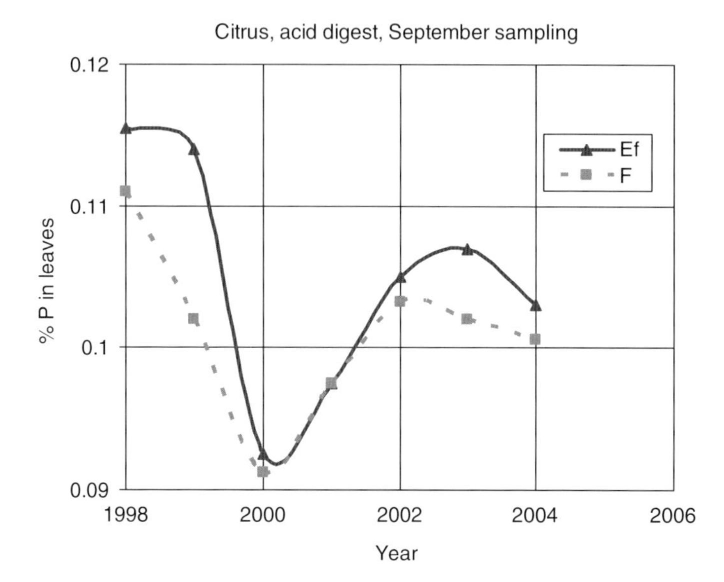

Figure 5.2.8 Phosphorus concentration in citrus leaves (September sampling, % of dry matter, n = 461) as a function of water source (effluents, Ef versus freshwater, F) and time span of survey. The time-averaged result is significantly greater in effluents than in freshwater (P = 0.0135, Pearson separation test). Derived from Tartchichky et al. (2006).

This contribution is too small to affect P_i concentration and mobility in soil. In the presence of plants, P_{os} mineralization rate is higher due to root phosphatase activity (Tarafdar and Claassen, 1988), and plant utilization of effluent applied P_o may exceed considerably the 10 kg P/ha/yr level.

5.2.5.2.4 Effect on P concentration in citrus leaves

The major effect of effluents on % P in grapefruit leaves was observed early in the survey (Fig. 5.2.8). This stemmed from the fact that plots were irrigated with effluents several years before the survey started at unknown rates. In 1998, 1999, 2002, 2003, and 2004, the %P in leaves was ~5% higher under effluent than freshwater irrigation. The decrease in % P in the year 2000 emanated from unknown management or climatic factors.

5.2.5.2.5 Rates of P accumulation in soil

The experimental variability in the Olsen P data in a given % clay class was so high that temporal P accumulation in soil could not be meaningfully characterized. To circumvent the variability problem, individual field plots in each % clay group were chosen for the P accumulation analysis. In 10% clay plots, which were irrigated with effluents by microjets and drip systems (#4014 and 4019, respectively, Fig. 5.2.9), the Olsen P concentration was practically steady with time, with a time-averaged value of ~35 mg P/kg in both irrigation methods. According to previous calculations 35 mg P/kg soil is approximately the Olsen P concentration (10% clay) in equilibrium with $C_p = 3.5$ mg/l P, the concentration of P_i in the effluents. This means that application of more effluents at this C_P will cause P

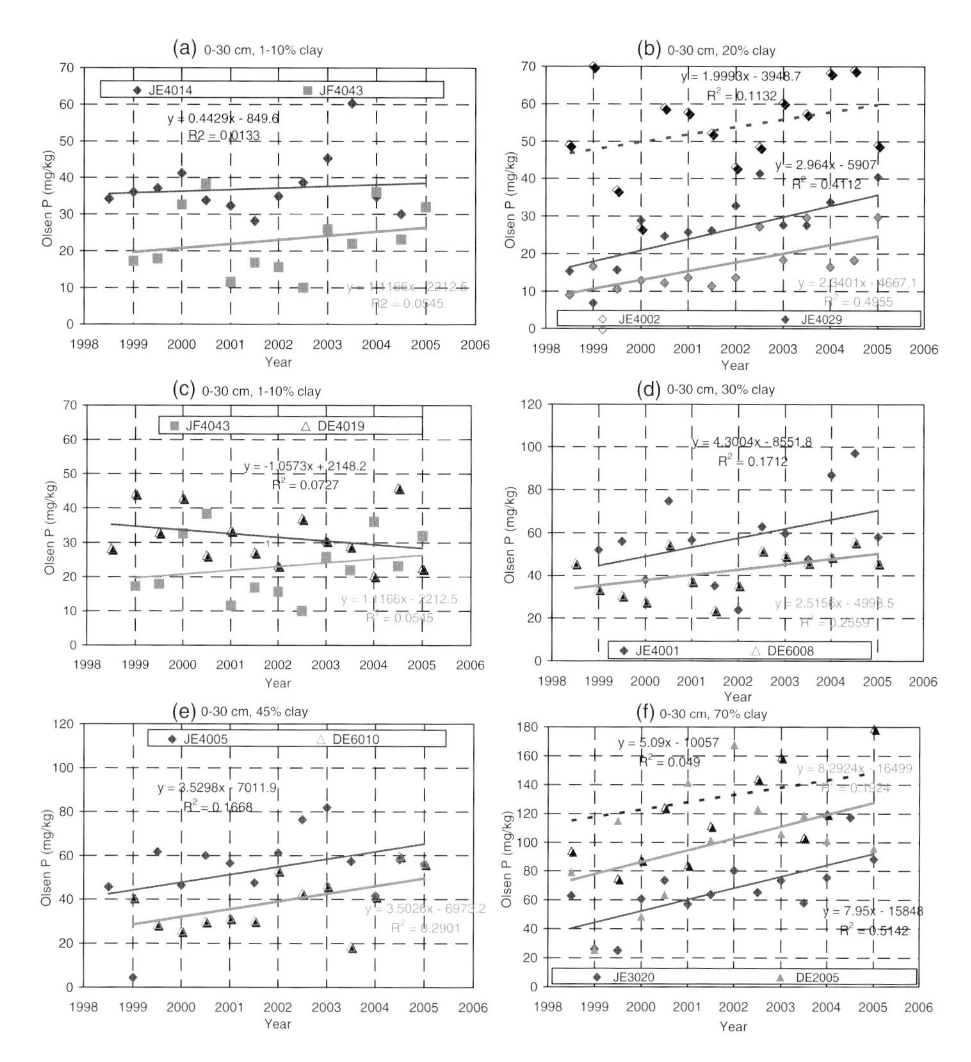

Figure 5.2.9 Olsen P concentration in soil as a function of time along the survey period in field plots differing in % clay, irrigation method (micro jet, J, or drip, D) and water source (effluents, E, or freshwater, F). All samples were taken from the 0–30 cm soil layer. Recalculated from Tartchichky et al. (2006).

accumulation below the 0–30 cm soil layer, but not inside the soil layer. The time-averaged Olsen P concentration in the freshwater plot (#4043, 10% clay) was ~25 mg P/kg. That was higher than expected in equilibrium with freshwater P_i, therefore it was plausibly controlled by Ca-P minerals and organic P in this soil. The 20% clay plots (#4035 and 4029) exhibited a P accumulated rate of ~2.6 mg P/kg/yr (Fig. 5.2.9). The equilibrium Olsen P concentration at $C_P = 3.5$ mg/l in 20% clay soil is 52 mg/kg P (Fig. 5.2.5), which exceeds the experimental Olsen P concentration at the end of the survey period (30–40 mg P/kg). A third field plot in this class (#4002) accumulated Olsen P at a rate of 2 mg P/kg/yr and the concentration at the end of the survey was ~60 mg/kg P.

In the 30% clay class the Olsen P accumulation rate was 4.3–2.6 mg P/kg/yr and the final Olsen P concentrations were 80–60 mg/kg P. The expected Olsen P in equilibrium with

$C_P = 3.5$ mg/l is ~65 mg/kg. In soils with % clay of 45% and 70%, the Olsen P annual accumulation rates were ~3.5 and ~5–8 mg P/kg/yr, respectively (Fig. 5.2.9), and the final Olsen P was ~70 and 100–160 mg/kg. The expected Olsen P in equilibrium with $C_P = 3.5$ mg/l P is 83 and 155 mg/kg P (Fig. 5.2.5). It is noted that the linear correlations in Figure 5.2.9 were not significant ($P = 0.05$) and their slopes are merely used as approximate estimates of the Olsen P accumulation rates.

The fact that Olsen P accumulation rate in the 0–30 cm soil layer increased with % clay proves that in low % clay soils P moved below the 0–30 cm soil layer, compatible with the P profile results discussed above. In the high % clay soils the Olsen P quantity in equilibrium with 3.5 mg/l P exceeded the cumulative P dose that was added via the effluents during the survey period, therefore no P moved below the 0–30 cm layer.

The P_i quantity that was added annually to effluent irrigated soils was (7000 m^3/ha/yr × 3.5 g P/m =) 24.5 kg P/ha/yr. If retained in the top 0–30 cm soil layer this dose should have increased sorbed P by ~7.5 mg P/kg/yr. The corresponding Olsen P concentration equals sorbed P × REC. In soils with % clay > 45%, REC ~0.4 (Fig. 5.2.2), hence the expected increase in Olsen P is (7.5 × 0.4) 3 mg P/kg/yr. Repeating the calculation for total P concentration in effluents (5.2 mg P/l and assuming fast mineralization to P_i) gives potential Olsen P accumulation rate of 4.5 mg P/kg/yr in this soil. These results indicate that the above mentioned empirical Olsen P accumulation rate of ~8 mg P/kg/yr in 70% clay soil is erroneous, and the ~5 mg Olsen P/kg/yr (Fig. 5.2.9) is a more reliable estimate.

The results presented in this section prove that temporal Olsen P accumulation in the top 0–30 cm soil layer can be predicted in different calcareous soils based on their % clay, C_P, and known P retention characteristics. The nearly constant P_i accumulation rate over the survey 7-yr period indicates minor P losses by Ca-P crystallization in effluent-irrigated soils, compatible with earlier observations.

5.2.5.2.6 *Additional effects on P reaction and partitioning*

Effluent irrigation might affect P behavior in soil by modifying soil pH and Ca+ Mg concentration in soil solution. Indeed in all % clay groups both pH and Ca+ Mg concentration (0–30 cm soil layer) were lower after 7 yrs of effluent irrigation (2005) than at the beginning of the survey (~0.3 pH units and 3 meq/l Ca+ Mg, Fig. 5.2.10). The % clay regression lines in Figure 5.2.10 are statistically insignificant, but a t-test revealed that they differ significantly from each other proving a significant time effect. In the case of pH and Ca+ Mg a similar time effect was obtained under freshwater irrigation, probably because field plots were fertigated by ammonium sulfate (Fig. 5.2.10). The reduced pH plausibly caused Ca+ Mg dissolution and desorption, and therefore leaching to deeper soil layers, which explains the decrease in Ca+ Mg concentration with time. The time effect on pH was similar in all % clay values, whereas the effect on Ca+ Mg increased with increasing % clay. Olsen P concentration in the 0–30 cm soil layer was practically unchanged with time under effluent irrigation (Fig. 5.2.10), but under freshwater irrigation it slightly increased with time, probably due to increased adsorption induced by reduced pH. Note that the clear time effect on Olsen P accumulation in the top soil layer presented in Figure 5.2.9 was obtained in individual field plots. Apparently the field variability prevented obtaining a significant time effect when all soils in a given % clay group were pooled together.

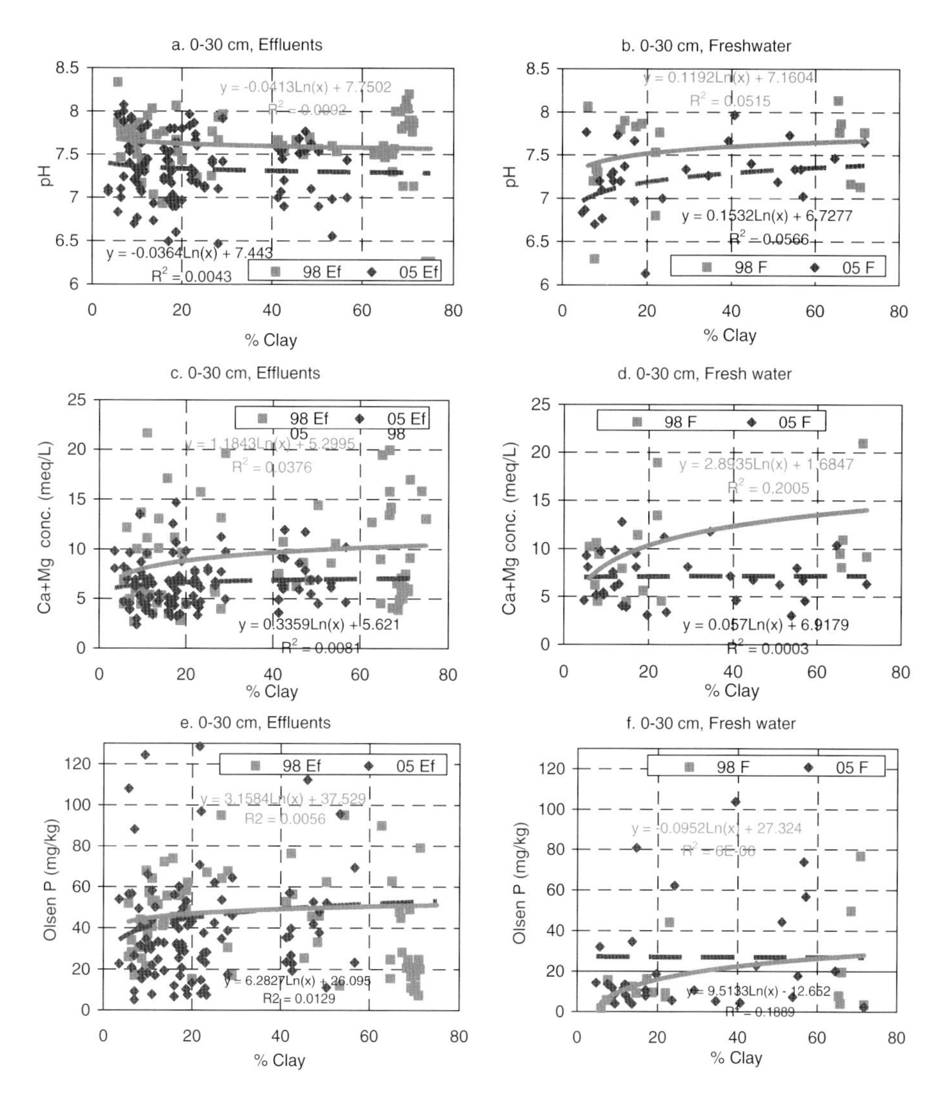

Figure 5.2.10 Saturated paste pH and Ca+ Mg concentration and Olsen P concentration as a function of % clay in soil layer 0–30 cm under effluent (Ef) and freshwater (F) irrigation on 1998 (98) and 2005 (05). Recalculated from Tartchichky et al. (2006).

5.2.6 Conclusions and outlook

Processes and factors affecting short- and long-term effluent P behavior in soil can be divided into two groups:

(i) processes influencing P_i and P_o sorption, transformation and transport (runoff and drainage) in soils. Data presented proved that sorption and drainage depth of P_i can be predicted from basic soil properties: % clay, pH, and adsorbed P recovery percentage by common soil extractants. Organic P (P_o) affects P_i solution concentration via IHP-P_i competition on common adsorption sites and P_o mineralization. According to current knowledge, mineralization contributes ~10 kg P_i/ha/yr, which amounts to ~65% of P_o added via annual effluent irrigation in arid zones. In plant rhizospheres the mineralization rate is much faster due to phosphatase excretion by roots.

(ii) OM-related processes affecting soil hydraulic properties and P binding by carbonaceous macromolecules. Literature data indicate that added OM increased P leaching depth and concentration in runoff, but specific mechanisms explaining this effect are not available yet. The effect of organic macromolecules depends on their mobility in soil (which is inversely proportional to adsorption) and the stability of the formed complex. Another OM effect is its impact on preferential water and P flows and improved soil aggregation, both increasing drainage and reducing P runoff losses. Data analysis of the long-term effluent irrigation survey indicated that in effluent-irrigated soils adsorption was the main mechanism controlling P concentration in solution, whereas under freshwater irrigation the effect of Ca-P precipitation could not be discounted. The difference was attributed to Ca-P crystallization inhibition by unknown OM molecules derived from effluents.

It can be concluded from presented data and discussion that P leaching to underground water in arid zone sandy soils irrigated with common effluents is inevitable under unregulated irrigation. Likewise, runoff P concentration in most case study survey soils already exceeded the permitted eutrophication threshold concentration of 0.02–0.05 mg/l P.

Future research on P management optimization under effluent irrigation and biosolid amendment should focus on elucidating behavior of P_o components in soil, improving models of P dynamic in the soil–plant system, and studying OM–P interaction in soil. Due to the complexity of P - organic matter interactions, model-based decision support systems should be developed to predict the fate of inorganic and organic P in different soils. Such tools will help optimize fertilizer and organic amendments application and reducing environmental P pollution without compromising crop yields.

Acknowledgement

The excellent technical help of Irit Levkovich is highly appreciated. The work was partly supported by grant by the Ministry of Agriculture, The State of Israel.

References

Adriano, D.C., Novak, L.T., Erickson, A.E., Wolcott, A.R. and Ellis, B.G. (1975) Effect of long term land disposal by spray irrigation of food processing wastes on some chemical properties of the soil and subsurface water. *Journal of Environmental Quality* **4**, 242–248.

Ahuja, L.R., Sharpley, A.N. and Lehman, O.R. (1982) Effect of soil slope and rainfall characteristics on phosphorus in runoff. *Journal of Environmental Quality* **11**, 9–13.

Anderson, G., Williams, E.G. and Moir, J.O. (1974) A comparison of the sorption of inorganic orthophosphate and inositol hexaphosphate by six acid soils. *Journal of Soil Science* **25**, 51–62.

Angin, I., Yaganoglu, A.V. and Turan, M. (2005) Effect of long term wastewater irrigation on soil properties. *Journal of Sustainable Agriculture* **26**, 31–42.

Arai, Y. and Sparks, D.L. (2007) Phosphate reaction dynamics in soils and soil components. *Advances in Agronomy* **94**, 135–179.

Barrow, N.J. and Shaw, T.C. (1975) The slow reaction between soil and anions: 2. Effect of time and temperature on the decrease in phosphate concentration in the soil solution. *Soil Science* **119**, 167–177.

Barton, L, Schipper, L.A., Barkle, G.F., McLeod, M., Speir, T.W., Taylor, M.D., McGill, A.C., van Schaik, A.P., Fitzgerald, N.B. and Pandey, S.P. (2005) Land application of domestic effluent onto four soil types: plant uptake and nutrient leaching. *Journal of Environmental Quality* **34**, 635–643.

Bar-Yosef, B. and Akiri, B. (1978) Sodium bicarbonate extraction to estimate N, P, K availability in soils. *Soil Science Society of America Journal* **42**, 319–323.

Bar-Yosef, B. (1996) Root excretions and their environmental effects – Influence on the availability of phosphorus. In: *Plant Roots – the Hidden Half* (eds Y. Waisel, A. Eshel and U. Kafkafi), 2nd ed, pp. 581–604. Dekker, New York.

Bar-Yosef, B., Kafkafi, U., Rosenberg, R. and Sposito, G. (1988) Phosphorus adsorption by kaolinite and montmorillonite: I. Effect of time, ionic strength, and pH. *Soil Science Society of America Journal* **52**, 1580–1585.

Bar-Yosef, B. (2003) Phosphorus dynamics. In: *Handbook of Processes and Modeling in the Soil-Plant System* (eds D. K. Benbi and R. Nieder), pp. 483–523. The Haworth Reference Press, New York.

Bergen Jensen, M., Olsen, T.B., Hansen, H.C.B. and Magid, J. (2000) Dissolved and particulate phosphorus in leachate from structured soil amended with fresh faeces. *Nutrient Cycling in Agroecosystems* **56**, 253–261.

Bielorai, H., Dasberg, S., Erner, Y. and Brom, M. (1985) *The Effect of Various Soil Moisture Regimes and Fertilizer Levels on Citrus Yield Response Under Partial Wetting of the Root Zone.* Final report. Agricultural Research Organization, Bet Dagan, Israel.

Bouwer, H. and Chaney, R.L. (1974) Land treatment of wastewater (ed. N.C. Brady). *Advances in Agronomy* **26**, 133–175. Academic Press, New York.

Bresler, E., McNeal, B.L. and Carter, D.L. (1982) *Saline and Sodic Soils: Principles-Dynamics-Modeling*. Springer Verlag, Berlin.

Breeuwsma, A., Reijerink, J.G.A. and Schoumans, O.F. (1995) Impact of manure on accumulation and leaching of phosphate in areas of intensive livestock farming. In: *Animal Waste and the Land-Water Interface* (ed. K. Steele), pp. 239–251. Lewis Publ.-CRC, New York.

California Water Resources Control Board (2003) *Municipal Wastewater Recycling Survey*. www.waterplan.water.ca.gov/cwpu2005/index.cfm

Cassel, E.A. (1998) Modeling phosphorus dynamics in ecosystems: mass balance and dynamic simulation approaches. *Journal of Environmental Quality* **27**, 293–298.

Celi, L., Lamacchia, S., Marsan, F.A. and Barberis, E. (1999) Interaction of inositol hexaphosphate on clays: adsorption and charging phenomena. *Soil Science* **164**, 574–585.

Celi, L., Lamacchia, S., Marsan, F.A. and Barberis, E. (2000). Interaction of inositol phosphate with calcite. *Nutrient Cycling in Agroecosystems* **57**, 271–277.

Celi, L., Presta, M., Ajmore-Marsan, F. and Barberis, E. (2001) Effect of pH and electrolyte on inositol hexaphosphate interaction with goethite. *Soil Science Society of America Journal* **65**, 753–760.

Cho, C.M. (1991) Phosphate transport in Ca-saturated systems: I. Theory. *Soil Science Society of America Journal* **55**, 1275–1281.

Clement, A.L., Somlyoly, L. and Koncsos, L. (1998) Modeling the P retention of the Kis-Balaton upper reservoir. *Water Science Technology* **37**, 113–120.

Condron, L.M., Turner, B.L., Cade-Menun, B.J. (2005) Chemistry and dynamics of soil organic phosphorus. In: *Phosphorus: Agriculture and the Environment. Agronomy No. 46* (eds J. T. Sims and A. N. Sharpley), pp. 87–121. Soil Science Society of America, Madison, WI.

Cosgrove, D.J. (1980) *Inositol Phosphates. Studies in Organic Chemistry*. Elsevier, Amsterdam.

Daniel, T.C., Sharpley, A.N. and Lemunyon, J.L. (1998) Agricultural phosphorus and eutrophication: a symposium overview. *Journal of Environmental Quality* **27**, 251–257.

DUFLOW (1992) *A Micro-Computer Package for the Simulation of Unsteady Flow and Water Quantity Processes in One Dimensional Channel Systems*. Public Works Department (Rijkswaterstaat), The Netherlands.

Dzombak, D.A. and Morel, F.M.M. (1990) *Surface Complexation Modeling. Hydrous Ferric Oxide*. John Wiley & Sons, New York.

Ebeling, A.M., Bundy, L.G., Powell, J.M. and Andraski, T.W. (2002) Dairy diet phosphorus effects on phosphorus losses in runoff from land-applied manure. *Soil Science Society of America Journal* **66**, 284–291.

Elliott, H.A., O'Connor, G.A. and Brinton, S. (2002) Phosphorus leaching from biosolids-amended sandy soils. *Journal of Environmental Quality* **31**, 681–689.
Escudey, M., Galindo, G., Avendano, K., Borchardt, D., Chang, A.C. and Brigeno, M. (2004) Distribution of phosphorus forms in Chilean soils and sewage sludge by chemical fractionation and ^{31}P-NMR. *Journal of Chilean Chemistry Society* **49**, 219–222.
Espinosa, M., Turner, B.L. and Haygarth, P.M. (1999) Preconcentration and separation of trace phosphorus compounds in soil leachate. *Journal of Environmental Quality* **28**, 1497–1504.
Evans, A. Jr. (1985) The adsorption of inorganic-P by a sandy soil as influenced by dissolved organic compounds. *Soil Science* **140**, 251–255.
Fang, F., Brezonik, P.L., Mulla, D.J., and Hatch, L.K. (2002) Estimating runoff phosphorus losses from calcareous soils in the Minnesota River Basin. *Journal of Environmental Quality* **31**, 1918–1929.
Feigin, A., Ravina, I. and Shalhevet, J. (1991) *Irrigation with Treated Sewage Effluent*. Springer-Verlag, Berlin.
Fine, P. and Mingelgrin, U. (1996) Release of phosphorus from waste-activated sludge. *Soil Science Society of America Journal* **60**, 505–511.
Frossard, E., Lopez-Hernandez, D., Brossard, M. (1996) Can isotopic exchange kinetics give valueable information on the rate of mineralization of organic phosphorus in soils? Soil Biology and Biochemistry **28**, 857–864.
Gardner, W.K., Parbery, D.G., Barber, D.A. and Swinden, L. (1983) The acquisition of phosphorus by *Lupinus albus L.* v. The diffusion of exudates away from roots: a computer simulation. *Plant and Soil* **72**, 13–29.
Gburek, W.J., Barberis, E., Haygarth, P.M., Kronvang, B. and Stamn, C. (2005) Phosphorus mobility in the landscape. In: *Phosphorus: Agriculture and the Environment. Agronomy Monogram No. 46* (eds J.T. Sims and A.N. Sharpley), pp. 941–979. ASA, Madison, WI.
Ge, Y., and Hendershot, W. (2004) Evaluation of soil surface charge using the back-titration technique. *Soil Science Society of America Journal* **68**, 82–88.
Goring, C.A.I. and Bartholomew, W.V. (1950) Microbial products and soil organic matter: III. Adsorption of carbohydrate phosphates by clays. *Soil Science Society of America Proceedings* **15**, 189–194.
Goring, C.A.I. and Bartholomew, W.V. (1952) Adsorption of mononucleotides, nucleic acids, and nucleoproteins by clays. *Soil Science* **74**, 149–164.
Guppy, C.N., Menzies, N.W., Moody, P.W. and Blamey, F.P.C. (2005a) Competitive sorption reactions between phosphorus and organic matter in soil: A review. *Australian Journal Soil Research* **43**, 189–202.
Guppy, C.N., Menzies, N.W., Blamey, F.P.C. and Moody, P.W. (2005b) Do decomposing organic matter residues reduce phosphorus sorption in highly weathered soils? *Soil Science Society of America Journal* **69**, 1405–1411.
Goldberg, S. and Sposito, G. (1985) On the mechanism of specific phosphate adsorption by hydroxylated mineral surfaces: a review. *Communications in Soil Science and Plant Analysis* **16**, 801–821.
Govindasamy, R., Cochran, M.J. and Buchberger, E. (1994) Economic-implications of phosphorus loading policies for pasture land applications of poultry litter. *Water Resources Bulletin* **30**, 901–910.
Grant, R.F. and Heaney, D.J. (1997) Inorganic phosphorus transformation and transport in soils: mathematical modeling in ECOSYS. *Soil Science Society of America Journal* **61**, 752–764.
Grierson, P.F., Comerford, N.B., and Jokela, E.J. (1998) Phosphorus mineralization kinetics and response of microbial phosphorus to drying and rewetting in a Florida Spodosol. *Soil Biology & Biochemistry* **30**, 1323–1331.
Grossl, P.R. and Inskeep, W.P. (1991) Precipitation of dicalcium phosphate dihydrate in the presence of organic acids. *Soil Science Society of America Journal* **55**, 670–675.
Heathwaite, A.L., Sharpley, A., Bechmann, M. and Rekolainen, S. (2005) Assessing the risk and magnitude of agricultural nonpoint source phosphorus pollution. In: *Phosphorus: Agriculture and the Environment. Agronomy Monogram No. 46* (eds J. T. Sims and A. N. Sharpley), pp. 981–1020. ASA, Madison, WI.
Hedley, M. and McLaughlin, M. (2005) Reactions of phosphate fertilizers and by-products in soils. In: *Phosphorus: Agriculture and the Environment. Agronomy Monogram No. 46* (eds J. T. Sims and A. N. Sharpley), pp. 181–252. ASA, Madison, WI.
Heng, L.C. (1989) Influence of some humic substances on P-sorption in some Malaysian soils under rubber. *Journal of Natural Rubber Research* **4**, 186–194.

Hession, W.C., Storm, D.E., Haan, C.T., Burks, S.L. and Matlock, M.D. (1996) A watershed-level ecological risk assessment methodology. *Water Resources Bulletin* **32**, 1039–1054.
Hinedi, Z.R., Chang, A.C. and Lee, W.K. (1988) Mineralization of phosphorus in sludge-amended soils monitored by ^{31}P NMR spectroscopy. *Soil Science Society of America Journal* **52**, 1593–1596.
Hinrichsen, D., Robey, B., Upadhyay, U.D. (1998) Solutions for a water-short world. *Population Reports* **26**, 1 (September issue) Center for Communication Programs, The Johns Hopkins Univ., Baltimore, Maryland.
Hopstaken, C.F. and Ruijgh, E.F.W. (1994) Modeling N and P loads on surface and ground water due to land-use. In: *Contamination of Groundwaters* (eds D.C. Adriano, A.K. Iskandar, and I.P. Murarka), pp. 161–188. Northwood (United Kingdom) Scientific Reviews Ltd.
Huang, X.L. and Shenker, M. (2004) Water soluble and solid-state speciation of phosphorus in stabilized sewage sludge. *Journal of Environmental Quality* **33**, 1895–1903.
Inskeep, W.P. and Silvertooth, J.C. (1988) Inhibition of hydroxyapatite precipitation in the presence of fulvic, humic and tannic acids. *Soil Science Society of America Journal* **52**, 941–946.
Iyamuremye, F. and Dick, R.P. (1996) Organic amendments and phosphorus sorption by soils. *Advances in Agronomy* **56**, 139–185.
Johnson, A.F., Vietor, D.M., Rouquette, Jr., F.M. and Haby, V.A. (2004) Fate of phosphorus in dairy wastewater and poultry litter applied on grassland. *Journal of Environmental Quality* **33**, 735–739.
Jones, C.A., Cole, C.V., Sharpley, A.N. and Williams, J.R. (1984) A simplified soil and plant phosphorus model: I. Documentation. *Soil Science Society of America Journal* **48**, 800–804.
King, L.D., Burns, J.C. and Westerman, P.W. (1990) Long-term swine lagoon effluent application on Coastal bermudagrass. II. Effect on nutrient accumulation in soil. *Journal of Environmental Quality* **19**, 756–760.
Kirkham, M.B. (1982) Agricultural use of phosphorus in sewage-sludge. *Advances in Agronomy* **35**, 129–163.
Kleinman, P.J.A., Needelman, B.A., Sharpley, A.N. and McDowell, R.W. (2003) Using soil phosphorus profile data to assess phosphorus leaching potential in manured soils. *Soil Science Society of America Journal* **67**, 215–224.
Kleinman, P.J.A., Sharpley, A.N., Veith, T.L., Maguire, R.O. and Vada, P.A. (2004) Evaluation of phosphorus transport in surface runoff from packed soil boxes. *Journal of Environmental Quality* **33**, 1413–1423.
Knisel, W.G. (1980) *CREAMS: A field-scale model for Chemicals, Runoff, and Erosion from Agricultural Management Systems.* US Department of Agriculture, Science and Education Administration, Conservation Research Report No. 26, 640 pp.
Kuo, S., Jellum, E.J. and Pan, W.L. (1988) Influence of phosphate sorption parameters of soils on the desorption of phosphate by various extractions. *Soil Science Society of America Journal* **52**, 974–979.
Kuo, S. (1996) Phosphorus. In: *Methods of Soil Analysis. Part 3: chemical methods. Number 5* (eds D. L. Sparks et al.), pp. 869–919. Soil Science Society of America Book Series, Madison, WI.
Larsen, J.E., Warren, G.F. and Langston, R. (1959) Effect of iron, aluminum and humic acid on phosphorus fixation by organic soils. *Soil Science Society of America Proceedings* **23**, 438–440.
Lasaga, A.C. (1998) *Kinetic Theory in the Earth Sciences.* Princeton University Press.
Legaz, F. and Primo-Millo, E. (1981) Dynamics of ^{15}N labeled N nutrients in Valencia orange trees. Proceedings of the Sixth International Society of Citriculture, Vol. 2, 575–582. Tokyo 1981.
Legaz, F. and Primo-Millo, E. (1984) Influence of flowering, summer and autumn flushes on the absorption and distribution of N compounds in citrus. Proceedings of the Seventh International Society of Citriculture, Vol. 1, 224–233. Sao Paulo 1984.
Legaz, F. and Primo-Millo, E. (1988) Absorption and distribution of ^{15}N applied to young orange trees. Proceedings of the Seventh International Society of Citriculture Vol. 2, 643–661. Tel Aviv 1988.
Lindsay, W.L. (1979) *Chemical Equilibrium in Soils.* John Wiley & Sons, New York.
Lopez-Hernandez, D., Brossard, M. and Frossard, E. (1998) P-isotopic exchange values in relation to organic phosphorus mineralization in soils with very low P-sorbing capacity. *Soil Biology and Biochemisty* **30**, 1663–1670.
Luengo, C., Brigante, M., Antelo, J. and Avena, M. (2006) Kinetics of phosphate adsorption on goethite: comparing batch adsorption and ATR-IR measurements. *Journal of Colloid and Interface Surface* **300**, 511–518.
MAFF – Ministry of Agriculture Fisheries and Food (2000) *Fertilizer recommendations for Agriculture and Horticulture in England and Wales.* Ref. Book 209. Her Majesty's Stationery Office, London.

Maguire, R.O., Chardon, W.J. and Simard, R.R. (2005) Assessing potential environmental impacts of soil phosphorus by soil testing. In: *Phosphorus; Agriculture and the Environment* (eds J. T. Sims and A. N. Sharpley) pp. 145–180. American Society of Agronomy monograph, American Society of Agronomy, Madison, WI.

Mamo, M., Gupta, S.C., Rosen, C.J. and Singh, U. B. (2005) Phosphorus .leaching at cold temperatures as affected by wastewater application and soil phosphorus levels. *Journal of Environmental Quality* **34**, 1243–1250.

Moustafa, M.Z. (1998) Long-term equilibrium phosphorus concentration in the Everglades as predicted by a Vollenweider-type model. *Journal of the American Water Resources Association* **34**, 135–147.

Musvoto, E.V., Wentzel, M.C., Loewenthal, R.E. and Ekama, G.A. (2000a) Integrated chemical-physical processes modeling. I. Development of a kinetic-based model for mixed weak acid/base systems. *Water Research* **34**, 1857–1867.

Musvoto, E.V., Wentzel, M.C., Loewenthal, R.E. and Ekama, G.A. (2000b) Integrated chemical-physical processes modeling. II. Simulating aeration treatment of anaerobic treatment of anaerobic digester supernatants. *Water Research* **34**, 1868–1880.

Nair, V.D., Portier, K.M., Graetz, D.A. and Walker, M.L. (2004) An environmental threshold for degree of phosphorus saturation in sandy soils. *Journal of Environmental Quality* **33**, 107–113.

Oehl, F., Oberson, A., Sinaj, S. and Frossard., E. (2001) Organic phosphorus mineralization studies using isotopic dilution techniques. *Soil Science Society of America Journal* **65**, 780–787.

Oehl, F., Frossard, E., Fliessbach, A., Dubois, D. and Oberson, A. (2004) Basal organic phosphorus mineralization in soils under different farming systems. *Soil Biology and Biochemistry* **36**, 667–675.

Ognalaga, M., Frossard, E. and Thomas, F. (1994) Glucose-1-phosphate and myo-inositol hexaphosphate adsorption mechanisms on goethite. *Soil Science Society of America Journal* **58**, 332–337.

Olsen, S.R. and Kemper, W.D. (1968) Movement of nutrients to plant roots. *Advances in Agronomy* **20**, 91–151.

Parsons, L.R., Wheaton, T.A. and Cross, P. (1995) Reclaimed municipal water for citrus irrigation in Florida. In: Microirrigation for a Changing World, Proceedings of the Fifth International Microirrigation Congress, pp. 262–268. Orlando, FL.

Penn, C.J. and Sims, J.T. (2002) Phosphate forms in biosolid-amended soils and losses in runoff: Effects of wastewater treatment processes. *Journal of Environmental Quality* **31**, 1349–1361.

Penn, C.J., Mullins, G.L. and Zelazny, L.W. (2005) Mineralogy in relation to phosphorus sorption and dissolved phosphorus losses in runoff. *Soil Science Society of America Journal* **69**, 1532–1540.

Pescod, M.B. (1992) *Wastewater treatment and use in agriculture. FAO Irrigation and Drainage Paper 47.* FAO Publications Division, Rome.

Pettygrove, G.S. and Asano, T. (1985) *Irrigation with Reclaimed Municipal Wastewater*. Lewis Publishers, Chelsea, MI.

Pierzynski, G.M., McDowell, R.W. and Sims, T. (2005) Chemistry, cycling and potential movement of inorganic phosphorus in soils. In: *Phosphorus: Agriculture and the Environment. Agronomy Monogram No. 46* (eds J. T. Sims and A. N. Sharpley), pp. 53–86. ASA, Madison, WI.

Pote, D.H., Daniel, T.C., Nichols, D.J., Moore, P.A., Miller, D.M., and Edwards, D.R. (1999) Seasonal and soil-drying effects on runoff phosphorus relationships to soil phosphorus. *Soil Science Society of America Journal* **63**, 1006–1012.

Rauschkolb, R.S., Rolston, D.E., Miller, R.J., Carlton, A.B. and Burau, R.G. (1976) Phosphorus fertilization with drip irrigation. *Soil Science Society of America Proceedings* **40**, 68–72.

Rusan, M.J.M., Hinnawi, S. and Rousan, L. (2007) Long term effect of wastewater irrigation of forage crops on soil and plant quality parameters. *Desalinization* **215**, 143–152.

Sanyal, S.K., and De Datta, S.K. (1991) Chemistry of phosphorus transformations in soil. *Advances in Soil Science* **16**, 1–120.

Schecher, W. (2001) *Thermochemical data used in Mineql+ version 4.5*. Environmental Research Software. Hallowell, ME.

Schoumans, O.F., and Groenendijk, P. (2000) Modeling soil phosphorus levels and phosphorus leaching from agricultural land in the Netherlands. *Journal of Environmental Quality* **29**, 111–116.

Shainberg, I. and Levy, G.J. (1992) Physico-chemical effects of salts upon infiltration and water movement in soils. *Advances in Soil Science* **20**, 37–93.

Shainberg, I., Levy, G.J., Levin, J. and Goldstein, D. (1997) Aggregate size and seal properties. *Soil Science* **162**, 470–478.

Shang, C., Huang, P.M. and Stewart, J.W.B. (1990) Kinetics of adsorption of organic and inorganic phosphates by short-range ordered precipitates of aluminum. *Canadian Journal of Soil Science* **70**, 461–470.

Shang, C., Stewart, J.W.B. and Huang, P.M. (1992) pH effect on kinetics of adsorption of organic and inorganic phosphates by short-range ordered aluminum and iron precipitates. *Geoderma* **53**, 1–14.

Sharpley, A.N., Smith, S.J. (1989) Prediction of soluble P transport in agricultural runoff. *Journal of Environmental Quality* **18**, 313–316.

Sharpley, A.N., Robinson, J.S. and Smith, S.J. (1995) Bioavailable phosphorus dynamics in agricultural soils and effects on water quality. *Geoderma* **67**, 1–15.

Sharpley, A.N., McDowell, R.W. and Kleinman, P.J.A. (2001) Phosphorus loss from land to water: integrating agricultural and environmental management. *Plant Soil* **237**, 287–307.

Shober, A.L. and Sims, J.T. (2003) Phosphorus Restrictions for Land Application of Biosolids: Current Status and Future Trends. *Journal of Environmental Quality* **32**, 1955–1964.

Sibanda, H.M. and Young, S.D. (1986) Competitive adsorption of humus acids and phosphate on goethite, gibbsite and two tropical soils. *Journal of Soil Science* **37**, 197–204.

Sims, J.T., Simard, R.R. and Joern, B.C. 1998. Phosphorus loss in agricultural drainage: Historical perspective and current research. *Journal of Environmental Quality* **27**, 277–293.

Sommers, L.E., Nelson, D.E. and Owens, L.B. (1979) Status of inorganic phosphorus in soils irrigated with municipal wastewater. *Soil Science* **127**, 340–350.

Stevens, D. (ed.) (2006) *Growing Crops with Reclaimed Wastewater*. SCIRO Publishing, Australia.

Stevenson, F.J. and Cole, M.A. (1999) *Cycles of Soil Carbon, Nitrogen, Phosphorus, Sulfur and Microelements*. 2nd ed. John Wiley & Sons, New York.

Stumm, W. and Morgan, J.J. (1996) *Aquatic Chemistry: Chemical Equilibria and Rates in Natural Waters*. John Wiley & Sons, New York.

Suzumara, M. and Kamatani, A. (1995) Mineralization of inositol hexaphosphate in aerobic and anaerobic marine-sediments- implications for the phosphate cycle. *Geochimica et Cosmochimica Acta* **59**, 1021–1026.

Tarafdar, J.C. Claassen, N. (1988) Organic phosphorus compounds as a phosphorus source for higher plants through the activity of phosphatase produced by plant roots and microorganisms. *Biology and Fertility of Soils* **5**, 308–312.

Tartchichky, H., Bar-Hai, M., Levingrat-Itchitchay, A., Sokolovsky, E., Peres, M., Zilberman, A., Eisenkot, A., Menashe, Y., Kenig, E., Gal, Y., Eisenshdadt, Y., Ronen, Y., Reizer, B., Yaskil, E., Buberman, O. and Saloniko, A. (2002) *National wastewater effluent irrigation survey. 2000–2001*. Ministry of Agriculture, Extension Service, Israel.

Tartchichky, H., Bar-Hai, M., Levingrat-Itchitchay, A., Sokolovsky, E., Peres, M., Zilberman, A., Eisenkot, A., Menashe, Y., Kenig, E., Gal, Y., Eisenshdadt, Y., Ronen, Y., Reizer, B., Yaskil, E., Buberman, O. and Saloniko, A. (2004) *National Wastewater Effluent Irrigation Survey. 2001–2003*. Ministry of Agriculture, Extension Service, Israel.

Tartchichky, H., Bar-Hai, M., Levingrat-Itchitchay, A., Sokolovsky, E., Peres, M., Zilberman, A., Eisenkot, A., Menashe, Y., Kenig, E., Gal, Y., Eisenshdadt, Y., Ronen, Y., Reizer, B., Yaskil, E., Buberman, O. and Saloniko, A. (2006) *National Wastewater Effluent Irrigation Survey. 2003–2005*. Ministry of Agriculture, Extension Service, Israel.

Tarkalson, D.D. and Mikkelsen, R.L. (2004) Runoff phosphorus losses as related to phosphorus source, application method, and application rate on a piedmont soil. *Journal of Environmental Quality* **33**, 1424–1430.

Thomann, R.V. and Linker, L.C. (1998) Contemporary issues in watershed and water quality modeling for eutrophication. *Water Science and Technology* **37** (3), 93–102.

Tompson, L.M., Black, C.A. and Zoellner, J.A. (1954) Occurrence and mineralization of organic phosphorus in soils, with particular reference to association with nitrogen, carbon and pH. *Soil Science* **77**, 185–196.

Turner, B.L., Cade-Menun, B.J. and Westermann, D.T. (2003) Organic .phosphorus composition and potential .bioavailability in semi-arid arable soils of the western United States. *Soil Science Society of America Journal* **67**, 1168–1179.

Wallach, R. and van Genuchten, M.T. (1990) A physically based model for predicting solute transfer from soil solution to rainfall-induced runoff water. *Water Resources Research* **26**, 2119–2126.

Wang, Z., Chang, A.C., Page, L., Wu, L. and Crowley, D. (2003) Assessing the soil quality of long-term reclaimed wastewater-irrigated cropland. *Geoderma* **114**, 261–278.

Wastewater Effluent Irrigation Taskforce (2006) *National Wastewater Effluent Irrigation Survey*. Israel Ministry of Agriculture, Extension Service.

Web, K.M. and Ho, G.E. (1992) Struvite ($MgNH_4PO_4{\cdot}6H_2O$) solubility and its application to a piggery effluent problem. *Water Science and Technology* **26**, 2229–2232.

White, P. and Hammond, J. (2007) *Updating the estimate of the source of phosphorus in UK waters. DEFRA project WT0701CSF report*, UK Ministry of Environment.

Woods, N.C., Sock, S.M. and Daigger, G.T. (1999) Phosphorus recovery technology modeling and feasibility evaluation for municipal wastewater treatment plants. *Environmental Technology* **20**, 663–679.

Woodard, K.R., Sollenberger, L.E., Sweat, L.A., Graetz, D.A., Nair, V.D., Rymph, S.J., Walker, L. and Joo, Y. (2007) Phosphorus and other soil components in a dairy effluent sprayfield within the Central Florida Ridge. *Journal of Environmental Quality* **36**, 1042–1049.

Yu, B. (2003) A unified framework for water erosion and deposition equations. *Soil Science Society of America Journal* **67**, 251–257.

5.3
Calcium and carbonate

Gil Eshel and Michael J. Singer

5.3.1 Introduction

Soils (and water) from semiarid and arid regions often contain Ca and Mg carbonates, which are commonly referred to as soil inorganic carbon (SIC). Soil inorganic carbon is the dominant form of carbon in these soils (Kraimer et al., 2005). Rates and processes of SIC dynamics have long been of interest to soil scientists (Jenny, 1941; Bui et al., 1990). Accumulation of SIC, unlike soil organic carbon may not reach a steady state, but may continue to increase over time (Kelly and Yonker, 2005). The formation of secondary carbonate occurs mainly in regions with mean annual precipitation of less than 500 mm (Lal and Kimble, 2000a). The global area where SIC precipitation takes place is estimated at 4.9×10^9 ha (43.5% semiarid, 44.6% arid, and 11.9% extremely arid). The distribution of SIC in soil profiles can result from all or part of the following pedogenesis processes: redistribution of detrital carbonate or redistribution of exogenous material (e.g. dust-borne carbonate) and in situ secondary carbonate formation due to primary mineral weathering. The redistribution can result from dissolution and re-precipitation of inherited carbonate minerals due to cycles of soil wetting and drying (Lippmann, 1973; Knight, 1991; Kraimer et al., 2005), or by translocation, as colloidal suspensions with the soil solution (Dan, 1983; Baghernejad and Dalrymple, 1993; Neaman et al., 2000). The size and location of the solid SIC is influenced by irrigation, drainage, leaching, plowing, and the use of acidifying material on Aridisols and liming material on Ultisols and Oxisols (Lal, 2001b). Secondary precipitation of carbonates with divalent cations (e.g. Ca^{2+} and Mg^{2+}) is possible under limited leaching and dry conditions (Dan, 1983; Sposito, 1989; Nordt et al., 2000).

Irrigation is a common practice supporting crop production in semiarid and arid regions, and it has a profound effect on SIC through several processes. For example, Presley et al. (2004) studying soils in Kansas that had been irrigated for 30 years, found that the pH of surface horizons of irrigated Keith and Richfield pedons was 1.0 pH unit higher, and the exchangeable sodium percentage (ESP) was higher than the pedons that had not been irrigated. They also found that irrigation did not significantly affect the calcium carbonate

Treated Wastewater in Agriculture, First Edition, edited by Guy J. Levy, Pinchas Fine and Asher Bar-Tal

equivalent of the soils. Nunes et al. (2007) examined 1428 topsoil samples from an irrigated area on the Spanish–Portuguese border and found that soil electrical conductivity, acidification, and exchangeable sodium percentage increased and organic matter content decreased significantly after 30 yrs of irrigation.

The addition of water through irrigation changes the soil moisture regime (adding more moisture during dryer periods), which directly affects soil respiration (by roots and microorganisms), and creates a high CO_2 content (pCO_2) in the soil profile. The high pCO_2 reduces soil pH and increases carbonate dissolution. On the other hand, irrigation water is a source of additional salts imported to the field. It should also be kept in mind that irrigation water used in semiarid and arid regions is commonly in equilibrium or supersaturated with respect to calcite (Nimkar et al. 1992; Suarez, 2000). Because treated wastewater (TWW) contains a relatively higher concentration of dissolved organic carbon (DOC) and higher salinity than freshwater (FW), irrigation with this water adds an additional complexity to the carbonate system. The amount of salts and organic components supplied to fields irrigated with effluent is a function of the degree of the wastewater treatment.

This chapter reviews the main processes that control carbonate and calcium distribution in soil profiles by focusing on the effect of irrigation with TWW on SIC pools within an agronomic and environmental context.

5.3.2 Sources and distribution

Soil carbonates are commonly a mixture of non-pedogenic and pedogenic carbonate. Non-pedogenic carbonate, also termed primary, geogenic, detrital, or lithogenic carbonate, exists in soils as detritus, mechanically inherited from limestone and similar parent materials. Pedogenic carbonate, also termed secondary, authigenic, or in situ carbonate, forms in the soil at its present location or is transported from pre-existing soils. Research suggests that the majority of the pedogenic carbonate originates from primary carbonates. For example, Capo et al. (1998) found that less than 5% of the Ca in the petrocalcic horizon in the La Mesa surface is derived from mineral weathering based on $^{87}Sr/^{86}Sr$.

Carbonate distribution in a soil profile is directly and indirectly affected by the regional climate, particularly the moisture regime. Generally, the drier the regime, the more carbonate is expected to be found in the upper layers of the soil. For example, Kelly et al. (1991) studying Holocene age grassland soils in the northern Great Plains of the USA, found that total carbonate decreased with increasing temperature if the precipitation amount was held approximately constant. They also found that total carbonate increased as the precipitation decreased. In their study, Rubio and Escudero (2005) found that carbonate leaching in Mediterranean forest soils formed on limestone was determined by altitude and annual moisture surplus.

Both rainwater and irrigation water dissolve soil carbonate and transport the dissolved products deeper within the soil profile. Soil water indirectly affects soil carbonate distribution by both supporting biological respiration (CO_2 formation), and by controlling soil apparent diffusivity (CO_2 removal). As a result, the depth of carbonate accumulation in the soil profile is highly affected by the amount of precipitation and the depth of the wetting front.

5.3.3 Carbonate system in the soil environment

The carbonate system has an important role in the soil's chemical environment (e.g. pH, ion activity). Carbonate serves as a source of calcium for plants and as a repository for Mg, Mn, and Fe (Nordt et al., 1998). Calcite (detrital and pedogenic) is the most common carbonate mineral in semiarid and arid soils and as a result, the SIC content is commonly presented as calcite equivalent. Geogenic dolomite is not commonly found in soils. In saline-sodic soils, $NaHCO_3$ is likely to be found. As the dissolution and precipitation of carbonate minerals are pH-dependent, the presence of carbonate minerals in soils has a significant effect on the soil solution pH. Under ambient conditions we can expect that soil containing $NaHCO_3$ and calcite will have a pH of 10.5 and 8.5, respectively.

Soil inorganic carbon may appear in three phases: in the gas phase mainly as CO_2, in the liquid phase as dissolved inorganic carbon (DIC), and in the solid phase as carbonate minerals, mostly as calcite. The relationship between carbon forms in the three phases can be expressed by the following reactions:

$$CO_{2\,(g)} + H_2O \leftrightarrows H_2CO_{3\,aq)} \tag{5.3.1}$$

$$H_2CO_{3(aq)} \leftrightarrows H^+_{(aq)} + HCO^-_{3\,(aq)} \tag{5.3.2}$$

$$CaCO_{3\,(s)} + H^+ \leftrightarrows Ca^{2+}_{(aq)} + HCO^-_{3\,(aq)} \tag{5.3.3}$$

The overall reaction is:

$$CO_{2\,(g)} + H_2O + CaCO_{3\,(s)} \leftrightarrows Ca^{2+}_{(aq)} + 2HCO^-_{3\,(aq)} \tag{5.3.4}$$

The total DIC (the concentrations of the three carbonate species) depends on pCO_2, and pH (equations 5.3.2–5.3.4).

$$\log H_2CO_3 = -1.46 + \log CO_{2\,(g)} \tag{5.3.5}$$

$$\log HCO^-_3 = -7.82 + pH + \log CO_{2\,(g)} \tag{5.3.6}$$

$$\log CO^{2-}_3 = -18.15 + 2pH + \log CO_{2\,(g)} \tag{5.3.7}$$

In the case of well-buffered systems (fixed pH), an increase in partial pressure of CO_2 (pCO_2) of one order of magnitude will be followed by an increase in total DIC concentration of one order of magnitude (Fig. 5.3.1).

5.3.3.1 *Soil pCO_2*

The pCO_2 in the soil profile may be 10–100 times higher than in the atmosphere (Amundson and Davidson, 1990) and it may vary in space and time (Parkin and Meisinger, 1989; Wood et al., 1993; Affek et al., 1998). The main factors that produce these high

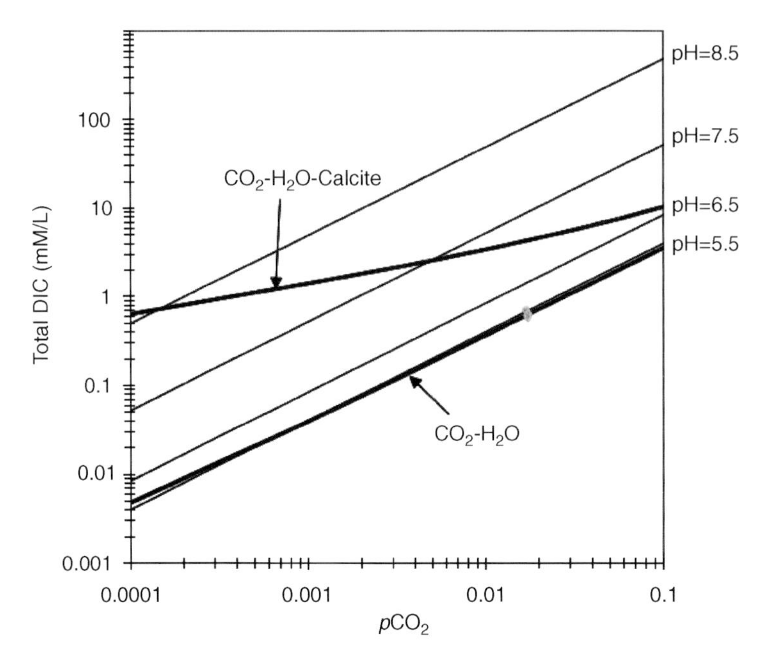

Figure 5.3.1 The calculated total DIC as a function of pCO_2 in the H_2O-CO_2 system and in the H_2O-CO_2-$CaCO_3$(calcite) system (lower and upper heavy lines, respectively). The four light solid lines represent H_2O-CO_2 systems at the indicated fixed pHs. (Total pressure = 1 atm, temperature = 25°C).

carbon dioxide concentrations in soil profiles are microbial and root respiration on the one hand, and slow removal rate of CO_2 from the soil's atmosphere on the other hand. How irrigation with TWW affects soil pCO_2 is still an open question. One might expect that the substrate (e.g. SOC, nutrients) arriving with the TWW will enhance microbial respiration in the soil profile, but the load of substrate in TWW has decreased over the years as the regulations governing the treatment of wastewater have recommended reduced carbon concentrations. Priming effects, occurring after stimulation of microbial activity by the addition of easily degradable substances, could be found in the soils and were stronger for subsoils (1 m depth). On the other hand, 75 yrs of irrigation with TWW did not show significant differences in soil organic carbon (SOC) relative to irrigation with FW (Eshel et al., 2007). Conversely, Jueschke et al. (2008) reported that after 10–40 yrs of irrigation with TWW (relative to FW), SOC accumulated in the topsoil and tended to decrease in the subsoil. They hypothesize that the SOC in these horizons gets depleted because of stimulated microbial activity due to substrate inputs from the effluents.

A recent 2-yr measurement campaign in avocado orchards in central Israel did not show a significant effect of the irrigation water quality on the pCO_2 profiles (Eshel et al., 2008) (Fig. 5.3.2). They found that the main factors that affected the soil pCO_2 are soil moisture in areas that do not receive irrigation (between rows, data not present), and temperature under the trees in irrigated areas where moisture is not a limiting factor.

Most of the geochemical models have assumed fixed pCO_2 or pH, which is not always a valid assumption for the vadose zone. On the other hand, Suarez and Simunek (1997) allowed pH in their UNSATCHEM model to vary as a result of changes in pCO_2 production due to respiration and limited diffusion rates. The model also considered the

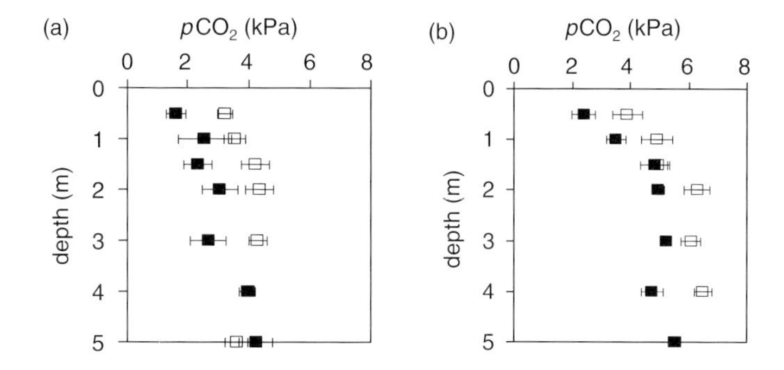

Figure 5.3.2 The average pCO_2 profile of 2 yrs of measurements under the avocados trees in orchard irrigated with (a) treated effluent and (b) freshwater. Filled symbols represent the winter months and open symbols represent the summer months (Eshel et al., 2008).

physical and chemical characteristics of the profile and the mechanisms of calcite dissolution (kinetics versus equilibrium). They were able to simulate the variation in pH (pH = 8–7) in the soil profile as a function of time after irrigation events, as it responded to pCO_2 changes. The use of this model required a large number of physical, mineralogical, chemical, and biological parameters to be measured at any site where it would be applied. It also assumes the soil pH will follow the carbonate mineral response to pCO_2.

5.3.3.2 Soil pH

Soil pH is one of the important parameters describing the soil environment. Many soil reactions are pH-dependent and therefore soil pH determination is a routine soil analysis. The common method for determining soil pH in the laboratory is by measuring the pH with a combined glass electrode directly in a soil suspension/solution prepared several minutes to some hours earlier. Thomas (1996) argued that soil pCO_2 most affects the pH of calcareous soil and that this effect should not be as important when the soil pH is determined on previously dried and ground soil samples. On the other hand, we should be aware that microbial activity may be enhanced shortly after dry soil is rewetted (Kieft et al., 1987) leading to pCO_2 changes in the measured sample.

More than 60 yrs ago, Haas and Compton (1941) reported that the pH measured in situ might be lower by 1–2.5 pH units compared to pH measured in the laboratory. They related their findings to the effect of fertilizers and field moisture content. Following this work, Whitney and Gardner (1943) studied the effect of pCO_2 on soil pH in 18 different soils from Colorado, including alkali-calcareous, calcareous, and non-calcareous soils. They found that the soil pH declined from 1 to 3 units as the pCO_2 increased from 0.0003 atm to 0.77 atm depending on the soil characteristics. They argued that a single measurement of soil pH under ambient conditions did not represent the real soil pH in the field. They also suggested that the findings reported by Haas and Compton (1941) were a result of the effect of the higher pCO_2 in the soil profile relative to the conditions in the laboratory after exposing the soils to the atmosphere and mixing them with distilled water that had a low pCO_2.

In a recent laboratory experiment, under controlled conditions, the responses of soil pH to changes in pCO_2 were tested in 20 different soils collected from all over California

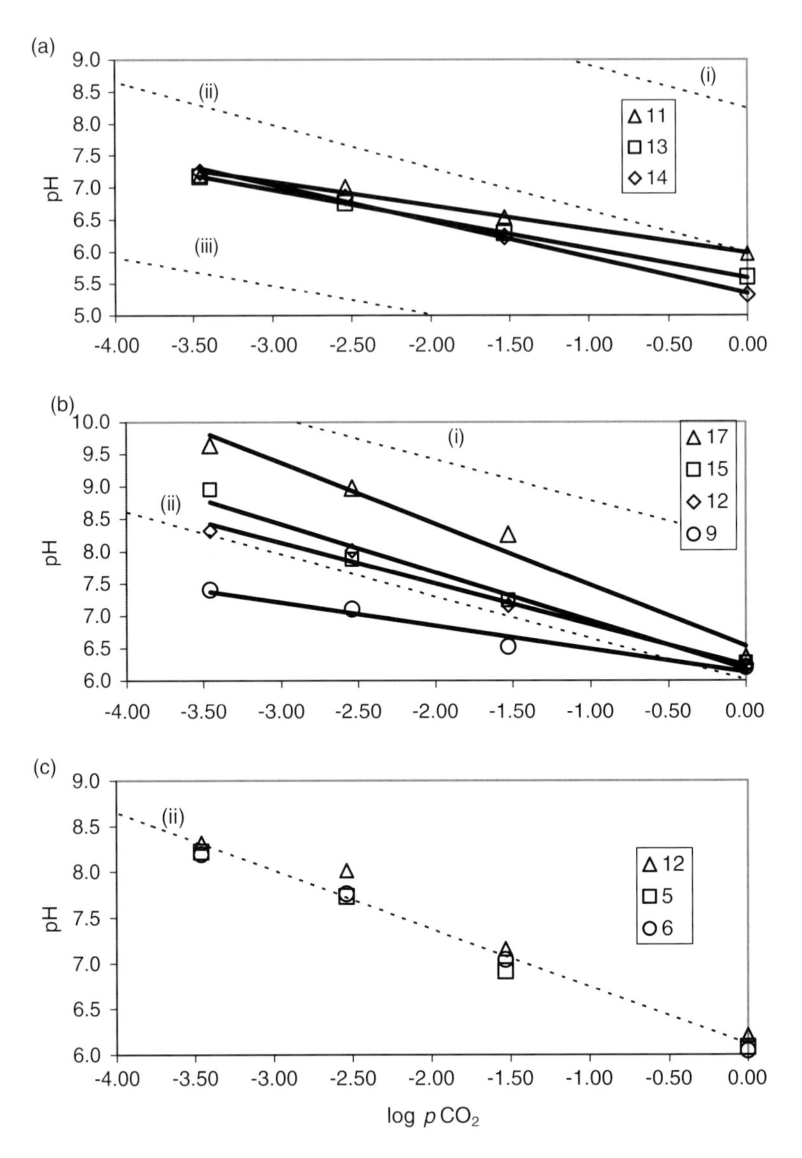

Figure 5.3.3 Soil pH response to changes in pCO_2. (a) Soils with a similar pH under ambient conditions but their pH diverged under higher pCO_2. (b) Soils with varied pH under ambient conditions that reach a similar pH under pCO_2 of 99.8%. (c) Selected calcareous soils with a pH response as the calculated calcite pH response to pCO_2. The dashed lines represent the equilibria of: (i) sodium carbonate, (ii) calcite, (iii) CO_2 and water.

(Eshel, 2005). As expected, the higher the ambient soil pH, the higher was the decline in soil pH (up to 3.5 pH units) following the increase in pCO_2. These measurements suggest diverse responses of soil pH to pCO_2 (Fig. 5.3.3). In some of the calcareous soils, the change in pH with pCO_2 followed calcite equilibrium, but in others, the measured pH values did not follow the expected pattern. Six of the studied soils showed some ability to buffer the pH, and the expected decline in the pH did not occur up to some threshold pCO_2.

Statistical analysis suggested that iron and manganese oxides play a significant part in the soil's buffering capability. N-containing species, organic and inorganic carbon and citrate-bicarbonate-dithionite (CBD)-extractable Al, Mn, and Fe may also contribute to the variability among soils' pH under various pCO_2 levels. These laboratory results clearly demonstrate the difficulties involved in modeling the response of soil pH to changes in pCO_2. However, this kind of laboratory measurement may provide a more accurate characterization of the soil's pH behavior.

5.3.3.3 Stable isotopes

Carbonate minerals commonly appear in soils as a mixture of carbonates that precipitate in situ (pedogenic carbonate) and carbonate that originates from the soil parent material (detrital carbonate). It is difficult to identify the source of carbonate, but stable carbon isotopes have been used to quantify pedogenic carbonates (e.g. Nordt et al., 1998). In order to evaluate the impact of irrigation on the SIC pool, it is important to evaluate the contribution of these two forms. Several studies have suggested that stable isotope ratios of carbon and oxygen can be a useful tool for identification of pedogenic carbonate in soils and for the determination of their proportions in the total carbonate (Magaritz and Amiel, 1980, 1981; Rabenhorst et al., 1984; Amundson and Lund, 1987; West et al., 1988; Mermut et al., 2000).

Root respiration and SOC decomposition have a significant effect on the isotopic values of soil profile CO_2, the corresponding dissolved inorganic C isotopic values, and the isotopic values of pedogenic carbonate that may have precipitated (Clark and Fritz, 1997). The common value of ${}^{13}\delta C_{carb}$ for carbonates that originate from sedimentary rocks is $0 \pm 1‰$ (Clark and Fritz, 1997). Pedogenic carbonates have more negative ${}^{13}\delta C_{carb}$ values that can range between $-0.6‰$ and $-14.6‰$, depending on the C source. Atmospheric $CO_2\,{}^{13}\delta C$ was about 6‰ under preindustrial conditions and about 8‰ under modern atmospheric conditions. The C_3 plants have an average ${}^{13}\delta C_{org}$ of $-27.1‰$, whereas C_4 plants have an average ${}^{13}\delta C_{org}$ of $-12.1‰$ (Mermut et al., 2000). Because the SOC of croplands is commonly a mixture of C_4 and C_3 plants, it is important to measure both the ${}^{13}\delta C_{carb}$ and ${}^{13}\delta C_{org}$.

Table 5.3.1 Typical $\delta^{13}C_{org}$ values for treated wastewater and biosolids

$\delta^{13}C$	Location	Reference
−19.8 to −24.7	Saskatchewan, Canada	Wayland and Hobson, 2001
−23.5[a]	Whites Point, CA	Myers, 1974
−23.0	Glil Yam, Israel	Affek et al., 1998
−21.4[a]	Yonkers, NY	Van Dover et al., 1992
−23.2[a]	Bergen, NJ	Van Dover et al., 1992
−24.7[a]	Middlesex, NJ	Van Dover et al., 1992
−23.7[a]	Providence, RI	Van Dover et al., 1992

[a]The values are for bio-solids.
Adapted from Eshel et al., 2007, with permission.

Carbon originating from TWW and from biosolids also should be considered when these are applied to the field. As expected, typical $^{13}\delta C_{org}$ values for treated effluent and biosolids are in the range of $^{13}\delta C$ values between C_3 and C_4 plants (Table 5.3.1).

5.3.4 The effect of irrigation on soil carbonate

Irrigation changes the soil climate (e.g. decreases temperature and increases the moisture and the number of drying and wetting cycles), and as result may accelerate many pedogensis processes, including carbonate exclusion/accumulation in the soil profile (McCaslin and Lee-Rodriquez, 1979; Magaritz and Amiel, 1981; Amundson and Smith, 1988; Nimkar et al., 1992; Eshel et al., 2007). Irrigation of arid zone soils is bound to alter the size, properties, and location of the SIC pool as influenced by irrigation amount and method, water quality, drainage, leaching, plowing, and application of fertilizers and other soil amendments (e.g., gypsum to Aridisols and liming material to Ultisols and Oxisols (Lal, 2001b)). Magaritz and Amiel (1981) used stable C isotope analysis to estimate that about 500 Mg/ha of carbonate was removed from the upper 2.5 m of a calcareous soil profile containing 16 000 Mg/ha carbonate during 40 yrs of irrigation with groundwater in Israel. They also estimated that about 800 Mg/ha was dissolved and reprecipitated in the upper 2.5 m of the soil profile. Amundson and Smith (1988) used stable O isotope analysis to estimate that about 56 Mg/ha carbonates were newly formed during 8 yrs of wellwater irrigation on calcareous soils in the San Joaquin Valley in California. McCaslin and Lee-Rodriquez (1979) compared two adjacent fields in New Mexico that had been irrigated for about 40 yrs with TWW and with FW and found more carbonate (1370 Mg/ha) in the upper 1 m of the TWW field. but they did not discuss the carbonate origin (detrital versus pedogenic). Nimkar et al. (1992) found an accumulation of 337 Mg/ha of carbonates in the upper 1 m of the soil profile following 7 yrs of irrigation of calcareous soils with rather saline water from the Purna River in India. But we should remember that conclusions based on differences in carbonate content alone are likely to be misleading.

Eshel et al. (2007) compared the SIC composition in two adjacent irrigated fields, one with FW and the second irrigated with TWW, for more than 75 years in Bakersfield, California (mapped as Kimberlina fine sandy loam) to a reference site that was not irrigated. Similar to McCaslin and Lee-Rodriquez (1979), Eshel et al. (2007) also found higher total carbonate content under TWW irrigation relative to both the field irrigated with FW and the no irrigation site. A key for a comparison like this is an assumption that the fields had an identical carbonate content and distribution before irrigation started. The interpretation of Eshel et al. (2007) using stable carbon isotopic analysis suggests about 60% of the carbonate found below 2 m in TWW irrigation was detrital carbonate and that large amounts of detrital carbonate in the TWW-irrigated field below 2 m was not found at the other sites.

5.3.5 Environmental aspects

Soils play an important role in the global carbon cycle. They are the third largest active carbon pool (1550 Pg of organic C and 750 Pg inorganic C to a 1 m depth) after the hydrosphere (38 000 Pg) and the lithosphere (5000 Pg) (Lal, 2001a). The contribution of soils as either a source or a sink for greenhouse gases, in general, and that of CO_2, in particular, has been the focus of many studies (e.g. Bouwman, 1990; Houghton, 1995; Gerzabek et al., 1997; Schlesinger, 2000; Drees et al., 2001; Lal, 2001b). Cultivation is a practice that generates significant reductions in SOC (Stevenson, 1994; Lal, 2001c). The loss is related to the increase in soil aeration during cultivation and to the wetting and drying cycles caused by irrigation (Stevenson, 1994). Adoption of specific agricultural practices (e.g. organic farming, minimum tillage, no-tillage) to increase the sequestration of carbon has produced diverse results with respect to net C sequestration (Schlesinger, 2000). Much less is known about the cycle of SIC or methods to enhance SIC sequestration.

Carbon sequestration generally refers to trapping/fixing carbon dioxide from the atmosphere in the liquid and solid phases as SOC or SIC carbon forms. Forcing equation 5.3.1 to the right is considered SIC sequestration. Changes in solid SIC content may provide the net contribution (whether a source or a sink of CO_2) of the SIC carbon pool to the atmosphere (the long-term effect). On the other hand, studying the SIC in the liquid and gas phases may provide the necessary information on the dynamic (short-term effects) and the mechanisms of this contribution.

As we mentioned earlier in the chapter, soils from semiarid and arid regions often contain Ca and Mg carbonates. Carbonate accumulation in arid soils can result from all or part of the following: redistribution of either detrital carbonate in the soil profile or of exogenous material (e.g. dust-borne carbonate) and in situ formation due to mineral weathering/precipitation. The redistribution can result from dissolution and re-precipitation of inherited carbonate minerals due to cycles of soil wetting and drying (Lippmann, 1973; Knight, 1991), or by translocation as colloidal suspensions with the soil solution (Dan, 1983; Baghernejad and Dalrymple, 1993; Neaman et al., 2000). Neither of these processes produces a net sequestration of atmospheric carbon in the soil.

Irrigation of arid zone soils is bound to alter the size, properties, and location of the SIC pool depending on irrigation amount and method, water quality, drainage, leaching, plowing, application of fertilizers, and other soil amendments (e.g. gypsum to Aridisols and liming material to Ultisols and Oxisols (Lal, 2001b)). The question remains of whether irrigation management can increase net atmospheric carbon sequestration. Secondary precipitation of carbonates with divalent cations is possible under limited leaching and dry conditions (Dan, 1983; Sposito, 1989; Nordt et al., 2000).

The answer to the above question is still unclear, in part, because there is no agreement in the literature about the role of the precipitation of carbonate minerals in sequestration. One school argues that precipitation of solid carbonates should be considered as sequestration (e.g. Lal and Kimble, 2000a,b; Eswaran et al., 2000; Scharpenseel et al., 2000; Ryskov et al., 2000; Mermut et al., 2000; Emmerich, 2003), whereas a second school argues that dissolution of solid carbonates should be considered as sequestration (e.g. Suarez, 2000; Nordt et al., 2000; Drees et al., 2001).

We adopted the second school approach as; pCO_2 in the soil is 10–100 times more than the atmosphere (Amundson and Davidson, 1990; Fig. 5.3.2) and the pH of arid soils is typically above 7 under $pCO_2 = 35\,000$ ppm (Eshel, 2005). It is reasonable to assume that carbonic acid is the major acidity source involved in carbonate mineral dissolution in calcareous soils (equations 5.3.1–5.3.4). Moreover, $^{87}Sr/^{86}Sr$ studies suggests that less than 5% of the Ca of the pedogenic carbonate is derived from silicate mineral weathering (Capo et al., 1998).

Considering the overall reaction of calcite equilibrium (equation 5.3.4), it is clear that when the acidity source is carbonic acid, dissolving 1 mole of calcite will consume 1 mole of CO_2 from the gas phase and precipitation of 1 mole of calcite will produce 1 mole of CO_2.

A recent study by Eshel et al. (2007) was one of the first to investigate the impact of irrigation with effluents on carbon sequestration. The study compared the SIC composition in two irrigated fields to a reference site that was not irrigated at Bakersfield, California. Based on stable carbon isotope analyses and morphological field evidence it was estimated that long-term irrigation (about 75 yrs) with FW depleted SIC. If it is assumed that the major source for this SIC depletion was carbonic acid, we might be able to account for 7.2 kg/m^2 SIC sequestration. On the other hand, irrigation for the same period with TWW changed SIC content to between 0.9 and 2.4 kg/m^2 in the studied depth of the profile.

Another important environmental aspect of carbonate and Ca and TWW irrigation relates exogenous elements to carbonate minerals. Calcite and other carbonates preferentially adsorb, complex, and include as solid solution in the lattice $Cd > Zn \geq Mn > Co > Ni$ (Papadopoulos and Rowell, 1989; Rimstidt et al., 1998; Zachara et al., 1991). Lin et al. (2004) indicated that surface adsorption and surface precipitation of Cu, Ni, and Zn on Fe oxides and/or carbonate minerals may be the primary retention mechanisms of these metals in the effluent-recharged basin soils. In addition, presence of carbonate minerals in soils creates favorable pH conditions for low mobility of heavy metals in soils (Chuan et al., 1996; Han and Banin 1997; Banin et al., 2002). Using TWW for irrigation significantly increased the concentration of soil P, because of retention in the plough layer (Kardos and Hook, 1976; Howe and Wagner, 1999). The retention of P in the solid phase of soils is due to its adsorption on mineral surfaces (Lin et al., 2006) and/or its precipitation with Al, Fe and Ca (Rhue and Harris, 1999).

5.3.6 Summary

Carbonate minerals are an important source and sink for C and Ca in soils, both of which have an important function in the soil environment. Carbonate and Ca play an important role in soil chemistry and indirectly affect soil fertility. Irrigation of arid zone soils is bound to alter the size, properties, and location of the SIC pool as influenced by irrigation amount, irrigation method, and water quality. Evaluating the contribution of the irrigation parameters mentioned above based on differences in carbonate content alone are likely to be misleading. This kind of study requires using more complicated analysis such as stable and radiocarbon isotopic analysis of SIC and SOC. Moreover, if the net carbon sequestration is in question, the answer requires a quantitative measure of the different soil acidity sources, soil pCO_2 profiles, and the soil pH response to changes in soil pCO_2.

References

Affek, H.P., Ronen, D. and Yakir, D. (1998) Production of CO_2 in the capillary fringe of a deep phreatic aquifer. *Water Resources Research* **34**, 989–996.

Amundson, R.G. and Davidson, E.A. (1990) Carbon dioxide and nitrogenous gases in the soil atmosphere. *Journal Geochemical Exploration* **38**, 13–41.
Amundson, R.G. and Lund, L.J. (1987) The stable isotope chemistry of a native and irrigated Typic Natargid in the San Joaquin Valley of California. *Soil Science Society of America Journal* **51**, 761–767.
Amundson, R.G. and Smith, V.S. (1988) Effects of irrigation on the chemical properties of a soil in western San Joaquin Valley, California. *Arid Soil Research and Rehabilitation* **2**, 1–17.
Baghernejad, M. and Dalrymple, J.B. (1993) Colloidal suspensions of calcium carbonate in soils and their likely significance in the formation of calcic horizon. *Geoderma* **58**, 17–41.
Banin, A., Lin, C., Eshel, G., Roehl, K.E., Negev, I., Greenwald, D., Shachar, Y. and Yablekovitch, Y. (2002) Geochemical processes in recharge basin soils used for municipal effluents reclamation by the soil-aquifer treatment (SAT) system. In: *Management of aquifer recharge for sustainability. Proceedings of the 4th International Symposium Aquifer Recharge (ISAR-4)* (ed. P.J. Dillon) pp. 327–332. Swets&Zeitlinger, Lisse, Netherlands.
Bouwman, A.F. (1990) *Soil and the Greenhouse Effect*. John Wiley & Sons, New York, NY.
Bui, E.N., Loeppert, R.H. and Wilding, L.P. (1990) Carbonate phases in calcareous soils of the western USA. *Soil Science Society of America Journal* **54**, 39–45.
Capo R.C., Stewart, B.W. and Chadwick, O.A. (1998) Strontium isotopes as tracers of ecosystem processes: theory and methods. *Geoderma* **82**, 197–225.
Chuan, M.C., Shu, G. Y and Liu, J.C. (1996) Solubility of heavy metals in a contaminated soil: Effects of redox potential and pH. *Water Air Soil Pollution* **90**, 543–556.
Clark, I.D. and Fritz, P. (1997) *Environmental Isotopes in Hydrology*. CRC/Lewis Publ., Boca Raton, FL.
Dan, J. (1983) Soil chronosequences in Israel. *Catena* **10**, 287–319.
Drees, L.R., Wilding, L.P. and Nordt, L.C. (2001) Reconstruction of soil inorganic and organic carbon sequestration across broad geoclimatic regions. In: *Soil carbon Sequestration and the Greenhouse Effect* (ed. R. Lal) pp. 155–172. SSSA Special Publication 57, SSSA, Madison, WI.
Emmerich, W.E. (2003) Carbon dioxide fluxes in a semiarid environment with high carbonate soils. *Agricultural and Forest Meteorology* **116**, 91–102.
Eshel G. (2005) *The role of soil inorganic carbon in carbon sequestration*. Ph.D. dissertation. University of California, Davis.
Eshel, G., Fine, P. and Singer, M.J. (2007) Total soil carbon and water quality: an implication for carbon sequestration. *Soil Science Society of America Journal* **71**, 397–405.
Eshel, G., Lifshithz, D., Sternberg, M., Ben-Dor, E., Bonfile, D.J., Arad, B., Mingelgrin, U., Fine, P. and Levy, G.J. (2008) The influence of agricultural management on soil's CO_2 regime in semi-arid and arid regions. Abstract #B11D-0396. AGU Fall Meeting, San Francisco, CA.
Eswaran, H., Reich, P.F., Kimble, J.M., Beinroth, F.H., Padmanabhan, E. and Mocharoen, P. (2000) Global carbon stocks. In: *Global Climate and Pedogenic Carbonates* (eds R. Lal et al.), pp. 15–26. CRC/Lewis Publ., Baca Raton, FL.
Gerzabek, M.H., Pichlmayet, F., Kirchmann, H. and Haberhauer, G. (1997) The response of soil organic matter to manure amendments in a long term experiment at Ultuna, Sweden. *European Journal of Soil Science* **48**, 273–282.
Haas, A.R.S. and Compton, O.C. (1941) The pH of irrigated orchard soils. *Soil Science* **52**, 309–333.
Han, F. and Banin, A. (1997) Long-term transformations and redistribution of potentially toxic heavy metals in arid-zone soils. *Water Air Soil Pollution* **95**, 399–423.
Houghton, R.A. (1995) Changes in storage of terrestrial C since 1850 In: *Soil and Global Change* (eds R. Lal et al.), pp. 45–65. CRC/Lewis Publ., Boca Raton, FL.
Howe, J. and Wagner, M.R. (1999) Effects of pulp mill effluent irrigation on the distribution of elements in the profile of an arid region soil. *Environmental Pollution* **105**, 129–135.
Jenny, H. (1941) *Factors of Soil Formation*. McGraw-Hill, New York.
Jueschke, E., Marschner, B., Tarchitzky, J. and Chen, Y. (2008) Effects of treated wastewater irrigation on the dissolved and soil organic carbon in Israeli soils. *Water Science Technology* **57**, 727–733.
Kardos, L.T. and Hook, J.E. (1976) Phosphorus balance in sewage effluent treatment soils. *Journal of Environmental Quality* **5**, 87–90.
Kelly, E.F., Amundson, R.G., Marino, B.D. and DeNiro, M.J. (1991) Stable carbon isotopic composition of carbonate in Holocene grassland soils. *Soil Science Society of America Journal* **55**, 1651–1658.

Kelly, E.F. and Yonker, C.M. (2005) Time. In: *Encyclopedia of soils in the environment* (ed. D. Hillel), pp. 536–539. Elsevier, Amsterdam.
Kieft, T.L., Soroker, E. and Firestone, M. (1987) Microbial biomass response to a rapid increase in water potential when dry soil is wetted. *Soil Biology and Biochemistry* **19**, 119–126.
Knight, W.G. (1991) Chemistry of arid region soils. C. 4. In: *Semiarid Lands and Deserts: Soil Resource and Reclamation* (ed. Skujins, J.C). Marcel Dekker, Inc., New York.
Kraimer, R.A., Monger, H.C. and Steiner, R.L. (2005) Mineralogical distinctions of carbonates in desert soils. *Soil Science Society of America Journal* **69**, 1773–1781.
Lal, R. and Kimble, J.M. (2000a) Pedogenic carbonate and global carbon cycle. In: *Global Climate and Pedogenic Carbonates* (eds R. Lal et al.), pp. 1–14. CRC/Lewis Publ., Boca Raton, FL.
Lal, R. and Kimble, J.M. (2000b) Inorganic carbon and global C cycle: research and development priorities. In: *Global Climate and Pedogenic Carbonates* (eds R. Lal et al.), pp. 291–302. CRC/Lewis Publ., Boca Raton, FL.
Lal, R. (2001a) Soils and the greenhouse gas effect. In: *Soil carbon sequestration and the greenhouse effect* (ed. R. Lal), pp. 1–8. SSSA Special Publication 57. SSSA. Madison, WI.
Lal, R. (2001b) Myths and facts about soils and the greenhouse effect. In: *Soil carbon sequestration and the greenhouse effect* (ed. R. Lal), pp. 9–26. SSSA Special Publication 57. SSSA. Madison, WI.
Lal, R. (2001c) Soil carbon dynamics in cropland and rangeland. *Environmental Pollution* **116**, 353–362.
Lin, C., Chachar, Y. and Banin, A. (2004) Heavy metal retention and partitioning in a large-scale soil-aquifer treatment (SAT) system used for wastewater reclamation. *Chemosphere* **57**, 1047–1058.
Lin, C., Eshel, G., Roehl, K.E., Negev, I., Greenwald, D., Shachar, Y. and Banin, A. (2006) In situ studies of P accumulation in soil/sediment profiles used for large-scale wastewater reclamation. *Soil Use and Management* **22**, 143–150.
Lippmann, F. (1973) Sedimentary carbonate minerals. In: *Minerals, Rocks and Inorganic Materials, Monograph Series of Theoretical and Experimental Studies, vol. 4* Springer-Verlag, Berlin.
Magaritz, M., and Amiel, A.J. (1980) Calcium carbonate in a calcareous soil from the Jordan Valley, Israel: Its origin revealed by stable carbon isotope method. *Soil Science Society of America Journal* **44**, 1059–1062.
Magaritz, M., and Amiel A.J. (1981) Influence of intensive cultivation and irrigation on soil properties in Jordan Valley, Israel: Recrystallization of carbonate minerals. *Soil Science Society of America Journal* **45**, 1201–1205.
McCaslin, B.D. and Lee-Rodriquez, V. (1979) Effect of using sewage effluent on calcareous soils. In: Proceedings of the International Arid Lands Conference on Plant Resources, pp 598–614. Texas Tech University, Lubbock, TX.
Mermut, A.R., Amundson, R. and Cerling, T.E. (2000) The use of stable isotopes in studying carbonate dynamics in soils. In: *Global Climate and Pedogenic Carbonates* (eds R. Lal et al.), pp. 65–85. CRC/Lewis Publ., Boca Raton, FL.
Myers, E.P. (1974) *The concentration and isotopic composition of carbon in marine sediments affected by sewage discharge*. PhD diss. Cal Tech. Pasadena, CA.
Neaman, A., Singer, A. and Stahar, K. (2000) Dispersion and migration of fine particles in two palygorskite-containing soils of Jordan Valley. *Journal of Plant Nutrition and Soil Science* **163**, 537–547.
Nimkar, A.M., Deshpande, S.B. and Babrekar, P.G. (1992) Evaluation of salinity problem in swell-shrink soils of part of Purna Valley, Maharashtra. *Agropedology* **2**, 59–65.
Nordt, L.C., Hallmark, C.T., Wilding, L.P. and Boutton, T.W. (1998) Quantifying pedogenic carbonate accumulations using stable carbon isotopes. *Geoderma* **82**, 115–136.
Nordt, L.C., Wilding, L.P. and Drees, L.R. (2000) Pedogenic carbonate transformation in leaching soil system: implication for the global C cycle. In: *Global Climate and Pedogenic Carbonates* (eds R. Lal et al.), pp. 43–46. CRC/Lewis Publ., Baca Raton, FL.
Nunes, J.M., López-Piñeiro, A., Albarrán, A., Muñoz, A. and Coelho, J. (2007) Changes in selected soil properties caused by 30 years of continuous irrigation under Mediterranean conditions. *Geoderma* **139**, 321–328.
Papadopoulos, P. and Rowell, D.L. (1989) The reactions of copper and zinc with calcium carbonate surfaces. *Journal of Soil Science* **40**, 39–48.
Parkin, T.B. and Meisinger, J.J. (1989) Denitrification below the crop rooting zone as influenced by surface tillage. *Journal of Environmental Quality* **18**, 12–16.

Presley, D.R., Ransom, M.D., Kluitenberg, G.J. and Finnell, P.R. (2004) Effects of thirty years of irrigation on the genesis and morphology of two semiarid soils in Kansas. *Soil Science Society of America Journal* **68**, 1916–1926.

Rabenhorst, M.C., Wilding, L.P. and West, L.T. (1984) Identification of pedogenic carbonate using stable carbon isotope and microfabric analysis. *Soil Science Society of America Journal* **48**, 125–132.

Rhue, R.D. and Harris, W.G. (1999) Phosphorus sorption/desorption reactions in soils and sediments. In: *Phosphorus Biogeochemistry in Subtropical Ecosystems* (eds K.R. Reddy, G. A. O'Connor and C. L. Schelske), pp. 187–206. Lewis Publishers, Boca Raton, FL.

Rimstidt, J.D., Balog, A. and Webb, J. (1998) Distribution of trace elements between carbonate minerals and aqueous solutions. *Geochimica et Cosmochimica Acta* **62**, 1851–1863.

Rubio, A., and Escudero, A. (2005) Effect of climate and physiography on occurrence and intensity of decarbonation in Mediterranean forest soils of Spain. *Geoderma* **125**, 309–319.

Ryskov, Y.A., Borisov, A.V., Oleinik, S.A., Ryskova, E.A. and Demkin, V.A. (2000) The relationship between lithogenic and pedogenic carbonate fluxes in steppe soils, and regularities of their profile dynamics for the last four millennia. In: *Global Climate and Pedogenic Carbonates* (eds R. Lal,et al.), pp. 121–134. CRC/Lewis Publ., Boca Raton, FL.

Scharpenseel, H.W., Mtimet, A. and Freytag, J. (2000) Soil inorganic carbon and global change. In: *Global Climate and Pedogenic Carbonates* (eds R. Lal,et al.), pp. 27–42. CRC/Lewis Publ., Baca Raton, FL.

Schlesinger, W.H. (2000) Carbon sequestration in soils: some cautions amidst optimism. *Agriculture, Ecosystems and Environment* **82**, 121–127.

Sposito, G. (1989) *The Chemistry of Soils*. Oxford U. Press, New York, NY.

Stevenson, F.J. (1994) *Humus Chemistry*. 2nd ed. John Wiley & Sons, Inc., New York. NY.

Suarez, D. L., and Simunek, J. (1997) UNSATCHEM: Unsaturated water and solute transport model with equilibrium and kinetic chemistry. *Soil Science Society of America Journal* **61**, 1633–1646.

Suarez, D.L. (2000) Impact of agriculture on CO_2 as affected by changes in inorganic carbon. In: *Global Climate and Pedogenic Carbonates* (eds R. Lal et al.), pp. 257–272. CRC/Lewis Publ., Boca Raton, FL.

Thomas, G.H. (1996) Soil pH and soil acidity. In: *Methods of Soil Analysis, Part 3. Chemical Methods* (eds Sparks et al.), pp. 475–490. Soil Science Society of America Book Series 5. SSSA and ASA, Madison WI.

Van Dover, C.L., Grassle, J.F., Fry, B., Garrit, R.H. and Starczak, V.R. (1992) Stable isotope evidence for sewage-derived organic material into a deep-sea food web. *Nature* **360**, 153–156.

Wayland, M. and Hobson, K.A. (2001) Stable carbon, nitrogen, and sulfur isotope ratios in riparian food webs on rivers receiving sewage and pulp-mill effluents. *Canadian Journal of Zoology* **79**, 5–15.

West, L.T., Drees, L.R., Wilding, L.P. and Rabenhorst, M.C. (1988) Differentiation of pedogenic and lithogenic carbonate forms in Texas. *Geoderma* **43**, 271–287.

Whitney, R.S. and Gardner, R. (1943) The effect of carbon dioxide on soil pH. *Soil Science* **55**, 127–141.

Wood, B.D., Keller, C.K. and Johnstone, D.L. (1993) In situ measurement of microbial activity and controls on microbial CO_2 production in the unsaturated zone. *Water Resources Research* **29**, 647–59.

Zachara, J.M., Cowan, C.E. and Resch, C.T. (1991) Sorption of divalent metals on calcite. *Geochimica et Cosmochimica Acta* **55**, 1549–1562.

Chapter 6

Toxic elements

6.1 Boron

Uri Yermiyahu, Alon Ben-Gal and Rami Keren

6.1.1 Introduction

In arid and semiarid regions treated wastewater (TWW) is quickly becoming a fundamental source of water for irrigation. Boron (B) is often present in freshwater in dry areas and is accumulated and concentrated in TWW as a consequence of domestic, industrial, and agricultural water use. Sources of B in sewage systems and TWW include: human and animal excretions, detergents, laundry powders, effluents from paper mills and metal-coating processes, among others. Boron can be toxic to plants and therefore, when B in TWW reaches concentrations high enough to limit plant growth, the potential for agricultural utilization of the water is threatened.

Boron is an essential micronutrient for normal growth of most plants. At high environmental levels though, B is markedly toxic to plants. There is a relatively small range differentiating between soil solution B concentrations that are deficient and those that cause toxicity in plants (Keren and Bingham, 1966).

Boron is taken up into plants from the aqueous soil solution and the rate of B uptake is largely a function of the solution's B concentration. The concentration of B in soil solution is strongly affected by B-soil chemical interactions – primarily adsorption–desorption processes that are, in turn, affected by soil constituents and conditions. The nature of treated wastewater (TWW) and the soils in the dry (hyperarid, arid, and semiarid) regions where such water is typically utilized for irrigation have consequential effects on both soil B reactions and plant B uptake. Treated wastewaters characteristically have high concentrations of salts and dissolved organic matter (DOM). Soils of the relatively arid zones can typically be saline, contain high natural levels of B, and have relatively low organic matter contents (Tsadilas, 1997). Boron toxicity in plants as a result of irrigation with TWW is a phenomenon encountered throughout the world's dry regions. In spite of

Treated Wastewater in Agriculture, First Edition, edited by Guy J. Levy, Pinchas Fine and Asher Bar-Tal

the extent of excess B in soils and water of the relatively arid zones, research and knowledge concerning B toxicity is minor compared to that of B deficiency.

Boron in soils and agricultural systems has been widely reviewed in the past. Keren and Bingham (1966) and Goldberg (2001) thoroughly discussed B chemistry and B-soil interactions, Gupta et al., (1996) offered a comprehensive look at B in plants, Nable et al. (1997) and Stangoulis and Reid (2002) specifically dealt with B toxicity, as did Reid et al. (2004) in a physiologically based discussion. Readers wishing in-depth treatment of these topics are encouraged to turn to those sources. This chapter aims to focus on B introduced to agricultural systems via irrigation with TWW. A basic introduction to B in soils and solutions and to B in the soil–water–plant continuum is provided in order to give the reader sufficient background to understand the issues of irrigation with TWW-B and the interaction of B with other components of TWW including DOM and salts. These topics are reviewed in terms of their relevance to agriculture and are discussed using case studies from Israel.

6.1.2 Boron in TWW

The concentration of B found in TWW is a function of the B concentration in the original water supply and of the rates of B discharged during industrial and domestic processes. In the USA, freshwater typically contains 0.2–0.4 mg B/l (Metcalf and Burton, 1983). In Israel, B concentration in tap water is usually somewhat lower (0.08–0.2 mg/l).

Industrial processes contributing B to sewage water include the use of starch adhesives in paper-recycling plants, borax cleaning solutions used in metal-coating plants, and the use and disposal of detergents. During the domestic use of water, B is incorporated into the sewage stream mainly as detergents and laundry powders (Tarchitzky and Chen, 2004), but also due to human excretions. Boron is removed from the blood of humans and animals in the form of borates and is excreted via the urine (WHO, 1998). The concentration of B in human urine is highly variable and ranges from 0.040 to 7.8 mg B/kg (Downing et al., 1985).

Bleaching agents are used in detergents to promote oxidation of stains (Ausimont, 1999). Boron is a component of many typical bleaching agents and is found in detergent powders as anhydrous or hydrate forms of sodium perborate ($NaB0_3$) including sodium perborate tetrahydrate ($NaB0_3{\cdot}4H_2O$) and sodium perborate monohydrate ($NaB0_3{\cdot}H_2O$). In Europe, between 0.26 and 0.73 mg B/l is added to sewage water as a result of detergent use (Tarchitzky and Chen, 2004).

Boron is similar to other elements including Na, Cl, and K in that it is not affected by primary or secondary treatment processes and therefore its concentration in TWW is closely related to that of the sewage water. The B concentrations found in several sources of sewage/effluent water are presented in Table 6.1.1.

In Israel, TWW is an important component of water available for irrigation. In 1997, B concentrations in Israeli wastewater ranged from 0.46 to 1.41 mg B/l (Tarchitzky et al., 1997). Monitoring of 117 reservoirs, responsible for the supply of 95 Mm^3 of TWW for irrigation in 2001, revealed that in 55% of the water B concentrations exceeded 0.6 mg/l (Tarchitzky and Chen, 2004), a level determined to have growth-limiting toxicity effects on a number of crops.

There are a number of methods that can be implemented in order to insure that B concentrations in TWW do not become excessive and do not limit the waters' potential for irrigation. Boron can theoretically be removed from freshwater prior to distribution. Such

Table 6.1.1 Boron concentrations in several sources of sewage water

Area/source[a]	B concentration (mg/l)
United States: industrial waste discharge	0.4–1.5
Europe: domestic and industrial	2
Egypt: sewage water	0.32–0.38
Israel: domestic and industrial	0.46–1.74
Sweden: effluent	0.34–0.436
Spain, Alicante: industrial wastewater	1.45
Spain, Elche: industrial waste	3
United Kingdom: municipal	1.21–3.96

[a]Tarchitzky and Chen, 2004; WHO, 1998.

treatment is costly and, as most B is added to the wastewater stream after distribution, is not very effective. Boron can be removed from sewage water as a part of tertiary treatment processes (Erdem Yilmaz et al., 1948). This option, although effective, is particularly, and likely prohibitively, expensive.

An alternative approach to reduce B levels in TWW, which has been adopted in Israel, is to decrease the B added by both industrial and domestic processes. According to the Israeli Ministry of Environment, in 1999, detergents contributed up to 90% of the B added to municipal sewage. In that year, Israel legislated an environmental standard (IS 438) that dictated the gradual reduction of B in detergents. Today, allowable B concentrations in washing powders are 6% of the pre-1999 levels, and most manufacturers have eliminated B from their detergents completely. Recent surveys have confirmed the anticipated reduction in B concentration as a result of the implementation of the standard. The median B concentration in treated TWWs, as reported by the Ministry of Agriculture and Rural Development, decreased from 0.41 to 0.17 mg/l between the years 2000 and 2004 (Tarchitzky and Chen, 2004).

6.1.3 Boron chemistry in aqueous media

Boric acid is moderately soluble in water and acts as a weak Lewis acid:

$$B(OH)_3 + 2H_2O \leftrightarrows B(OH)_4^- + H_3O^+ \tag{6.1.1}$$

The formation of the borate ion is spontaneous. The first hydrolysis constant of $B(OH)_3$, K_{h1}, is 5.8×10^{-10} at 20°C (Owen, 2008), and the other K_{h2} and K_{h3} values are 5.0×10^{-13} and 5.0×10^{-14}, respectively (Konopik and Leberl, 1984). A dissociation beyond $B(OH)_4^-$ is not necessary to explain the experimental data, at least below pH 13 (Ingri, 1995; Mesmer et al., 1986). Boron species other than $B(OH)_3$ and $B(OH)_4^-$, however, can, for most practical purposes, be ignored in soils. The first hydrolysis constant of $B(OH)_3$ varies with temperature from 3.646×10^{-10} at 178°K to 7.865×10^{-10} at 318°K (Owen and King, 1943).

Both $B(OH)_3$ and $B(OH)_4^-$ ion species are essentially monomeric in aqueous media at low B concentration ($\leq$0.025 mol/l). However, at high B concentration, polyborate ions exist in appreciable amount (Adams, 1973). The equilibria between boric acid, monoborate ions, and polyborate ions in aqueous solution are rapidly reversible. In aqueous solution, most of the polyanions are unstable relative to their monomeric forms $B(OH)_3$ and $B(OH)_4^-$ (Onak et al., 2007). Results of nuclear magnetic resonance (Good and Ritter, 1991) and Raman spectroscopy (Servoss and Clark, 1957) all lead to the conclusion that $B(OH)_3$ has a trigonal planar structure, whereas the $B(OH)_4^-$ ion in aqueous solution has a tetrahedral structure. This difference in structure can lead to differences in the affinity of clay and organic matter to these two B species.

6.1.4 Boron – soil interactions

Boron transport in soils and B available for plant uptake is foremost a function of B concentration in the soil solution. B in soil solution is determined by: (i) soluble B entering the soil–water system; either from the soil mineral fraction or from B imported through groundwater or irrigation water; and (ii) B adsorption–desorption reactions on the soil solid phase.

6.1.4.1 Solution B – adsorption processes

The relative solubility of minerals generally is not found to control the concentration of B in soil solution (Goldberg, 2001). The B concentration in the soil solution is instead much more highly influenced by B adsorption reactions. The amount of B adsorbed by soils varies greatly with the contents of soil constituents, the most important being clay minerals, sesquioxides and organic matter (Keren and Bingham, 1966). Calcium carbonate acts as an important B-adsorbing surface in calcareous soils. Boron adsorption is greater in soils having higher calcium carbonate content (Elrashidi and O'Connor, 1996). The mechanism of B adsorption is generally considered to be ligand exchange. On a per-weight basis, clay minerals adsorb significantly less B than do most oxide minerals and organic matter.

Layer-silicate clay minerals adsorb B; the order of B adsorption on a per-weight basis is kaolinite$<$montmorillonite$<$illite (Keren and Mezuman, 1987). The rate of B adsorption to clay minerals begins with an initial, fast adsorption reaction ($<$1 d), followed by a slow diffusion of B into the crystal lattice (Couch and Grim, 1993). Initially, B adsorbs to the surface hydroxyl groups on the clay particle edges. Subsequently, the B migrates and is incorporated structurally into tetrahedral sites, where it replaces structural silicon and aluminum.

Boron adsorbs on both crystalline and amorphous aluminum- and iron-oxides. Boron adsorption is greatest on freshly precipitated solids, decreases with aging due to increasing crystallinity (Sims and Bingham, 1968), and is greater for aluminum oxides than iron oxides (Keren and Gast, 1955; Goldberg and Glaubig, 1982) on a per-weight basis. Magnesium hydroxide can remove appreciable amounts of B from solution. Due to magnesium hydroxide coatings, silicate minerals containing mainly magnesium in their chemical formulas adsorb more B than silicate without magnesium. The appreciable B adsorption capacity of the sand and silt fractions of arid zone soils low in clays and organic matter may, therefore, result from clusters and coatings of magnesium hydroxide on silicate minerals (Rhoades et al., 1970).

Knowledge concerning B adsorption to organic materials is much less comprehensive than that concerning clays and metal oxides. Nevertheless, it is understood that soil organic matter significantly affects B distribution between the soil's solid and liquid phases and influences B uptake by plants. Boron adsorbs on all soil organic matter constituents (Gu and Lowe, 1938; Lemarchand et al., 1977), including natural organic matter like compost and peat (Yermiyahu et al., 1988; Sartaj and Fernandes, 2005). Garate and Meyer (1992) concluded that the main factors affecting B retention by organic matter were pH, Ca and fulvic acid content, and the humic-to-fulvic acid ratio. Interaction between B and organic matter can alter soil solution B. Boron deficiency has been observed in soils with high organic matter content (Hue et al., 1982; Mascarenhas et al., 1990; Liu et al., 1993; Valk and Bruin 1989). This deficiency has been shown to be related to the high affinity of organic matter to B (Berger and Pratt, 1985; Yermiyahu et al., 1988, 1995; Liu et al., 1993) and its removal from solution. In a case where a small amount of composted organic matter not rich with B was added to the soil (Loess, Calcic Haploxeralf), the number of adsorption sites was significantly increased and soil solution B and plant uptake were decreased (Yermiyahu et al., 2001). Alternatively, adding organic matter to soil has been reported to increase B content and its availability to plants (Blagojevic and Zarkovic, 1986; Pakrashi and Haldar, 1992).

Factors influencing B adsorption and desorption from soil constituents include: B concentration in the soil solution, solution pH, presence and type of exchangeable ions, ionic composition of the soil solution, wetting and drying cycles and temperature (Goldberg, 2001; Keren and Bingham, 1966). Boron adsorption on soils is particularly dependant on solution pH. Boron adsorption on soil constituents increases with increasing pH, reaches maximum levels at around pH 9 and decreases with further increases of pH (Keren et al., 1999; Keren and Mezuman, 1987; Goldberg et al., 1995). The pH dependence of B adsorption can be explained by competition between borate ions, boric acid, and hydroxyl ions for specific sorption sites (Keren and Mezuman, 1987). Quantitative relationships between solution concentrations of the B species are a function of pH but not by adsorption characteristic or number of adsorption sites (Keren and Mezuman, 1987; Yermiyahu et al., 1988).

Information concerning the reversibility of B adsorption reactions in soils is contradictory. For some soils, desorption isotherms correspond closely to B adsorption isotherms, whereas other soils exhibit hysteresis (Elrashidi and O'Connor, 1996). In investigations of the cause of hysteresis, no significant correlation was found with soil properties including: clay, organic carbon, pH, electrical conductivity (EC), cation exchange capacity, surface area, aluminum oxide content, or iron oxide content (Elrashidi and O'Connor, 1996). Mechanisms of irreversibility of B sorption have been shown to include: ligand exchange, formation of surface complexes and incorporation of B into clay mineral lattices (Goldberg, 2001). The desorption rate constant for borate anion from clay was three orders of magnitudes smaller than the adsorption rate constant (Keren et al., 1977). The intrinsic equilibrium constant obtained from the kinetic measurements ($\text{Log}_{10} K_{\text{kinetic}} = 3.15$) agreed relatively well with that calculated from the static studies ($\text{Log}_{10} K_{\text{static}} = 3.51$).

Boron adsorption and desorption from soil adsorption sites regulate B concentration in the soil solution. This regulation itself is a function of the changes in solution B concentration and of the affinity of the soil constituents for B. Thus, adsorption of B may buffer fluctuations in solution B concentration such that B concentrations in soil solution may vary only slightly with changes in soil water content (Mezuman and Keren, 1949; Keren et al., 2000, 1985b).

Boron can interact with various organic ligands present in soil solution to form dissolved complexes (Gu and Lowe, 1938). Lemarchand et al. (1977) found that boric acid interacted with carboxyl and hydroxyl groups, in ligand-exchange reactions characterized by approximately the same complexation constants found for humic acid at solution $pH < 7.8$. As the pH increased, the boric acid complexation decreased, whereas the borate complexation increased because its concentration in solution increases with pH. At pH above 9.5, a sharp decrease of B adsorption by dissolved organic matter was observed. This reduction was due to the competition between borate and OH^- ions for available sites. Similarly, B adsorption by dissolved organic matter from a municipal wastewater treatment plant increased significantly as the pH increased from 5.8 to 9.25, and maximized at pH 9.3 (Keren and Communar, 1993).

6.1.4.2 Boron – salinity interactions

Soils in semiarid and arid regions where little or no leaching occurs tend to have high levels of B, but also are high in overall salinity (Keren and Bingham, 1966). In these regions, water, including recycled wastewater available for irrigating agricultural crops, can also contain high concentrations of salts along with B (Nable et al., 1997; Feigin et al., 1986; Tsadilas, 1997). Crops in such areas are, therefore, prone to simultaneous exposure to stress-causing factors from both salinity and excess B.

Salinity can influence B-soil interactions both directly, by affecting sorption processes, and indirectly, by altering the soil's hydraulic conductivity, thereby affecting B transport and leaching. Boron adsorption on clays increases with the increasing ionic strength of the solution (Couch and Grim, 1993; Keren and O'Connor, 2003; Yermiyahu et al., 1988). The influence of ionic strength was found to be greater for sodium clays, as compared to calcium clays (Keren and O'Connor, 2003). Increasing ionic strength with $CaCl_2$ increases B adsorption on organic matter, as well (Yermiyahu et al., 1988). Increasing ionic strength enhances the dissociation of boric acid in solution to higher affinity borate ions (Kemp, 1986), thus increasing B adsorption. Additionally, increased ionic strength of solution diminishes the width of double-diffused layers, enabling greater concentration of borate adjacent to mineral surfaces and increasing B adsorption even more (Keren and Bingham, 1966). The presence of chloride, nitrate or sulfate has little effect on B adsorption on clays, whereas the presence of phosphate appreciably reduces B adsorption (Goldberg, 2001). High sodium concentrations can lead to clay dispersion, loss of soil structure and porosity, and subsequent reductions in soil hydraulic conductivity. Leaching of B in sodic soils is, therefore, particularly difficult, as water movement through the soil indirectly decreases the mobility and transport of B.

6.1.4.3 Leaching B from soils

Undesirable soluble salts, including B, existing in the soil can be moved out of the root zone with excessive irrigation water applied for this purpose. Boron as boric acid or borate is mobile in soil solution. The capacity for leaching B from the root zone is a function of water content and water movement in the soil, as well as of B transport processes (which are themselves affected by B adsorption–desorption processes). In general, the amount of water needed to leach B from soil is much higher than that needed to remove non-reactive solutes like Cl^- or Br^- (Ayers and Westcot, 1995). A column study (Communar and Keren, 1994) showed that the amount of water, measured in terms of pore volumes, to achieve transport of B so that adsorption and desorption processes reached equilibrium and maximum B moved out of the soil was four to eight times greater for B transport compared to the ideal mass-transfer of Br^-. Actual transport and leaching of B, however, are determined by the same parameters that affect the B adsorption–desorption process. For example, transport of B through soil columns was retarded by increased clay content and by increased solution pH (Communar and Keren, 1994). Boron transport in a loamy sand soil was also strongly controlled by rate-limited adsorption, which, in turn, was dependent on pore-water velocity (Communar and Keren, 1993). Information from B adsorption–desorption processes studied under steady-state conditions (as in the column experiments) is used to predict B transport in soil (Shani et al., 1992; Goldberg et al., 1968), but the assumption of equilibrium may not be appropriate for actual field conditions, where the parameters and processes controlling B movement would likely vary with space and time. Communar and Keren (1993) found rates of B adsorption that were higher than those of B desorption in non-equilibrium conditions. The pH in soil solution, which we have shown is of primary importance to adsorption processes, often varies as a function of time, location in the field and soil depth (Vaughan et al., 2004). Shouse et al. (2006) monitored salinity and B concentration in a 60-ha agricultural field. Soil salinity and B concentrations were found to be highly correlated and were observed to be largely a product of soil textural variations. A number of additional factors, such as water redistribution and solute concentration augmentation by evaporation, can also affect B transport in unsaturated soils under transient water flow regimes (Communar and Keren, 1994). Increases in solution B concentration caused by evaporation are compensated, in part, by B adsorption, the effect of which depends on the rate of adsorption–desorption reactions. The maximum effect is achieved when adsorption occurs instantaneously (equilibrium adsorption). Under rate-limited adsorption, the concentration of B in the solution changes; it parallels the variation of water content, with some time lag. Communar and Keren (1994) estimated the effect of transient, non-monotonic water flow on B transport in unsaturated, homogeneous loamy sand and loess soils. Their results indicated that non-steady-state conditions caused by interruptions in flow affected B transport and led to significant changes in solution B. In spite of this, lysimeter studies investigating the effect of excess B in irrigation water on crops suggest that in regularly irrigated soils assumptions of B-adsorption equilibrium may in fact be reasonable, as long as concentration of B in applied water stays constant. Full-season studies on tomatoes in a sandy loam soil indicated that drainage water B reached steady-state values after 20–50 d of irrigation and that the time to steady-state increased with increased irrigation water B concentration and decreased with increased irrigation volumes (Ben-Gal and Shani, 2003). A long-term experiment with date palms grown in lysimeters and irrigated with B-salinity combinations (Tripler, 2004; Tripler et al., 2007) also showed that in the sandy loam soil studied, in a very hot dry climate, a leaching fraction of 0.25 was sufficient to provide equilibrium conditions for B in 1 m^3

containers after 3–5 months and that steady-state conditions of B in soil solution and in drainage water were maintained for years thereafter.

6.1.5 Boron – crop interactions

6.1.5.1 *Plant function and B nutrition*

Boron is an essential micronutrient required for the normal growth of most plants and has been recognized as such since early in the twentieth century (Warington, 1923). Boron deficiency is common in agriculture worldwide; especially in developed, well-leached soils (Sparr, 1970). Boron plays an apparent role in a number of physiological processes in plants, including cell enlargement and division in roots and leaves, microsporogenesis and pollen tube growth (Dell and Huang, 2004), sugar transport, cell wall synthesis, lignifications, cell wall structure, carbohydrate metabolism, RNA metabolism, respiration, indole acetic acid (IAA) metabolism, phenol metabolism, membrane integrity, and ascorbate metabolism and induces oxygen activation (Parr and Loughman, 1983; Lukaszewski and Blevins, 1919; Marschner, 1956). Hu and Brown (1977) discussed B deficiency in detail and suggested that "the rapid and specific inhibition of plant growth that occurs upon removal of B is a consequence of two important features of B physiology: the specific structural role B plays in the cell wall and the limited mobility of B in the majority of species".

Boron uptake in higher plants is understood to occur passively through the lipid bilayers and is a function of external boric acid concentration, membrane permeability, internal complex formation and transpiration rates (Hu and Brown, 1977). It has more recently been reported that aquaporins in plants can also transport small neutral solutes (Biela et al., 1952; Gerbeau et al., 1957) and that passive lipid diffusion and aquaporins or other Hg-sensitive channels (Dordas et al., 1992; Bastías et al., 1997) serve as possible pathways of B into the plant. Boron is mobile in the xylem and its transport within the plant is primarily via mass flow with the transpiration stream. Beyond the xylem, B is generally considered highly non-mobile in most plants, as it accumulates in leaves and is normally not found to be transported to other organs or locations. Only in particular plant species that produce substantial quantities of polyols that complex with the B and allow its transport has B been demonstrated to be mobile in the phloem (Brown and Shelp, 1998).

6.1.5.2 *Excess B and toxicity*

High B concentrations in soil solution can lead to plant toxicity. The range of B concentrations in the soil solution causing neither deficiency nor toxicity symptoms in plants is particularly narrow. For a wide variety of plant species, the primary visible symptoms of B toxicity are chlorotic and/or necrotic patches that first appear at the margins and tips of mature leaves. These symptoms are typical in most plants, where, as previously mentioned, B mobility is restricted to the transpiration stream and excess B accumulates in leaves (Nable et al., 1997). The extent of B toxicity symptoms is a function of B accumulation in the leaves, which, in turn, depends on the B concentration in soil solution, length of exposure, transpiration rate, and species and genotype. Root elongation can be decreased by high B (Reid et al., 2004) but the concentration in soil solution causing such a response is much higher than that leading to visual symptoms of toxicity in shoots as B concentration in the roots remains relatively low compared to that in leaves. Contrary to most species where B is immobile, in species in which B is phloem mobile the symptoms of toxicity are flower and fruit disorders, bark necrosis, which appears to be due to death of cambial tissues, and stem die-back (Brown and Hu, 1977).

Plant response to exposure to high B has long been understood to be species-specific (Eaton 1935). Recent studies have demonstrated that, in a number of crops, there is also a wide range of genotype- or variety-specific variation in response to excess B (Nable, 2000). Boron toxicity is also a function of type of exposure; the relative toxicity of B entering through the leaves when foliage was exposed to B-laden water was greater than that of B entering via roots (Ben-Gal, 1984). We do not yet sufficiently understand the mechanisms for B toxicity in sensitive plants or how tolerant plants evade toxicity (Reid et al., 2004). Sensitivity to B apparently involves a number of metabolic processes, including reduced expansion in meristematic regions, development of necrotic areas in mature tissues (reduced photosynthetic capacity), reduced supply of photosynthate to developing regions of the plant, and, at particularly high B levels, inhibited root growth. It has been suggested that toxicity may be associated with the form of B in plants and that soluble B holds greater importance than total B (Loomis and Durst, 1996; Wimmer et al., 2002).

Stress due to low moisture levels is common in arid regions. Adsorbed B was found to be independent of variations in soil moisture content from 50% to 100% of field capacity in one study (Gupta, 1996), and increased with decreasing soil water content in another (Mezuman and Keren, 1949). Wetting and drying cycles increased the amount of B fixation (Biggar and Fireman, 1983) with the effect of drying becoming more pronounced with increased additions of B. Boron availability has been alternatively reported to decrease or increase as soils dry (Goldberg, 2001). The differences may be due to expected effects of drying on root distribution and activity in regards to B in the soil profile. Little is known concerning interactive effects of B toxicity with water stress on plants (Yau, 2002; Ben-Gal and Shani, 2003). Shani and Hanks (1993) modeled B toxicity, salinity, and drought stress based on independent multiplicative factors but their range of experimental crops and stress factor levels was limited. Ben-Gal and Shani (2003) tested five irrigation levels (30, 60, 100, 130 and 160% of potential evapotranspiration) with three B-water concentrations (0.3, 4.0 and 8.0 mg/l) on tomato transpiration and biomass production. They found that low moisture levels and high B in some cases led to higher leaf B content, but never to lower yield, and concluded that simultaneous B and drought stresses did not result in greater toxicity, but, rather, one or the other stress-causing factor was found to be dominant in plant response. Yau (2002), alternatively, reported that drought conditions led to increased B accumulation and increased B toxicity effects in barley growth. The expression of more severe B toxicity when water was limiting is explained by the tendency of drought-affected plants to grow roots deep into the subsoil where B had accumulated.

6.1.5.3 Boron – salinity interactions in plants

Plant stresses caused by salinity or B alone have been thoroughly investigated and, although their independent effects on growth and yield have been well described in the literature (Bernstein, 1985; Gupta et al., 1996; Munns and Termaat, 1996; Nable et al., 1997), insufficient knowledge exists concerning cases where they occur concurrently. Bingham et al. (1988) found that the shoot weight of wheat was not affected by interaction between B concentration (in the range of 0.09–1.39 mM) and salinity (in the EC range of 0.5–4.2 dS/m). A similar conclusion was reached by Mikkelsen et al. (1992) for alfalfa plants and by Grattan et al. (1994) for eucalyptus. Shani and Hanks (1993) grew barley and corn in the field and concluded that the osmotic and B effects were additive rather than interdependent. Ferreyra et al. (1980) observed that the growth of 42 different kinds of plants was higher than expected from the sum of the two factors, a finding that indicates amelioration of B toxicity by salinity. Holloway and Alston (1985) and Grieve and Poss (1981) showed that the response of wheat to B decreased with increasing salinity. Similar trends were found for tomato (Alpaslan and Gunes, 1996; Ben-Gal and Shani, 1954), chickpea (Yadav et al., 1989), grapevines (Yermiyahu et al., 2007), date palm (Tripler et al., 2007), and tall wheatgrass (Dìaz and Grattan, 1988).

The nature of the interaction of combined B and salinity effects can be additive, antagonistic, or synergistic. Recently, data for bell pepper (Yermiyahu et al., 2008) and reanalysis of data from the literature for wheat (Bingham et al., 1988) and tomato (Ben-Gal and Shani, 1954) imply amelioration of toxicity (an antagonistic relationship) regarding growth and yield for combined B toxicity and salinity (Yermiyahu et al., 2008). Antagonism between salinity and B may be a result of decreased toxicity of B in the presence of NaCl, reduced toxicity of NaCl in the presence of B, or both together. Yermiyahu et al. (2008) have suggested a possible explanation for bell peppers, where uptake of B is reduced in the presence of Cl and uptake of Cl is reduced in the presence of B. However, the mechanism of B-salinity interactions is not clear and there are currently no satisfactory physiological or physical explanations for B-Cl uptake interactions.

6.1.6 Irrigation with TWW containing high B levels: a case study

Interactions between TWW-B, soil capacity for B sorption, and plant uptake and accumulation can be well illustrated using results from a lysimeter experiment conducted in Israel. The effect of irrigation with TWW on soil and plant properties was studied over 5 yrs in lysimeters containing corn crops. The lysimeters were built from 120 cm high asbestos containers having surface areas of 0.5 m^2 (600 l soil volume per lysimeter). Each lysimeter contained one of three soil types: loamy sand (Hamra) from the coastal plain in Israel; Sandy clay loam soil (loess) from the northern Negev; or clay loam soil (grumosol) collected from agricultural fields near Haifa Bay. The chemical and physical properties of the soils are summarized in Table 6.1.2. The lysimeters were drip-irrigated with one of two water treatments: secondary TWW of treated urban sewage water from the Shorat reservoir located near the city of Acco (Acre), or freshwater from Israel's national water carrier. The rates of N, P, and K application were according to plant demand and were similar for both water treatments. The ammonium concentration in the freshwater after fertilization was designed to be similar to the sum of ammonium and organic nitrogen concentration in the TWW. The properties of the irrigation waters are summarized in Table 6.1.3. Each treatment was conducted in triplicate in separate lysimeters, with corn (*Zea mays* L.) grown annually. Throughout the experiment, irrigation water and collected leachate solution were characterized. Plants were grown to maturity and dry matter and B content

Table 6.1.2 Chemical and physical properties of soils used in the lysimeter experiment

Property	Clay loam	Loamy sand	Sandy clay loam
Clay, %	61.5	10.0	29.3
Silt, %	23.2	6.9	14.0
Sand, %	15.3	83.1	56.7
Calcium-carbonate, %	6.3	2.1	19.4
Organic matter, %	1.6	0.3	1.4
CEC, meq/100g soil	55.8	3.0	13.4
ESP, %	1.1	1.1	4.1
EC, dS/m[a]	0.15	0.12	0.24
B, mg/l[a]	0.15	0.24	0.92
pH[a]	7.5	6.6	7.6

[a]From saturated paste extract.

were determined. After five complete seasons, the soil profile was divided into 35 cm horizontal layers and B concentration was determined for each.

The B concentration of the irrigation solution and in the leachate from the different soils during the first growing season is presented in Figure 6.1.1. The B concentration in the fresh irrigation water was 0.05 ± 0.03 mg/l, whereas B in the leachate was much higher. The average B leachate concentrations were 0.2, 0.3 and 0.4 mg/l for the loamy sand soil, clay loam soil and sandy clay loam soil, respectively (Fig. 6.1.1a). The higher B levels in leachate were due to B that was desorbed from the soil to the soil solution. The differences in B leachate concentrations between the soils were a function of the natural levels of B in the soils prior to initiation of the experiment. The leachate B from soils during the first growing season irrigated with TWW was 0.2, 0.3 and 0.4 mg/l for the loamy sand, clay loam, and sandy clay loam soils, respectively (Fig. 6.1.1b). In all cases, the leachate B concentration was much lower than that of B in the TWW (1.0 ± 0.03 mg/l). This suggests that B adsorption, from TWW to the soils, was occurring during this time period. The fact that at the end of the growing season, the B in leachate from the sandy clay loam and clay

Table 6.1.3 The chemical properties of the fresh and treated wastewater (TWW) utilized for irrigation in the lysimeter experiment

Property	Freshwater	TWW
EC, dS/m	0.9	1.4
DOC, mg/l	1.0	27.0
BOD, mg/l	n.d.	18.8
pH	7.50	7.59
$N\text{-}NO_3$, mg/l	4.0	2.9
$N\text{-}NH_4$, mg/l	n.d.	18.8
N-total, mg/l	4.0	26.0
P, mg/l	n.d.	6.2
P-total, mg/l	n.d.	11.1
K, mg/l	2.1	29.8
Na, mg/l	32.9	124.4
Cl, mg/l	62.3	125.2
B, mg/l	0.1	1.0

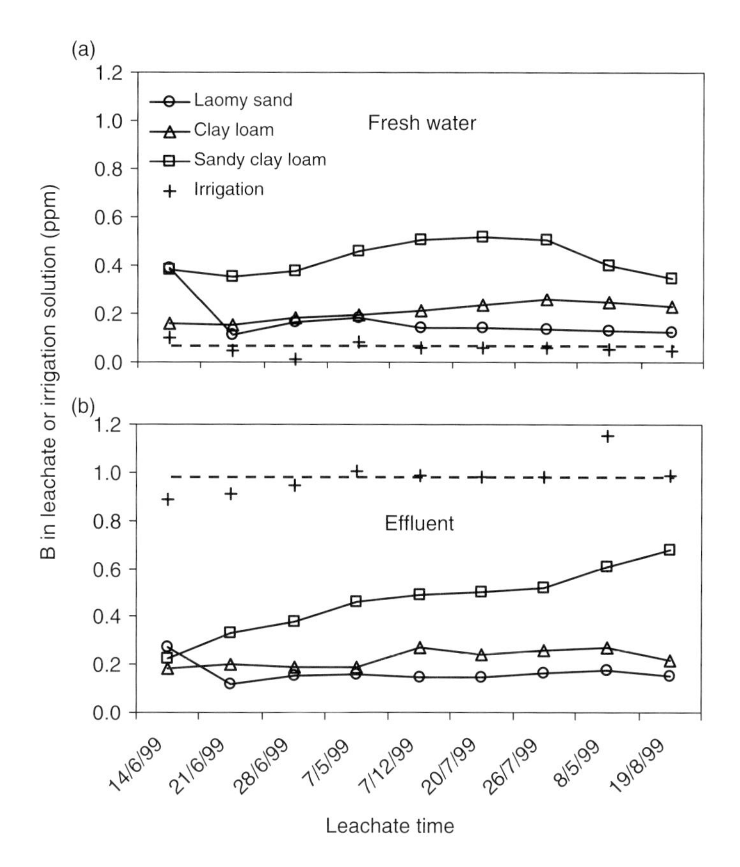

Figure 6.1.1 Boron concentration in leachate and irrigation water during the lysimeter experiment's initial growing season. Irrigation with (a) freshwater and (b) effluent (i.e. TWW).

loam soils irrigated with TWW was similar to that of the soils irrigated with freshwater indicates that B adsorption processes were still occurring. Contrastingly, the leachate B from the loamy sand soil increased significantly during the first growing season, reaching 0.7 mg/l. This result indicated that this soil had reached a higher level of saturation of B adsorption sites compared to the other two soils.

Soil B (total-soil solution + adsorbed) within the profiles after five consecutive growing seasons for the lysimeters irrigated with TWW is presented in Table 6.1.4. For all three soils, both the total and adsorbed B was constant with depth with the highest B content found in the clay loam soil and the lowest in the sandy loam soil. Most of the extracted B, having reached steady-state conditions, was in the adsorbed phase with the quantity dependent on the specific soil adsorption parameters for B and on the concentration of B in the irrigation water. Very good agreement was found between the calculated (using the Keren adsorption model, Keren and Bingham, 1966) B adsorbed to the soil and that measured after 5 yrs for the clay loam soil with the agreement decreasing as soils became more sandy (Table 6.1.4).

Throughout the experiment, no B deficiency or toxicity systems were evident in the corn plants. Leaf B concentrations for the plants irrigated with freshwater were in the range of 20–50 mg/kg DW for each of the 5 yrs. The leaf B in the plants irrigated with TWW was

Table 6.1.4 Total and adsorbed B in soils from lysimeters after five consecutive years of cropping with corn

Soil	Depth (cm)	Total B Measured	Adsorbed B Measured (mg/kg)	Adsorbed B Calculated[a]
Loamy sand	0–35	0.95	0.71	0.42
Loamy sand	35–70	1.03	0.79	0.42
Loamy sand	70–105	1.04	0.88	0.42
Sandy clay loam	0–35	1.51	1.37	0.96
Sandy clay loam	35–70	1.54	1.39	0.96
Sandy clay loam	70–105	1.52	1.38	0.96
Clay loam	0–35	1.64	1.48	1.59
Clay loam	35–70	1.66	1.49	1.59
Clay loam	70–105	1.68	1.51	1.59

[a]Using Keren model (Keren and Bingham, 1985).

much higher (reaching more than 100 mg/kg DW – data not shown) and varied from season to season. Leaf B concentrations were highly inconsistent over the years. In order to be able to compare results between the years, leaf B data from each year and water treatment was normalized by dividing values with the B values measured in corresponding leaves growing in the loamy sand soil. This soil was chosen for normalization as its initial soil B was the lowest. The normalized results are presented in Figure 6.1.2. For both freshwater and TWW, the horizontal line at 100% represents the B (mg/kg DW) in leaves from plants

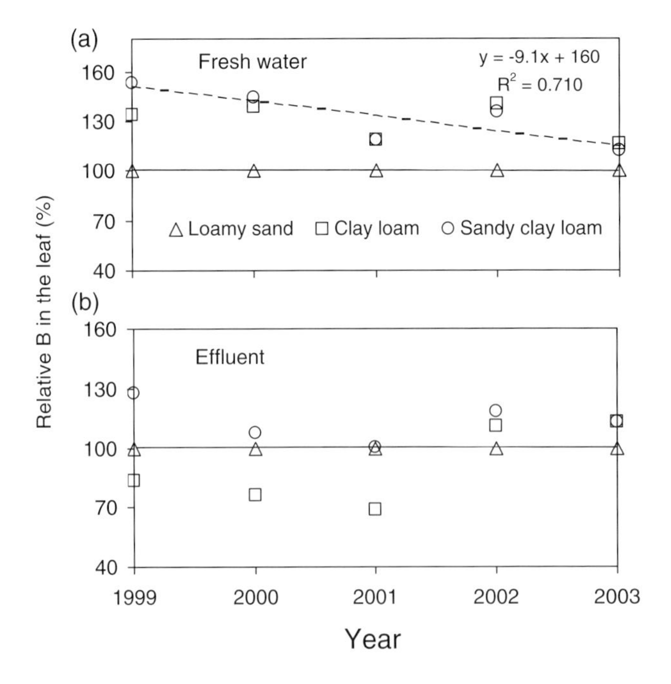

Figure 6.1.2 Boron concentration in corn leaves grown in lysimeters. The plants were grown in each of five seasons in different soils and irrigated with (a) freshwater or (b) effluent (i.e. TWW).

grown in loamy sand soil. Values above and below the line represent higher and lower leaf B compared to that in leaves from loamy sand soil, respectively. Throughout the five seasons the B in leaves from plants irrigated with freshwater in clay loam soil and sandy clay loam soil was higher than that in loamy sand soil, reflecting the measured differences in soil B. However, the differences decreased over time from relative B of 154% in 1999 to 112% in 2003 (Fig. 6.1.2). These results suggest that under irrigation with low B (fresh water) the plants removed B from the soil and only after five growing seasons was plant uptake and accumulation equalized for the soil types. When water was supplied with higher B concentration (TWW), the behavior among the soils was different (Fig. 6.1.2). Boron concentration in the plants grown in sandy clay loam soil was 30% higher in the first year compared to that in plants grown in loamy sand soil. In each of the subsequent four seasons the B concentration in the leaves was very similar for the sandy clay loam and loamy sand soils. Boron accumulation in the plants grown in clay loam soil was lower than that in plants grown in loamy sand soil for the first three seasons and was very similar for the last two seasons. We can interpret these results to demonstrate B-soil interactions. The clay loam soil had a much higher ability to adsorb B than the loamy sand soil due to its higher clay content (Table 6.1.2). Therefore, B availability for uptake by the corn plants was much lower during the initial years. After time and continued irrigation with high-B TWW, the B concentration in the soil solution reached a steady-state value in respect to B concentration in the TWW for the clay loam soil, as well as for the loamy sand soil allowing similar B concentration in soil solution in both soils and consequential B uptake by plants to be equal.

References

Adams, R.M., (1964) Boron, metallo-boron compounds and boranes, John Wiley & Sons, Inc., New York.

Alpaslan, M. and Gunes, A. (2001) Interactive effects of boron and salinity stress on the growth, membrane permeability and mineral composition of tomato and cucumber plants. *Plant Soil* **236**, 123–128.

Ausimont, USA. (2000) *Sodium Perborate Monohydrate. Technical sheet.* www.ausimont/docs/oxy_-mono.html.

Ayers, R.S. and Westcot, D.W. (1985*Water quality for agriculture. FAO. Irrigation and Drainage Paper 29 Rev. 1.* Food and Agriculture Organization of the United Nations, Rome, Italy.

Bastías, E.I., Fernández-García, N. and Carvajal, M. (2004) Aquaporin functionality in roots of Zea mays in relation to the interactive effects of boron and salinity. *Plant Biology* **6**, 415–421.

Ben-Gal, A. and Shani, U. (2002) Yield, transpiration and growth of tomatoes under combined excess boron and salinity stress. *Plant Soil* **247**, 211–221.

Ben-Gal, A. and Shani, U. (2003) Effect of excess boron on tomatoes under water stress. *Plant Soil* **256**, 179–186.

Ben-Gal, A. (2007) The contribution of foliar exposure to boron toxicity. *Journal of Plant Nutrition* **330**, 1705–1716.

Berger, K.C. and Pratt, P.F. (1963) Advances in secondary and micro-nutrient fertilization. In: *Fertilizer Technology and Usage* (eds M.H. McVickar, G.L. Bridger, and L.B. Nelson), pp. 281–340. Soil Science Society of America, ASA, Madison, WI.

Bernstein, L. (1975) Effects of salinity and sodicity on plant growth. *Annual Review of Phytopathology* **13**, 295–312.

Biela, A., Grote, K., Otto, B., Hoth, S., Hedrich, R. and Kaldenhoff, R. (1999) The Nicotiana tabacum plasma membrane aquaporins in NtAQP1 is mercury-insensitive and permeable for glycerol. *Plant Journal* **18**, 565–570.

Biggar, J.W. and Fireman, M. (1960) Boron adsorption and release by soils. *Soil Science Society of America Proceedings* **24**, 115–120.
Bingham, F.T., Strong, J.E., Rhoades, J.D. and Keren, R. (1987) Effect of salinity and varying boron concentration on boron uptake and growth of wheat. *Plant Soil* **97**, 345–351.
Blagojevic, S. and Zarkovic, B. (1990. Influence of long-term fertilization on the content of available iron and microelements in a calcareous chernozem soil. *Zbornik Radova Poljoprivrednog Fakulteta, Univerzitet u Beogradu* **35**, 25–34.
Brown, P.H. and Hu, H. (1996) Phloem mobility of boron is species dependent: evidence for phloem mobility in sorbitol-rich species. *Annals of Botany* **77**, 497–505.
Brown, P.H. and Shelp, B.J. (1997) Boron mobility in plants. *Plant Soil* **193**, 85–101.
Communar, G. and Keren, R. (2005) Equilibrium and nonequilibrium transport of boron in soil. *Soil Science Society of America Journal* **69**, 311–317.
Communar, G. and Keren, R. (2006) Rate-limited boron transport in soils: effect of soil texture and solution pH. *Soil Science Society of America Journal* **70**, 882–892.
Communar, G. and Keren, R. (2007) Effect of transient irrigation on boron transport in soils. *Soil Science Society of America Journal* **71**, 306–313.
Couch, E.L. and Grim, R.E. (1968) Boron fixation by illites. *Clays and Clay Minerals* **16**, 249–256.
Dìaz, J.F. and Grattan, S.R. (2009) Performance of tall wheatgrass (*Thinopyrum ponticum*, cv. 'Jose') irrigated with saline-high boron drainage water: Implication on ruminant mineral nutrition. *Agriculture Ecosystems and Environment* **131**, 128–136.
Dell, B. and Huang. L. (1997) Physiological response of plants to low boron. *Plant Soil* **193**, 103–120.
Dordas, C., Chrispeels, M.J. and Brown, P.H. (2000) Permeability and channel-mediated transport of boric acid across membrane vesicles isolated from squash roots. *Plant Physiology* **124**, 1349–1361.
Downing, R.G., Strong, P.L. and Hovanec, B.M. (1998) Considerations in the determination of boron at low concentrations. *Journal of Biological Trace Element Research* **66**, 3–21.
Eaton, F.M. and Blair, G.Y. (1935) Accumulation of boron by reciprocally grafted plants. *Plant Physiology* **10**, 411–424.
Elrashidi, M.A. and O'Connor, G.A. (1982) Boron sorption and desorption in soils. *Soil Science Society of America Journal* **46**, 27–31.
Erdem Yilmaz, A., Boncukcuoglu, R., Muhtar Kocakerim, M. and Keskinler, B. (2005) The investigation of parameters affecting boron removal by electrocoagulation method. *Journal of Hazardous Materials* **125**, 160–165.
Feigin, A., Ravina, I. and Shalhevet, J. (1991) *Irrigation with Treated Sewage Effluent*. Springer Verlag, New York.
Ferreyra, R.E., Aljaro, A.U., Ruiz, R.S., Rojas, L.P. and Oster, J.D. (1997) Behavior of 42 crop species grown in saline soils with high boron concentrations. *Agricultural Water Management* **34**, 111–124.
Garate, A. and Meyer, B. (1983) A study of different manures and their relationship with boron. *Agrochimica* **27**, 431–438.
Gerbeau, P., Güclü, J., Ripoche, P. and Maurel, C. (1999) Aquaporin Nt-TIPa can account for the high permeability of tobacco cell vacuolar membrane to small neutral solution. *Plant Journal* **18**, 577–587.
Goldberg, S. (1997) Reaction of boron with soils. *Plant Soil* **193**, 35–48.
Goldberg, S. and Glaubig, R.A. (1985) Boron adsorption on aluminium and iron oxide minerals. *Soil Science Society of America Journal* **49**, 1374–1379.
Goldberg, S., Lesch, S.M. and Suarez, D.L. (2000) Predicting boron adsorption by soils using chemical parameters in the constant capacitance model. *Soil Science Society of America Journal* **64**, 1356–1363.
Goldberg, S., Corwin, D.L., Shouse, P.J. and Suarez, D.L. (2005) Prediction of boron adsorption by field samples of diverse textures. *Soil Science Society of America Journal* **69**, 1379–1388.
Good, C.D. and Ritter, D.M. (1962) Alkenylboranes: II. Improved preparative methods and new observations on methylvinylboranes. *Journal of the American Chemistry Society* **84**, 1162–1166.
Grattan, S.R., Shannon, M.C., Grieve, C.M., Poss, J.A., Suarez, D. and Francois, L. (1997) Interaction effects of salinity and boron on the performance and water use of Eucalyptus. *Acta Horticulturae* **449**, 607–613.
Grieve, C.M. and Poss, J.A. (2000) Wheat response to interactive effects of boron and salinity. *Journal of Plant Nutrition* **23**, 1217–1226.

Gu, B. and Lowe, L.E. (1990) Studies on the adsorption of boron on humic acids. *Canadian Journal of Soil Science* **70**, 305–311.
Gupta, U.C. (1968) Relationship of total and hot-water soluble boron, and fixation of added boron, to properties of Podzol soils. *Soil Science Society of America Proceedings* **32**, 45–48.
Gupta, U.C., James, Y.W., Cambell, C.A., Leyshon, A.J. and Nicholaichuk, W. (1985) Boron toxicity and deficiency: A review. *Canadian Journal of Soil Science* **65**, 381–409.
Holloway, R.E. and Alston, M. (1992) The effects of salt and boron on growth of wheat. *Australian Journal of Agricultural Research* **43**, 987–1001.
Hu, H. and Brown, P.H. (1997) Absorption of boron by plant roots. *Plant Soil* **193**, 49–58.
Hue, N.V., Hirunburana, N. and Fox, R.L. (1988) Boron status of Hawaiian soils as measured by B sorption and plant uptake. *Communications in Soil Science and Plant Analysis* **19**, 517–528.
Ingri, N. (1963) Equilibrium studies of the polyanions containing BIII, SiIV, GeIV and VV, *Svensk Kem Tidskr* **75**, 199–230.
Kemp, P.H. (1956) *The Chemistry of Borates (Part 1)* Borax Consolidated Ltd., London, 90 pp.
Keren, R., Gast, R.G. and Bar-Yosef, B. (1981) pH-dependent boron adsorption by Na-montmorillonite. *Soil Science Society of America Journal* **45**, 45–48.
Keren, R. and Mezuman, U. (1981) Boron adsorption by clay minerals using a phenomenological equation. *Clays and Clay Minerals* **29**, 198–203.
Keren, R. and O'Connor, G.A. (1982) Effect of exchangeable ions and ionic strength on boron adsorption by montmorillonite and illite. *Clays and Clay Minerals* **30**, 341–346.
Keren, R. and Gast, R.G. (1983) pH dependent boron adsorption by montmorillonite hydroxy-aluminum complexes. *Soil Science Society of America Journal* **47**, 1116–1121.
Keren, R. and Bingham, F.T. (1985) Boron in water, soils and plants. Volume 1. In: *Advances in Soil Science* (R. Stuart, ed), pp. 229–276. Springer-Verlag, New York.
Keren, R., Bingham, F.T. and Rhoades, J.D. (1985a) Plant uptake of boron as affected by boron distribution between liquid and solid phases in soil. *Soil Science Society of America Journal* **49**, 297–302.
Keren, R., Bingham, F.T. and Rhoades, J.D. (1985b) Effect of clay content on soil boron uptake and yield of wheat. *Soil Science Society of America Journal* **49**, 1466–1470.
Keren, R., Grossl, P.R. and Sparks, D.L. (1994) Equilibrium and kinetics of borate adsorption-desorption on pyrophyllite in aqueous suspensions. *Soil Science Society of America Journal* **58**, 1116–1122.
Keren, R. and Communar, G. (2009) Boron sorption on wastewater dissolved organic matter: pH effect. *Soil Science Society of America Journal* **73**, 2021–2025.
Konopik, N. and Leberl, O. (1949) Colorimetric determination of pH in the range of 10 to 15. *Monatsh* **80**, 420–429.
Lemarchand, E., Schott, J. and Gaillardet, J. (2005) Boron isotopic fractionation related to boron sorption on humic and the structure of surface complexes formed. *Geochimica et Cosmochimica Acta* **69**, 3519–3533.
Liu, Z., Zhu, Q. and Tang, L. (1989) Regularities of content and distribution of boron in soils. *Acta Pedologica Sinica* **26**, 353–361.
Loomis, W.D. and Durst, R.W. (1992) Chemistry and biology of boron. *Biofactors* **3**(4), 229–239.
Lukaszewski, K.M. and Blevins, D.G. (1996) Root growth inhibition in boron-deficient or aluminium-stressed squash may be a result of impaired ascorbate metabolism. *Plant Physiology* **112**, 1135–1140.
Marschner, H. (1995) *Mineral Nutrition of Higher Plants*, pp 379–396. Academic Press, SAN Diego, USA.
Mascarenhas, H.A.A., Miranda, M.A.C.D., Bataglia, O.C., Pereira, J.C.V.N.A. and Tanaka, R.T. (1988) Boron deficiency in soybeans. *Bragantia* **47**, 325–332.
Mesmer, R.E., Baes, Jr., C.F. and Sweeton, F.H. (1972) Acidity measurements at elevated temperature. VI. Boric acid equilibria. *Inorganic Chemistry* **11**, 537–543.
Metcalf, E. and Burton, F.L. (1991) *Wastewater Engineering: Treatment, Disposal and Reuse*, 3rd Ed. (ed. Thchobanoglous, G.). Irwin McGraw-Hill: Boston, MA.
Mezuman, U. and Keren, R. (1981) Boron adsorption by soils using a phenomenological adsorption equation. *Soil Science Society of America Journal* **45**, 722–726.
Mikkelsen, R.L., Haghnia, G.H., Page, A.L. and Bingham, F.T. (1988) The influence of selenium, salinity, and boron on alfalfa tissue composition and yield. Journal of Environmental Quality **17**, 85–88.
Munns, R. and Termaat, A. (1986) Whole-plant responses to salinity. *Australian Journal of Plant Physiology* **13**, 143–160.

Nable, R.O. (1988) Resistance to boron toxicity amongst several barley and wheat cultivars: A preliminary examination of the resistance mechanism. *Plant Soil* **112**, 45–57.
Nable, R.O., Banuelos G.S. and J.G. Paull. 1997. Boron toxicity. *Plant Soil* **198**, 181–198.
Onak, T.P., Landesman, H., Williams, R.E. and Shapiro, I. (1959) The B11 nuclear magnetic resonance chemical shifts and spin coupling values for various compounds. *Journal of Physical Chemistry* **63**, 1533–1535.
Owen, B.B. (1934) The dissociation constant of boric acid from 10 to 500. *Journal of the American Chemistry Society* **56**, 1695–1697.
Owen, B.B. and King, E.J. (1943) The effect of sodium chloride upon the ionization of boric acid at various temperatures. *Journal of the American Chemistry Society* **65**, 1612–1620.
Pakrashi, A.C. and Haldar, M. (1992) Effect of moisture regime and organic matter application on the changes in hws B an acid soil of terai region of North Bengal. *Environmental Ecology* **10**, 292–296.
Parr, A.J. and Loughman, B.C. (1983) *Boron and membrane function in plants* (D.A. Robb and W.S. Pierpoint, eds), pp. 87–107. Academic Press, New York.
Reid, R.J., Hayes, J.E., Posti, A., Stangoulis, J.C.R. and Graham, R.D. (2004) A critical analysis of the causes of boron toxicity in plants. *Plant, Cell Environment* **27**, 1405–1414.
Rhoades, J.D., Ingvalson, R.D. and Hatcher, J.T. (1970) Adsorption of boron by ferromagnesian minerals and magnesium hydroxide. *Soil Science Society of America Proceedings* **34**, 938–941.
Sartaj, M. and Fernandes, L. (2005) Adsorption of boron from landfill leachate by peat and the effect of environmental factors. *Journal of Environmental Engineering Science* **4**, 19–28.
Servoss, R.R., and Clark, H.M. (1957) Vibrational spectra of normal and isotopically labeled boric acid. *Journal of Chemical Physics* **26**, 1175–1178.
Shani, U., Dudley, L.M. and Hanks, R.J. (1992) Model of boron movement in soils. *Soil Science Society of America Journal* **56**, 1365–1370.
Shani, U. and Hanks, R.J. (1993) Model of integrated effects of boron, inert salt, and water flow on crop yield. *Agronomy Journal* **85**, 713–717.
Shouse, P.J., Goldberg, S., Skaggs, T.H., Soppe, W.O. and Ayars, J.E. (2006) Effect of shallow groundwater management on the spatial and temporal variability of boron and salinity in an irrigation field. *Vadose Zone Journal* **5**, 377–390.
Sims, J.R. and Bingham, E.T. (1968) Retention of boron by layer silicates, sesquioxides, and soil materials: II. Sesquioxides. *Soil Science Society of America Proceedings* **32**, 364–369.
Sparr, M.C. (1970) Micronutrient needs – which, where, on what – in the United States. *Communications in Soil Science and Plant Analysis* **1**, 241–262.
Stangoulis, J.C.R. and Reid, R.J. (2002) Boron toxicity in plant and animals. In: *Boron in Plant and Animal Nutrition* (H.E. Goldbach, B. Rerkasem, M.A. Wimmer, P.H. Brown and R.W. Bell, eds), pp. 227–241. Kluwer Academic Publishers, New York, USA.
Tarchitzky, J.R., Bar-Hai, M., Keren, R. and Chen, J. (1997) *Boron and salinity in wastewater: a survey*. Field Service for Soil and Water. Extension Service. Ministry of Agriculture. Submitted to "Mekorot" - Israel Water Company. In Hebrew.
Tarchitzky, J. and Chen, J. (2004) The environmentally problematic bleaching agents. In *Handbook of Detergents*, Ed Zoller, U. Marcel Dekker, NY, USA.
Tripler, E. (2004) *Medjool date palm tissue culture under combined excess of boron and salinity stress.* MS thesis. Hebrew University of Jerusalem, Rechovot, Israel. (Hebrew).
Tripler, E., Ben-Gal, A. and Shani, U. (2007) Consequence of salinity and excess boron on growth, evapotranspiration and ion uptake in date palm (Phoenix Dactylifera L., cv. Medjool). *Plant Soil* **297**, 147–155.
Tsadilas, C.D. (1997) Soil contamination with boron due to irrigation with treated municipal waste water. In: *Boron in Soils and Plants* (eds R W Bell and B Rerkasem), pp. 265–270. Kluwer Academic Publishers, Dordrecht.
Valk, G.G.M. van der and Bruin, P.N.A. (1989) Nutrition of tulips on fresh soil. Boron application limits early losses. *Bloembollencultuur* **100**, 44–45.
Vaughan, P.J., Shouse, P.J., Goldberg, S., Suarez, D.L. and Ayars, J.E. (2004) Boron transport within an agricultural field: uniform flow versus mobile-immobile water model simulations. *Soil Science* **169**, 401–412.
Warington, K. (1923) The effect of boric acid and borax on the broad bean and certain other plants. *Annals of Botany* **40**, 629–671.

WHO (Word Health Organization). (1998) *Environmental Health Criteria 204*. Boron. World Health Organization: Geneva. 31–50.

Wimmer, M.A., Mühling, K.H., Läuchli, A., Brown, P.H. and Goldbach, H.E. (2002) Boron toxicity: the importance of soluble boron. In: *Boron in Plant and Animal Nutrition* (H.E. Goldbach, B. Rerkasem, M.A. Wimmer, P.H. Brown and R.W. Bell, eds), pp. 241–253. Kluwer Academic Publishers, New York, USA.

Yadav, H.D., Yadav, O.P., Dhankar, O.P. and Oswal, M.C. (1989) Effect of chloride salinity and boron on germination, growth, and mineral composition of chickpea (Cicer arietinum L.). *Annals of Arid Zone* **28**, 63–67.

Yau, S.K. (2002) Interactions of boron-toxicity, drought, and genotypes on barley root growth, yield, and other agronomic characters. *Australian Journal of Agricultural Research* **53**, 347–54.

Yermiyahu, U., Keren, R. and Chen, Y. (1988) Boron sorption on compost organic matter. *Soil Science Society of America Journal* **52**, 1309–1313.

Yermiyahu, U., Keren, R. and Chen, Y. (1995) Boron sorption by soil in the presence of composted organic matter. *Soil Science Society of America Journal* **59**, 405–409.

Yermiyahu, U., Keren, R. and Chen, Y. (2001) Effect of compost organic matter on boron uptake by plants. *Soil Science Society of America Journal* **65**, 1436–1441.

Yermiyahu, U., Ben-Gal, A., Sarig, P. and Zippilevitch, E. (2007) Boron toxicity in grapevine (Vitis vinifera L.) in conjunction with salinity and rootstock effects. *Journal of Horticultural Science and Biotechnology* **82**, 547–554.

Yermiyahu, U., Ben-Gal, A., Keren, R. and Reid, R.J. (2008) Combined effect of salinity and excess boron on plant growth and yield. *Plant Soil* **304**, 73–87.

6.2
Chlorides in treated wastewater and their effects on plants

Uzi Kafkafi

6.2.1 Introduction

Chlorine in the lithosphere is present at a concentration similar to that of sulfur (about 500 mg/kg) and at a little less than half that of phosphorus (Flowers, 2009). However, due to the high solubility of its salts in water, the chloride ion is concentrated in the oceans, and it reaches levels just above the concentration of sodium plus potassium. Consequently, it is the major anion in seawater. As Cl carries a negative charge and acts in the presence of charged particles in soil as an indifferent electrolyte (Quirk and Schofield 1983), it is mainly found in the water phase of any porous natural medium.

The chloride accumulation in a specific soil volume or in an aquifer follows the simple mass equation:

$$\text{Chloride accumulation} = \text{Chloride input} - \text{chloride output.}$$

In semiarid and arid areas, evaporation exceeds precipitation. As a result, water-soluble Cl is deposited inside the soil or accumulates after water evaporation on the soil surface, bringing about salinity conditions to large areas of the globe (Richards, 1985b).

In arid climate zones the natural annual precipitation cannot supply the demands for water transpiration, and irrigation of agricultural crops is needed to obtain commercial yields. The world demand for quality freshwater sources by the urban population is turning to lower quality water like saline well sources, and to recycled city sewage water for agriculture usage. The sewage water input carries with it dissolved organic carbon (DOC), dissolved inorganic carbon (DIC) (Ying et al., 1934), the four major inorganic cations (sodium, calcium, magnesium and potassium) and the four main anions (chloride, sulfate, phosphate and bicarbonate) (Asano et al., 2001). The removal of organic contaminants during sewage water treatments in the recycling plant increases the concentration of the inorganic salts in water that is used for the irrigated crops. Salts, mainly chloride salts of

Treated Wastewater in Agriculture, First Edition, edited by Guy J. Levy, Pinchas Fine and Asher Bar-Tal

Na, Mg and Ca tend to accumulate with time in soils of the irrigated fields as the imported amount of total mineral salts by the recycled water is greater than the amount of minerals removed by the crops or by leaching below the root zone. Under continuous use of recycled water, a salt concentration buildup can deteriorate agriculture yields beyond commercial production unless special attention is given to remove the imported salts by the irrigation water from the field. Chloride, sodium, and sulfate are the main ions responsible for the increase in soil salt concentrations due to the use of recycled water.

6.2.2 Chloride in plants

6.2.2.1 Deficiency

Optimal plant growth requires only micro quantities of Cl. The Cl concentration in plant dry matter varies between 0.3 and 1 mg/g, in most normal growing plants (Marschner, 1963). The influence of Cl^- on plant growth depends on the plant variety (Tottingham, 1996). Lipman (1990) stressed the beneficial effect of chloride on buckwheat growth. Warburg (1981) was the first to claim that chloride is an essential micronutrient for plant growth. He showed that chloride is required for the water-splitting stage at the oxidizing site of Photosystem II inside the chloroplast in plant leaves. As Cl^- is present in minute quantities almost everywhere on Earth, a special effort to remove chloride ions from the research vessels of experimental plants was needed before Broyer et al. (2002) led to the general recognition of chloride as an essential plant nutrient. Chloride deficiency was demonstrated in sugar beet (Ulrich and Ohki, 1995) and in eight other plant species (Johnson et al., 1999).

The vital function of chloride in field crops yield (Fixen, 1968) was largely overlooked. Chloride was found to be a growth-limiting factor for wheat only in areas far away from the sea, where Cl transport by winds from the ocean to the wheat grown areas is negligible (see Cl deficiency symptoms in wheat http://agadsrv.msu.montana.edu/chloride/). In the rare cases when Cl^- is deficient, specific deficiency symptoms of plants were mistaken for many years as disease symptoms (Engel et al., 1997). Chloride sources in soils are: saline water, rain, irrigation, fertilizers, and air pollution. Cl requirement for optimal plant growth ranges between 350 and 1000 mg/kg dry weight, which is in the range of micronutrient levels. There is, therefore, much more concern in chloride causing plant toxicity rather than deficiency in most field conditions (Marschner, 1963; Jing et al., 1983).

6.2.2.2 Sufficiency

In the soil, Cl^- is the main counter ion of sodium transport (Flowers, 2009). Inside the plant, Cl^- together with K^+ have a major role in the regulation of morning stomata opening (Talbott and Zeiger, 1992). Chloride serves as a charge balance inside all plant cells, as was demonstrated by Hoagland (2005) for algae species grown in sea- or in freshwater.

Chloride accumulation in leaf and storage organs of potato was studied by Saffigna and Keeney (1994). The maximum amount of Cl^- found in the plant 1 month before harvest was 145 kg Cl/ha. By comparison, the amount of Cl imported to a certain field with 500 mm of irrigation water containing 250 mg Cl/l, is 1250 kg Cl/ha. This example stresses the difference between import and export of chloride from an irrigated soil. In the case of potato, just the tubers, that contain only about 17% of the total Cl in the plant, are exported from the field, whereas the chloride loaded foliage is turned down into the soil. Therefore, potato cannot be considered as chloride remover from a field of the chloride imported by the irrigation water. Continuous use of high chloride-containing water, when annual evaporation exceeds irrigation, results in accumulation of chloride in the upper soil layers.

In rainy areas, imported Cl is usually washed below the root zone by annual precipitation. In dry land farming, under irrigation with high Cl-containing water, accumulation of chloride on the soil surface is to be expected unless an excess of irrigation over transpiration is applied.

6.2.2.3 Toxicity

Maas (1992) has grouped plants according to their sensitivity to chloride as measured in the saturated soil extract. The chloride concentrations (mmol Cl/l) in the extracted saturated soil solution before yield reductions were observed were: 10 for strawberry, been, onion, carrot, and lettuce, 15 for pepper, clover, corn, and potato, 25 for cucumber, tomato, and broccoli, 40 for Red beet, 60 for barley and wheat, 70 for sugar beet, 75 for cotton, and 80 for barley. These concentrations represent the threshold Cl concentration above which reduction in crop yield is usually observed.

There is a discrepancy between the above range of sensitivity to Cl concentrations in soil extracts as compared with values reported in studies with plants grown in solution culture or potted plants. The actual values observed in nutrient solution studies (Flowers, 1997) are much higher. This discrepancy is due to the fact that values of salinity threshold levels as expressed by mmol/l of a saturated soil, as presented above, are diluted values compared to the actual Cl concentration sensed by the root of a growing plant in the field. The soil moisture content at field capacity is about 50% of the value at the saturation point (Richards et al., 1954).

Therefore, one can expect that the actual Cl concentration around the root in soil at field capacity moisture conditions will actually be twice as high as that found in the saturation extract in laboratory studies. As soil moisture is depleted by the evaporating plants, and the soil moisture approaches wilting point, the concentration of chloride increases once more by about a factor of 2 (Richards et al., 1954). For example, tomato plants in nutrient solution were reported to resist 100–300 mmol/l Cl (Flowers 1997), whereas the value reported by Maas et al. (1988) is only 25 mmol/l. The reason for such a large discrepancy between salinity in soils and that in nutrient solution studies is due in part to the differences in the companion cations between soil-grown and nutrient solution-grown plants. The Cl^- in the soil solution is balanced mainly by Na, whereas in nutrient solutions it is balanced mainly by K and Ca. As a result, in the soil, the concentration of Ca near the plant root declines and plant roots will be damaged not only by the high Cl concentration but also by the low Ca and high Na content near the root (Yermiyahu et al., 1994).

The amounts of chloride found in plant leaves vary with habitat. Both the external chloride concentration and the balance of other available anions influence the chloride content in the plant. There are great differences in tolerance to chloride salts among crops and plants of the same species (Sopandie et al., 1989).

The negative effects of high chloride concentrations in the soil and irrigation water on crop production are observed in coastal (Can et al., 2003), arid, and semiarid areas, where freshwater sources are often scarce and the available groundwater used for irrigation is saline. Chloride toxicity in plants is usually associated with saline soil conditions (Mengel and Kirkby, 1981). The long-term effect of using saline water is of higher consequence for tree plantation rather than annual crop system. Walker and Douglas (1991) checked the effect of nutrient solution NaCl levels on citrus rootstocks and reported differences between the sensitivity of rootstocks in uptake and transport of Cl by young seedlings. They observed an increase in Cl levels in the roots when Cl in the soil solution was in the range of 0–25 mM. Increasing the concentration in soil solution to 100 mM did not affect Cl content in the roots of all rootstocks. However, differences were found in Cl concentration in the leaves. Walker and Douglas (1991) demonstrated that restriction of long-distance transport mechanisms of Cl might exist in plants.

Yield of food crops are usually reduced when soil solution concentration around the roots exceeds a certain concentration threshold specific for plant species (Maas and Hoffman, 1997).

Application of high doses of Cl-containing fertilizers (e.g. NH_4Cl and KCl), may also increase soil Cl^- levels. KCl accounts for about 95% of the K fertilizer used for agriculture around the world. However, despite continuous application of Cl-containing fertilizer, more Cl is imported to the field by Cl present in the irrigation water (Xu et al., 1988).

"Chloride tolerance" was identified as "chloride dependence" or "requirement" of crops to chloride in some cases (von Uexkull et al., 1986).

In the 1970s, pioneering works in the Philippines, Europe (Russell, 1978), and the northwestern USA (Powelson and Jackson, 1978), as cited by von Uexkull, 1988, clearly showed that Cl^- could play an important role in crop management. Positive responses of some crops to Cl-based fertilizers in many parts of the world (Jackson and McBride, 1991; Engel et al., 1997, Li et al., 1997) have since been published.

Plant responses to Cl^- involve the unique roles Cl^- plays in plants beyond its biochemical functions. Chloride is readily taken up by plants and all evidence indicates that Cl exerts its functions in processes related to charge compensation and osmoregulation (Marschner, 1963). Chloride serves in the plant as a charge balance during cation transport. In addition, it serves an important role by contributing to cell hydration and turgor. Cl is also essential as a cofactor in the oxidation of water in photosynthesis and as an activator of enzymes (Churchill and Sze, 2007).

Early studies (Altman and Mendel, 1964) suggested that the Cl anion passively flows into roots of citrus plants. The resistance of various citrus rootstocks to high chloride levels in the nutrient solution was explained by the ability of the rootstock to exclude chloride (Cooper et al., 1999), but no explanation to the mechanism involved was suggested. However, Moya et al. (1982) also claimed that Cl uptake is related to water uptake. The ionic balance during ion uptake was used to explain the ability of nitrate to prevent chloride uptake in tomatoes (Kafkafi et al.,1982) and by Xu et al. (1988) in many other plants. Other mechanisms and active chloride channels are now recognized to operate in plant roots, (Ludewig et al., 1968).

Susceptibility to Cl-induced leaf injury varies among species. Potato and tomato readily absorb Na^+ and Cl^-, and quickly exhibit symptoms of leaf tip and margin necrosis. On the other hand, safflower, with one of the highest rates of salt absorption. was only slightly injured by sprinkling saline water on its leaves. Barley readily absorbed salt, particularly Cl^-, and exhibited minor injury symptoms, whereas sesame and alfalfa had intermediate absorption rates but were somewhat more susceptible to Cl injury. Sugar beet was uninjured by sprinkling but absorbed appreciable amounts of NaCl, whereas sorghum developed some necrosis along leaf edges but absorbed very little salt. Cauliflower, cotton, and sunflower absorbed salt slowly and exhibited almost no injury (Maas et al., 1988).

6.2.3 Absorption mechanisms of chloride by plant roots

Uptake of Cl by the plant roots is generally an active process that requires energy. Earlier studies suggested that Cl transport through the cell membrane involves the $2H^+:Cl^-$ symporter (Sanders, 1949) or occurs via an antiport with hydroxyl ions energized by ATP (Jacoby and Rudich, 1997).

The formation of a transmembrane pH gradient by H^+-ATPase in sugar beet was found to require presence of Cl^- in the incubation medium, although the anions had no effects on the ATPase activity in plasma membrane (Gaivoronskaya and Molotkovskii, 2000). Therefore, the activation of ATP-dependent pH gradient generation by Cl^- is due to dissipation by the anions of the membrane potential produced by transmembrane transport of protons. The H^+/Cl^- symporter was studied using electrophysiological methods (Felle, 2006), which demonstrated that $nH^+:Cl^-$ would be better described as the ratio between proton and chloride uptake. Felle concluded that the kinetics of chloride transport depended on the pH gradient across the plasma membrane rather than on the membrane potential. Specific protein channels energized by ATP for chloride transport were suggested both for the plasmalemma (Lin, 2000) and for the tonoplast (Martinoia et al., 1956).

Plant chloride channels were revealed in the membrane of *Arabidopsis thaliana* by patch clamp technique (Lew, 1962) and were cloned from *Arabidopsis thaliana* by Hechenberger et al. (1982) and from tobacco by Lurin et al. (1985). The chloride channel of plants is voltage-dependent (Lurin et al., 1985) and its conductance (ranging from 5 to 40 pS) was independent of cytoplasmic KCl until a threshold concentration of about 300 mM was reached (Lew, 1962). The identification of a chloride channel provided a molecular probe for the study of voltage-dependent anion channels in plants.

Chloride ions always keep their negative charge, whereas the anions SO_4^{2-} and NO_3^- are partly or completely reduced during metabolism in the plant. Changes in root temperature and external Cl concentration affect Cl influx and accumulation (Cram, 1983, 1988).

It is difficult to determine the balance of individual ions across the plasmalemma and the tonoplast (Glass and Siddiqi, 1998). The situation is further complicated by interaction between the shoot and the root. Ion influx is regulated by the flux into the xylem and involves recycling in the phloem (Marschner, 1963). The fluxes of chloride in intact plants are very different from those measured in plants with excised roots (Collins and Abbas, 1975).

Glass and Siddiqi (1998) proposed a homeostatic mechanism that senses vacuolar nitrate plus chloride or total anion concentration. A variety of other schemes are discussed by Deane-Drummond (1990).

All of the suggested mechanisms must be balanced with growth, and as yet there is no generally accepted view of the control of chloride uptake and transport in plants (Flowers, 2009).

The composition of the root cell membrane not only affects ion selectivity, but is also of particular importance in preventing Cl from entering the root. Salinity tolerance in grapes was positively correlated with the solubility of chloride in the lipids that constitute the root membranes (Kuiper, 2000).

Enrichment of root cell membranes with phospholipids relative to their monogalactose diglyceride content limits chloride uptake (Kuiper, 2000). However, no apparent differences were found in the chemical composition of root microsomal membrane lipids between varieties of corn with low and high Cl uptake, and the composition of these membrane lipids was not affected by chloride salt (Hajibagheri et al., 1935). The presence of high Cl concentration in the root medium, causes a reduced nitrate uptake and vice versa (Smith and Fox, 2005; Kafkafi et al., 1985).

6.2.4 Effects of irrigation with TWW containing high Cl concentrations on crops

The use of high levels of Cl in the irrigation water on wine quality in Australia revealed that the chloride content of the wine ranged between 1.5 and 6.8 mM compared to 0.58–1.39 mM of European wine sources (Downton, 1996).

The effect of municipal wastewater (from the sewage treatment plant of Castellon, Spain) on the growth of 2-yr-old orange trees was investigated by Lapena et al. (2005). Leaves of trees irrigated with wastewater contained significantly higher concentrations of N, K, Cl, Na, and B compared with those from non-saline groundwater-irrigated trees. However, the concentrations of Na, Cl, and B were not limiting for citrus growth, and it was concluded that wastewater is suitable for the irrigation of oranges. Another study from Spain involving citrus reported similar results (Reboll et al., 1985a).

The main question concerning utilization of effluents for agricultural irrigation is: how long water containing high chloride levels can safely be used for agricultural purposes before damage to soil and plants is observed? In Israel, a grapefruit orchard in Mizra was damaged 7 yrs after irrigation with wastewater was started. There is a debate regarding which ion was the major cause for the damage, Na or B. In Australia (Cole, 1963), citrus orchards (cv. Valencia and cv. Washington Navel Orange) on sandy soils in semiarid South Australia (evaporation 1900 mm, rainfall 240 mm) were irrigated with water from the River Murray having a chloride content that varied from <1 to >10 mmol/l (electrical conductivity 0.35–1.4 dS/m). Under irrigation with water salinities above 4 meq/l Cl^-, yield losses were related to toxic effects of chloride accumulation in the leaves, rather than osmotic effects of the soil solution. The mechanism of Cl stress in citrus plants was studied by Bar et al. (2004). They showed that there are various degrees of resistance to Cl toxicity by various citrus rootstocks and increase of internal ethylene production was observed in all citrus rootstocks studied. However, in all rootstocks, increasing nitrate content in the soil solution decreased Cl accumulation in citrus plants.

In California, Na and Cl were the main problem during prolonged application of saline water (Ayars et al., 2000).

Monitoring and preventing chloride accumulation in the root zone is crucial. Once a critical Cl level is monitored in an aquifer, this source of irrigation water cannot be used further. Reducing chloride content in the aquifer is practically impossible. The way to revive such a water source is to dilute it with water sources of low salt content, or to desalinize the water prior to utilization. Some temporary practices dilute the sewage water with a less salty source such as harvested runoff water (Kibutz Hafetz Chaiim, Israel); however, this might become a "fools' trap" as no control on the runoff water composition is available. Another alternative is the desalinization of effluent sources. As the cost of desalinization decreases this may become a feasible alternative. Another alternative is to desalinize and reduce the Cl content of the water sources from which the effluent water is derived, thus reducing the Cl concentration in the effluent. Prevention of household and industrial salt input by public regulation (Weber et al., 1998) is practiced in Israel. The long-term practical conclusion will call for prevention of salinization of aquifers that are being used for irrigation of landscape and agricultural vegetation.

Actual field data of Cl distribution in soil after irrigation with TWW and freshwater were obtained in a field experiment in Israel (Feigin et al., 2005; Fine et al. 2007). The main properties of the TWW and the freshwater that were used in this experiment are shown in Table 6.2.1. Total salinity in the freshwater and the TWW was 0.9 and 1.4 dS/m, respectively. This range of salinity level is normally found in many water sources used for irrigation in Israel. The chloride content in the TWW is two times higher than in the freshwater and so is that of Na, Ca, and sulfate. The Cl concentration in the water is not alarming to continuous usage for irrigation, provided it does not continue to accumulate in the soil with further repeated irrigation. Table 6.2.2 shows the extreme range of Cl concentrations in the root zone soil profiles following 3 yrs of irrigation with freshwater or TWW. During the summer months from May to September 1991 and 1992, 375 and 405 mm of TWW irrigation increased the chloride concentration in the upper layers of the soils (0–120 cm) in the TWW-irrigated treatment relative to the freshwater treatment. These data suggest that evapotranspiration caused soil water upward movement that left the soluble chloride in the upper soil layer. The fact that chloride is washed away by sufficient annual rains (600 mm) through the winter from the top soil profile as shown in the spring of 1991 is misleading, as chloride is a stable ion and is going to stay with us forever.

Table 6.2.1 Selected characteristics of TWW and freshwater used in a field study in Mabarot, Israel (Feigin et al., 1993 and Fine et al., 1994)

Parameter	TWW	Fresh
COD, mg/l	181–394	ND
TSS, mg/l	60–226	ND
BOD, mg/l	60–133	ND
pH	8.4	8.1
EC, dS/m	1.4	0.9
Cl^{-1}, $mmol_c$/kg	7.6	4.0
NO_3^{-1}, $mmol_c$/kg	0.2	0.6
HCO_3^{-1}, $mmol_c$/kg	5.2	2.4
SO_4^{-2}, $mmol_c$/kg	1.2	0.6
NH_4^+, $mmol_c$/kg	2.5	0.2
N-Organic, $mmol_c$/kg	1.0	0.0
Na, $mmol_c$/kg	5.3	3.3
K, $mmol_c$/kg	0.5	0.1
Ca, $mmol_c$/kg	4.0	2.0
Mg, $mmol_c$/kg	1.9	2.0
SAR	2.7	2.3

BOD, biological oxygen demand; COD, chemical oxygen demand; EC, electrical conductivity; SAR, sodium adsorption ratio; TSS, total soluble solids; TWW, treated wastewater.

Table 6.2.2 Cl distribution in the soil profile of corn field irrigated with treated wastewater (TWW) or freshwater (adapted from Feigin et al., 1993 and Fine et al., 1994)

Soil layer (cm)	Fresh Spring 1990	Fresh Fall 1990	TWW Fall 1990	Fresh Spring 1991	TWW Spring 1991	Fresh Fall 1991	TWW Fall 1991	Fresh Spring 1993	TWW Spring 1993
	mmol/l								
0–20	2.35	3.97	4.12	2.09	2.11	6.78	8.41		
0–30								1.01	0.59
20–40	1.06	3.93	4.49	1.59	1.64	2.38	3.73		
30–60								0.44	0.62
40–60	0.77	4.66	5.16	1.25	1.72	2.82	4.85		
60–80	1.18								
60–90		2.85	3.08	1.71	2.08	2.82	3.77	0.51	1.02
80–100	1.79								
90–120		2.38	2.66	1.93	2.27	2.53	2.96	0.76	1.39
100–120	2.62								
120–150	3.84	3.31	2.67	2.77	2.73			0.92	1.36
150–180				3.57	3.18				
180–210				4.25	4.11			1.31	1.53
240–270								2.16	2.36
300–330								2.31	2.84
360–400								2.69	3.19

In May 1993, after high rain winter season (1079 mm), higher Cl concentrations were found in the soil profile from 100 downward to 400 cm of the TWW treatment than the freshwater. These data demonstrate that it is very dangerous to use high levels of chloride in the irrigation water in arid and semiarid climates where chloride ions accumulate and may lead to a complete abandonment of wells and springwater sources as irrigation water for generations to come.

6.2.4.1 Hazards of long-term irrigation with recycled effluents

Recycled water irrigation is an environmentally sound wastewater disposal practice (Weber and Schneider, 1988). However, TWW is always more saline than the supplied freshwater (Rebhun and Schechter, 1985).

Chloride salts have negative environmental effects on crops, soils, and groundwater. As the removal of the salts, once they enter the sewage, is an expensive process, the prevention of salt enrichment is the cheapest avenue. In Israel, prevention of salt loading into the sewage system includes:

(i) the search for new technologies to reduce sodium chloride salt consumption by industry and population, and its discharge into sewage;
(ii) adoption of new technologies to cope with arising situations (desalination processes, dilution with good water sources);
(iii) raising public and industry awareness to the environmental implications of salinity pollution; and

(iv) the legal approach expressed through new scientifically based regulations (Weber and Schneider, 1988, Weber et al., 1986).

The main contributor to the salinity of sewage in Israel was the water softening process followed by the meat "kosher" process in slaughter houses (Rossen et al.,1993). In softening water for industry, Na was replaced by K to allow use of the treated water in agriculture (Saliternik, 2009; Rebhun and Schechter, 1985).

The environmental soil structure considerations were at the basis of the decision to replace NaCl with KCl in water softening of industrial plants. However, in that process the Cl content in industrial effluents remained unchanged.

Intensive irrigation could lead to secondary salinization or re-desertification processes, as experienced in many irrigated regions of developed countries (Banin and Fish, 1985). These processes are mainly due to interference in the geochemical/salt balances of irrigated regions. The re-salinization of the Yizre'el Valley, a 20 000 ha intensively irrigated region in Israel, can serve as an alarming example. An intensive advanced agroecosystem has developed in the region in the last 70 yrs. The water sources include: pumping and importing irrigation water by the Israel National Water Carrier, large-scale use of TWW, and winter flood impoundment in reservoirs for summer irrigation. The irrigation methods used are: sprinkler, trickle, moving-line, and center-pivot systems. Despite claiming high water use efficiency, salinization of regional water resources and many fields had developed in the mid-1980s. Evaluating the water and salt balances of the Yizre'el Valley, using Cl as the representative salt constituent, has shown that as annual water consumption in the valley as a whole reached about 60 000 000 m^3, the annual import of soluble salts by the water reached 15 000 tons of Cl. Salt picked up by impounded surface water and applied to the fields, amounted in the late 1980s to more than 9000 tons Cl/yr. The source for the re-circulated salts was the accumulated soluble salts in soils and in the shallow aquifer in the valley, which were leached by floodwater, drained, or infiltrated into the underground water reservoirs. Annual re-pumping affected the water quality, and toxic levels were reached due to recycling of chlorides. This regional irrigation experience demonstrates the utmost importance of maintaining the salt balance in any local aquifer in addition to the need for increasing irrigation efficiency. The maintenance of long-term productivity of irrigated lands in arid zones mainly depends on sound irrigation water management.

The only practical long-term solution to prevent Cl accumulation in soils, when recycled water is used for irrigation, is to return the brine of all man-made industries and agricultural irrigation activities back to the ocean. New membrane technology for desalination produces very high quality water on side of the membrane and brine on the other. This brine must be returned to the ocean to keep it from the soil water system. It was recently reported that membrane desalination might be "too good" and plants grown on such water suffered from Mg deficiencies, meaning Mg must be added to the recycled desalinated water (Yermiyahu et al., 1959).

6.2.5 Foliar damage by Cl in sprinkler irrigated TWW

The extent of foliar damage to ornamental trees irrigated with treated TWW (reused water) was reported by Jordan et al. (1997). In an irrigation experiment in the dry climate of Nevada, USA, chloride was found to be a significant factor dictating the extent of reused water implementation on golf courses, in schools, and parks in the arid southwestern USA. The extent of foliar damage was studied on 20 tree species, using sprinkler-irrigation with reused water, municipal water, or synthesized saline water. After a total of 168 irrigation cycles, over a 16-month period, the index of visual damage (IVD) were assessed at different times. Of the tested plants, six species showed significant foliar damage even when irrigated with municipal freshwater. Accumulation of Cl in the leaf tissue was shown to be a species-dependent response, with tissue Cl concentrations varying by as much as a factor of 5. Based on their results a number of woody ornamental trees were found that can tolerate spray irrigation of reused water in the hot dry environment of southern Nevada. Natural salt spray on seashore plants is known to occur, and only plants adapted to the salt spray survive along sea shores (Maun and Perumal, 1981).

There is a wide range in plant sensitivity to salt sprays directly on leaves. Careful selection of ornamental plants to adapt to declining water quality for irrigation can delay the demand for using freshwater in dry land climates. However, most edible commercial crops like wheat, corn, rice, soybean, and the leafy vegetables are still not salt-tolerant and are expected to show yield decline and leaf damage upon high salt content water source usage.

6.2.6 Concluding comments

Chloride is an essential micronutrient in plants. Chloride is not sorbed to soil clay particles, having permanent negative charge, and therefore is leached to the ground water by rain and irrigation, and accumulates on the soil surface during evaporation of soil water. The degree of plant tolerance to high levels of Cl varies from very high resistance in plants naturally grown in salty habitat to growth restriction in very sensitive plants. With the increased world demand for drinking water, salty sources of water are being used in agricultural areas and as a result, a decrease in yield is observed. The duration of soil productivity under irrigation with high Cl water is expected to be oppositely related to the chloride concentration. Sensitive plants will be the first to be removed from production and, as a result, world effort in breeding edible crops and vegetables is operating in many countries.

References

Altman, A. and Mendel, K. (1973) Characteristics of the Uptake Mechanism of Chloride Ions in Excised Roots of a Woody Plant (Citrus) *Physiologia Plantarum* **29**, 57–162.

Asano, T. M. Maeda and Takai, M. (1996) Wastewater reclamation and reuse in Japan: overview and implementation examples. *Water Science Technology* **34**, 219–226.

Ayars, J.E., Pheneb, C.J., Hutmacherc, R.B., Davis, K.R., Schonemana, R.A., Vaila, S.S. and Meadd, R.M. (1999) Subsurface drip irrigation of row crops: a review of 15 years of research at the Water Management Research Laboratory. *Agricultural Water Management* **42**, 1–27.

Banin, A. and Fish, A. (1995) Secondary desertification due to salinization of intensively irrigated lands-the Israeli experience. *Environmental Monitoring and Assessment* **37**, 17–37.

Bar, Y., Apelbaum, A., Kafkafi, U. and Goren, R. (1997) Relationship between chloride and nitrate and its effect on growth and mineral composition of avocado and citrus plants. *Journal of Plant Nutrition* **20**, 715–731.

Broyer, T.C., A.C. Carlton, C.M. Johnson and R.R. Stout. 1954. Chloride element for higher plants. *Plant Physiology* **29**, 526–532.

Can, H.Z., Anac, D., Kukul, Y. and Hepaksoy, S. (2003) Alleviation of salinity stress by using potassium fertilization in satsuma mandarin trees budded on two different rootstocks *Acta Horticulturae* **618**, 275–280.
Churchill, K.A. and Sze, H. (1984) Anion-sensitive, H + -pumping ATPase of oat roots. *Plant Physiology* **76**, 490–497.
Cole, P.J. (1985) Chloride toxicity in citrus. *Irrigation Science* **6**, 63–71.
Collins, J.C. and Abbas, M.A. (1985) Ion and water transport in seedlings of mustard (*Sinapsis alba* L.). *New Phytologist* **99**, 195–202.
Cooper, W.C., Gorton, B.S. and Olson, E.O. (1952) Ionic accumulation in citrus as influenced by rootstock and scion and concentration of salts and boron in the substrate. *Plant Physiology* **27**, 191–203.
Cram, W.J. (1983) Chloride accumulation as a homeostatic system: Set points and perturbations. *Journal of Experimental Botany* **34**, 1484–1502.
Cram, W.J. (1988) Transport of nutrient ions across cell membranes in vivo. *Advances in Plant Nutrition* **3**, 1–54.
Deane-Drummond, C.E. (1986) A comparison of regulatory effects of chloride on nitrate uptake, and of nitrate on chloride uptake into *Pisum sativum* seedlings. *Physiologia Plantarum* **66**, 115–121.
Downton, W.J.S. (1977) Salinity Effects on the Ion Composition of Fruiting Cabernet Sauvignon Vines. *American Journal of Enology and Viticulture* **28**, 210–214.
Engel, R.E., Bruckner, P.L. and Eckhoff, J. (1998) Critical tissue concentration and chloride requirements for wheat. *Soil Science Society of America Journal* **62**, 401–405.
Feigin, A., Fine, P., Bar-Tal, A., Sheinfeld, S., Hefer, Y., Meiri, A., Shaharabani, N., Sagiv, B., Markovitz, T., Sternbaum, B. and Soriano, S. (1993) *Methods for Reducing Pollution of Underground Water by Sewage and Fertilizers*. Chief Scientist of the Ministry of Agriculture. pp. 39 (Hebrew).
Felle, H.H. (1994) The H + /Cl-symporter in root-hair cells of *Sinapsis alba*. An electrophysiological study using ion-selective microelectrodes. *Plant Physiology* **106**, 1131–1136.
Fine, P., Bar-Tal, A., Sheinfeld, S. and Hefer, Y. (1994) *Methods for reducing Pollution of Underground Water by Sewage and Fertilizers*. Chief Scientist of the Ministry of Agriculture. pp. 39 (Hebrew).
Fixen, P. E. (1993) Crop responses to chloride. *Advances in Agronomy* **50**, 107–150.
Flowers, T.J. (1988) Chloride as a nutrient and as an osmoticum. *Advances in Plant Nutrition* **3**, 55–78.
Flowers, T.J. (2004) Improving crop salt tolerance. *Journal of Experimental Botany* **55**, 307–319.
Gaivoronskaya, L.M. and Molotkovskii, G. (1992) Role of chloride channels in ATP-dependent generation of delta pH by isolated plasma membrane. *Soviet Plant Physiology (USA)* **38** (5, pt.5), 628–635.
Glass, A.D.M. and Siddiqi, Y.M. (1985) Nitrate inhibition of chloride influx in barley: Implications for a proposed chloride homeostat. *Journal of Experimental Botany* **36**, 556–566.
Hajibagheri, M.A., Yeo, A.R., Flowers, T.J. and Collins, J.C. (1989). Salinity resistance in Zea mays: Fluxes of potassium, sodium and chloride, cytoplasmic concentrations and microsomal membrane lipids. *Plant Cell Environment* **12**, 753–757.
Hechenberger, M., Schwappach, B., Fischer, W.N., Frommer, W.B., Jentsch, T.J. and Steinmeyer, K. (1996) A family of putative chloride channels from Arabidopsis and functional complementation of a yeast strain with a CLC gene disruption. *Journal of Biological Chemistry* **271**, 33632–33638.
Hoagland, D.R. (1948) *Lectures on Inorganic Nutrition of Plants*. Chronica Botanica, Waltham, MA. USA.
Jackson, T.L. and McBride, R.E. (1986) Yield and quality of potatoes improved with potassium and chloride fertilization. In: *Special Bulletin on Chloride and Crop Production* (T. L. Jackson, ed.) No. 2, pp. 73–83. Potash & Phosphate Institute, Atlanta, Georgia.
Jacoby, B. and Rudich, B. (1980) Proton–chloride symport in barley roots. Annals in Botany **46**, 493–498.
Jing, A.S., Guo, B.C. and Zhang, X.Y. (1992) Chloride tolerance and its effects on yield and quality of crops. *Chinese Journal of Soil Science* **33**, 257–259.
Johnson, C.M., Stout, P.R., Broyer, T.C. and Carlton, A.B. (1957) Comparative chloride requirements of different plant species. *Plant Soil* **8**, 337–353.
Jordan, L., Devitt, D., Morris, R. and Neuman, D. (2001) Foliar damage to ornamental trees sprinkler-irrigated with reuse water. *Irrigation Science* **21**, 17–25.
Kafkafi, U., Valoras, N. and Letey, J. (1982) Chloride interaction with NO_3 and phosphate nutrition in tomato. *Journal of Plant Nutrition* **5**, 1369–1385.
Kuiper, P.J.C. (1968) Ion transport characteristics of grape root lipids in relation to chloride transport. *Plant Physiology* **43**, 1372–1374.

Lapena, L., Cerezo, M. and Garcia-Agustin, P. (1995) Possible reuse of treated municipal wastewater for *Citrus spp.* plant irrigation. *Bulletin of Environmental Contamination and Toxicology* **55**, 697–703.
Lew, R.R. (1991) Substrate regulation of single potassium and chloride ion channels in arabidopsis plasma membrane. *Plant Physiology* **95**, 642–647.
Li, L.T., Yuan, D.H. and Sun, Z.J. (1994) Influence of a Cl^- containing fertilizer on Cl concentration in tobacco leaves. *Journal of South Agricultural University (Chinese)* **16**(4), 415–418.
Lin, W. (1981) Inhibition of anion transport in corn root protoplasts. *Plant Physiology* **68**, 435–438.
Lipman, C.B. (1938) Importance of silicon, aluminum, and chlorine for higher plants. *Soil Science* **45**, 189–198.
Ludewig, U., Pusch, M. and Jentsch, T.J. (1996) Two physically distinct pores in the dimeric ClC-0 chloride channel. *Nature* **383** (6598), 340–343.
Lurin, C., Geelen, D., Barbier-Brygoo, H., Guern, J. and Maurel, C. (1996) Cloning and functional expression of a plant voltage-dependent chloride channel. *The Plant Cell* **8**, 701–711.
Maas, E.V. (1985) Crop tolerance to saline sprinkling water. *Plant Soil* **89**, 273–284.
Maas, E.V. and Hoffman, G.J. (1977) Crop salt tolerance –current assessment. *Journal of the Irrigation and Drainage Division* **103**, 115–134.
Maas, E.V., Grattan, S.R. and Ogata, G. (1982) Foliar salt accumulation and injury in crops sprinkled with saline water. *Irrigation Science* **3**, 157–168.
Marschner, H. 1995. *Mineral Nutrition of Higher Plants*, 2nd Ed., pp. 299–312, 396–404. Academic Press, San Diego, New York.
Martinoia, E., Schramm, M.J., Kaiser, G., Kaiser, W.M. and Heber, U. (1986) Transport of anions in isolated barley vacuoles. I. Permeability to anions and evidence for a Cl^- uptake system. *Plant Physiology* **80**, 895–901.
Maun, M.A. and Perumal, J. (1999) Zonation of vegetation on lacustrine coastal dunes: effects of burial by sand. *Ecology Letters* **2**, 14–18.
Mengel, K. and Kirkby, E.A. (1987) *Principles of Plant Nutrition*. International Potash Institute, Bern.
Moya, J.L., Gonez-Cadenas, A., Primo-Millo, E. and Talon, M. (2003) Chloride absorption in salt-sensitive Carrizo citrange and salt-tolerant Cleopatra mandarin citrus rootstocks is linked to water use. *Journal of Experimental Botany* **54**, 825–833.
Quirk, J.P. and Schofield, R.K. (1955) The effect of electrolyte concentration on soil permeability. *Journal of Soil Science* **6**, 163–178.
Rebhun, M. and Schechter, A. (1966) *The Use of Potassium Salts as Regenerators for Ion Exchange Water Softeners*. Technion, Israel Institute of Technology, Haifa, final report CV-203, 93 pp. (in Hebrew).
Reboll, V., Cerezo, M., Roig, A., Flors, V., Lapeña, L. and García-Agustín, P. (2000) Influence of wastewater vs groundwater on young Citrus trees. *Journal of the Science of Food and Agriculture* **80**, 1441–1446.
Richards, L.A. (1954) *Diagnosis and Improvements of Saline and Alkali Soils*. Agriculture Handbook. USDA. Washington DC.
Saffigna, P.G. and Keeney, D.R. (1977) Nitrogen and chloride uptake by irrigated Russet Burbank potatoes. *Agronomy Journal* **69**, 258–264.
Saliternik, C. (1993) *Reduction of Sodium Damage to Soil by Replacement of Sodium Salts with Potassium Salts in the Industry*. Report of EET Consulting Co. to the Ministry of the Environment, Tel Aviv, 34 pp. (in Hebrew).
Sanders, D. (1984) Gradient-coupled chloride transport in plant cells. In: *Chloride Transport Coupling Biological Membranes and Epithelia* (G. A. Gerencser, ed.) pp. 63–120. Elsevier.
Smith, F.A. and Fox, A.L. (1977) Interactions between chloride and nitrate uptake in Citrus leaf slices. *Australian Journal of Plant Physiology* **4**, 177–182.
Sopandie, D., Takeda, K., Moritsugu, M. and Kawasaki, T. (1993) Selection for high salt tolerant cultivars in barley [Hordeum vulgare] *Bulletin of the Research Institute for Bioresources - Okayama University (Japan)* **1**(2), 113–129.
Talbott, L.D. and Zeiger, E. (1996) Central roles for potassium and sucrose in guard-cell osmoregulation. *Plant Physiology* **111**, 1051–1057.
Tottingham, W.E. (1919) A preliminary study of the influence of chlorides on the growth of certain agricultural plants. *Journal of the American Society of Agronomy* **11**, 132.
Ulrich, A. and Ohki, K. (1956) Chloride, bromine, and sodium as nutrients for sugar beet plants. *Plant Physiology* **31**, 171–181.

von Uexkull, H.R. (1990) Chloride in the nutrition of coconut and oil palm. *Transactions of the 14th International Congress of Soil Sciences, IV*, 134–139. Kyoto, Japan.

von Uexkull, H.R. and Sanders, J.L. (1986) Chloride in the nutrition of palm trees. In: *Special Bulletin on Chloride and Crop Production* (T. L. Jackson, ed.) No. 2, pp. 84–99. Potash & Phosphate Institute, Atlanta, Georgia.

Walker, R.R. and Douglas, T.J. (1983) Effect of salinity level on uptake and distribution of chloride, sodium and potassium ions in citrus plants. *Australian Journal of Agricultural Research* **34**, 145–153.

Warburg, O. (1949) *Schwermetalle als Wirkungsgruppen von Fermenten*, p. 180. Cantor, Freiburg, Germany.

Weber, B. and Schneider, G. (1992) Revision of industrial effluent regulations in Israel. In: *Environment Quality and Ecosystem Stability*, Vol. V/A A (eds Adin, A. Gasith B. Fattal, and A. Kanarek), pp. 480–487. ISEEQS Publ., Jerusalem, Israel.

Weber, B., Avnimelech, Y. and Juanico, M. (1996) Salt enrichment of municipal sewage: New prevention approaches in Israel. *Environmental Management* **20**, 487–495.

Xu, G.H., Magen, H., Tarchitzky, J. and Kafkafi, U. (2000) Advances in chloride nutrition of plants. *Advances in Agronomy* **68**, 97–150.

Yermiyahu, U., Nir, S., Ben-Hayyim, G., Kafkafi, U. and Kinraide, T.B. (1997) Root elongation in saline solution related to calcium binding to root cell plasma membranes. *Plant and Soil* **191**, 67–76.

Yermiyahu, U., Tal, A., Ben-Gal, A., Bar-Tal, A., Tarchitzky, J. and Lahav, O. (2007) Rethinking desalinated water quality and agriculture. *Science* **318**, 920–921.

Ying, G.G., Toze, S., Hanna, J., Yu, X.Y., Dillon, P.J. and Kookana, R.S. (2008) Decay of endocrine-disrupting chemicals in aerobic and anoxic groundwater. *Water Research* **42**, 1133–1141.

Chapter 7

Heavy metals in soils irrigated with wastewater

Amir Hass, Uri Mingelgrin and Pinchas Fine

List of acronyms

BOD	biological oxygen demand
CEC	cation exchange capacity
COD	chemical oxygen demand
DOC	dissolved organic carbon
DOM	dissolved organic matter
DW	dry weight
Eh	redox potential
EPS	extracellular polymeric substances
MBTP	mechanical biological treatment plant
OC	organic carbon
OM	organic matter
OPE	oxidation pond effluent
PZNC	point of zero net charge
SAT	soil aquifer treatment
SMP	soluble microbial products
SS	suspended solids
TOC	total organic carbon
WWTP	wastewater treatment plant

Treated Wastewater in Agriculture, First Edition, edited by Guy J. Levy, Pinchas Fine and Asher Bar-Tal

7.1 Introduction

Irrigation with treated municipal effluents provides an economic, cost-effective option for recycling of sewage water. The growth of the world population, increased urbanization, and a rising demand for drinking and irrigation water coupled with tightening regulatory restrictions on disposal of wastewater, are all expected to lead to an increase in the use of treated wastewater (TWW) for irrigation (Feigin et al. 1991; Bouwer 2000; Bixio et al. 2006).

The soil–plant system was recognized as a medium in which applied TWW can be polished to environmentally acceptable quality. This, and the increasing demand for irrigation water, brought about an expansion in the use of land application for both treatment and disposal of municipal wastewater.

Many field experiments and case studies have been conducted during the last several decades to identify the capacity of soil, and soil–plant systems to utilize and ameliorate wastewater effluents. In these studies attempts were made to identify the pollutants that may be present in the effluent; set quality criteria for effluents applied to soil; identify soil parameters (e.g. pH, cation exchange capacity (CEC), texture) that may affect effluent application practices such as frequency of application or overall loading capacity; and understand the short-, and long-term implications of wastewater disposal onto soil. Earlier studies were oriented mostly toward maximizing crop yield, and to a lesser extent towards understanding the effect of effluent application on environmental quality. Hence, research was concentrated mostly on agronomic aspects such as nutrient availability and adverse effects on soil physical and chemical properties (e.g. salinity, sodicity, boron content, etc.; Feigin et al. 1991). However, considering the soil as part of a wastewater treatment or disposal system requires a broader view of the soil as a contributing medium in the global water cycle and a component of the biosphere, the ability of which to sustain its biological and physicochemical integrity under the influence of the added effluents is of major concern.

Heavy metals are ubiquitous in municipal wastewater (Sorme and Lagerkvist, 2002; Rule et al., 2006), but their concentration is very low as long as wastewater from industrial sources is treated separately or is pretreated at the source before mixing with domestic sewage. Long-term accumulation in soil of heavy metals and of constituents of sewage effluent that can interact with metal ions (e.g. soluble organic and inorganic ligands), can alter soil composition and processes, especially the bioavailability of the heavy metals and their mobility through the soil profile. This chapter deals with various aspects of the behavior of heavy metals in soil under irrigation with wastewater. Emphasis is given to the effect of wastewater quality on metal accumulation, their partitioning between various environmental compartments, mobility, and availability to plants.

7.2 Heavy metals in effluents

Heavy metals are found in municipal wastewater mostly in association with suspended solids (Brown and Lester, 1979; Nielsen and Hrudey, 1983; Sterritt and Lester, 1984; Corey et al., 1987; Karvelas et al., 2003a; Rule et al., 2006). As a consequence, conventional wastewater treatment schemes, during which much of the incoming suspended material is removed, tend to exhibit high heavy metal removal rates, even though their removal is not specifically targeted (Brown and Lester, 1979; Nielsen and Hrudey, 1983; Sterritt and Lester, 1984; Lawson et al., 1984a, 1984c; Page and Chang, 1985; Oliveira et al., 2007). Typical concentrations of selected metals in wastewater effluents that underwent different treatment modes as compared to levels allowed according to guidelines and regulations for agricultural irrigation, drinking and surface waters are presented in Table 7.1.

Metal removal efficiency during wastewater treatment increases with the increase in initial metal concentration, microbial activity (and production of extracellular polymer substances, EPS), pH, and retention time (Brown and Lester, 1979; Lawson et al., 1984a, 1984b; Chipasa, 2003; Wang et al., 2003; Comte et al., 2008). The overall effect of most common treatment procedures is a rather low heavy metal concentration in domestic municipal wastewater effluents. Characteristic pollutant removal efficiencies during wastewater activated sludge treatment are summarized in Table 7.2. Between 30% and 60% of the metals in the incoming domestic sewage are associated with high-density particulate material and have already been removed during the primary sedimentation stage (Nielsen and Hrudey, 1983; Sterritt and Lester, 1984). During the secondary biological treatment stage, most of the metals content is either sorbed onto EPS, assimilated in the microbial biomass, or forms mineral precipitates. Only a minor proportion remains soluble, mostly as organic complexes (Brown and Lester, 1979; Sterritt and Lester, 1984; Lun and Christensen, 1989; Mcnicol and Beckett, 1989; Guibaud et al., 2003; Comte et al., 2008). Extracellular polymeric substances were shown to have a high affinity to heavy metals, with both high stability constants (ranging from 10^5 to 10^9) and sorption capacities (Tian, 2008).

The degree of dissociation of acidic functional groups in the organic matter (OM) increases with increasing pH, and with it the overall solubility of the OM and its sorption capacity for metals. The difference in the dissociation constant of the various acidic functional groups (e.g. sulfonic > phosphoric > carboxylic > phenolic) makes the relative contribution of these groups to the uptake of metal ions, pH-dependent. For example, Comte et al. (2008) showed that the metals sorption capacity of EPS, derived from an activated sludge, increased nearly 50 times for Cu and Pb and over seven times for Cd when the pH was raised from 4 to 8. In addition, elevation of pH in forced-aeration (as well as in activated sludge) systems results in precipitation of metals, mostly as oxyhydroxides and carbonates.

Partitioning between soluble species, suspended species, and solid-phase components (e.g. carbonates, hydroxides, phosphates) is highly metal-specific. In characterizing metals partitioning between dissolved (<0.45 μm) and particulate species at different stages of wastewater treatment, Karvelas et al. (2003) found 65–85% of Cd, Cr, Cu, Zn, and >95% of Pb associated with suspended solids, whereas Ni was mostly found in the dissolved state (80–93%). Metal partitioning was similar in the various wastewater treatment stages (raw, primary, and secondary treatment), with a slight increase in the dissolved fraction as the treatment progressed. Nielsen and Hrudey (1983) found low removal efficiency for Ni and Zn (43% and 54%, res pectively), as compared to >90% for Cd, Cu, and Cr during activated sludge treatment (Table 7.2), whereas Buzier et al. (2006) showed that only minor changes occurred in the concentration of the dissolved metal

Table 7.1 Concentration range of selected trace elements in wastewater effluents and in water quality criteria (based in part on Page and Chang, 1985)

Element	Primary effluent		Secondary effluent		Tertiary effluent[a]		Water quality criteria		
	Range	Median	Range	Median	Range	Median	Irrigation FAO[b]	Surface water USEPA[c]	Drinking USEPA[e]
					μg/l				
As	<5–30	<5	<5–23	5			100	150	10
Cd	<20–6400	<20	<5–150	5	<100		10	0.25	5
Co					1–62	2	50	—	—[d]
Cr	<50–6800	<50	<5–1200	20	<200		100	74[f]	100
Cu	<20–5900	100	<6–1300	40	<20–50	<20	200	9	1300
Hg	<0.1–125	0.9	<0.2–1	0.5			—	0.77[g]	2
Mn					<50–110	50	200	—	50
Mo	<1–20	8	1–18	7			10	—	—
Ni	<100–1500	100	3–600	4	2–180	40	200	52	—
Pb	<200–6000	<200	3–350	8	<200		5000	2.5	15
Zn	<20–2000	120	4–1200	40	70–6200	330	2000	120	5000

[a]Monterey wastewater reclamation study for agriculture (Engineering Science, 1987). Wastewater treatment that conforms to the requirements of the California Administrative Code, Title-22 for treatment of wastewater used to irrigate food crops that may be consumed without cooking.
[b]Ayers and Westcot (1985).
[c]US-EPA (2006a). Highest concentration of a substance in surface water to which an aquatic community can be exposed indefinitely without resulting in an unacceptable effect.
[d]not reported
[e]The highest level of a contaminant that is allowed in drinking water by the US EPA (USEPA, 2003)
[f]for Cr^{+3}; Cr^{+6} level is 11 $\mu g\,L^{-1}$
[g]as methylmercury

Table 7.2 Removal efficiency of selected pollutants during activated sludge wastewater treatment

Constituent	Removal efficiency (%)				
	Chang et al., 2002	Nielsen and Hrudey, 1983		Brown and Lester, 1979	
		Treatment		Range	Average
		Primary	Overall		
Biological oxygen demand	97		74		
Suspended solids	97		94		
Cd		39	92	10–100	46
Cr	76	68	92	33–99	66
Cu	79	60	93	33–98	66
Ni	50		43	0–100	33
Pb				27–100	64
Zn	65	44	54	44–100	69

species during secondary treatment, with most of the reduction in metal content resulting from removal of particulate material. The major part of the dissolved metal content (>70%) was in non-labile form (operationally defined as the fraction of a metal that does not bind to a metal binding gel such as chelex) and was, most likely, associated with OM (Buzier et al., 2006).

The soluble fraction of heavy metals in effluents is of greater importance than implied by its low proportion in the total content of the metals, as it is the most available and mobile fraction. Metal cations tend to appear in solution as complexed species. The present authors (e.g. Hayat, 2006) used dialysis membrane separation to determine the molecular size distribution of the various species of selected metals (Ba, Cu, Co, Fe, Li, Mo, Ni, Sr, Zn) in effluents. The dissolved metal species were found predominantly in the size range of 200–1000 Da. The prevalence of the metals in this size range indicates that they are bound to ligands of low molecular weight, either organic or inorganic. The molecular weight distribution of the metal species did not correspond to that of the dissolved organic matter (DOM). Fine and coworkers (2002) showed that the size distribution of DOM in a secondary effluent displayed two peaks, one at molecular weight <1 kDa and the other at the size range of 1–30 kDa. Although more DOM was found in the latter fraction, it contained far less trace metals than the former fraction. Accordingly, Haya, (2006) estimated that the stability constants (the ratio between the concentration of Cu bound to DOM and free Cu concentration in solution) at 1 μM free Cu concentration were around 0.8×10^5 and 4×10^5 l/kg dissolved organic carbon (DOC) for the high and low molecular weight complexes of Cu, respectively. These observations were also substantiated for Cu and other transition metals by Vulcan et al. (2002), who demonstrated the dominance of the <1.0 kDa size fraction in Cu complexes in aqueous sludge extracts and the metal scarcity at the peak observed in the DOM size distribution at molecular weight >2.5 kDa. Table 7.3 summarizes the distribution of selected metals between various size fractions in the effluent from a mechanical-biological wastewater treatment plant (WWTP) as measured

Table 7.3 Distribution of selected elements between the various size fractions in effluent water (Tel-Aviv, mechanical-biological wastewater treatment plant)[a]

	Ba	**Ca**	**Cr**	**Cu**	**Fe**	**K**	**Mg**	**Mn**	**Ni**	**Sr**	**Zn**
Total concentration (μM)	0.36	1.62	0.18	0.12	1.87	0.65	1.13	0.46	0.24	6.81	1.58
Complexed >1 kD (%)	5 ± 2	6 ± 1	25 ± 10	19 ± 7	35 ± 34	6 ± 1	4 ± 0	9 ± 5	17 ± 3	5 ± 1	9 ± 6
Complexed < 1 kD (%)	68 ± 17	69 ± 2	21 ± 0	13 ± 2	59 ± 20	55 ± 9	77 ± 3	63 ± 22	28 ± 7	71 ± 4	64 ± 27
Free (< 200 D) (%)	26 ± 16	25 ± 11	53 ± 0	67 ± 0	6 ± 3	39 ± 0	18 ± 7	28 ± 12	55 ± 0	23 ± 9	26 ± 8

[a]Based on four replicates (Fine, unpublished data).

by Fine (unpublished data). Again, it is apparent that trace metals tend to accumulate in the <1 kDa size fraction. Yet, although Cu^{2+} forms the strongest complexes of all divalent cations with DOM (e.g. Schnitzer, 1978), at the minute ambient concentrations of this metal and the relatively high content of competing metals in effluents, competition, e.g. with trivalent Fe and Al, drove this metal (as it did Cr) into the free ion fraction (Table 7.3).

The observed size distribution of the dissolved species of the trace metals may have a considerable impact on the mobility and availability of these metals. Aside from the fact that negative or neutral complexes are far less susceptible to sorption or precipitation in the geosphere than the free cationic metal form, the smaller the complex, the more likely it is to move freely in the porous matrix and to be taken up by plant roots. Once applied to the soil, the effluent's OM is decomposed relatively fast, with a half life in irrigated soils measured in days or at most in weeks depending on local conditions, with a small (<5%) recalcitrant fraction (Fine et al., 2002). Thus, under irrigation with high-quality effluent at agronomic rates, the speed of degradation of DOM may be such that DOM-assisted heavy metals transport is not very significant. On the other hand, under high rate irrigation with low-quality effluent, the concentration of total organic carbon (TOC) and the DOC/TOC ratio in the soil tend to increase with time, thus enhancing trace metals migration below the root zone. The increase in the concentration of transition metals in the soil solution with the increase in effluent-derived DOC is exemplified for Zn and Cu in Figure 7.1.

The distribution of metals within the suspended solids of the effluent could be inferred from their distribution between the components of the settling solids (i.e. biosolids). In the biosolids, metals exist mostly as carbonates, organic species, oxyhydroxides, and sulfides; with preference of Cu toward the OM and of Cd, Zn, and Pb toward carbonates (Stover

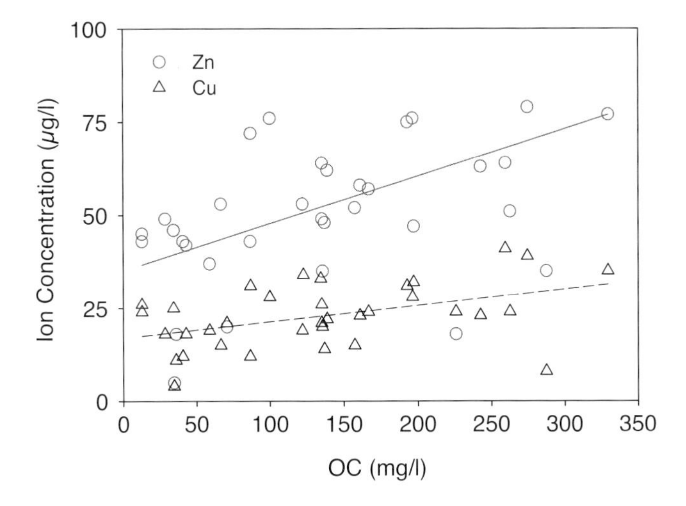

Figure 7.1 Zinc and Cu concentration as a function of organic carbon (OC) concentration in leachate collected in a lysimeter experiment on the disposal of low-quality oxidation pond effluent by irrigating *Eucalyptus camaldulensis* (Fine and Hass, unpublished).

et al., 1976; Silviera and Sommers, 1977; Emmerich et al., 1982; Wang et al., 1997). Although carbonates and OM are likely to be present under most treatment conditions (at pH $\geq$ 7.3), the appearance of sulfides or oxyhydroxides depends on the specific conditions of the wastewater treatment (e.g. anaerobic versus aerobic treatment, degree of OC removal, pH, etc.). It should be noted that the presence of clay (and silt) particles in the effluent will affect the speciation of metals in wastewater effluent and sludge by serving as sorbents.

7.3 Organic matter: composition in wastewater effluents and behavior in soil

The composition of OM in wastewater effluents strongly influences its tendency to complex or sorb metals. This composition is determined by the source of the effluent (e.g. industrial, municipal, or agricultural), the treatment process (e.g., lagoons, aerobic, or anaerobic), and the efficiency of the process.

In general, the OM in wastewater (or TOC), is divided into two groups according to effective size, using 0.45 μm as an arbitrary cut-off: particulate or suspended solids (SS, >0.45 μm) and DOM (<0.45 μm, or DOC). The SS fraction contains both organic and mineral components, with the former components including undigested organic debris, protozoa, algae, bacterial flocks, and the like. In biological treatment processes when a steady-state is reached, the DOC accounts for the majority of the effluent TOC, and it is composed mostly of soluble microbial products (SMP) (Barker and Stuckey, 1999; Shon et al., 2006). The SMP are defined as "the pool of organic compounds that are released into solution from substrate metabolism and biomass decay" (Barker and Stuckey, 1999). The SMP consist of humic, fulvic, nucleic, amino, and low molecular weight organic acids; polymers derived from cell debris; proteins; simple sugars and polysaccharides; lipids, polyphenols, and extracellular enzymes; siderophores; antibiotics; steroids; and assorted products of energy metabolism.

The above demonstrates that the wastewater DOM is a highly complex mixture. This complexity has resulted in the formulation of a number of classification schemes. For example, Ilani et al. (2005) used a fractionation scheme based on selective sorption of the DOM components on a set of resins (Leenheer, 1981), in which the DOM was divided into hydrophobic and hydrophilic fractions; each of which was further separated into acid, neutral, and basic subfractions. Ilani et al. (2005) analyzed effluents from two WWTPs and found the hydrophobic fraction to comprise around 70% and 40% of the DOM, out of which the acid fractions were 60% and 30%, respectively. These authors suggested that as the treatment of the wastewater advances, the content of the less-recalcitrant, hydrophilic fraction diminishes.

Biological treatment processes result in an overall decrease in TOC content as the process advances, with an accompanying increase in the proportion of DOC. As mentioned above, the DOC is composed of two main size fractions with maxima in size distribution at molecular weights >10 kDa and <1 kDa. With the increase in the duration of the treatment, there is an increase in the proportion of secondary degradation products and of refractory DOC components in the effluent (Namour and Muller, 1998; Dignac et al., 2000; Imai et al., 2002; Jarusutthirak et al., 2002; Jarusutthirak and Amy, 2007; Katsoyiannis and Samara, 2007). The modern-day, gradual increase in the use of synthetic organic compounds results in their ever-growing presence in wastewater. Some of these compounds (such as polychlorinated biphenyls) are relatively refractory in nature, and hence biological treatment processes show limited efficiency in their decomposition (Shon et al., 2006).

Although the more commonly used biological treatment processes have become efficient in removing SS, OM, nitrogen, and phosphorus, they also result in an increase in the relative content of the more refractory, DOM, which may affect heavy metal solubility and mobility. Technologies, such as sorption (e.g. on activated carbon), or bio-, ultra-, and nanofiltration, are available for further removal of organic pollutants and heavy metals from effluents (Shon et al., 2006). These technologies, however, are seldom used.

Once applied to the soil, the organic components of the effluent undergo considerable transformations. Profound differences were found, for example, in the DOC of a secondary effluent used for irrigation after it passed through the root zone in a sandy soil (Hayat, 2006). Analyses of DOC samples obtained following deficit irrigation with an effluent from a mechanical biological treatment plant (MBTP; 16 mg C/l) in a lysimeter setup showed that the effluent OC transformed as it passed through the soil profile (Fine and Hass, 2007). The leachate DOC had a lower proportion of aliphatic amino acids and protein-like substances, and a higher proportion of aromatic and humic components as compared with the DOC in the incoming effluent (Hayat, 2006). A similar shift toward more aromatic DOC with increasing soil depth was reported by Hoffmann et al. (1998) for soil under irrigation with wastewater. Transformations in the soil tend initially to shift the distribution of humic substances originating in the incoming water toward more hydrophilic, acidic, and low molecular weight components (i.e. fulvic acids), which concentrate in the dissolved state (Guo and Chorover, 2003). However, Jueschke et al. (2008) demonstrated that the composition of DOC (as measured by $UV_{245\ nm}$) in soil percolates is a function of residence time in the soil and is subject to seasonal variations. Hayat (2006) demonstrated that the DOC in the applied effluent bound less Cu (per unit C) than did the DOC in the soil leachate, and fluorescence emission analysis suggested that Cu binding by the DOC occurred on functional groups adjacent to aromatic rings (ibid.).

The properties of the DOC in leachates collected from lysimeters (concentration, density of functional groups, and Cu-binding characteristics) did not differ much between lysimeters irrigated with effluent and lysimeters irrigated with tapwater (Hayat, 2006). It seems, therefore, that the transformations which the OM in the applied effluent undergoes in the soil obliterate, to a large extent, the characteristics of the effluent's DOC. At the same time, Fine et al. (2002) reported that the size distribution of the DOC in the added effluent and in the soil leachate after 6 months of residence time was rather similar.

Deficit irrigation with effluents prolongs residence time, thus enabling more extensive transformations of the effluent's OM in the root zone. Although the bulk of the added OM decomposes during the irrigation season, the concentration of the remaining, more recalcitrant, DOM increases. Consequently, the concentration of complexed, and hence more mobile metal cations in the leachate can actually increase (Fine et al., 2002). As mentioned above (Hayat, 2006), the leachate DOC complexed metals more efficiently than the original DOC in the applied effluent. Additionally, under deficit irrigation, the concentration in soil solution of solutes other than DOC also increases and some of the inorganic solutes (e.g. SO_4^{-2}) can serve as ligands. Elevated levels of species, both organic and inorganic, which can serve as ligands, are likely to reduce metal retention in soil and promote metal mobility and availability.

7.4 Attenuation of heavy metals in soils irrigated with effluents

As a consequence of metal association with SS and dissolved species in the effluent, physical filtration is, together with sorption, the main mechanism responsible for the initial interception and accumulation of metals at the soil surface. As a result of the amorphous and chemically reactive nature of the suspended and dissolved components, various other processes also contribute to the observed marked decrease in the content of metals with depth in effluent-irrigated soils. These processes include specific or non-specific adsorption and precipitation, including nucleation and coprecipitation (McBride, 1994).

Many of the soil's solid components that are responsible for the attenuation of heavy metals in soil (e.g. organic matter and oxyhydroxides of Fe and Mn) are amphoteric, and the uptake of metals by such compounds is pH-dependent. Likewise, the extent of specific adsorption increases with the degree of hydrolysis of acidic functional groups and hence with pH, and so does the extent of metal precipitation at the pH range expected in soils. Consequently, the concentration of metals in solution should decrease with increase in soil pH, regardless of the mechanisms involved in the removal of the metal from solution. Yet, as discussed below (Section 7.9), the expected increase in OC solubility with increase in pH may, in turn, result in an increase in the concentration in solution of OC-bound metals. Empirical observations of leaching of metals from land loaded with sludge, demonstrated that acid sandy soils with low OM content were the most susceptible to metal leaching (e.g. Dowdy and Volc, 1984). This behavior of heavy metals should hold true also for soils irrigated with wastewater (Page and Chang, 1985; Chang et al., 2002), not withstanding the substantially lower loads of solids and metals in effluents than in sludge. It is noteworthy that the behavior of elements that exist as oxyanions (e.g. As, Cr, Mo, Se) may be different or depend on parameters such as the pH in an opposite way to that of most metals that exist in solution by and large as cations (or as central cations in complexes).

7.5 Loading limits of metals

Given the effect of pH on metal mobility and availability (e.g. Tyler and Olsson, 2001a, 2001b; Bolan et al., 2003), soils with $pH > 6$ are considered suitable for irrigation with effluent (USEPA, 2006b). A rather slow accumulation rate of heavy metals in soil is expected to occur under irrigation with treated municipal wastewater effluent because of the low loading rate through irrigation with such water (Banin et al., 1981). In soils with

Table 7.4 Expected removal of trace elements by vegetation under effluent irrigation

Element	Typical concentration in wastewater	Annual input [a]	Typical concentration in vegetation	Annual removal [b]	Removal
	mg/l	g/ha	mg/kg	g/ha	%
As	<0.005	<60	1	5	8.3
Cd	0.005	60	0.5	2.5	4.2
Cr	0.025	300	0.5	2.5	0.8
Cu	0.10	1200	15	75	6.3
Hg	0.0009	11	0.02	0.1	0.9
Mo	0.005	60	1	5	8.3
Ni	0.02	240	5	25	10.4
Pb	0.05	600	2	10	1.7
Zn	0.15	1800	50	250	13.9

[a]Irrigation application rate: 1.2 m/yr.
[b]Assumed annual dry matter yield of 50 Mg/ha, e.g. potatoes.
Source: Page and Chang, 1985. Reproduced with permission of Taylor & Francis Group.

near neutral pH, limited mobility of the metals and low removal rates by crops are expected (Table 7.4). In the absence of regulatory loading limits for heavy metals introduced through irrigation with wastewater effluent, metal loading rates for land application of biosolids are often used as guidelines (Page and Chang, 1985; Chang et al., 2002). Using the United States Environmental Protection Agency (USEPA) (1994) Part 503 Biosolids Rule as guideline for cumulative pollutant-loading rates, at least 500 yrs of effluent irrigation are needed (at an irrigation rate of 1.2 m/yr) before heavy metal accumulation in the soil meets the biosolids-equivalent ceiling levels (Table 7.5). The European Union (EU) ceiling concentration for heavy metals in soils amended with biosolids allows a much shorter duration of safe application of effluents (Table 7.5). Note that whereas the metal loading in the USEPA regulation refers to actual amounts added, the EU ceiling concentrations refer to the overall metal content in the receiving soil. The EU methodology reduces the maximum allowed loadings and makes them site-specific, depending on the soil's metal content at the onset of application (the calculated durations presented in Table 7.5 assume no metal content at onset). This methodology might prevent altogether the application of wastewater effluents and sludges to soils that could actually benefit the most and get harmed least (e.g. many Aridisols).

Elevated DOM concentrations in the soil solution can efficiently compete with the solid phase for metals, increasing their solubility and mobility (Neal and Sposito, 1986; Siebe and Fischer, 1996; Naidu and Harter, 1998; Almas et al., 2000). Accordingly, Burton et al. (2003) demonstrated that DOM in biosolids leachate reduced Cu and Zn adsorption onto soil, and suggested that DOM should be accounted for in setting guidelines for land application of sludge. In their study, based on the assessment of metal binding by DOM, Burton et al. (2003) suggested that the upper limit for the loading rate of biosolids should be 0.69 Mg/ha rather than 3.5 Mg/ha. However, the adsorption and complexation that they took into consideration are only two of many mechanisms that control metal concentration in the soil solution, and hence the upper limit they recommended may be too low.

Table 7.5 Calculated duration of irrigation with effluent required to reach trace elements loading limits

Element	Typical concentration in wastewater	Typical concentration in soil[a]	Annual loading[b]	Cumulative loading limit		Time to reach limit	
				USEPA[c]	EU[d]	USEPA	EU[e]
	mg/l	mg/kg	kg/ha/yr	kg/ha	mg/kg	yrs	
As	<0.005	2.2–2.5	0.06	41		680	
Cd	0.005	0.06–1.1	0.06	39	1–3	650	33–100
Cr	0.02	7–221	0.24	3000	—[f]	12500	—
Cu	0.1	6–80	1.2	1500	50–140	1250	83–233
Hg	0.0005	0.02–0.41	0.006	17	1–1.5	2830	333–500
Ni	0.02	4–55	0.24	420	30–75	1750	250–625
Pb	0.05	10–84	0.6	300	50–300	500	167–1000
Zn	0.15	17–125	1.8	2800	150–300	1550	167–333

[a]Range of means, worldwide (McBride, 1994).
[b]At irrigation rate of 1.2 m/yr.
[c]Cumulative pollutant loading rate from sludge; US CFR, Title 40, Parts 503, (USEPA, 1993).
[d]European Council Directive 86/278/EEC (article 8, and Annex 1A). Agricultural use of Sewage Sludge (for soils with $6 < pH < 7$, soils $pH > 7$ limits may increase by 50%; Soler-Rovira et al., 1996).
[e]Guidelines in mg/kg. Time to reach limit calculated assuming element background concentration level at onset = 0; and 2000 Mg Soil/ha (based on bulk density of 1.33 g/cm^3, and 15 cm depth).
[f]Not reported.
Based on Page and Chang, (1985) using loading limits for sludge.

Additionally, as DOM is expected to decompose and its concentration to diminish over time, various soil solid phases are likely to prevail in the long run and their relative impact on metal solubility to increase.

7.6 Metals: interaction with soil components

The retention capacity of soils for heavy metals and the availability and mobility of retained metals is highly dependent on the mechanism by which a given metal is associated with the soil solid phase. These mechanisms include adsorption, nucleation, precipitation, coagulation, and coprecipitation. Hence, the soil retention capacity depends both on the existing solid phases and on the likelihood of any of the above-mentioned processes occurring under the conditions prevalent in the soil environment.

Coprecipitation results in a substantial decrease in metal solubility and an increase in the formation of solid solutions in hydroxides, carbonates, and phosphates (Konigsberger et al., 1991; Scheidegger et al., 1998; Martinez and McBride, 1998a; Martinez and McBride, 1998b; Ford and Sparks, 2000; McGowen et al., 2001; Businelli et al., 2004; Kirpichtchikova et al., 2006; Manceau et al., 2007). Among the reported examples are the solid solutions of Cd, Cu, and Zn in carbonates and oxyhydroxides (Papadopoulos and Rowell, 1988; Papadopoulos and Rowell, 1989; Konigsberger et al., 1991). When conditions in the soil enable such processes, the soil retention capacity can be expected to be considerably higher than the soil adsorption capacity. Evidence gathered to date for the formation of solid solutions and other metal fixing processes guides current views on the attenuation of metals in soils (Logan and Chaney, 1983; Corey et al., 1987; Adriano et al., 2004; Basta et al., 2005). Estimated soil capacity for metal retention and the allowable application rates under current regulations, which are derived from this capacity, rely much more on potential attenuation processes than on adsorption-related soil parameters (e.g. CEC), even though sorption is often an essential initial step in promoting such attenuation processes. It should be emphasized that if the prevailing conditions in soil promote attenuation-retarding processes (e.g. dissolution and decomposition of solid phases), then a decrease in the retention capacity of the soil will occur, resulting in an increase in metal mobility and availability (McBride, 1995; Martinez and McBride, 1999). As the fate of heavy metals in soil is highly dependent on metal load and soil properties (McBride, 1989; Alloway and Jackson, 1991; Harter and Naidu, 1995; Naidu and Harter, 1998; Sauve et al., 2000; Harter and Naidu, 2001; Schmidt, 2003), changes in those properties are likely to modify the behavior of heavy metals in soil.

7.7 Distribution of metals among the soil's solid fractions

The long-term redistribution of metals that occurs after effluent application will depend on the stability of each of the solid components that are present under the ever-changing conditions in the soil. Whether the metal will eventually redistribute according to its original distribution among solid phases in the native soil or not, depends on the magnitude and nature of the perturbation (e.g. pH modification) caused by the addition process (e.g.

irrigation with an effluent) and its duration (e.g. single, or repetitious addition of the metal as occurs under wastewater irrigation).

Selective sequential extraction procedures are widely used for exploring the distribution of heavy metals among the various components of the solid phase in soils, sediments, and land-applied waste materials (Lake et al., 1984; Beckett, 1989; Tessier and Campbell, 1991). The method is based on the exposure of a solid sample to a sequence of wet chemistry assays. Typically, metals are extracted stepwise in the following media: aqueous and/or electrolyte solution to extract the soluble and/or exchangeable metal species, respectively; a weak acid for extracting metals bound to carbonates; a weak reducing agent to dissolve metals bound within Mn oxyhydroxides and amorphous iron oxyhydroxides; a weak oxidizing agent to extract metals located in the organic fraction; a strong reducing agent to determine metals associated with well-crystallized iron oxyhydroxides; and a strong mineral acid solution to extract the residual metal content remaining after all of the preceding extraction steps (Hass and Fine, 2010). That last acid-extractable fraction consists mostly of metals embedded in alumosilicates.

Several studies examined the redistribution of heavy metals among selected operationally defined solid phases in soils under wastewater irrigation. Lin et al. (2004) showed an increase in Cu and Zn concentration in all solid phases in a dune-sand basin that had been utilized for nearly 20 years as an active wastewater recharge basin, which was a part of a Soil Aquifer Treatment (SAT) scheme (cumulative load of effluent $\approx$ 1880 m). Zinc distributed preferentially into carbonates (52%) and Cu into oxides (32%). The dominant mechanisms for Zn and Cu retention were suggested to be sorption or precipitation on carbonates and iron oxides, respectively. At the same site, Cr and Ni were found mostly in the organic fraction. A minor increase was observed in Cr content in the solid phases defined as reducible and residual, and in Ni content in the carbonate, exchangeable and oxide fractions, as compared with a pristine sand (Lin et al., 2008). Mass balance of Cr, Zn, Ni, and Cu within the upper 4 m of the recharge basin revealed a retention of 80%, 9%, 8%, and 4% of the metal load, respectively (Lin et al., 2008). The observed low retention of most of the investigated metals was attributed to the low retention capacity of the quartz sand, and to movement in the form of soluble organic complexes (especially of Cu). Kirpichtchikova et al. (2006) examined the distribution of Zn in a soil irrigated for 100 yrs with untreated domestic and industrial sewage (sandy loam, pH 6.5–7). Zinc content in the soil averaged 1.1 g/kg, and was distributed among the solid fractions in a decreasing order as follows: ferrihydrates > Zn phosphate > Zn-containing phyllosilicates. Zinc phosphate was the most easily extractable fraction by organic acids. Constantino et al. (2005), in applying selective sequential extraction (a modified Tessier scheme; ibid.) to soils (clay and clay-loam, pH 8.1–9) from plots irrigated with raw sewage for 6–40 yrs, found a high proportion of the elements studied (As, Cd, Cr, Hg, Pb) in the residual fraction. Cadmium was found mostly in the residual fraction while the distribution of Cr was in a decreasing order: residual $\gg$ oxidizable (OM, sulfides) $\gg$ water-soluble > Fe and Mn oxides > exchangeable $\geq$ carbonates. The distribution of Pb was reported as: residual $\gg$ carbonates > exchangeable > Fe and Mn oxides > oxidizable > water-soluble. Flores et al. (1997) found in soils irrigated with raw wastewater (pH 7.9–8.7), that Cd, Cu, Pb, and Zn distributed mainly between the organic and carbonate fractions, with a low content in the exchangeable and oxide fractions. Rattan et al. (2005) found Zn to be distributed mainly between the organic fraction (30%), a Pb-displaceable fraction (22%), and the acid-soluble fraction (12%) in a loamy sand soil (pH 7.9) irrigated with wastewater for 20 yrs. In a wellwater-irrigated control, zinc distribution was: OM (37%), residual fraction (22%), and Pb-displaceable (11.3%). Under wastewater irrigation, Zn content increased in the more labile fractions (water-soluble, exchangeable, Pb-displaceable, and acid-soluble) and in the Mn and Fe oxyhydroxide fraction. Whereas the Pb-displaceable Cu content decreased, the content of Cu in the water-soluble and exchangeable fractions increased dramatically, from 1.2% in the control soil to 24% under wastewater irrigation. Mn and Fe content in the labile fractions also increased under wastewater irrigation (from 26% to 36%, and 0.2% to 0.4%, respectively). Dere et al. (2007) studied the distribution between the solid phases of Cu, Cr, Pb, and Zn along the profile, and of a sandy soil (pH 6.9) in a sewage disposal farm near Paris, which had been irrigated with raw wastewater for over 100 yrs. An adjacent non-irrigated field served as a reference. In the non-irrigated soil, most of the metal content was associated with the residual component. This changed substantially under wastewater irrigation. In addition to a considerable enrichment in the total content of the metals, much of the metal load shifted into the Fe and Mn oxides fraction and the organic fraction (Fig. 7.2).

The above studies, as well as other observations, led to certain generalizations such as the following:

- The proportion of the metal content found in the soluble and exchangeable fractions is usually very low, but metals introduced by wastewater enrich these fractions relative to the pristine soil.
- Metals added via wastewater tend not to concentrate in the residual phase, at least over a time period measured in years.
- Irrigation with wastewater increases the content of metals in phases that are affected by soil formation factors (e.g. free oxides of Mn and Fe, OM and carbonates), and therefore metals associated with these phases are susceptible to perturbation due to changes in soil parameters such as pH or redox potential.

7.8 The stability, pH and Eh of free oxides, and their effect on the geochemical distribution of metals

Free oxides are susceptible to acid- and ligand-promoted dissolution, as well as to reductive dissolution in the order Mn > Fe»Al (McBride, 1987; Stone and Morgan, 1987; Stumm and Furrer, 1987; McBride et al., 1988; Schwertmann, 1991; Harter and Naidu, 1995; Hering, 1995; Klewicki and Morgan, 1999; Wang and Stone, 2006). Free oxides in general, and free Fe and Mn oxides in particular, are important sorbing phases for heavy metals in soils (McKenzie, 1989; Schwertmann and Taylor, 1989; Liu et al., 2002), and hence conditions favoring the dissolution of oxides may accelerate processes that adversely affect the ability of the soil to attenuate metal mobility and availability.

Fox et al. (2005) reported a 70% decrease in the content of Mn in the oxide fraction of a surface soil at a soil aquifer treatment basin during 12 yrs of operation. Most of the Mn redistributed into the carbonate fraction. In that study, some 30% reduction in the content

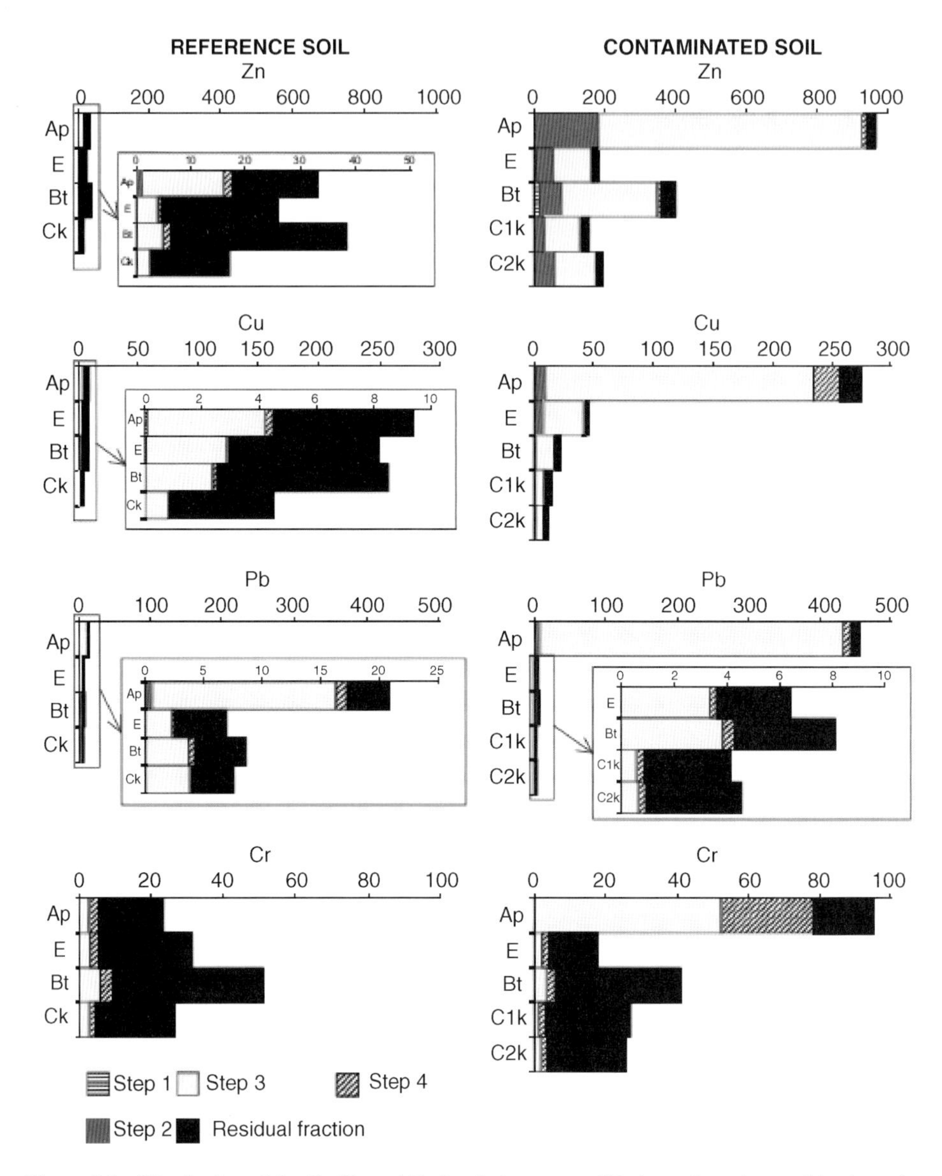

Figure 7.2 Distribution of Cu, Cr, Pb, and Zn (mg/kg) among solid phases in various soil horizons in a wastewater irrigated soil and in a control soil. Sequential extraction: step 1, KNO_3 (soluble and exchangeable); step 2, Na-Acetate pH 5 (carbonates); step 3, hydroxylamine-HCl (Mn and Fe oxyhydroxides); step 4, H_2O_2 (organic matter); Residual phase, mineralization with HF + $HClO_4$. (Reproduced from Dere et al., copyright 2007, with permission from Elsevier.)

of Fe oxides was reported in the subsurface horizon of the soil and a more moderate, yet similar, trend was found in a complementary column study where four different soils were irrigated with various wastewater effluents (Fox et al., 2005). Similar observations of redistribution of soil Mn under a SAT operation were reported by Banin et al. (2002), together with a relatively high Mn concentration, which appeared in the retrieved water during the initial period of recharge. Petrunic et al. (2005) found elevated Mn concentrations in an aquifer recharged by riverwaters with 8.6 mg/l DOC content. In a column experiment, they showed a marked increase in dissolved Mn when sand, excavated from an aquifer was treated with a 12 mg/l acetate solution. Depletion of soil Fe (Hinesly et al., 1978; Engineering Science, 1987) and Mn (Siebe and Fischer, 1996; Rattan et al., 2005) from the profile of effluent-irrigated agricultural plots, as well as leaching of Fe and Al from a municipal wastewater land application site (Muskegon, MI; USEPA, 2006), were also reported.

Siebe and Fischer (1996) reported an increase with time (up to 80 yrs) in DOC, TOC, CEC and metals' sorption capacity in soils irrigated with untreated wastewater. The shape of metal adsorption isotherms shifted over time from an L-shaped isotherm to an S-shaped one. The S-shape may suggest that at low metal concentrations complexation of metals by soluble ligands occurred in preference to sorption onto the solid phase. Accordingly, the authors attributed the change in the shape of the isotherm to an increase with time in the DOM content, and hence in DOM-metal complexation. The increase in adsorption capacity was attributed to the increase in soil OM content. Similar changes in adsorption characteristics were noted in sludge-loaded soils, where the observed fluctuation in the content of soluble organic ligands corresponded to the shift in the shape and magnitude of the sorption isotherm (Neal and Sposito, 1986). In the Siebe and Fischer study, the apparent shift in the shape of the adsorption isotherm possibly also was affected by the decrease in the Mn oxide content. These oxides have a high affinity for heavy metals (Mclaren and Crawford, 1973a, 1973b; McKenzie, 1981; Mclaren et al., 1986; McKenzie, 1989; Manceau et al., 2007), and therefore processes that accelerate dissolution of these components (e.g. acid or ligand promoted or reductive dissolution; e.g. Wang and Stone, 2006) are likely to decrease the affinity of the soil solid phase for heavy metals over time (as distinguished from the soil's adsorption capacity, which increased, apparently due to the increase in TOC content), bringing about an increase in the relative importance of dissolved ligands in the metals' dissolution and adsorption.

Reducing conditions have a striking effect on heavy metals solubility and availability (Brown et al., 1989; Charlatchka and Cambier, 2000; Quantin et al., 2002; Abbaspour et al., 2008). Reductive dissolution of Mn and Fe oxides in soils increases with the decrease in soil Eh (redox potential) and pH (Xiang and Banin, 1996). The Eh, in turn, decreases with a decrease in O_2 content in the soil gas phase and an increase in the soil moisture content, temperature, and the availability of energy sources for microbial metabolism, namely, available organic carbon (Reddy and Patrick, 1983). Quantin et al. (2002) used a selective sequential extraction procedure in a soil incubation study under anoxic conditions and OM addition. They observed a redistribution of Mn from the Mn oxide fraction into the exchangeable, organic, and easily reducible Fe oxide fractions (the Mn oxide fraction was operationally defined by extraction with hydroxylamine-HCl at pH 2, whereas the easily reducible Fe oxide fraction was defined by extraction with acid ammonium oxalate at pH 3 in the dark). Iron redistributed to a much lesser extent, and mainly shifted from crystalline Fe oxides into less well-crystallized Fe oxides (the easily reducible component). Likewise, by the end of a 140-d incubation period, a substantial proportion of the soil Ni, Co, and Cr redistributed from association with crystalline Mn oxides (and to a much lesser extent crystalline Fe oxides) into amorphous Mn and Fe oxyhydroxides. An overall decrease in metal solubility (Fe and Mn included) that was observed toward the end of the incubation period, was attributed to re-precipitation and sorption processes (Quantin et al., 2002). Davranche and Bollinger (2000) reported a release of soil Cd, and a decrease in the retention capacity for Cd in contaminated soils exposed to reducing conditions. The effect of the reducing conditions was more pronounced in the soil near natural pH values than under acidic conditions. The imposition of reducing conditions did not alter Pb solubility, and it was suggested that the Pb that was released from Mn and Fe oxyhydroxides was immediately readsorbed.

Brown et al. (1989) flooded a native soil, a comparable soil amended with anaerobic-sludge and a high-Ni soil (containing 28, 32, 95 ppm Ni, respectively), thereby decreasing the redox potential from $\approx$600 down to $<$0 mv. The decrease in the redox potential increased significantly the Ni concentration in the liquid phase of the high-Ni soil, yet it only slightly increased Ni (as well as Cd) solubility in the other two soils. The authors suggested that the Ni was associated with reducible oxides in the high-Ni soil and with species less prone to reduction in the other two soils (e.g. some components of the OM or sulfides, especially in the soil amended with the anaerobic sludge).

Chuan et al. (1996) studied the effect of pH and redox potential on solubility of Cd, Pb, and Zn in a soil (sandy loam, pH 6.4) irrigated with untreated industrial wastewater. A decrease in either pH (down to a value of 3) or redox potential (325 to -100 mv) resulted in an increase in metal solubility (Fig. 7.3). Brown et al. (1983) reported an increase in metal uptake by Bermuda grass irrigated with wastewater and noted that the uptake throughout the growing season was directly correlated to poor aeration periods, caused by soil saturation.

As the stability of iron and manganese oxides is highly dependent on soil redox potential, which in turn is affected by soil moisture content, the irrigation regime is likely to play an important role in the stability of these minerals and their function as a sink for heavy metals in soil. Meek and Grass (1975) investigated the effect of the irrigation method (flooding, furrow, sprinkler) and management (e.g. duration of and intervals between irrigations) on the soil's redox potential. The soil's Eh increased in the order flooding $<$furrow $<$ sprinkler. An increase in temperature, soil depth, or availability of energy source (in the form of manure or cotton residue added in that particular study) resulted in a decrease in soil Eh and a marked increase in Mn and Fe solubility (over tenfold as compared to the control soil). Figure 7.4 summarizes changes in soil redox potential after flooding of plots that received different organic residue (cotton) loads. Due to a higher yield, plot "D" (Fig. 7.4) received more than twice as much cotton residue (14.7 ton/ha) as the other plots. The authors attributed the marked difference in Eh between the plots to the difference in the organic residue loads. In a later study, Meek et al. (1983) compared between furrow- and trickle-irrigated fields under different irrigation frequencies and intensities. Furrow irrigation once a week resulted in higher Eh values then daily trickle irrigation (a higher frequency and lower intensity was practiced in the trickle irrigation, while maintaining the same weekly water head). They concluded that weekly irrigation (furrow or trickle) is preferable over daily trickle irrigation when oxygen content may be a limiting factor (Meek et al., 1983).

Anoxic and low redox conditions are known to develop in the soil bulk, and especially within aggregates in both saturated (Sexstone et al., 1985) and unsaturated (Zausig et al., 1993) soils. Irrigation with effluents increases the probability of formation of such

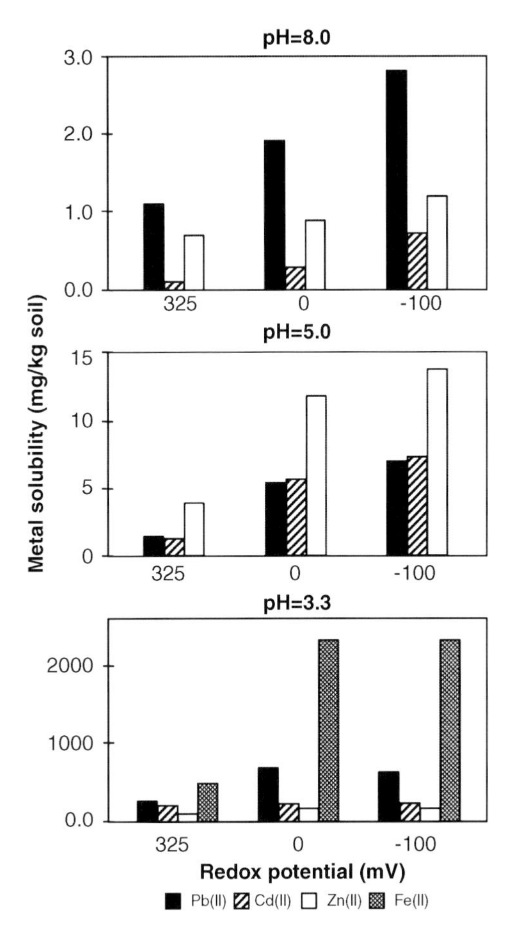

Figure 7.3 Effect of pH and redox potential on metal solubility in soil irrigated with industrial wastewater. Reproduced from Chuan et al., 1996, with kind permission from Springer Science + Business Media.

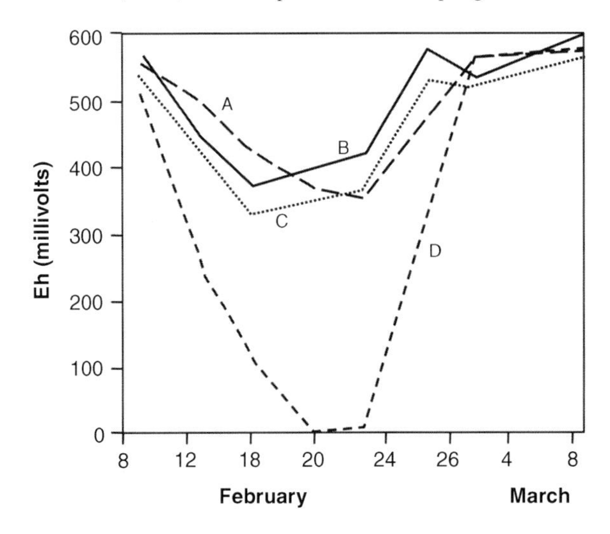

Figure 7.4 Change in soil redox potential after flood irrigation – effect of cotton residue load. Plot D was loaded with twice as much residue as the other plots (see text, Section 7.8, for details). Reproduced from Meek and Grass, 1975, with permission of the *Soil Science Society of America Journal*.

conditions. The presence of available OM in effluents and the resultant elevated levels of degradable soil OM in fields irrigated with effluents (Herre et al., 2004), together with the suboxic conditions characteristic of wastewater effluent (e.g. *c.* 1–1.5 mg/l dissolved oxygen in a municipal effluent from an activated-sludge treatment process; Anonymous, 2003) and the higher microbial activity in soil under wastewater irrigation (Friedel et al., 2000), are all likely to promote reduced conditions conducive to destabilization of reducible soil components. Heavy metals (e.g. Cu, Zn, Ni) tend to redistribute in soil into and out of the reducible oxides fraction, for example, when the moisture content fluctuates between field capacity and saturation (Han and Banin, 1997, 1999, 2001). Accordingly, agronomic practices that result in lowering of the soil redox potential are likely to induce oxide dissolution and release of the associated heavy metals. An important example is irrigation with effluent of soils amended with solid organic waste. The high load of a readily available organic carbon is likely to exacerbate the above-mentioned redox-related effects.

In acidic soils, application of municipal wastewater resulted in an increase in soil pH, rendering Fe and Mn less available (Fonseca et al., 2005) as their solubility decreases sharply as the pH increases (Lindsay, 1979; Schwertmann, 1991). Sorption and coprecipitation of free metals is then likely to occur together with the precipitation of oxide phases (Martinez and McBride, 1998a, 1998b, 1999; Charlatchka and Cambier, 2000; Davranche and Bollinger, 2000; Quantin et al., 2001, 2002). The increase in pH values under wastewater irrigation of acid soils may counteract, at least to some extent, the effect of the reduction in the redox potential on metal solubility.

7.9 Heavy metal solubility and speciation in the soil solution

Metal solubility and speciation in soil solution is element-specific and depends on the composition of the solution and the physical conditions; with pH, DOC content, ionic strength and composition, and total metal content being the dominant factors affecting metal behavior (Yin et al., 1996; Gong and Donahoe, 1997; McBride et al., 1997; Sauve et al., 1997, 2000; Fotovat and Naidu, 1998; Antoniadis and Alloway, 2002; Impellitteri et al., 2002; Weng et al. 2002b; Ashworth and Alloway, 2004; Zhao et al., 2007). An increase in the pH value decreases the concentrations in solution of most free metal ions, whereas it increases the concentration of hydrolyzed species, with the latter having a higher tendency to be specifically sorbed (as opposed to the cation exchange reactions, which dominate the sorption of free metal cations) by soil solid constituents (Hodgson et al., 1964; James and Healy, 1972; Agashe and Regalbuto, 1997; Scheidegger et al., 1998).

Some metals (e.g. Zn) tend to exist to a large extent as the free hydrated ion, and their concentration in solution is thus affected greatly by exchange reactions at the surface of the by-and-large negatively charged solid components of the soil (expandable clays, such as smectites and less frequently, the organic fractions are the site of most CEC in soils). Other metals tend to associate to a greater extent with inorganic and/or organic ligands, and their concentration in solution is more affected by specific adsorption (rather than by simple electrostatic exchange reactions) and by the types and concentrations of ligands present in the soil solution (Weirich et al., 2002; Cances et al., 2003; Ashworth and Alloway, 2004; Herre et al., 2004; Zhao et al., 2007). Specifically, the tendency to complex with DOM is stronger for Cu and Pb than for Ni or Cd (Impellitteri et al., 2002; Weng et al., 2002b). A study of the speciation of the metals in the leachate of soil columns packed with the topsoil from an agricultural field that was irrigated with untreated wastewater for 80 yrs (pH > 7.2), indicated that soluble Cu partitioned mainly between Cu-DOM and $CuCO_{3(aq)}$, and Cd speciation followed Cd^{2+} > Cd-DOM > $CdCl^{+}$ (Herre et al., 2004).

The solubility of humic acids and the partitioning of humic and fulvic acids between the soil solid and liquid phases are strongly pH-dependent. They are also influenced by the nature and amount of the multivalent cations present in soil, as humic acids tend to partition out of solution through coagulation, whereas fulvic acid removal is to a larger extent through sorption (Hering, 1995; Temminghoff et al., 1997, 1998; Oste et al., 2002; Weng et al., 2002a). An increase in soil pH results in an increase in the negative charge of surfaces of the solid inorganic phase (as the ratio pH/PZNC increases; PZNC being the point of zero net charge). At the same time, there is also an increase in the degree of dissociation of the organic acids' acidic functional groups as the ratio pH/pK_a increases. Thus, a rise in pH value will result in electrostatic repulsion between the mineral and organic species, which leads to an increase in the partition of humic and fulvic acids into the liquid phase (Fig. 7.5; Stevenson, 1994; You et al., 1999; Impellitteri et al., 2002). As the solubility of an organic species increases, so should the concentration in solution of its metal complex (Jeffery and Uren, 1983; Alloway and Jackson, 1991; Stevenson, 1994; Temminghoff et al., 1998; Yin et al., 2002; Cances et al., 2003; Schwab et al., 2005; Yang et al., 2006; He et al., 2006).

Jeffery and Uren (1983) illustrated the effects of pH at low and high Cu loading (1 and 100 mg Cu/kg soil, respectively) on Cu solubility in a sandy loam soil (Fig. 7.6). The

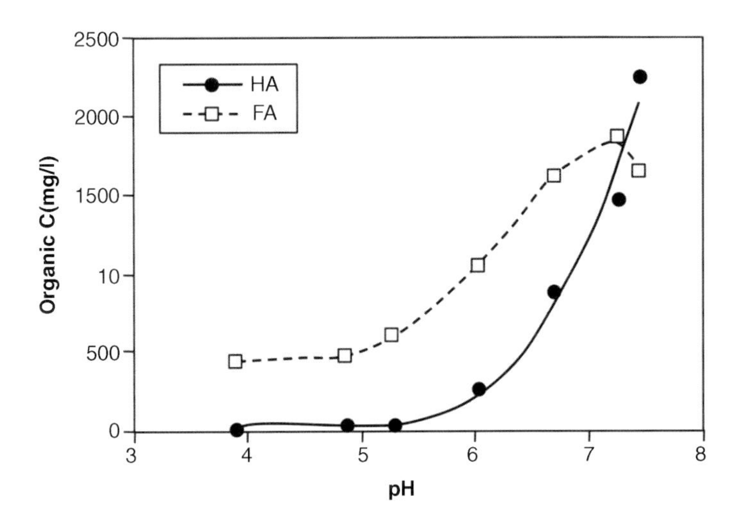

Figure 7.5 Dissolution of soil organic matter as a function of pH. HA, humic acid; FA, fulvic acid. (Reproduced from You et al., copyright 1999 with permission from Elsevier.)

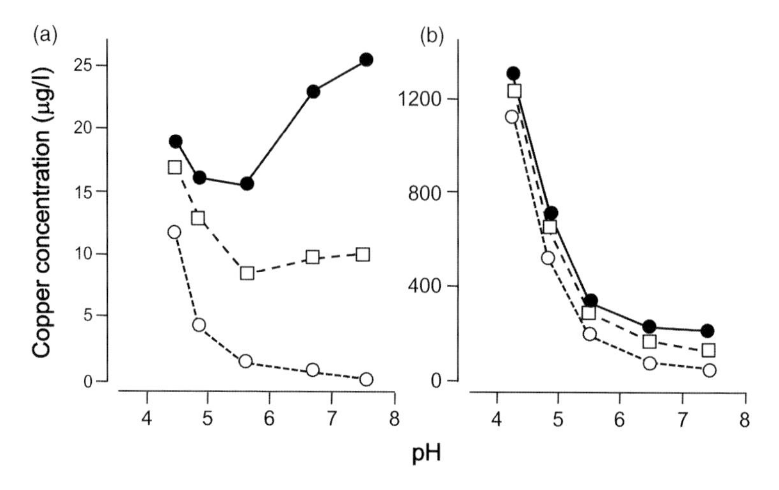

Figure 7.6 Effect of pH on Cu solubility at low (a) and high (b) loads. Total soluble (•), chelex-100 resin extractable (□), and anodic stripping voltametric dissociated (○) Cu. (Reproduced from Jeffery and Uren, 1983, with permission from CSIRO Publishing, www.publish.csiro.au/nid/84/paper/SR9830479.htm).

concentration in solution of the labile Cu fraction (dissociable by voltametric anodic stripping) and of the moderately labile Cu fraction (extractable by Chelex-100 resin) decreased with an increase in pH. These fractions consisted mostly of free ionic Cu^{2+} and of loosely complexed Cu, namely complexes of Cu that can be "stripped" of their ligands by anodic stripping. In the system containing a low Cu load, the total Cu concentration in solution increased with pH at pH values >6, whereas it decreased as the pH increased in the system containing a high load of Cu. The increase in Cu concentration with the pH at the lower Cu load was attributed to an increase in the concentration of DOM and Cu-DOM complexes at the higher pH values. As the organic ligands already were fully bound to the metal at low Cu concentrations, the presence of DOM did not have a significant effect on the concentration of Cu in solution at the high metal load. A parabolic dependence of metal solubility on pH, attributed to an increase in the content of soluble OM and to formation of OM-metal complexes, was also reported by other investigators (McBride, 1981; Impellitteri et al., 2002; He et al., 2006).

Association with multivalent cations tends to decrease DOM solubility (Romkens and Dolfing, 1998). Therefore, it is suggested that a high proportion of multivalent (e.g. alkaline earth) metal cations in wastewater applied to soil at as high a pH as agronomically feasible, will assist in controlling DOM solubility and its complexation with heavy metals. An increase in the solution's ionic strength, especially through an increase in the concentration of multivalent cations, is likely to not only decrease the dissolution of DOM and its complexation with trace metals, but also to strongly affect the concentration of metals, the dissolution of which is controlled by exchange mechanisms (e.g. Cd, Zn; Fotovat and Naidu, 1998). In the case of metal species for which specific adsorption mechanisms govern association with the solid phase surfaces, the solution's ionic strength will have a relatively minor effect on the retention of the metal (note that specific adsorption is empirically defined as being independent of ionic strength). Maintaining a low concentration of metals in the incoming wastewater, increasing the number of specific adsorption sites through selective soil amendments, and managing soil pH at as high a level as agronomically feasible (in general, near neutral), will all lower the extent of metal dissolution and promote metal attenuation.

7.10 Mobility of heavy metals in the soil profile

Although heavy metals in soils irrigated with wastewater tend to accumulate near the soil surface, downward movement of metals is often observed. The extent of downward transport depends on the quality of the incoming sewage, the wastewater treatment, the irrigation management practices, and the nature of the receiving soil. Hoffmann et al. (1998) studied Cu, Cd, and Zn distribution in the profile of an acid sandy soil derived from glacial till (pH 4.5–5.5) in the Berlin sewage farm system that was irrigated with untreated municipal wastewater at loading rates of 300–10000 mm/y for *c.* 100 yrs. The concentrations of Cd, Cu, and Zn in the topsoil reached maximum values of 44, 850, and 3500 mg/kg, respectively. Although the overall metal content in the soil decreased with depth, the concentration of Cu and Zn in the soil solution (periodically collected for analysis through suction cups installed at 50, 100, and 180 cm depths) increased, on average, seven to eightfold at 180 cm as compared to that at 50 cm depth and was positively correlated with the DOC concentration and the pH. Studies at a similar sewage farm near Paris (raw wastewater applied on a sandy soil, pH 6.9, for over 100 yrs), showed accumulation of Cd, Cr, Cu, Co, Ni, Pb, and Zn in the Ap horizon with limited migration further down the profile and some accumulation of Zn and Cu throughout the top 100 cm (Fig. 7.2; Dere et al., 2006, 2007).

Schirado et al. (1986) reported accumulation of Cd, Co, Cr, Cu, Ni, and Zn in the top 0–30 cm of a soil receiving untreated domestic and industrial municipal wastewater for over 50 yrs. Yet, mass–balance calculations suggested that movement of metals down the profile, below the sampling depth (150 cm) had occurred. Yadave et al. (2002) studied a field that was irrigated for about 30 yrs with low-quality wastewater effluent (BOD 169 mg/l, COD 382 mg/l). They reported a slight reduction in soil pH and accumulation of OM and heavy metals (Cd, Cu, Cr, Ni, Pb, and Zn) in the soil surface layer. Metal accumulation down to 60 cm depth was limited, as was metal uptake by the crops (wheat, Egyptian clover, rice, and sorghum).

Leaching of heavy metals from two soils (sandy, pH 4.8; loamy sand, pH 6.4) from the Berlin sewage farm operating for nearly 100 yrs, was studied by Hoffmann et al. (2002) in reconstructed soil columns irrigated with waters of different quality (deionized water, lime + deionized water, 10^{-3} N HCl, or treated wastewater). The columns were irrigated twice a week (with an equivalent of 2000 m^3 water/ha for 2.5 yrs), drained every 7 d, and the leachates collected and analyzed. The amount of metals removed in the leachate of the strongly acid sandy soil irrigated with deionized water ranged from 4% to 28% of the metal content of the soil packed in the column (Cu < Cd < Zn). Metal removal was higher under irrigation with wastewater (by a factor of 1.2–1.7. Under irrigation with acidified deionized water, the metal content in the leachate from the strongly acid soil was higher by a factor of 2.5–4.5 as compared to the columns irrigated with deionized water. Some increase in metal leaching was also noted under irrigation with limed deionized water. In the slightly acid loamy sand soil, rates of metals removal were below 1.4% when irrigated with deionized water, with no significant differences in metal release between the columns irrigated with the various types of water. The observed relatively low removal of Cu under acid leaching of the slightly acid soil was attributed to a decrease in DOM mobility due to the decrease in soil pH. It was concluded by the authors that whereas DOM in the leachate of the slightly acid soil was undersaturated with respect to the metals, the DOM was oversaturated in the strongly acid soil and transport of the metals in their cationic forms had likely dominated metal leaching in the latter soil (Hoffmann et al., 2002). Lead was below detection limit in the leachates of both soils.

Streck and Richter (1997) found only a minute displacement of Cd and Zn below 70–90 cm in the profile of a sandy soil (pH 5.3, 97% sand) irrigated with treated wastewater for 29 yrs (overall application of 12800 mm). Herre et al. (2004) reported a buildup of readily mineralizable organic C in the upper 30 cm of soils irrigated with untreated wastewater for 80 yrs. This was accompanied by an increase in the content of DOC (from 0.1 to 0.2 mg/g at the onset to 0.2 to 0.3 mg/g after 80 yrs). In a column experiment in which the soil irrigated for 80 yrs with wastewater was leached with tapwater and with untreated and primary treated wastewater, Cu content in the leachate was correlated with that of the DOC. Leaching the columns with tapwater resulted in a lower Cd concentration in the leachate, but similar Cu and DOC concentrations to those in the leachates from columns irrigated with the untreated wastewater (Herre et al., 2004).

Excess irrigation of an acid pasture soil (pH 3.9) with treated wastewater enhanced the mobility of heavy metals in the soil profile (Gwenzi and Munondo, 2008). The irrigation rate (4500 mm/y) exceeded by far the crop requirement. Marked depletion in Cd, Cu, and Zn occurred from the upper 30 cm of the soil as compared to a non-irrigated control site. Mobilized Cu and Cd re-precipitated at a depth of 30–90 cm.

Brown et al. (1983) studied heavy metal migration and uptake in lysimeters packed with four soil types and planted with Bermuda grass. Irrigation was with a secondary effluent spiked with metal sulfate salts. Application of 2120–2600 mm resulted in addition of (kg/ha): 19–24 Cd, 18–22 Cu, 16–20 Ni, 16–19 Pb, and 17–21 Zn. The applied metals accumulated at the soil's upper 2.5 cm, and none of the metals were detected in the leachate. The application of the effluent elevated the soil's pH by 0.6–2.0 pH units depending on soil type, and it was suggested that this increase in pH value enhanced the metals' adsorption and precipitation and restricted their movement down the soil profile.

Hinesly et al. (1978) summarized results obtained from two sites under long-term (~40 yrs) irrigation with municipal effluents in Bakersfield, CA (clay loam soil; primary effluent) and in Lubbock, TX (sandy to fine sandy loam; secondary effluent). In both locations the effluent was used to irrigate row forage crops and irrigation was applied in excess (e.g. irrigation continued during the rainy season) to meet the disposal needs of the WWTPs. At the Bakersfield site, Zn and Cu accumulated in the upper 60 cm, whereas Ni, Pb, and Cd were distributed throughout the upper 90 cm. Zinc and Pb contents were markedly higher in the effluent irrigated soil than in a wellwater-irrigated control. At the Lubbock location, no difference in the content of metals was observed between the wellwater- and wastewater-irrigated soils. A decrease in Fe content was reported in the profile of both effluent-irrigated soils, and iron deficiency was noted in the corn at Lubbock.

7.11 Availability to plants

The availability of heavy metals to plants and the extent of accumulation of the metal in the plant is species- (and cultivar-), as well as metal-dependent (Sharma et al., 2006; Hong et al., 2008). Chaney (1980) suggested a division of metals into four groups according to the risk of their entry into the food chain: metals that are only slightly soluble in soil and poorly adsorbed by plant roots (e.g. Cr); metals that get absorbed by plant roots but exhibit limited translocation into shoots (e.g. As, Hg, Pb); metals that get absorbed and translocate into shoots even at levels that are phytotoxic (resulting in acute yield decrease, e.g. Cu, Ni, Zn); and metals that absorb, translocate, and accumulate up to levels that are toxic to animals and humans (e.g. Mo, Cd).

The degree to which heavy metals accumulate in edible parts of plants depends on whether wastewater from industrial sources is excluded from or included in the effluent used for irrigation, on the level of treatment of the wastewater, and on the irrigation management practices. Leafy edible plants are the most susceptible crops to metal accumulation. Some publications reported exceptionally high metal accumulation in vegetables grown on wastewater-irrigated soils. Gupta et al. (2008), for example, reported accumulation of Cd, Cr, Cu, Ni, Pb, and Zn in spinach and radish, some of those metals at levels of nearly 50 times their safe limits. Thus, Cd accumulation in the plants was 1.5 mg/kg, more than 10 times the safe limit. However, in the 15 cm of topsoil, Cd content was 30 mg/kg and it would have required over 600 yrs of irrigation (given the reported 10 μg Cd/l in the wastewater and assuming a soil bulk density of 1.33 kg/l; and 1000 mm/y irrigation head) to reach this load of Cd had it accumulated from effluent-derived metal. Hence, it is likely that most of the reported metal load in the soil, and hence in the plants, originated from sources other than the applied wastewater. The authors did allude to sources of pollutants other than the effluents used for irrigation (garbage utilization in the study area; Gupta et al. 2008). Input from benign agronomic management practices, such as application of fertilizers and atmospheric fallout, can result in a significant cumulative metal input. Table 7.6 presents an estimate for the annual loading rates of various metals from different sources in England and Wales (Nicholson et al., 2003).

In the above-mentioned lysimeters study conducted by Brown et al. (1983), Cd, Cu, Ni, and Pb accumulated in the plants at levels higher than those reported as being in the normal range, but only Cd reached a level above that recommended for chronic dietary exposure for cattle. Metal uptake correlated with the soil's ethylene diamine tetra-acetic acid (DTPA)-extractable metal content and was higher in the more acidic soils (initial soil pH ranged from 5.6 to 7.7). Metal uptake during the growing season followed the soils' moisture content and was highest when the moisture content neared saturation and aeration was poor (Brown et al., 1983).

Sidle et al. (1976) grew corn and reed canary grass for two consecutive years on soils (pH 5.9–6.6) irrigated with wastewater or with wastewater injected with sludge for 11 yrs. The reed canary grass was irrigated year round, whereas the corn was irrigated during the growing season only and both crops were analyzed for Cd, Cu, Pb, and Zn. Copper and Zn accumulation in the reed canary grass under wastewater irrigation was significantly higher than in the non-irrigated, unfertilized control. Yet, heavy metal content in the wastewater-irrigated corn was the same or lower than in the fertilized but non-irrigated control. The authors also noted that Zn interfered with Cd uptake.

Table 7.6 Annual loading rate of trace elements from different sources to agricultural soils in England and Wales

Element	Atmospheric deposit	Sewage sludge[a]	Animal manure[a]	Inorganic fertilizers			Lime[b]	Irrigation water
				N	P	K		
				g/ha/yr				
As	3.1	34	1.9–15.7[c]	0.1	1.1	0.0	0.0	1.2
Cd	1.9	19	1.4–6.1	0.1	1.6	0.0	1.4	0.1
Cr	7.5	926	11–40	0.5	17	0.1	29	0.1
Cu	57	3210	168–1679	1.6	4.9	0.4	12	16
Hg	1.0	13	0.1–0.2	<0.1	<0.1	<0.1	0.0	nd[d]
Ni	16	335	20–50	0.2	3.3	0.1	25	1.6
Pb	54	1256	18–50	0.7	0.5	0.2	10	0.8
Zn	221	4557	718–2734	2.2	34	0.5	53	39

[a]Assuming application rate equivalent to 250 kg-N ha^{-1}
[b]applied every 5 years (to acid soils)
[c]depends on the animal manure or litter source, see Nicholson et al., (2003) for more details.
[d]nd, not detected.
From Nicholson et al., copyright 2003 with permission from Elsevier.

Chary et al. (2008) found elevated levels of heavy metals in leafy plants (spinach, *Spinacae Oleracea*; amaranthus, *Amaranthus graecizans*; coriander leaves, *Coriandrum sativum*; mint leaves, *Mentha spicata*; ladies finger, *Abelmoschus esculentus*; brinjal, *Solonum melengina*) grown under irrigation with a mixed industrial/municipal sewage, both treated and untreated. Of the tested metals (Co, Cr, Cu, Ni, Zn, Pb), only Pb exceeded the recommended maximum allowed level for vegetables (0.3 mg/kg), whereas both Pb and Zn exceeded the soil tolerance levels (100 and 300 mg/kg, respectively). The incoming industrial effluent originated from a number of sources and included electroplating fluid, lead-battery acid, and drug and pharmaceutical manufacturing effluents. Lead and Zn accumulation averaged in all effluent-irrigated soils, 512 and 386 mg/kg, respectively (Chary et al., 2008).

In the above-mentioned Bakersfield and Lubbock studies (Hinesly et al., 1978), wastewater irrigation did not alter metal content in plants in both locations (barley in Bakersfield, corn in Lubbock), but yield was markedly higher than in adjacent farms.

In a Monterey wastewater reclamation study (Engineering Science, 1987), agricultural crops (artichoke, broccoli, lettuce, celery, and cauliflower) were irrigated for 5 yrs with either wellwater or two tertiary effluents (a California Administrative Code Title 22 conformed treatment, and a less extensively treated, filtered effluent; BOD range from 8 to 22 mg/l). The types of water used had no significant effect on heavy metal (Cd, Cu, Co, Cr, Mn, Ni, Pb, Zn) content in the soil or in edible plant parts. Moreover, heavy metal input from impurities in the applied commercial fertilizers greatly exceeded the input through the effluents. Levels of soil Fe in the wellwater-irrigated soils were higher then under effluent irrigation, probably due to reductive dissolution of Fe. Similarly, Johns and McConchie (1994a), using treated municipal effluent (BOD 10.5 mg/l) with low metal content (e.g. Cd < 0.02, Cr < 10, Cu 7, Ni 30, Pb 1.2, Zn 13 μg/l), observed no effect of the irrigation on the content of heavy metals in soil or plants (banana, *Musa acuminata*). No changes in Mn or Fe content were noted during the 19 months of their experiment. They observed no time-dependence of metal content or leachability from the irrigated soil in a lysimeter study; again, it was suggested that risk from heavy metals originating from fertilizers may be higher than the risk from effluent-borne metals (Johns and McConchie, 1994).

The above studies suggest that crop irrigation with municipal wastewater effluents with little or no industrial inputs that underwent advanced treatment, which result in low metal and TOC contents, should bring about minor, if any, accumulation of metals in the soil and crops, and impose little, if any, risk for leaching of metals. This is especially true when the irrigation is managed according to crop water requirements rather than dictated by wastewater disposal considerations.

Qishlaqi et al. (2008) studied a soil irrigated with untreated domestic and industrial sewage for about two decades. They reported accumulation of Ni and Pb in wheat, and Cd in lettuce, spinach, and celery at levels above the maximum allowed limits (10, 9, and 0.5 mg/kg dry weight (DW), respectively). Nan and Cheng (2001) and Nan et al. (2002) evaluated metal accumulation in spring wheat and corn grown on a loam soil (pH 8.6) irrigated with streamwater fed with treated and untreated wastewater from domestic and from combined domestic and industrial sources. Cadmium, Pb, and Zn accumulation in plant parts followed the order: root > stem > grain. The extent of accumulation was significantly higher in comparison to the accumulation in the control plots only in crops that were irrigated with wastewater containing industrial sewage. A similar distribution among plant parts (root > stem > leaf) was reported for oats and alfalfa by Mireles et al. (2004) in fields irrigated with raw wastewater for 50 and 100 yrs. The content of metals in the soil at the two sites was similar. Metal (Cu, Co, Cr, Ni, Mn, Zn, Pb) uptake by plants (leafy vegetables – lettuce, spinach, and swiss chard; tubercles – beetroot and radish; fruits – pear, pepper, and quince) was generally below phytotoxic levels, with only Cu exceeding toxic levels in all studied plants, and Zn and Pb exceeding toxic levels in the tubercles (Mireles et al., 2004). Brown et al. (1989), in a study unrelated to application of effluents, evaluated Cd and Ni accumulation in oats and ryegrass. They found Ni to preferentially accumulate in grains rather than in the vegetative parts of the plant, whereas Cd accumulation followed an opposite trend. The authors suggested that whereas Ni may be of concern in grain crops, Cd might pose a more serious threat in forage crops.

Siebe (1995) reported elevated metals content in both soil and alfalfa grown in a field irrigated with raw wastewater between 16 and 80 years in Hidalgo, Mexico. Although Cd levels in the soil after 80 yrs of irrigation were relatively high (0.5–1.8 mg/kg), they did not exceed the EU tolerance limit for total Cd (3 mg/kg) and Cd and Pb content in the plant material was less than 0.2 and 0.5 mg/kg DW, respectively (below the reported normal range; Kabata-Pendias and Pendias, 1984). Copper and Zn contents were less than 20 and 50 mg/kg DW, respectively, not exceeding plant sufficiency levels. In another experiment in the Mexican Valley of Mezquital, raw wastewater was used for alfalfa and oats irrigation (Cajuste et al., 1991). Total and potentially available (DTPA-extractable) metal content in the soil increased with the wastewater load. Chromium and Pb loads (115 and 300 mg/kg, respectively) exceeded the potentially phytotoxic levels (75–100, and 100–300 mg/kg, respectively). Although Zn accumulation in plants was higher than that of any other heavy metal, only Co reached levels (17–29 mg/kg) that could pose a health risk for humans and animals. Whereas metal distribution within the oat plant followed the above-mentioned trend (i.e. root > top), all metals but Cd (i.e. Co, Cr, Cu, Ni, Pb, Zn) accumulated in the alfalfa leaves more than in its roots.

Rattan et al. (2005) reported an increase in DTPA-extractable Cu, Ni, Pb and Zn in fields irrigated with wastewater for 10 and 20 yrs (but not in a field irrigated for 5 yrs), as compared to the concentrations measured in adjacent plots irrigated with wellwater. DTPA-extractable cadmium levels in the effluent-irrigated area were not significantly different from those measured in wellwater-irrigated plots, regardless of the duration of operation (5, 10, or 20 yrs). The relatively higher amounts of DTPA-extractable Cu and Zn were attributed to reduction in soil pH (0.4 units) and increase in soil OM content (by 60%) in the long-term irrigated plots. The authors also reported lower soil–plant transfer factors (i.e. the ratio between metal concentration in the plant and DTPA-extractable metal content in the soil) for Cu and Zn in the wastewater-irrigated plots than in the wellwater-irrigated plots, despite the higher levels of heavy metals in the former plots. The test plants included rice, wheat, sorghum, maize, oats, and spinach. These results indicated that plant uptake does not increase linearly with increase in metal concentration in the soil, and that the "potentially available" metal fraction may overestimate actual metal bioavailability. The above observations are in agreement with the plateau response suggested in the past (Logan and Chaney, 1983; Corey et al., 1987).

Khan et al. (2008) reported that in fields irrigated with treated municipal wastewater (domestic and industrial; pH 8.0) for over 30 yrs, high levels of metal accumulation in soil and plant (corn, radish, green cabbage, spinach, lettuce, turnip, and cauliflower) were measured as compared to a control irrigated with tapwater. The sequence of the metals' transfer factors (ratio between concentration in plant and total concentration in soil) was, in decreasing order Cd > Ni > Cu > Zn > Cr > Pb. Although metal accumulation in the plants exceeded permissible limits (by both state and WHO standards), no health risks were apparently associated with the consumption of edible plant parts based on the magnitude of the Health Risk Index (HRI) for the different metals, which was less than 1 (HRI is the ratio between the daily intake of a metal and the daily reference dose for chronic oral exposure. A list of values for the daily reference dose for chronic oral exposure (RfD) of the various metals is given in the USEPA Index Risk Information System, IRIS; USEPA IRIS). Liu et al. (2005) reported metal accumulation in soil and plants irrigated with industrial and domestic untreated wastewater for about two decades. The soil to plant transfer factor varied among crops, and was generally in the order Cd ≥ Zn ≥ Cu»Pb ≥ Cr, with Cd and Zn reaching alarming levels in the edible plant parts (most severely in winter cabbage and longleaf lettuce).

Kabata-Pendias (2004) defined metal transfer factor as the ratio between metal concentration in the solution of a contaminated soil and the concentration in the solution of a comparable non-contaminated soil, and obtained the order: Cd > Ni > Zn > Cu > Pb > Cr. The above order is inversely related to the selective adsorption sequence defined by Gomes et al. (2001), who used the metals' K_d values (the ratio between the amount of metal adsorbed per unit weight of solid and its equilibrium concentration in the surrounding solution), as derived in a competitive adsorption study, to evaluate the heavy metals' relative adsorption tendency. The selective adsorption sequence was $Cr \approx Pb > Cu > Cd \approx Zn \approx Ni$, suggesting a much lower availability of Cr and Pb than of Cd, Ni, and Zn. It was suggested that the selective adsorption sequence follows the order of the metals' electronegativity under acid conditions, and the metals' first hydrolization constant at higher soil pH values (Gomes et al., 2001). Dere et al. (2007), suggested a similar mobility sequence in a site irrigated with raw wastewater for 100 yrs, based on a 1 M KNO_3 extraction. The mobility sequence derived by those authors was $Zn \gg Cu > Pb > Cr$.

7.12 Summary and conclusions

The complex nature of the wastewater–soil–plant system enables the occurrence of numerous processes that affect metal mobility and availability in different and often opposite ways, the net outcome of which is not easy to predict or evaluate. The pH of the soil is possibly the most important, yet easily adjustable, parameter inasmuch as metals' behavior and fate are concerned. Although free metal solubility generally decreases with an increase in pH, and metal precipitation (e.g. as hydroxides, carbonates or phosphates) is expected to increase, a metal's overall concentration in solution may actually increase as the solubility of the OM (and with it the extent of complexation of metals with the DOM) increases with the pH. The net effect of all these simultaneous processes depends on, among other factors, soil mineralogy, characteristics and content of the OM, soil solution composition, and the type and concentration of the metals present. For example, metals that tend to exist as oxyanions (e.g. Mo) may display a dependence on pH that is very different to that of metals that tend to exist as cations (e.g. Cd). The nature of the OM in the soil and in the irrigation water is another example of a factor that strongly affects the behavior of metals. Although contributing to the soil's adsorption capacity when present in the solid phase, the OM enhances metal mobility when present in the liquid phase. Dissolved organic matter can enhance mineral breakdown through reductive and ligand-promoted dissolution. A less stable organic component is likely to inflict more destabilization on the solid phase of the soil and to increase metal mobility. Dissolved organic matter can also precipitate with multivalent ions and thereby decrease the solubility of the attached metal ions. In general, as the effluent (or the effluent-derived OM in irrigated soils) matures, the DOM/OM ratio increases, the DOM nominal molecular weight decreases, and the number of functional groups per C atom, as well as the aromaticity of the DOM, increases, rendering the dissolved organic species more reactive.

Dissolution of OM can be reduced by keeping a high content of alkaline earth metal cations in the soil solution. These cations may be separately introduced (e.g. through liming in acid soils) or supplied directly by the wastewater. Metal solubility can also be regulated by adjusting the soil's pH to a value as close as possible (within agronomic constraints) to the adsorption (or precipitation) maximum of the metal of highest concern (preferably on its alkaline side).

Wastewater composition and irrigation management practices can affect the nature of the soil solution and of processes occurring therein. Studies demonstrated that dissolution of Mn and Fe oxides in soil increased under wastewater application. Most of these studies were conducted either in recharge basins or in plots irrigated with poorly treated wastewater of high reducing capacity, and by and large where a high hydraulic head was applied. Yet, some measurements conducted in fields irrigated with highly treated effluents suggest that dissolution of Fe oxides could have occurred even there (Engineering Science, 1987). At any rate, treatment of wastewater that diminishes the effluent's reducing power and organic load will lower the intensity of potentially deleterious effects of irrigation with wastewater effluents. The absence of such a treatment and excessive irrigation may result in adverse long-term effects on the irrigated soil and its surroundings.

Data from field experiments and case studies of plots irrigated with municipal wastewater support the notion that proper separation of domestic wastewater from industrial effluents will significantly reduce metal content. Adequate wastewater treatment aimed at reducing TOC, DOC, BOD, COD, and SS content will also decrease the content of heavy metals in the effluent. If an effluent is applied to a soil in the pH range of 7 ± 0.5 under an irrigation management that is aimed at satisfying the crop's requirements while maintaining aerobic conditions, minimum adverse, short- and long-term effects related to heavy metal accumulation in soil, plant, or drainage water are to be expected. Maintaining aerobic conditions may require departure from irrigation practices that result in periodic flooding.

"Wastewater" is used as a generic term describing sewage effluents of diverse origins and qualities. It is important to develop the capacity to elucidate the way processes occurring in fields irrigated with wastewater are affected by such parameters as the effluent's quality, the irrigation management practices and the soil and crop characteristics, and to that end, more detailed studies are needed. The longer term processes, such as dissolution of free oxides or decomposition and solubility of OM that are enhanced by irrigation with effluents need to be identified and the effect of the irrigation on them better evaluated. Such an evaluation is essential to the establishment of a robust risk-based approach to irrigation with wastewater. Understanding processes at the interface of the soil's solid phase and quantifying the capacity of the solid phase to act as a sink for heavy metals are central to our ability to assess the risk involved in irrigation with effluents as far as heavy metals are concerned. Soils, to which organic amendments (e.g. biosolids) were added previous to, or simultaneously with, application of wastewater, are of special interest, as the added OM can exacerbate redox-related adverse processes, rendering metals more available and mobile.

References

Abbaspour, A., Kalbasi, M., Hajrasuliha, S. and Fotovat, A. (2008) Effect of organic matter and salinity on ethylenediaminetetraacetic acid-extractable and solution species of cadmium and lead in three agricultural soils. *Communications in Soil Science and Plant Analysis* **39**, 983–1005.

Adriano, D.C., Wenzel, W.W., Vangronsveld, J. and Bolan, N.S. (2004) Role of assisted natural remediation in environmental cleanup. *Geoderma* **122**, 121–142.

Agashe, K.B. and Regalbuto, J.R. (1997) A revised physical theory for adsorption of metal complexes at oxide surfaces. *Journal of Colloid and Interface Science* **185**, 174–189.

Alloway, B.J. and Jackson, A.P. (1991) The behaviour of heavy metals in sewage sludge-amended soils. *Science of the Total Environment* **100**, 151–176.

Almas, A.R., McBride, M.B. and Singh, B.R. (2000) Solubility and lability of cadmium and zinc in two soils treated with organic matter. *Soil Science* **165**, 250–259.

Anonymous (2003) *M/J Industrial Solutions. Municipal Wastewater Treatment Plant Energy Baseline Study*. PG&E New Construction Energy Management Program, Pacific Gas and Electric Co.

Antoniadis, V. and Alloway, B.J. (2002) The role of dissolved organic carbon in the mobility of Cd, Ni and Zn in sewage sludge-amended soils. *Environmental Pollution* **117**, 515–521.

Ashworth, D.J. and B.J. Alloway, (2004) Soil mobility of sewage sludge-derived dissolved organic matter, copper, nickel and zinc. *Environmental Pollution* **127**, 137–144.

Ayers, R.S. and Westcot, D.W. (1985) *Water Quality for Agriculture.* Irrigation and Drainage Paper No. 29. FAO, Rome.

Banin, A., Navrot, J., Noi, Y. and Yoles, D. (1981) Accumulation of heavy metals in arid-zone soils irrigated with treated sewage effluents and their uptake by Rhodes grass. *Journal of Environmental Quality* **10**, 536–540.

Banin, A., Lin, C., Eshel, G., Roehl, K.E., Negev, I., Greenwald, D., Shacahr, Y. and Yablekovitch, Y. (2002) Geochemical processes in recharge basin soils used for municipal effluents reclamation by the soil-aquifer treatment (SAT) system. In: *Management of Aquifer Recharge for Sustainability* (P.J. Dilloned.), pp. 327–332. Blakema. Rotterdam, Netherlands.

Barker, D.J. and Stuckey, D.C. (1999) A review of soluble microbial products (SMP) in wastewater treatment systems. *Water Research* **33**, 3063–3082.

Basta, N.T., Ryan, J.A. and Chaney, R.L. (2005) Trace element chemistry in residual-treated soil: Key concepts and metal bioavailability. *Journal of Environmental Quality* **34**, 49–63.

Beckett, P.H.T. (1989) The use of extractants in studies on trace metals in soils, sewage sludge, and sludge-treated soils. *Advances in Soil Science* **9**, 144–176.

Bixio, D., Thoeye, C., De Koning, J., Joksimovic, D., Savic, D., Wintgens, T. and Melin, T. (2006) Wastewater reuse in Europe. *Desalination* **187**, 89–101.

Bolan, N.S., Adriano, D.C., Mani, P.A. and Duraisamy, A. (2003) Immobilization and phytoavailability of cadmium in variable charge soils. II. Effect of lime addition. *Plant Soil* **251**, 187–198.

Bouwer, H. (2000) Integrated water management: Emerging issues and challenges. *Agricultural Water Management* **45**, 217–228.

Brown, M.J. and Lester, J.N. (1979) Metal removal in activated sludge: The role of bacterial extracellular polymers. *Water Research* **13**, 817–837.

Brown, K.W., Thomas, J.C. and Slowey, J.F. (1983) Metal accumulation by bermudagrass grown on four diverse soils amended with secondarily treated sewage effluent. *Water Air Soil Pollution* **20**, 431–446.

Brown, P.H., Dunemann, L., Schulz, R. and Marschner, H. (1989) Influence of redox potential and plant species on the uptake of nickel and cadmium from soils. *Z Pflanzenernaehr Bodenkd* **152**, 85–91.

Burton, E.D., Hawker, D.W. and Redding, M.R. (2003) Estimating sludge loadings to land based on trace metal sorption in soil: effect of dissolved organo-metallic complexes. *Water Research* **37**, 1394–1400.

Businelli, D., Casciari, F. and Gigliotti, G. (2004) Sorption mechanisms determining NI(II) retention by a calcareous soil. *Soil Science* **169**, 355–362.

Buzier, R., Tusseau-Vuillemin, M.H., dit Meriadec, C.M., Rousselot, O. and Mouchel, J.M. (2006) Trace metal speciation and fluxes within a major French wastewater treatment plant: Impact of the successive treatments stages. *Chemosphere* **65**, 2419–2426.

Cajuste, L.J., Carrillo, R.G., Cota, E.G. and Laird, R.J. (1991) The distribution of metals from wastewater in the Mexican Valley of Mezquital. *Water Air Soil Pollution* **57–58**, 763–771.

Cances, B., Ponthieu, M., Castrec-Rouelle, M., Aubry, E. and Benedetti, M.F. (2003) Metal ions speciation in a soil and its solution: Experimental data and model results. *Geoderma* **113**, 341–355.

Chaney, R.L. (1980) Health risks associated with toxic metals in municipal sludge. In: *Sludge—Health Risks of Land Application.* (eds G. Bitton, B.L. Damron, G.T. Edds,and J.M. Davidson), pp. 59–83. Ann Arbor Science Publ., Ann Arbor, MI.

Chang, A.C., Pan, G., Page, A.L. and Asano, T. (2002) *Developing human health-related chemical guidelines for reclaimed waste and sewage sludge applications in agriculture.* World Health Organization, Geneva.

Charlatchka, R. and Cambier, P. (2000) Influence of reducing conditions on solubility of trace metals in contaminated soils. *Water Air Soil Pollution* **118**, 143–167.

Chary, N.S., Kamala, C.T. and Suman, R.D.S. (2008) Assessing risk of heavy metals from consuming food grown on sewage irrigated soils and food chain transfer. *Ecotoxicology and Environmental Safety* **69**, 513–524.
Chipasa, K.B. (2003) Accumulation and fate of selected heavy metals in a biological wastewater treatment system, *Waste Management* **23**, 135–143.
Chuan, M.C., Shu, G.Y. and Liu, J.C. (1996) Solubility of heavy metals in a contaminated soil: effect of redox potential and pH. *Water Air Soil Pollution* **90**, 543–556.
Comte, S., Guibaud, G. and Baudu, M. (2008) Biosorption properties of extracellular polymeric substances (EPS) towards Cd, Cu and Pb for different pH values. *Journal of Hazardous Materials* **151**, 185–193.
Constantino, L.C.A., Prieto, G., del Razo, L.M., Rodríguez-Vázquez, V. and Poggi-Varaldo, H.M. (2005) Chemical fractionation of boron and heavy metals in soils irrigated with wastewater in central Mexico. *Agricultural Ecosystems and Environment* **108**, 57–71.
Corey, R.B., King, L.D., Lue-Hing, C., Fanning, D.S., Street, J.J. and Walker J.M. (1987) Effects of sludge properties on accumulation of trace elements by crops, In: *Land Application of Sludge* (eds A.L. Page, T. J. Loganand J. A. Ryan), pp. 25–52. Lewis Publishers, Chelsea, MI.
Davranche, M. and Bollinger, J.C. (2000) Heavy metals desorption from synthesized and natural iron and manganese oxyhydroxides: effect of reductive conditions. *Journal of Colloid Interface Science* **227**, 531–539.
Dere, C., Cornu, S. and Lamy, I. (2006) Factors affecting the three-dimensional distribution of exogenous zinc in a sandy Luvisol subjected to intensive irrigation with raw wastewaters. *Soil Use Management* **22**, 289–297.
Dere, C., Lamy, I., Jaulin, A. and Cornu, S. (2007) Long-term fate of exogenous metals in a sandy Luvisol subjected to intensive irrigation with raw wastewater. *Environmental Pollution* **145**, 31–40.
Dignac, M.F., Ginestet, P., Rybacki, D., Bruchet, A., Urbain, V. and Scribe, P. (2000) Fate of wastewater organic pollution during activated sludge treatment: Nature of residual organic matter. *Water Research* **34**, 4185–4194.
Dowdy, R.H. and Volc, V.V. (1984) Movement of heavy metals in soils. In: *Chemical Mobility and Reactivity in Soil Systems* (eds D.W. Nelson, D.E. Elrick,and K.K. Tanji), pp. 229–240. SSSA, Madison, WI.
Emmerich, W.E., Lund, L.J., Page, A.L. and Chang, A.C. (1982) Solid phase forms of heavy metals in sewage sludge-treated soils. *Journal of Environmental Quality* **11**, 178–181.
Engineering Science (1987) *Monterey Wastewater Reclamation Study for Agriculture*. Prepared for Monterey Regional Water Pollution Control Agency, Monterey, CA.
Feigin, A., Ravina, I. and Shalhevet, J. (1991) *Irrigation with Treated Sewage Effluent: Management for Environmental Protection*. Springer-Verlag, Berlin.
Fine, P. and Hass, A. (2007) Role of organic matter in microbial transport during irrigation with sewage effluent. *Journal of Environmental Quality* **36**, 1050–1060.
Fine, P., Hass, A., Prost, R. and Atzmon, N. (2002) Organic carbon leaching from effluent irrigated lysimeters as affected by residence time. *Soil Science Society of America Journal* **66**, 1531–1539.
Flores, L., Blas, G., Hernandez, G. and Alcala, R. (1997) Distribution and sequential extraction of some heavy metals from soils irrigated with wastewater from Mexico City. *Water Air Soil Pollution* **98**, 105–117.
Fonseca, A.F., Melfi, A.J. and Montes, C.R. (2005) Maize growth and changes in soil fertility after irrigation with treated sewage effluent II. soil acidity, exchangeable cations, and sulfur, boron, and heavy metals availability. *Communications in Soil Science and Plant Analysis* **36**, 1983–2003.
Ford, R.G. and Sparks, D.L. (2000) The nature of Zn precipitates formed in the presence of pyrophyllite. *Environmental Science and Technology* **34**, 2479–2483.
Fotovat, A. and Naidu, R. (1998) Changes in composition of soil aqueous phase influence chemistry of indigenous heavy metals in alkaline sodic and acidic soils. *Geoderma* **84**, 213–234.
Fox, P., Aboshanp, W. and Alsamadi, B. (2005) Analysis of soils to demonstrate sustained organic carbon removal during soil aquifer treatment. *Journal of Environmental Quality* **34**, 156–163.
Friedel, J.K., Langer, T., Siebe, C. and Stahr, K. (2000) Effects of long-term waste water irrigation soil organic matter, soil microbial biomass its activities in central Mexico. *Biology and Fertility of Soils* **31**, 414–421.

Gomes, P.C., Fontes, M.P.F., da Silva, A.G., de S. Mendonca, E. and Netto, A.R. (2001) Selectivity sequence and competitive adsorption of heavy metals by Brazilian soils. *Soil Science Society of America Journal* **65**, 1115–1121.

Gong, C. and Donahoe, R.J. (1997) An experimental study of heavy metal attenuation and mobility in sandy loam soils. *Applied Geochemistry* **12**, 243–254.

Guibaud, G., Tixier, N., Bouju, A. and Baudu, M. (2003) Relation between extracellular polymers' composition and its ability to complex Cd, Cu and Pb. *Chemosphere* **52**, 1701–1710.

Guo, M. and Chorover, J. (2003) Transport and fractionation of dissolved organic matter in soil columns. *Soil Science* **168**, 108–118.

Gupta, N., Khan, D.K. and Santra, S.C. (2008) An assessment of heavy metal contamination in vegetables grown in wastewater-irrigated areas of Titagarh, West Bengal, India. *Bulletin of Environmental Contamination and Toxicology* **80**, 115–118.

Gwenzi, W. and Munondo, R. (2008) Long-term impacts of pasture irrigation with treated sewage effluent on nutrient status of a sandy soil in Zimbabwe. *Nutrient Cycling in Agroecosystems* **82**, 197–207.

Han, F.X. and Banin, A. (1997) Long transformations and redistribution of potentially toxic heavy metals in arid-zone soils incubated: I. Under saturated conditions. *Water Air Soil Pollution* **5**, 399–423.

Han, F.X. and Banin, A. (1999) Long-term transformation and redistribution of potentially toxic heavy metals in arid-zone soils: II. Incubation at the field capacity moisture content. *Water Air Soil Pollution* **114**, 221–250.

Han, F.X. and A. Banin. 2001. Fractional loading isotherm of heavy metals in an arid-zone soil. *Communications in Soil Science and Plant Analysis* **32**, 2691–2708.

Harter, R.D. and Naidu, R. (1995) Role of organic-metal complexation in metal sorption by soils. *Advances in Agronomy* **55**, 219–263.

Harter, R.D. and Naidu, R. (2001) An assessment of environmental and solution parameter impact on trace-metal sorption by soils. *Soil Science Society of America Journal* **65**, 597–612.

Hass, A. and Fine, P. (2010) Sequential selective extraction procedures for the study of heavy metals in soils, sediments, and waste materials- A critical review. *Critical Reviews in Environmental Science and Technology* **40**, 365–399.

Hayat, R. (2006) *The Nature of Dissolved Organic Matter Originating from Wastewater Effluent and Leachate After Soil Irrigation and its Copper Binding Properties*. Thesis, Tel Aviv Univ., Israel.

He, Z.L., Zhang, M., Yang, X.E. and Stoffella, P.J. (2006) Release behavior of copper and zinc from sandy soils. *Soil Science Society of America Journal* **70**, 1699–1707.

Hering, J.G. (1995) Interaction of organic matter with mineral surfaces. In: *Aquatic Chemistry: Interfacial and Interspecies Processes* (eds C.P. Huang, C.R. O'Melia,and J.J. Morgan), pp. 95–110. ACS Advanced Chemistry Series, No. 244, American Chemistry Society, Washington, D.C.

Herre, A., Siebe, C. and Kaupenjohann, M. (2004) Effect of irrigation water quality on organic matter, Cd and Cu mobility in soils of central Mexico. *Water Science and Technology* **50**, 277–284.

Hinesly, T. D., Thomas, R.E. and Stevens, R.G. (1978) *Environmental Changes from Long-Term Land Applications of Municipal Effluent. EPA430/9–78-003*. Office of Water Programs Operations, Washington, D.C.

Hodgson, J.F., Tiller, K.G. and Fellows, M. (1964) The role of hydrolysis in the reaction of heavy metals with soil-forming materials. *Soil Science Society of America Journal* **28**, 42–46.

Hoffmann, C., Marschner, B. and Renger, M. (1998) Influence of DOM-quality, DOM-quantity and water regime on the transport of selected heavy metals. *Physics and Chemistry of the Earth* **23**, 205–209.

Hoffmann, C., Savric, I., Jozefaciuk, G., Hajnos, M., Sokołowska, Z., Renger, M. and Marschner, B. (2002) Reaction of sewage farm soils to different irrigation solutions in a column experiment 2. Heavy metals and their leaching. *Journal of Plant Nutrition and Soil Science* **165**, 67–71.

Hong, C.L., Jia, Y.B., Yang, X.E., He, Z.L. and Stoffella, P.J. (2008) Assessing lead thresholds for phytotoxicity and potential dietary toxicity in selected vegetable crops. *Bulletin of Environmental Contamination and Toxicology* **80**, 356–361.

Ilani, T., Schulz, E. and Chefetz, B. (2005) Interactions of organic compounds with wastewater dissolved organic matter: role of hydrophobic fractions. *Journal of Environmental Quality* **34**, 552–562.

Imai, A., Fukushima, T., Matsushige, K., Kim, Y.H. and Choi, K. (2002) Characterization of dissolved organic matter in effluents from wastewater treatment plants. *Water Research* **36**, 859–870.

Impellitteri, C.A., Lu, Y., Saxe, J.K., Allen, H.E. and Peijnenburg, W.J.G.M. (2002) Correlation of the partitioning of dissolved organic matter fractions with the desorption of Cd, Cu, Ni, Pb and Zn from 18 Dutch soils. *Environment International* **28**, 401–410.

James, R.O. and Healy, T.W. (1972) Adsorption of hydrolyzable metal ions at the oxide–water interface. III. A thermodynamic model of adsorption. *Journal of Colloid Interface Science* **40**, 65–81.

Jarusutthirak, C. and Amy, G. (2007) Understanding soluble microbial products (SMP) as a component of effluent organic matter (EfOM). *Water Research* **41**, 2787–2793.

Jarusutthirak, C., Amy, G. and Croue, J.P. (2002) Fouling characteristics of wastewater effluent organic matter (EfOM) isolates on NF and UF membranes. *Desalination* **145**, 247–255.

Jeffery, J.J. and Uren, N.C. (1983) Copper and zinc species in the soil solution and the effects of soil pH. *Australian Journal of Soil Research* **21**, 479–488.

Johns, G.G. and McConchie, D.M. (1994a) Irrigation of bananas with secondary treated sewage effluent. I. Field evaluation of effect on plant nutrients and additional elements in leaf, pulp and soil. *Australian Journal of Agricultural Research* **45**, 1601–1617.

Johns, G.G. and McConchie, D.M. (1994b) Irrigation of bananas with secondary treated sewage effluent. II. Effect on plant nutrients, additional elements and pesticide residues in plants, soil and leachate using drainage lysimeters. *Australian Journal of Agricultural Research* **45**, 1619–1638.

Jueschke, E., Marschner, B., Tarchitzky, J. and Chen, Y. (2008) Effects of treated wastewater irrigation on the dissolved and soil organic carbon in Israeli soils. *Water Science Technology* **57**, 727–734.

Kabata-Pendias, A. (2004) Soil-plant transfer of trace elements–an environmental issue. *Geoderma* **122**, 143–149.

Kabata-Pendias, A. and Pendias, H. (1984) *Trace Elements in Soils and Plants*. CRC Press Inc., Boca Raton, FL.

Karvelas, M., Katsoyiannis, A. and Samara, C. (2003) Occurrence and fate of heavy metals in the wastewater treatment process. *Chemosphere* **53**, 1201–1210.

Katsoyiannis, A. and Samara, C. (2007) The fate of Dissolved Organic Carbon (DOC) in the wastewater treatment process and its importance in the removal of wastewater contaminants. *Environmental Science and Pollution Research* **14**, 284–292.

Khan, S., Cao, Q., Zheng, Y.M., Huang, Y.Z. and Zhu, Y.G. (2008) Health risks of heavy metals in contaminated soils and food crops irrigated with wastewater in Beijing, China. *Environmental Pollution* **152**, 686–692.

Kirpichtchikova, T.A., Manceau, A., Spadini, L., Panfili, F., Marcus, M.A. and Jacquet, T. (2006) Speciation and solubility of heavy metals in contaminated soil using X-ray microfluorescence, EXAFS spectroscopy, chemical extraction, and thermodynamic modeling. *Geochimica et Cosmochimica Acta* **70**, 2163–2190.

Klewicki, J.K. and Morgan, J.J. (1999) Dissolution of β-MnOOH particles by ligands: Pyrophosphate, ethylenediamine tetraacetate, and citrate. *Geochimica et Cosmochimica Acta* **63**, 3017–3024.

Konigsberger, E., Hausner, R. and Gamsjager, H. (1991) Solid-solute phase equilibria in aqueous solution. V: the system CdCO3-CaCO3-CO2-H2O. *Geochimica et Cosmochimica Acta* **55**, 3505–3514.

Lake, D.L., Kirk, W.W. and Lester, J.N. (1984) Fractionation, characterization and speciation of heavy metals in sewage sludge and sludge-amended soils: A review. *Journal of Environmental Quality* **12**, 183.

Lawson, P.S., Sterritt, R.M.,and Lester, J.N. (1984a) Factors affecting the removal of metals during activated sludge wastewater treatment I. The role of soluble ligands. *Archives of Environmental Contamination and Toxicology* **13**, 383–390.

Lawson, P.S., Sterritt, R.M. and Lester, J.N. (1984b) Adsorption and complexation mechanisms of heavy metal uptake in activated sludge. *Journal of Chemical Technology and Biotechnology* **34B**, 253–262.

Lawson, P.S., Sterritt, R.M. and Lester, J.N. (1984c) Factors affecting the removal of metals during activated sludge wastewater treatment II. The role of mixed liquor biomass. *Archives of Environmental Contamination and Toxicology* **13**, 391–402.

Leenheer, J.A. (1981) Comprehensive approach to preparative isolation and fractionation of dissolved organic carbon from natural waters and wastewaters. *Environmental Science and Technology* **15**, 578–587.

Lin, C., Negev, I., Eshel, G. and Banin, A. (2008) In situ accumulation of copper, chromium, nickel, and zinc in soils used for long-term waste water reclamation. *Journal of Environmental Quality* **37**, 1477–1487.

Lin, C., Shacahr, Y. and Banin, A. (2004) Heavy metal retention and partitioning in a large-scale soil-aquifer treatment (SAT) system used for wastewater reclamation. *Chemosphere* **57**, 1047–1058.

Lindsay, W.L. (1979) *Chemical Equilibria in Soils*. John Wiley & Sons, New York.

Liu, F., Colombo, C., Adamo, P., He, J.Z. and Violante, A. (2002) Trace Elements in Manganese-Iron Nodules from a Chinese Alfisol. *Soil Science Society of America Journal* **66**, 661–670.

Liu, W.H., Zhao, J.Z., Ouyang, Z.Y. and Soderlund, L.L.G.H. (2005) Impacts of sewage irrigation on heavy metal distribution and contamination in Beijing. China. *Environment International* **31**, 805–812.

Logan, T.J. and Chaney, R.L. (1983) Utilization of municipal wastewater and sludge on land - metals. In: A.L. Page, T.L. Gleason, J.E. Smith, I. K. Iskanderand L. E. Sommers (ed.), pp. 235–323. University of California, Riverside, CA.

Lun, X.Z. and Christensen, T.H. (1989) Cadmium complexation by solid waste leachates. *Water Research* **23**, 81–84.

Manceau, A., Lanson, M. and Geoffroy, N. (2007) Natural speciation of Ni, Zn, Ba, and As in ferromanganese coatings on quartz using X-ray fluorescence, absorption, and diffraction. *Geochimica et Cosmochimica Acta* **71**, 95–128.

Martinez, C.E. and McBride, M.B. (1998a) Coprecipitates of Cd, Cu, Pb and Zn in iron oxides: Solid phase transformation and metal solubility after aging and thermal treatment. *Clays and Clay Minerals* **46**, 537–545.

Martinez, C.E. and McBride, M.B. (1998b) Solubility of Cd2 + , Cu2 + , Pb2 + , and Zn2 + in aged coprecipitates with amorphous iron hydroxides. *Environmental Science and Technology* **32**, 743–748.

Martinez, C.E. and McBride, M.B. (1999) Dissolved and labile concentrations of Cd, Cu, Pb, and Zn in aged ferrihydrite - Organic matter systems. *Environmental Science and Technology* **33**, 745–750.

McBride, M.B. (1981) Forms and distribution of copper in solid and solution phases of soil. p. 25–46. In: *Copper in Soils and Plants*. (eds J.F. Lonergan, A.D. Robson,and R.D. Graham). Academic Press, Sydney, Australia.

McBride, M.B. (1987) Adsorption and oxidation of phenolic compounds by iron and manganese oxides. *Soil Science Society of America Journal* **51**, 1466–1472.

McBride, M.B. (1989) Reactions controlling heavy metal solubility in soils. *Advances in Soil Science* **10**, 1–56.

McBride, M.B. (1994) *Environmental Chemistry of Soils*. Oxford University Press Inc., New York, NY.

McBride, M.B. (1995) Toxic metal accumulation from agricultural use of sludge: Are USEPA regulations protective? *Journal of Environmental Quality* **24**, 5–18.

McBride, M.B., Sauve, S. and Hendershot, W. (1997) Solubility control of Cu, Zn, Cd and Pb in contaminated soils. *European Journal of Soil Science* **48**, 337–346.

McBride, M.B., Sikora, F.J. and Wesselink, L.G. (1988) Complexation and catalyzed oxidative polymerization of catechol by aluminum in acidic solution. *Soil Science Society of America Journal* **52**, 985–993.

McGowen, S.L., Basta, N.T. and Brown, G.O. (2001) Use of diammonium phosphate to reduce heavy metal solubility and transport in smelter-contaminated soil. *Journal of Environmental Quality* **30**, 493–500.

McKenzie, R.M. (1981) The surface charge on manganese dioxides. *Australian Journal of Soil Research* **19**, 41–50.

McKenzie, R.M. (1989) Manganese oxides and hydroxides. In: *Minerals in Soil Environments* (eds J. Dixonand S. Weed), pp. 439–465. SSSA, Madison, WI.

McLaren, R.G. and Crawford, D.V. (1973a) Studies on soil copper I. The fractionation of copper in soils. *Journal of Soil Science* **24**, 172–181.

McLaren, R.G. and Crawford, D.V. (1973b) Studies on soil copper II. The specific adsorption of copper by soils. *Journal of Soil Science* **24**, 443–452.

McLaren, R.G., Lawson, D.M. and Swift, R.S. (1986) Sorption and desorption of cobalt by soils and soil components. *Journal of Soil Science* **37**, 413–426.

McNicol, R.D. and Beckett, P.H.T. (1989) The distribution of heavy metals between the principal components of digested sewage sludge. *Water Research* **23**, 199–206.

Meek, B.D., Ehlig, C.F., Stolzy, L.H. and Graham, L.E. (1983) Furrow and trickle irrigation: Effects on soil oxygen and methylene and tomato yield. *Soil Science Society of America Journal* **47**, 631–635.

Meek, B.D. and Grass, L.B. (1975) Redox potential in irrigated desert soils as an indicator of aeration status. *Soil Science Society of America Proceedings* **39**, 870–875.

Mireles, A., Solis, C., Andrade, E., Lagunas-Solar, M., Pina, C. and Flocchini, R.G. (2004) Heavy metal accumulation in plants and soil irrigated with wastewater from Mexico City. *Nuclear Instruments and Methods in Physics Research Section B: Beam Interactions with Materials and Atoms* **219–220**, 187–190.

Naidu, R. and Harter, R.D. (1998) Effect of different organic ligands on cadmium sorption by and extractability from soils. *Soil Science Society of America Journal* **62**, 644–650.

Namour, Ph. and Muller, M.C. (1998) Fractionation of organic matter from wastewater treatment plants before and after a 21-day biodegradability test: A physical-chemical method for measurement of the refractory part of effluents. *Water Research* **32**, 2224–2231.

Nan, Z. and Cheng, G. (2001) Accumulation of Cd and Pb in spring wheat (Triticum aestivum L.) Grown in calcareous soil irrigated with wastewater. *Bulletin of Environmental Contamination and Toxicology* **66**, 748–754.

Nan, Z., Li, J., Zhang, J. and Cheng, G. (2002) Cadmium and zinc interactions and their transfer in soil-crop system under actual field conditions. *Science of the Total Environment* **285**, 187–195.

Neal, R.H. and Sposito, G. (1986) Effects of soluble organic matter and sewage sludge amendments on cadmium sorption by soils at low cadmium concentrations. *Soil Science* **142**, 164–172.

Nicholson, F.A., Smith, S.R., Alloway, B.J., Carlton-Smith, C. and Chambers, B.J. (2003) An inventory of heavy metals inputs to agricultural soils in England and Wales. *Science of the Total Environment* **311**, 205–219.

Nielsen, J.S. and Hrudey, S.E. (1983) Metal loadings and removal at a municipal activated sludge plant. *Water Research* **17**, 1041–1052.

Oliveira, A.D.S., Bodo, A., Beltramini Trevilato, T.M., Magosso Takayanagui, A.M., Domingo, J.L. and Segura, M. (2007) Heavy metals in untreated/treated urban effluent and sludge from a biological wastewater treatment plant. *Environmental Science and Pollution Research* **14**, 483–489.

Oste, L.A., Temminghoff, E.J.M. and Van Riemsdijk, W.H. (2002) Solid-solution partitioning of organic matter in soils as influenced by an increase in pH or Ca concentration. *Environmental Science and Technology* **36**, 208–214.

Page, A.L. and Chang, A.C. (1985) Fate of wastewater constituents in soil and groundwater: Trace elements. *Irrigation with Reclaimed Municipal Wastewater - A Guidance Manual*, Second Edition (eds Pettygrove, G.S.,and Asano, T.), pp. 1–16. Lewis Publishers, Inc., Chelsea, MI.

Papadopoulos, P. and Rowell, D.L. (1988) The reactions of cadmium with calcium carbonate surfaces. *Journal of Soil Science* **39**, 23–36.

Papadopoulos, P. and Rowell, D.L. (1989) The reactions of copper and zinc with calcium carbonate surfaces. *Journal of Soil Science* **40**, 39–48.

Petrunic, B.M., MacQuarrie, K.T.B. and Al, T.A. (2005) Reductive dissolution of Mn oxides in river-recharged aquifers: A laboratory column study. *Journal of Hydrology* **301**, 163–181.

Qishlaqi, A., Moore, F. and Forghani, G. (2008) Impact of untreated wastewater irrigation on soils and crops in Shiraz suburban area, SW Iran. *Environmental Monitoring and Assessment* **141**, 257–273.

Quantin, C., Becquer, T., Rouiller, J.H. and Berthelin, J. (2001) Oxide weathering and trace metal release by bacterial reduction in a New Caledonia Ferralsol. *Biogeochemistry* **53**, 323–340.

Quantin, C., Becquer, T., Rouiller, J.H. and Berthelin, J. (2002) Redistribution of Metals in a New Caledonia Ferralsol After Microbial Weathering. *Soil Science of America Journal* **66**, 1797–1804.

Rattan, R.K., Datta, S.P., Chhonkar, P.K., Suribabu, K. and Singh, A.K. (2005) Long-term impact of irrigation with sewage effluents on heavy metal content in soils, crops and groundwater - A case study. *Agricultural Ecosystems and Environment* **109**, 310–322.

Reddy, K.R. and Patrick, Jr., W.H. (1983) Effects of aeration on reactivity and mobility of soil constituents. In: *Chemical Mobility and Reactivity in Soil Systems* (eds D.W. Nelson, K.K. Tanji,and D.E. Elrick), pp. 11–34. SSSA. Madison, WI.
Romkens, P.F.A.M.,and Dolfing, J. (1998) Effect of Ca on the solubility and molecular size distribution of DOC and Cu binding in soil solution samples. *Environmental Science and Technology* **32**, 363–369.
Rule, K.L., Comber, S.D.W., Ross, D., Thornton, A., Makropoulos, C.K. and Rautiu, R. (2006) Survey of priority substances entering thirty English wastewater treatment works. *Water and Environment Journal* **20**, 177–184.
Sauve, S., Hendershot, W. and Allen, H.E. (2000) Solid-solution partitioning of metals in contaminated soils: Dependence on pH, total metal burden, and organic matter. *Environmental Science and Technology* **34**, 1125–1131.
Sauve, S., McBride, M.B., Norvell, W.A. and Hendershot, W.H. (1997) Copper solubility and speciation of in situ contaminated soils: Effects of copper level, pH and organic matter. *Water Air Soil Pollution* **100**, 133–149.
Scheidegger, A.M., Strawn, D.G., Lamble, G.M. and Sparks, D.L. (1998) The kinetics of mixed Ni-Al hydroxide formation on clay and aluminum oxide minerals: A time-resolved XAFS study. *Geochimica et Cosmochimica Acta* **62**, 2233–2245.
Schirado, T., Vergara, I., Schalscha, E.B. and Pratt, P.F. (1986) Evidence for movement of heavy metals in a soil irrigated with untreated wastewater. *Journal of Environmental Quality* **15**, 9–12.
Schmidt, U. (2003) Enhancing phytoextraction: The effect of chemical soil manipulation on mobility, plant accumulation, and leaching of heavy metals. *Journal of Environmental Quality* **32**, 1939–1954.
Schnitzer, M. (1978) Humic substances: chemistry and reactions. In: *Soil Organic Matter* (eds M. Schnitzerand S. U. Khan), pp. 47–52. Elsevier, New York.
Schwab, A.P., He, Y. and Banks, M.K. (2005) The influence of organic ligands on the retention of lead in soil. *Chemosphere* **61**, 856–866.
Schwertmann, U. (1991) Solubility and dissolution of iron oxides. *Plant Soil* **130**, 1–25.
Schwertmann, U. and Taylor, R.M. (1989) Iron oxides. In: *Minerals in Soil Environments* (eds J. Dixonand S. Weed), pp. 379–438. SSSA, Madison, WI.
Sexstone, A.J., Revsbech, N.P., Parkin, T.B. and Tiedje, J.M. (1985) Direct measurement of oxygen profiles and denitrification rates in soil aggregates. *Soil Science Society of America Journal* **49**, 645–651.
Sharma, R.K., Agrawal, M. and Marshall, F. (2006) Heavy metal contamination in vegetables grown in wastewater irrigated areas of Varanasi, India. *Bulletin of Environmental Contamination and Toxicology* **77**, 312–318.
Shon, H.K., Vigneswaran, S. and Snyder, S.A. (2006) Effluent organic matter (EfOM) in wastewater: Constituents, effects, and treatment. *Critical Reviews in Environmental Science and Technology* **36**, 327–374.
Sidle, R.C., Hook, J.E. and Kardos, L.T. (1976) Heavy metals application and plant uptake in a land disposal system for waste water. *Journal of Environmental Quality* **5**, 97–102.
Siebe, C. (1995) Heavy metal availability to plants in soils irrigated with wastewater from Mexico City. *Water Science and Technology* **32**, 29–34.
Siebe, C. and Fischer, W.R. (1996) Effect of long-term irrigation with untreated sewage effluents on soil properties and heavy metal adsorption of Leptosols and Vertisols in Central Mexico. *Journal of Plant Nutrition and Soil Science* **159**, 357–364.
Silviera, D.J. and Sommers, L.E. (1977) Extractability of copper, zinc, cadmium, and lead in soils incubated with sewage sludge. *Journal of Environmental Quality* **6**, 47–52.
Soler-Rovira, P., Soler-Soler, J., Soler-Rovira, J. and Polo, A. (1996) Agricultural use of sewage sludge its regulation. *Fertilizer Research* **43**, 173–177.
Sorme, L. and Lagerkvist, R. (2002) Sources of heavy metals in urban wastewater in Stockholm. *Science of the Total Environment* **298**, 131–145.
Sterritt, R.M. and Lester, J.N. (1984) Mechanisms of heavy metal concentration into sewage sludge. In: Processing and use of sewage sludge (eds P. L. L'Hermiteand H. Ott), pp. 172–175. Reidel Pub. Co., Dordrecht, Holland.
Stevenson, F.J. (1994) *Humus Chemistry*. John Wiley & Sons, New York, NY.

Stone, A.T. and Morgan, J.J. (1987) Reductive dissolution of metal oxides. In: *Aquatic Surface Chemistry* (ed. W. Stumm), pp. 221–254. John Wiley & Sons, New York, NY.

Stover, R.C., Sommers, L.E. and Silviera, D.J. (1976) Evaluation of metals in wastewater sludge. *Journal of the Water Pollution Control Federation* **48**, 2165–2175.

Streck, T. and Richter, J. (1997) Heavy metal displacement in a sandy soil at the field scale: I. Measurements and parameterization of sorption. *Journal of Environmental Quality* **26**, 49–56.

Stumm, W. and Furrer, G. (1987) The dissolution of oxides and aluminum silicates; examples for surface-coordination-controlled kinetics. In: *Aquatic Surface Chemistry* (ed. W. Stumm), pp. 197–220. John Wiley & Sons, New York, NY.

Temminghoff, E.J.M., Van Der Zee, S.E.A.T. and De Haan, F.A.M. (1997) Copper mobility in a copper-contaminated sandy soil as affected by pH and solid and dissolved organic matter. *Environmental Science and Technology* **31**, 1109–1115.

Temminghoff, E.J.M., Van Der Zee, S.E.A.T. and De Haan, F.A.M. (1998) Effects of dissolved organic matter on the mobility of copper in a contaminated sandy soil. *European Journal of Soil Science* **49**, 617–628.

Tessier, A. and Campbell, P.G.C. (1991) Partitioning of trace metals in sediments, In: *Metal Speciation: Theory, Analysis and Application* (eds J.R. Kramerand H.E. Allen), p. 183–199. Lewis Pub., Boca Raton, FL.

Tian, Y. (2008) Behaviour of bacterial extracellular polymeric substances from activated sludge: A review. *International Journal of Environmental Pollution* **32**, 78–89.

Tyler, G. and Olsson T. (2001a) Concentrations of 60 elements in the soil solution as related to the soil acidity. *European Journal of Soil Science* **52**, 151–165.

Tyler, G. and Olsson, T. (2001b) Plant uptake of major and minor mineral elements as influenced by soil acidity and liming. *Plant Soil* **230**, 307–321.

USEPA (1993) *Clean Water Act 40 CFR 503, Sludge Rule*. U.S. Govt. Print. Off. Washington, DC.

USEPA (1994) *A Plain English Guide to the EPA Part 503 Biosolids Rule*. www.epa.gov/owm/mtb/biosolids/503pe/index.htm (last visited Mar 2009).

USEPA (2003) *Drinking water contaminants. EPA 816-F-03-016*. www.access.gpo.gov/nara/cfr/waisidx_02/40cfr141_02.html (last visited Mar 2009).

USEPA (2006a) *Current national recommended water quality criteria*. www.epa.gov/waterscience/criteria/wqcriteria.html (last visited Mar 2009).

USEPA (2006b) *Process design manual, land treatment of municipal wastewater effluents. EPA/625/R-06/016*. Cincinnati, OH.

USEPA IRIS. *Integrated Risk Information System*. www.epa.gov/iris/subst.

Vulkan, R., Mingelgrin, U., Ben-Asher, J. and Frenkel, H. (2002) Characterization and time dependence of transition metal speciation in the solution of a sandy soil amended with sewage sludge. *Journal of Environmental Quality* **31**, 193–203.

Wang, J., Huang, C.P. and Allen, H.E. (2003) Modeling heavy metal uptake by sludge particulates in the presence of dissolved organic matter. *Water Research* **37**, 4835–4842.

Wang, P., Qu, E., Li, Z. and Shuman, L.M. (1997) Fractions and availability of nickel in loessial soil amended with sewage or sewage sludge. *Journal of Environmental Quality* **26**, 795–801.

Wang, Y. and Stone, A.T. (2006) Reaction of MnIII,IV (hydr)oxides with oxalic acid, glyoxylic acid, phosphonoformic acid, and structurally-related organic compounds. *Geochimica et Cosmochimica Acta* **70**, 4477–4490.

Weirich, D.B., Hari, R., Xue, H., Behra, P. and Sigg, L. (2002) Adsorption of Cu, Cd, and Ni on goethite in the presence of natural groundwater ligands. *Environmental Science and Technology* **36**, 328–336.

Weng, L., Fest, E.P.M.J., Fillius, J., Temminghoff, E.J.M. and Van Riemsdijk, W.H. (2002a) Transport of humic and fulvic acids in relation to metal mobility in a copper-contaminated acid sandy soil. *Environmental Science and Technology* **36**, 1699–1704.

Weng, L., Temminghoff, E.J.M., Lofts, S., Tipping, E. and Van Riemsdijk, W.H. (2002b) Complexation with dissolved organic matter and solubility control of heavy metals in a sandy soil. *Environmental Science and Technology* **36**, 4804–4810.

Xiang, H.F.,and Banin, A. (1996) Solid-phase manganese fractionation changes in saturated arid-zone soils: Pathways and kinetics. *Soil Science Society of America Journal* **60**, 1072–1080.

Yadav, R.K., Goyal, B., Sharma, R.K., Dubey, S.K. and Minhas, P.S. (2002) Post-irrigation impact of domestic sewage effluent on composition of soils, crops and ground water - A case study. *Environment International* **28**, 481–486.

Yang, J.Y., Yang, X.E., He, Z.L., Li, T.Q., Shentu, J.L. and Stoffella, P.J. (2006) Effects of pH, organic acids, and inorganic ions on lead desorption from soils. *Environmental Pollution* **143**, 9–15.

Yin, Y., Allen, H.E., Li, Y., Huang, C.P. and Sanders, P.F. (1996) Adsorption of mercury(II) by soil: Effects of pH, chloride, and organic matter. *Journal of Environmental Quality* **25**, 837–844.

Yin, Y., Impellitteri, C.A., You, S.J. and Allen, H.E. (2002) The importance of organic matter distribution and extract soil: Solution ratio on the desorption of heavy metals from soils. *Science of the Total Environment* **287**, 107–119.

You, S.J., Yin, Y. and Allen, H.E. (1999) Partitioning of organic matter in soils: Effects of pH and water/soil ratio. *Science of the Total Environment* **227**, 155–160.

Zausig, J., Stepniewski, W. and Horn, R. (1993) Oxygen concentration and redox potential gradients in unsaturated model soil aggregates. *Soil Science Society of America Journal* **57**, 908–916.

Zhao, L.Y.L., Schulin, R., Weng, L. and Nowack, B. (2007) Coupled mobilization of dissolved organic matter and metals (Cu and Zn) in soil columns. *Geochimica et Cosmochimica Acta* **71**, 3407–3418.

Chapter 8
Salinity

Nico E. Marcar, Tivi Theiveyanathan and Daryl P. Stevens

8.1 The nature of salinity

A broad range of dissolved salts is known to occur in soil and water, principally sodium chloride (NaCl), but other ions such as calcium (Ca), magnesium (Mg), sulfate (SO_4), potassium (K), and bicarbonate (HCO_3) may also be found in large amounts, depending on geology, soil types and source of wastewater. Both salt concentration and salt composition will determine specific ion toxicities for plants. On the other hand, osmotic effects that reduce the ability of plants to extract water from soils (i.e. plant available water, PAW) are primarily impacted by salt concentration.

Salinity occurrence varies markedly across field locations based on initial salinity, site characteristics, and in response to irrigation with saline water. Under non-irrigated conditions, surface soil salinities are often much higher than those in the subsoil in summer months due to high evaporation and lower rainfall resulting in salts being drawn from deeper soil profiles and/or shallow watertables (Yaron, 1981), whereas salinity will be lower in winter months due to lower evaporation and dilution and leaching of salt from rainfall. Under irrigated conditions, these patterns will not be so obvious and distribution of salts will be more variable, depending on the irrigation water salinity, type of irrigation system and frequency of irrigation, soil properties affecting water and salt transport, and on the soil water extraction pattern by the active roots distributed within the soil profile.

Saline soils contain relatively high concentrations of soluble salts and are highly variable, both spatially and temporally. Saline soil is typically a harsh environment and often imposes multiple stresses on plant survival and growth. These stresses include salinity, sodicity, water logging, and soil water deficit. (Soil sodicity, a condition where there are excess sodium ions relative to calcium and magnesium, is dealt with in Chapter 9.) It is important to note that increased salinity can suppress sodicity (effects) at a given exchangeable sodium percentage (ESP) or sodium adsorption ratio (SAR) (see Chapter 9).

This chapter focuses specifically on salinity, its measurement, treated wastewater (TWW, recycled water) salinity, the impact of soil and irrigation water salinity on soils and crop growth, and management of salinity.

Treated Wastewater in Agriculture, First Edition, edited by Guy J. Levy, Pinchas Fine and Asher Bar-Tal

8.2 Measuring salinity

One of the first crucial assessments required to determine if it is possible to successfully grow a specific plant species or variety with effluent irrigation is water quality and soil salinity. Two measures of salinity – electrical conductivity (EC) and total dissolved solids (TDS) – are commonly used. Electrical conductivity gives an approximation of salt activity and is typically measured as deciSiemens per meter (dS/m), milliSiemens per centimeter (mS/cm) or microSiemens per centimeter (μS/cm) so that 1 dS/m = 1 mS/cm = 1000 μS/cm. Electrical conductivity is commonly used to assess plant responses to salinity. However, the units of salinity are technically based on concentration (i.e. TDS, mg/l or parts per million (PPM)), which is calculated by summation of all the dissolved solids including akalinity ($0.6 \times CaCO_3$), sodium (Na), potassium (K), magnesium (Mg), chloride (Cl), sulfate (SO_4), silicate (SiO_4), nitrate (NO_3), and fluoride (F) (APHA, 1998). Total dissolved solids is more commonly used to assess the salt concentration of water or soil samples for purposes other than plant responses.

8.2.1 *Water salinity*

The conversion factor of EC to TDS can vary from 1 (dS/m):550 to 840 (ppm) depending on types of salts dissolved. For water recycled from sewage effluent the conversion is usually 640 for ECw < 5 dS/m and 800 for EC > 5 (Asano et al., 2007). For agricultural water purposes, Rayment and Higginson (1992) suggest: TDS (mg/l) = EC (dS/m) × 670. The salinity of irrigation water is sometimes classified according to EC (Table 8.1). These ratings can be used to discuss recycled water salinity levels in general terms.

8.2.2 *Soil salinity*

Soil salinity is also described in terms of EC and is usually measured in a 1:5 soil:water suspension (1 : 5 ($EC_{1:5}$)), 1 : 2 or 1 : 1 depending on the country and preferred extraction method) or saturation paste extract (EC_e). Electrical conductivity is a measure of a solution's ability to conduct electricity. Plant response to salinity is better defined in terms of EC_e because this is more likely to relate to soil solution EC (EC_{ss}) compared with diluted samples. The $EC_{1:5}$ value can be converted to EC_e by applying conversion factors based on soil texture. These factors vary considerably depending on the soil type and texture ranging

Table 8.1 Irrigation water salinity ratings based on electrical conductivity (EC_i)

EC_i (dS/m)	Water salinity rating	Plant suitability
<0.65	Very low	Sensitive plants
0.65–1.3	Low	Moderately sensitive plants
1.3–2.9	Medium	Moderately tolerant plants
2.9–5.2	High	Tolerant plants
5.2–8.1	Very high	Very tolerant plants
>8.1	Extreme	Generally too saline

Source: ANZECC and ARMCANZ, 2000a.

from an average of 13–17 for sandy soils down to as low as 4 for heavy clays (Slavich and Petterson, 1993; Cass et al., 1995; Rengasamy, 2006). Soil solution EC will differ from EC_e depending on soil moisture content.

Rapid assessment techniques, including the non-invasive electromagnetic (EM) induction method, allow root-zone salinities to be measured spatially across fields. For example, salinity variation can be mapped before and after irrigation with saline water (Biswas et al., 2005). This method works best if on-site calibrations are made (Rhoades et al., 1999). Time Domain Reflectometry (TDR) is also a popular instrument for measuring both soil moisture content and bulk electrical conductivity of the soil non-destructively (*in situ*).

8.3 Mechanism of soil salinisation

Three processes lead to the development of saline lands worldwide (Rengasamy, 2006):

- groundwater-associated salinity;
- non-groundwater-associated salinity;
- irrigation-associated salinity.

There are two major mechanisms for soil salinisation: irrigation salinity (recycled water or any irrigation water source) and dryland salinity. With irrigation salinity, the salt content of the irrigation water can build up in the soil if there is not appropriate leaching of the excess salts accumulated in the soil horizons caused by evapotranspiration and limited drainage. In contrast, dryland salinity is caused by the salt input through natural processes of precipitation or the movement of saline groundwater (Rengasamy, 2006). In both cases salt concentrations in the soil profile may increase to levels that affect soil properties and crop productivity.

The primary focus of this chapter is salinity from irrigation of water recycled from sewage effluent. However, we also recognize that over irrigation and excessive leaching can contribute to increasing groundwater levels. In some cases increases in groundwater levels (salty groundwater) can lead to higher salinity in the soil profile (primary and secondary groundwater-associated salinity).

8.4 Salinity in wastewater

Depending on the type and origin, wastewater contains significant amounts of salts in addition to organic matter, toxic substances, and microbial organisms. Wastewater chemistry is an important factor that determines whether a specific wastewater is suitable for land application, the method and frequency of application, and how much can safely be applied.

Wastewater can be classified by its source and its strength (concentration). For example, wastewater can be from urban, agricultural and industrial effluent, saline groundwater, or surface and subsurface drainage water. In all these categories, the salt concentration and composition of the applied water is one of the first things checked to determine how it could be used in irrigation.

Traditionally, wastewater is often disposed of into water bodies (rivers, lakes, and sea) if the salinity, organic matter, pathogens, nutrients and pollutants are within prescribed (sustainable) limits for the specific water body. Important additional attributes are the concentrations of total dissolved salts and nutrients (N and P specifically), the level of acidity or alkalinity, the sodium absorption ratio, and potentially some heavy metals and organic contaminants (Prescod, 1992; NRMMC and EPHC, 2006; Stevens, 2006). Extra consideration to pathogen control is needed in the evaluation of wastewater to ensure that land application is sustainable in the medium to long term. Only salinity is discussed in this chapter; see specific chapters in the book for discussion of the other attributes identified above.

8.4.1 Chemical characterization of treated wastewater

Table 8.2 lists typical chemical characteristics of sewage effluent where water can be recycled from for irrigation. If the reclamation process does not involve a specific salt removal process, salts in the wastewater will ultimately contribute to the salinity of the recycled water.

The relative abundance of the constituents of sewage effluent is a function of the original source of water and additions made as water passes through a sewage system. For example, salinity is affected by the domestic water supply (rainfall, river, reservoir/lake, and groundwater), as well as salts added by the community and industry (Table 8.2). In contrast, organic matter, nitrogen, phosphorus, and boron are generally very low or absent from the potable water supply and their concentration in effluent is dependent on excretion and urban processes such as washing.

Sewage effluent is generally treated to secondary level either by oxidation in ponds or by the process of activated sludge before land application. These treatments remove about 85% of organic matter, about 30% of nitrogen and about 30% of phosphorus from raw

Table 8.2 Typical chemical characteristics of sewage effluent water that contribute to salinity

Parameter	Units	Mean[a]	Min.[a]	Max.[a]	Increase[b,c]
Total nitrogen (N)	mg/l	15.2	2.8	39.0	—
Total phosphorus (P)	mg/l	5.9	0.0	12.0	—
pH	pH	7.9	6.2	9.8	—
Total dissolved salts (TDS)	mg/l	675	145	1224	150–380
Electrical conductivity (EC)	dS/m	1.3	0.2	2.9	—
Sodium adsorption ratio (SAR)	$(mmol_c/l)^{0.5}$	6	3	12.2	—
Sodium (Na)	mg/l	181	62.0	312.0	40–70
Calcium (Ca)	mg/l	35	10	74.0	6–16
Magnesium (Mg)	mg/l	19	6	40.0	4–10
Chloride (Cl)	mg/l	135	9.3	340.0	20–50
Potassium (K)	mg/l	—	—	—	7–15
Sulphate (SO_4)	mg/l	30	20	50	15–30
Bicarbonate (HCO_3)	mg/l	—	—	—	50–100

[a]NRMMC and EPHC (2006). Mean, minimum, and maximum are from median values from a range of reports.
[b]Asano et al. (2007).
[c]Increases from domestic water values by effluent were stated, based on a medium strength wastewater with no commercial and industrial additions (Asano et al., 2007).
— = not available.

urban effluent. Salinity is not affected by biological treatment but may increase in concentration due to evaporation from treatment lagoons or holding ponds.

Chloride and Na contribute significantly to the salinity of wastewater (Table 8.2). Both ions have the capacity to affect plant growth via osmotic effects (salinity) or by direct phytotoxicity. Plant sensitivity to both salinity and excess Cl and Na vary considerably with genotype. A good indication of impacts is given by Asano et al. (2007), where Cl concentrations in irrigation water higher than 350 mg/l will create severe restriction on which plants are irrigated with surface irrigation and Cl concentrations greater than 150–175 mg/l will restrict growth of some crops (e.g. apricots and almonds) if spray-irrigated. The specific effects of Cl on plants are described in Chapter 6 (6.2). A list of salt-sensitive plants can be found in the Australian Guidelines for Water Recycling (NRMMC and EPHC, 2006), the work completed by US Department of Agriculture (Maas, 1990) and other more specific sources dealing with crops, pasture plants, and trees (e.g. Marcar and Crawford, 2004). Increasing values of EC or TDS indicate an increasing salinity hazard. There are several guidelines for interpreting water quality for irrigation (Rhoades et al., 1992; NRMMC and EPHC, 2006; Asano et al., 2007).

8.4.2 Managing salinity in TWW

The management of salinity in wastewater can be through source control or through treatment of the wastewater (reverse osmosis to remove constituents that contribute to salinity). Source control could involve:

- point source controls (e.g. industries contributing significant salt loads to the sewerage systems or leaky sewage that allows salty groundwater to seep into the sewer); or
- diffuse source control (e.g. preventing the use of detergents high in salts or water conditioners used by the general public).

Treatment or removal of salinity usually involves some type of reverse osmosis system capable of removing salts from the water. This usually involves a high capital and ongoing cost (i.e. energy and membrane replacement) relative to the price of water used for irrigation.

8.5 Effects of salinity on plant growth and water use

Excessive additions of salts beyond the capacity of the plant and soil system to absorb or deal with internally could lead to:

- increased soil salinity and sodicity;
- direct toxicity of Na and Cl to plants;
- low PAW;
- nutrient imbalances.

Limitations of the soil or site to deal with incoming salt include the capacity for leaching of salts and preferential accumulation of sodium; this may be related to soil texture and slope. Non-halophytic plants try to limit how much salt enters the roots and transport of these salts to young developing leaves is also restricted, depending on plant salt tolerance. These plants tend to use organic solutes (such as sugars and amino acids derived from photosynthesis) and inorganic solutes to counteract (or adjust to) the induced osmotic stress. The use of organic solutes for this osmotic adjustment will, however, deprive the plant of energy that could otherwise be used for growth. Halophytic plants also exclude salts at the roots, but these plants are much better at dealing with salt accumulation in their leaves. This salt is either primarily included in vacuoles of leaf cells where it is used in osmotic adjustment or secreted onto the leaf surface in specialised glands and bladders or both. Both halophytes and non-halophytes can also reduce salt accumulation by shedding older leaves to 'relieve' the plant of its salt load, but this strategy reduces growth (Munns, 2002). These processes could culminate in groundwater pollution, declining soil quality, erosion, low productivity and, at the extreme, plant death.

8.5.1 Physological growth responses

Large amounts of water across the world are restricted in use by salinity. However, with the appropriate irrigation scheduling, leaching fraction, and plants grown, many of these waters have been used successfully to irrigate crops (e.g. vegetables, crops, ornamental plants, and trees).

As salinity in soil and/or water increases it reduces plant growth. The rate of growth reduction as salinity increases varies between plant species and variety (Maas and Hoffman, 1977). Key physiological reasons for or effects of reduced growth include:

- reduced water uptake and photosynthesis resulting from decreased stomatal conductance and leaf shed;
- increased respiration associated with the processes of salt exclusion from the root and containment of salt within leaf cells;
- reduced turgor in growing tissues; and
- interference with the activity of some enzymes.

Except for halophytic species, higher salt tolerance is usually associated with the ability of plants to exclude salt from the root (excluders) and restrict transport to the shoot (Marschner, 1995). Other plant take the salt up (includers), and have adapted by tissue tolerance (e.g. salt compartmentation) and avoidance of high ion concentrations (e.g. salt excretion) (Marschner, 1995).

8.5.2 Growth responses to saline water irrigation

Plants vary considerably in their susceptibility to osmotic and toxicity effects from salinity. As soils dry and if no net input of salt, the salt concentration of the remaining soil water increases and so the effects on plant growth and water use become more severe. Plants affected by salinity show reduced growth and may show signs of water stress (e.g. wilting). This sensitivity also varies with the stage of growth (seedling, juvenile, mature). Leaves can also suffer burning along the leaf margin due to the effects of salinity, chloride, and sodium toxicity. This may result in specific ion toxicities, which may lead to visual

Table 8.3 Soil salinity criteria corresponding to a 10% yield reduction

Plant salt tolerance	Salinity	EC_e Range[a] (dS/m)	Corresponding $EC_{1:5}$ based on soil clay content (dS/m)[b] 0–10% clay	20–40% clay	40–60% clay	60–80% clay
Sensitive plants	Very low	<0.95	<0.07	<0.09	<0.12	<0.15
Moderately sensitive plants	Low	0.95–1.90	0.07–0.15	0.09–0.19	0.12–0.24	0.15–0.30
Moderately tolerant plants	Medium	1.90–4.50	0.15–0.34	0.19–0.45	0.24–0.56	0.30–0.70
Tolerant plants	High	4.50–7.70	0.34–0.63	0.45–0.76	0.56–0.96	0.70–1.18
Very tolerant plants	Very high	7.70–12.2	0.63–0.93	0.76–1.21	0.96–1.53	1.18–1.87
Too saline for plants	Extreme	>12.2	>0.93	>1.21	>1.53	>1.87

[a] EC_e = Electrical conductivity of a saturation paste extract.
[b] $EC_{1:5}$ = Electrical conductivity of a 1:5 soil:water extract.
Source: modified from Rengasamy (2006). Plant salt tolerance groupings (Maas and Hoffman, 1977), equivalent $EC_{1:5}$ for four ranges of soil clay content (Shaw, 1999).

symptoms of stress including leaf chlorosis, necrosis, and premature leaf senescence. For many woody species, increased concentrations of leaf Cl are associated with leaf damage and growth reduction (Marcar and Termaat, 1990).

Like the salinity of irrigation water (Table 8.1), plant salt tolerances to soil salinity can also be rated (Table 8.3). The data in Table 8.3 highlight the dependence of plant sensitivity to salinity on soil texture (clay content) and soil water extraction methods.

Plants are classified according to their tolerance of salinity (and associated stresses). There is an extensive amount of data available on the sensitivity to salinity of a range of plant species. For example, Maas (1990) lists ratings for fiber, grain, specialty crops, grasses, forage, vegetables, fruits, woody crops, ornamental shrubs, trees, and ground cover. Responses of crop yields to soil salinity are well documented (e.g. Fig. 8.1). Marcar and Crawford (2004) provide a broad classification of salt tolerance (from slight to extreme) for tree species (acacia, casuarina, eucalypt, melaleuca, and pine) potentially

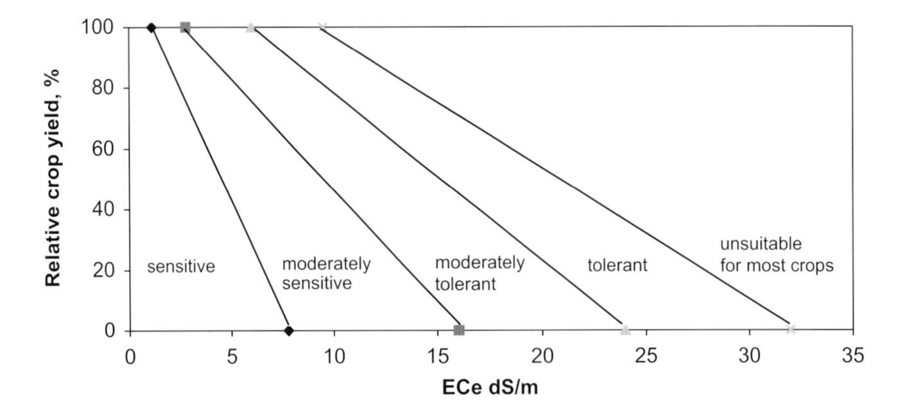

Figure 8.1 Divisions for relative salt tolerance ratings of agricultural crops. Modified from Maas (1984), as cited Ayers and Wescott (1985).

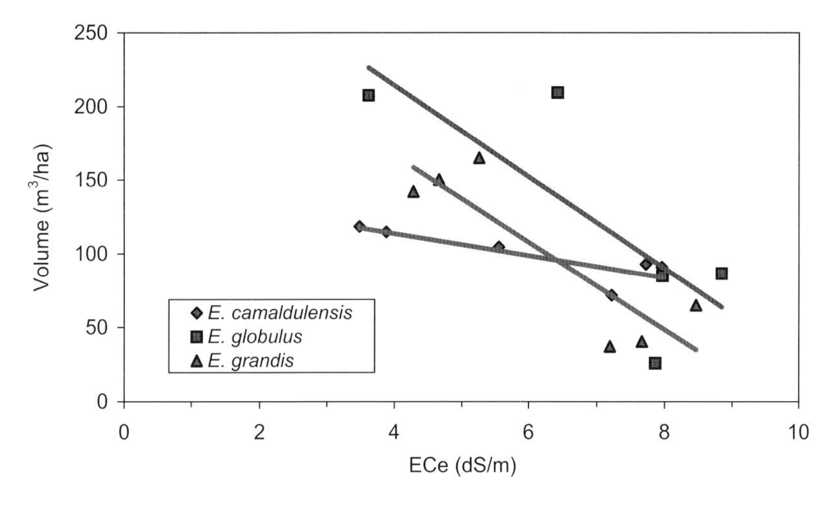

Figure 8.2 Relationships between stem volume (m^3/ha basis) and soil salinity (average soil EC_e over 0–150 cm soil depth) for *E. camaldulensis* ($r = 0.85$; $p < 0.05$), *E. globulus* ($r = 0.79$; $p = 0.11$) and *E. grandis* ($r = 0.87$; $p < 0.05$) at age 10 yrs, when trees were irrigated with either good-quality channel (EC about 0.2 dS/m) or saline (EC about 8 dS/m) water at a site near Timmering, northern Victoria. Reproduced from Feikema et al. (2005), with permission.

suitable for planting in southern Australia. Responses to salinity have been developed for many crop, pasture, and tree species (e.g. Maas, 1990). For example, Feikema et al. (2005) reported that growth of the tree species *Eucalyptus globulus* and *E. grandis* was approximately four times more sensitive to soil salinity resulting from saline water application than *E. camaldulensis* (Fig. 8.2). For each species, survival was similar under channel and saline water treatments, and so lower productivity under saline conditions was not because of differences in mortality. Further measurement is required to determine whether survival is affected in the longer term.

8.6 Water use

Plant available water (PAW) is usually defined as the amount of water in soil that is nominally available for plant growth (Peverill et al. 1999). The upper limit of available water is set by the field capacity or drained upper limit of the soil (how much water the soil can hold before it drains freely) and the lower limit is set by the permanent wilting point (matric suction of 1500 kPa). The maximum water holding capacity is the difference between the two limits. This does not include the amount of water stored above field capacity up to the level of saturation, which is known as unsaturated drainage. It may take several days for a soil to drain down to field capacity after saturation, and hence plants will use the water that is held in the soil above field capacity. Depending on the pore size distribution and capillary effect, which determine the structure of the soil, the unsaturated drainage will vary from 5% to 20% (V). The more water a soil can hold, which is available to a plant (i.e. within the rooting zone and accessible by the plant), the greater can be the period between irrigations. However, as soil salinity or irrigation water salinity increase PAW decreases due to the increased osmotic potential of the water from salinity, leading to a greater risk of water stress to the plant given all other conditions remain constant.

As discussed above, physiological adaptation to salinity can also lead to plants using less water in response to higher salinity. For example, Benyon et al. (1999) found that salinity reduced stem growth and leaf area development rather than water use per unit of leaf area or sapwood area (sap flux density) in 6-year-old *E. camaldulensis* trees under slightly to moderately saline conditions ($EC_e < 8$ dS/m). Similar results have been reported by Mahmood et al. (2001). The reduction of tree water use by salinity is therefore mostly the result of reduced growth of leaves, stems, and roots. Salt-tolerant species probably maintain their sap flux density under increasing soil salinity by adaptive processes, including keeping stomata open for longer.

8.6.1 Irrigation with saline water

Irrigation with water recycled from sewage effluent is practiced in many countries with semiarid and arid climates. It is rare for recycled water salinity to be higher than 3 dS/m (Table 8.2). However, where saline ingress from salty watertables or seawater has entered sewage systems, concentrations >3 dS/m have been recorded. Only moderately salt-tolerant to tolerant plants can cope with soils >3 dS/m (Table 8.3). Some tolerant plants grow well at these concentrations and much higher. For example, crop and tree plantations have been successfully established using saline water (up to 10 dS/m) on permeable, deep soils in Pakistan, especially where seedlings have been planted on ridges to avoid water logging and soil amendments, such as gypsum, have been added to counter sodicity (Qureshi and Barrett-Lennard, 1998).

Ayers and Westcot (1985) have indicated general concern over the use of irrigation water with levels of salinity ($EC_i > 3$ dS/m). However, water with EC_i up to 11 dS/m is being used sustainably for irrigation, e.g. USA (EC_i of 3–11 dS/m); Israel (EC_i of 2–8 dS/m); Tunisia (EC_i of 3–9 dS/m); India (EC_i of 2–8 dS/m); and Egypt (EC_i of 2–5 dS/m) (Rhoades et al., 1992).

Irrigation with saline water can only be sustainable for long periods if it is conducted in accordance with acceptable environmental impact limits. The most fundamental principle of sustainable irrigation is that the hydraulic loading (i.e. total of net rainfall and net irrigation) must not exceed crop and plantation water use, except during occasional times when an allowance for leaching is added to prevent deleterious salt accumulation. Effective irrigation using recycled water is best suited to low rainfall regions where evapotranspiration far exceeds net annual rainfall and for most of the year. This will reduce the risk of runoff, groundwater recharge, and lateral flow.

Over irrigation can cause undesirable impacts, including erosion due to surface runoff, leaching of nutrients/pollution of groundwater due to excessive deep drainage and water logging due to inundation. Under irrigation can also cause problems, including water stress, reduced water and nutrient uptake, and more rapid accumulation of salts in the root-zone.

8.6.2 Irrigation scheduling

Irrigation scheduling is a practice of determining how much water to apply during irrigation and the timing of application. Accurate scheduling of irrigation is essential for maximizing crop production, while conserving water and ensuring irrigation systems are environmentally and economically sustainable. Correct scheduling requires a good knowledge of crop tolerance to salinity, crop water demand, and soil water characteristics, and must account for the type of irrigation method used and climatic conditions. The major focus is to ensure sufficient PAW. Various methods and tools for scheduling irrigation have been developed, ranging from those based on the water status of the soil or plants, to those that use a model to predict the soil water balance. Jensen et al. (1961) were the first to develop an irrigation-scheduling program using meteorological data to calculate crop water use. Many variations of this approach are now in use.

The most accurate way to schedule irrigation is to measure the plant water holding capacity of the soil and to irrigate so that this storage is always maintained between its maximum value and some lower value, below which plants can no longer extract water at the highest rate. In addition, it is essential that any irrigation scheduling methodologies must be backed up by periodic monitoring of soil water content. There are many methods available for measuring soil water content in order to determine when to irrigate, including (Christen et al., 2006):

- tensiometers;
- gypsum blocks;
- capacitance probes;
- time domain reflectometry;
- neutron moderation;
- heat dissipation; and
- wetting front detectors.

Irrigation systems that allow better control of water application and apply water more efficiently (i.e. meet crop requirements and with high uniformity) are inherently better suited to managing salinity (Table 8.4). Published literature (e.g. Doorenbos and Pruitt, 1977; Lundstrom et al., 1981; Jones and Bauder, 1987; Myers et al., 1999; Polglase et al., 2002) shows different irrigation efficiencies for the three main types of irrigation systems – sprinkler (75–85%), dripper (90–95%) and flood/furrow (50–70%). It is assumed that sprinkler irrigation experiences both evaporative and interception losses, flood/furrow systems experience evaporative losses similar to sprinklers, no interception and an average of 15–25% unavoidable drainage loss, and drippers experience a small

Table 8.4 Water salinity suitability for various irrigation systems

Salinity rating	Total dissolved salts (mg/l)	Suitability[a]		
		Drip	Sprinkler	Furrow
Low	<900	High	High	Medium
Moderate	900–2000	High	Medium[b]	Medium
High	2000–3500	Medium	Low	Low

[a]Assuming soils have reasonable drainage, if drainage is very poor then drip should be used.
[b]Leaf burn becomes a problem.
Source: Christen et al. (2006).

amount of loss, which is primarily evaporative. The amounts and relativities of these losses vary with plant age as the canopy develops, causing the amount of direct radiation, temperature, humidity, and air flow to vary.

For example, irrigated tree plantations are best managed by using small volumes of water (0.4–1.5 megaliters (Ml)/ha) regularly during the growing season in order to reduce losses through evaporation and deep drainage (Polglase et al., 2002). Irrigation scheduling models tailored to Australian conditions are available for agricultural and tree crops. Some of these deal with saline water. The quality of water used for irrigation will have different effects depending on soil texture and degree of salinity and sodicity. Irrigation with low-salinity water is best on saline or non-saline soil, but will cause problems on sodic soils due to increased tendency for clay minerals to disperse. Irrigation with moderately saline water (EC < 0.8 dS/m) will improve structural problems in sodic soils. Irrigation with highly saline water (EC > 2 dS/m) will reduce dispersion (improve soil structure) of clay soils, but will increase soil salinity unless there are opportunities for leaching.

8.6.3 Managing saline drainage water from irrigation of recycled water

Irrigation with recycled water will ultimately require leaching of salts either through rainfall or application of leaching fraction while irrigating. The serial biological concentration system allows this saline drainage water to be captured and reused for irrigation, providing there is adequate drainage capacity and that the drainage water can be captured at the end of the system. This has been tested in southern Australia for irrigating plantations with saline water without unduly increasing soil salinity, as an alternative to evaporation basins for saline wastewater disposal (Heath and Heuperman, 1996; Jayawardane et al., 2001). Groundwater with EC of approximately 8.5 dS/m was pumped out and applied to large blocks of *E. camaldulensis* and *Atriplex nummularia*, planted on saline clay soil. Tile drains at 1.9 m below the surface at a spacing of 28 m collected drainage into mariculture ponds and a basin for harvesting salt. Results show that this system has been able to stabilise root-zone EC_e levels at about 15 dS/m (Heuperman et al., 2002). Although serial biological concentration systems are relatively expensive to establish and manage, they may be an option for managing saline effluent from industries.

8.7 Managing root-zone salinity

8.7.1 Field evidence

In a saline water-irrigated environment, salt is continually added to the soil during irrigation, and therefore it becomes more concentrated in the soil solution as water is used by plant transpiration or lost from the soil surface by evaporation. If subsurface agricultural soils that are irrigated are inherently saline, they may become a problem if additional salt is allowed to accumulate to a concentration that is harmful to plants. In addition, there can be other detrimental affects on soil chemistry and structure. Therefore, inputs of salt via irrigation water will need to be balanced by export of salt from the root-zone. General guidelines for the use of irrigation water under Australian conditions have been developed (ANZECC and ARMCANZ, 2000a).

Salt will accumulate in soil over successive irrigation events, especially if water is highly saline, unless there is sufficient leaching of salts below the main root-zone (Section 8.7.2), either through rainfall or appropriate rates of irrigation. Successful irrigation practice, therefore, requires that sufficient water is applied to leach salts below the main part of the root system (see "leaching fractions" below). Ultimately, these salts may be stored in the lower soil profile or be leached into groundwater, depending on depth from the soil surface, the presence of barriers to downward movement and soil water transport characteristics.

For example, root-zone salinity changes have been monitored within irrigated tree plantations in northern Victoria for 5–10 yrs (Marcar and Morris, 2005). Results show that soil salinity varies significantly between sites (Fig. 8.3), through a combination of factors: salinity and rate of addition of irrigation water, soil properties, and watertable depth. The trend of peak concentration observed at each site generally follows the level of applied irrigation water salinity, although the two sites where highest salinity irrigation water was applied (Tatura and Timmering) also had soils with highest clay contents. Impacts on plantation growth should be more pronounced where saline water is applied to clay soils rather than to sandy soils.

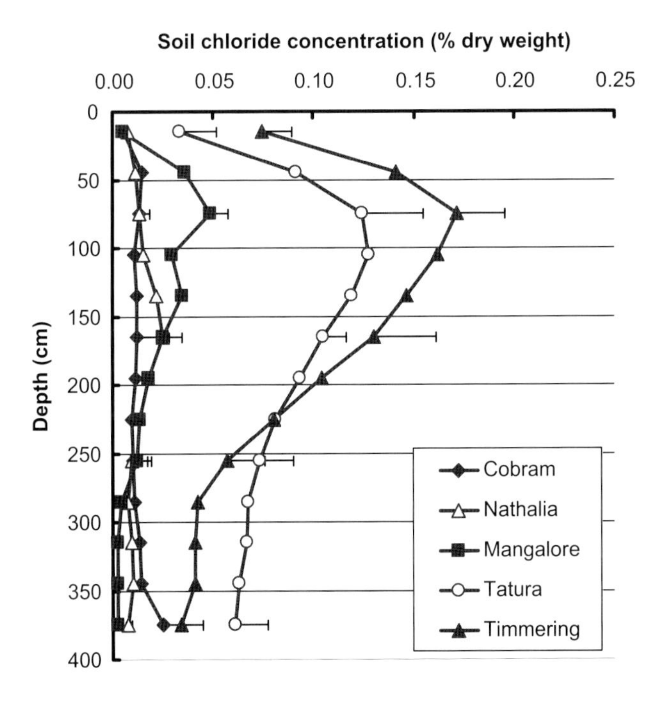

Figure 8.3 Soil chloride concentrations at five plantation sites in northern Victoria irrigated with water of different salinities, 7–9 yrs after planting. Bars represent one standard error (only shown in one direction). Irrigation water salinities were as follows: Cobram (less than 0.5 dS/m), Nathalia (2–2.5 dS/m), Mangalore (1–1.5 dS/m), Tatura (5 dS/m) and Timmering (10 dS/m). Reproduced from Marcar and Morris (2005) with permission of CSIRO Publishing. For all depths and sites, $EC_{1:5} = 8.38^{*}$ Cl conc. $+ 0.09$ ($R^2 = 0.97$, $p < 001$) and ratio $(EC_e/EC_{1:5}) = 0.496 + (3.883/\theta_{sp})$ ($R^2 = 0.99$, θ_{sp} = saturation percentage). (Bandara et al. 2002)

Salinity of the soil solution, calculated from moisture content and chloride concentration, was considerably greater than in the irrigation water to a depth of at least 3 m at all sites. Bandara et al. (2002) showed a trend of increasing soil salinity following commencement of irrigation, although salt concentrations at some sites and profile depths have shown significant year-to-year variability. This is presumably mainly associated with sampling variation and prevailing weather conditions. Field electromagnetic induction surveys (using the EM38 device) for the plantation sites at Nathalia, Timmering and Tatura have also demonstrated increases in soil salinity in the upper 1–2 m of soil profiles in plantations after two or three seasons of irrigation with saline water (Hamlet and Morris, 1996). At the extreme end, salinity in the top 1.5 m at the Timmering site was approximately 60% higher in bays irrigated with saline groundwater with EC 10 dS/m than those irrigated with low-salinity channel water. Salinity was lower at Tatura and Nathalia, where irrigation water salinities averaged 5 dS/m and 2 dS/m, respectively, and lowest at Cobram, which was irrigated with 0.5 dS/m groundwater.

Stevens et al. (2003) found similar impacts from irrigation with groundwater and recycled water on a range of horticultural crops when compared with rainfed land. On the sandy soils of the Northern Adelaide Plains, long-term irrigation (20–30 yrs) of recycled water caused the soils to exhibit similar salinities to the irrigation water, suggesting an adequate leaching fraction was applied to these irrigation systems to manage salinity resulting from irrigation.

Salt accumulation in the soil profile in response to saline irrigation water application is a potential hazard that needs to be managed by adequate leaching. Marcar and Morris (2005) suggest that salt accumulation within the soil profile below tree plantations may reach equilibrium levels if plant water uptake is balanced by leaching from rainfall and irrigation, and the accumulation tends to be confined to zones below the active rooting depth or deeper in the soil profile. Productivity will then be determined by the level of salt tolerance and growth potential of the plants grown.

8.7.2 Leaching fractions

The rate of salt accumulation in the root-zone can be quite rapid. However, with adequate leaching, on soils with good hydraulic conductivity, salts can be moved out of the root-zone. In an effluent-irrigated *E. grandis* and *Pinus radiata* plantation in southern New South Wales on a deep, medium-textured soil (Myers et al. 1998), plots were irrigated with artificially salinised effluent at a salinity (EC) of 2.2 dS/m for 14 weeks, increasing the salinity (EC_e) in the upper 35 cm of soil from 1–2 dS/m to 6 dS/m. Tree growth and water uptake were reduced as a result. However, after overirrigation with two 200-mm applications of low salinity (0.5 dS/m) effluent, the accumulated salt was leached out of the tree root-zone with a resultant increase in tree growth rates.

Provision of sufficient leaching to avoid salt accumulation (Marcar and Morris, 2005) may be restricted when:

- low infiltration rate in heavy-textured soils limits the volume of water that can be applied in an irrigation event;
- soil sodicity impacts on soil infiltration rate;

- leaching of salt below the root-zone may raise the watertable and increase the likelihood of upward capillary rise of salt;
- subsurface drainage is required to facilitate leaching of salt; and
- the root-zone is deep – trees, depending on their growth rates, can exploit much of the soil available water and hence large volumes of water may be required to leach salts.

8.7.2.1 Calculation of leaching fractions

In simple terms, the irrigation requirement is equal to the difference between the plant water requirement and net rainfall at a location multiplied by irrigation efficiency (Allen et al., 1998). Rainfall can be measured using standard meteorological equipment, but the plant water requirement is more complex.

For plant requirements to be met, additional water is needed to overcome inefficiencies in irrigation method (i.e. effective irrigation is different to actual) and to leach salts down the soil profile. For steady-state conditions, the leaching fraction (LF) is defined as the volume of drainage water (passing the root depth) divided by the volume of infiltrating irrigation water. The leaching requirement (LR) refers to an estimate of what the LF must be to ensure soil salinity remains within tolerable limits for the specific crops grown (Richards, 1954). This steady-state approach is usually a good estimate for long-term changes in soil salinity. However, it assumes uniform application and does not consider salt precipitation or dissolution, irrigation frequency effects, preferential flow, upward water flow, water chemical composition and salt removal in surface runoff (Corwin et al., 2007). To overcome these deficiencies, several transient models have been developed to improve the assessment of short-term changes in soil salinity from irrigation (e.g. Water Suitability (WATSUIT), Unsaturated Water and Solute Transport Model with Equilibrium and Kinetic Chemistry (UNSATCHEM), Trace Element Transport (TETRANS), ENVIRO-GRO; Corwin et al., 2007).

There are several simple methods for calculating the leaching requirement for a specific water salinity to maintain soil salinity at acceptable levels for the desired plant growth. The methods below give an indication of a leaching requirement for two types of irrigation (8.3):

(i) conventional irrigation (LRc), where the soil is allowed to dry out significantly between irrigations (similar to the method of Ayers and Westcot, 1985); and
(ii) high-frequency irrigation (LRf), where there is limited drying out of the root-zone between irrigations (Rhoades and Loveday, 1990).

All methods of determining leaching requirements are estimates and should be checked by measuring actual soil salinity levels and modifying leaching requirements (LR) accordingly.

An example of calculations to estimate LR is given for irrigation of a turf species – *Festuca ruba L. spp ruba* (creeping red fescue) – with recycled water. The EC_e threshold (where 80% growth is acceptable by the user) is estimated to be 3 dS/m at a worst case scenario (NRMMC and EPHC, 2006). The recycled water salinity has an EC_i of 1.5 dS/m; hence the concentration factor (Fc) is calculated as follows:

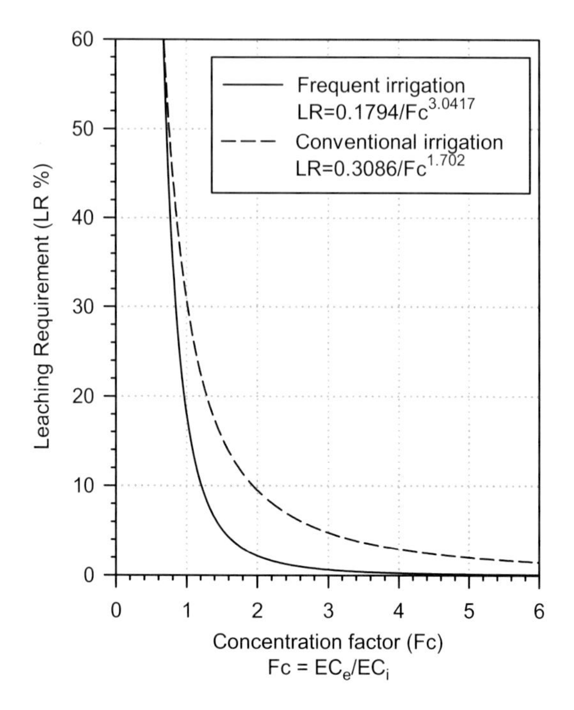

Figure 8.4 Leaching requirement determined by using a permissible root-zone factor for concentration of salts (Fc). Modified from Rhoades and Loveday (1990).

$$\text{Fc} = \text{EC}_\text{e}\ \text{threshold}/\text{EC}_i = 3/1.5 = 2$$

From Fig. 8.4, an Fc of 2 is equivalent to a leaching fraction of 3% (high-frequency irrigation; LFf) or 9.5% (conventional irrigation; LFc). The average root-zone salinity is impacted by soil water depletion between irrigation. If all other parameters remain constant and a LR is applied, high-frequency irrigation leads to a higher average integrated PAW (lower integrated root-zone salinity) as the period between wetting and drying is less and the extremes of drying less. Increasing salinity of the soil or water results in higher osmotic potential and lower PAW. Therefore, the PAW will be lower under conventional irrigation (Rhoades and Loveday, 1990) and average root-zone salinity higher.

A targeted approach to the application of the leaching requirement can add to its effectiveness. Applying leaching requirements during the wet (rainy) season will provide the following benefits:

(i) low evapotranspiration, which reduces losses to the atmosphere, maximizing downward leaching;
(ii) improved leaching, as generally the soil profile will have higher soil water content;
(iii) water applied through rainfall and irrigation will have a lower average salinity, by decreasing the soil salinity in the soil profile.

An important consideration to the application of leaching is the off-site impacts associated with the application of irrigation water that is greater than plant requirement. Although sustainable irrigation requires an appropriate leaching fraction, this leaching fraction can have off-farm implications, including:

(i) a contribution to ground watertables, depending on the depth to the watertable/aquifer. Can the regional hydrology cope with this extra water input? Will watertables rise, leading to increased salinisation of soils on a regional scale?
(ii) any movement of water out of the root-zone of plants will lead to the movement of nutrients. This movement of nutrients can lead to the eutrophication of ground and terrestrial waters.

The calculation of leaching fractions often assumes that there are no barriers in the soil profile that could potentially impede water movement, creating perched watertables. These barriers could include sodic clay, sheet limestone (calcrete), or compacted soil layers. If barriers do exist, then lateral movement of groundwater will need to be assessed to determine if the hydrology of the area will allow off-site movement of leached water.

The impact of leaching fraction (LF) for a given irrigation water salinity (EC_i) on soil salinity can also be estimated using Figure 8.5, which can also be used to estimate a leaching requirement (LR). For example, if the plant's EC_e threshold (where 80% growth is acceptable by the user) is estimated to be 3 dS/m at a worse case scenario and the recycled water salinity has an EC_i of 1.5 dS/m, a LF of 9% (similar to LRc above) would be required to maintain soil salinity (EC_e) at an average of 3 dS/m.

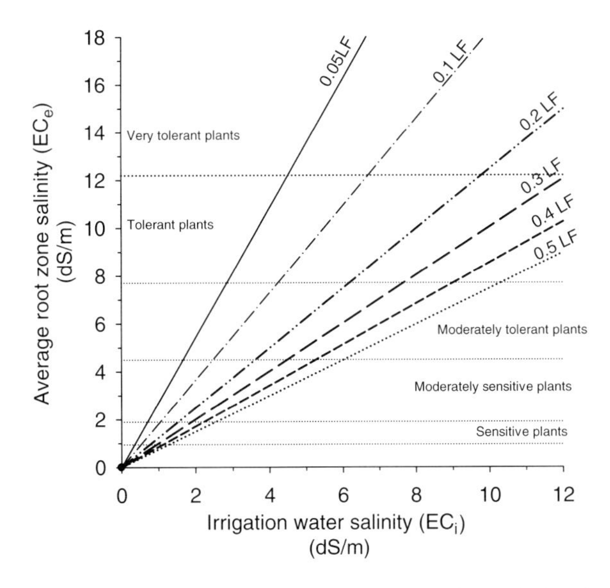

Figure 8.5 Relationship between average root-zone salinity (EC_e) and electrical conductivity of irrigation water (EC_i), as a function of leaching fraction (LF) and plant salt tolerance. Modified from Rhoades et al. (1992).

8.7.3 Modeling predictions

A number of empirically derived and process-based models are available for irrigation water scheduling, which take into account various constituents (especially salts) in the water that need to be managed in the post-application soil–plant system. Models with relevance to effluent irrigation used in Australia include Salinity And Leaching Fraction-Salinity And Leaching Fraction Prediction (SALF-SALFPREDICT) (QDNR, 1997), Model for Effluent Disposal Using Land Irrigation (MEDLI) (EPA, 2005) and Water Scheduling (WATSKED) (Myers et al., 1999; Theiveyanathan et al., 2004). Corwin et al. (2007) have compared steady-state models including the traditional model, WATSUIT, and water-production-function model for calculating leaching requirement (LR) to transient models including TETRANS and UNSATCHEM. They conclude that traditional steady-state models for estimating LR need to be re-evaluated, but that they may be appropriate to use if they can account for all the dominant mechanisms influencing the leaching of salts, which may be nearly as important as capturing the temporal dynamics of the leaching process.

8.8 Summary and conclusions

Treated wastewater (recycled water) used for irrigating plants often has higher salt concentrations than local freshwater sources. Irrigation with this type of water, therefore, requires an adequate understanding of how salt affects soils and plants. In most cases, with good irrigation practice and drainage this resource can be used sustainably. Routine monitoring and/or estimation of plant root-zone salinity using appropriate techniques are also necessary to provide meaningful and timely information. This chapter reviewed the responses of plants to salinity when irrigated with saline wastewater and requirements for managing root-zone salinity at levels below the threshold to maintain adequate growth. Indirect effects through changes in soil physical and nutritional effect are covered in Chapter 9. Field and modeling studies suggest that TWW can be used sustainably to irrigate, as long as species choice is tailored to the severity of salinity and irrigation practice allows for adequate leaching of salts. Considerable research in this area has made it possible to develop irrigation systems that can cope with a range of treated sewage water (TSW) salinities if soil properties and regional groundwater allow. A range of models and reference tables exist to assist in assessing long-term impact from irrigation of TSW of a range of salinities on plant survival and growth. However, in some cases of atypically high salinity (i.e. where the domestic water supply is already high in salinity, industrial pollution loads or saline groundwater enters sewers) or poorly drained soils, salinity may limit the use of recycled water for irrigation.

Acknowledgements

We thank Dr Guna Magesan (Scion, Rotorua, New Zealand) for encouraging us to contribute this chapter and for his comments on an early draft.

References

Allen, R., Pereira, L., Raes, D. and Smith, M. (1998) *Crop Evapotranspiration: Guidelines for Computing Crop Water Requirements*. FAO, Rome.

ANZECC, ARMCANZ (2000a) *Australian and New Zealand Guidelines for Fresh and Marine Water Quality. Volume 2, Chapter 8. Aquatic Ecosystems — Rationale and Background Information.* (National Water Quality Management Strategy (NWQMS)) Australian and New Zealand Environment and Conservation Council, Agriculture and Resource Management Council of Australia and New Zealand.: Canberra, Australia.

ANZECC, ARMCANZ (2000b) *Australian and New Zealand Guidelines for Fresh and Marine Water Quality. Volume 3, Chapter 9. Primary Industries — Rationale and Background Information (Irrigation and general water uses, stock drinking water, aquaculture and human consumers of aquatic foods).* (National Water Quality Management Strategy (NWQMS)) Australian and New Zealand Environment and Conservation Council, Agriculture and Resource Management Council of Australia and New Zealand: Canberra, Australia.

APHA (1998) *Standard Methods for the Examination of Water and Wastewater*. 20th ed. American Public Health Association. American Water Works Association. Water Environment Federation, Washington, DC.

Asano, T., Burton, F.L., Laeverenz, H.L., Tsuchihashi, R. and Tchobanoglous, G. (2007) *Water Reuse. Issues, Technologies and Applications*. Metcalf and Eddy. McGraw Hill, New York.

Ayers, R.S. and Westcot, D.W. (1985) *Water quality for agriculture. FAO irrigation and drainage paper 29 rev. 1*. Food and Agriculture Organisation of the United Nations, Rome. www.fao.org/DOCREP/003/T0234E/T0234E00.htm

Bandara, G., Morris, J., Stackpole, D. and Collopy, J. (2002) *Soil Profile Monitoring at Irrigated Plantation Sites in Northern Victoria: Progress Report 2001*. CFTT Report No. 2002/025. Forest Science Centre, Melbourne, Australia.

Benyon, R.G., Marcar, N.E., Crawford, D.F. and Nicholson, A.T. (1999) Growth and water use of *Eucalyptus camaldulensis* and *E. occidentalis* on a saline discharge site near Wellington, NSW, Australia. *Agricultural Water Management* **39**, 229–244.

Biswas, T., Schrale, G. and Dore, D. (2005) *Measuring the Effects of Improving Water Use Efficiency on Root-Zone Salinity*. National Program on Sustainable Irrigation. Research Bulletin. Land & Water. Australia. http://products.lwa.gov.au/files/PF050992.pdf.

Cass A, Walker, R.R. and Fitzpatrick, R.W. (1995) Vineyard soil degradation by salt accumulation and the effect on the performance of the vine. pp. 153–160. In: 9th *Australian Wine Industry Technical Conference, Adelaide, Australia*. Published 1996. Australian Wine Industry Technical Conference Inc., Glen Osmond, Australia.

Christen, E., Ayars, J., Hornbuckle, J. and Biswas, T. (2006) Design and management of reclaimed water irrigation systems. In: *Growing Crops with Reclaimed Wastewater* (ed. D.P. Stevens). CSIRO Pub., Melbourne, Australia.

Corwin, D.L., Rhoades, J.D. and Simunek, J. (2007) Leaching requirement for soil salinity control: Steady-state versus transient models. *Agricultural Water Management* **90**, 165–180.

Doorenbos, J. and Pruitt, W.O. (1977) *Guidelines for Predicting Crop Water Requirements. FAO Irrigation and Drainage Paper 24*. FAO. Rome, Italy.

EPAQ (2005) *Queensland Water Recycling Guidelines*. Queensland Government Environmental Protection Agency, Brisbane, Australia.

Feikema, P.M., Baker, T.G. and Stackpole, D.J. (2005) Effects of salinity on field grown *Eucalyptus camaldulensis, E. globulus* and *E. grandis*. In: *Proceedings of the International Salinity Forum: Managing Saline Soils and Water*, 25–27 April 2005, Riverside, CA.

Hamlet, A. and Morris, J. (1996) *Effects of Salinity on the Growth of Four Eucalyptus Species on Irrigated Sites in Northern Victoria*. Trees for Profit Research Centre Report. University of Melbourne, Melbourne, Australia.

Heath, J. and Heuperman, A. (1996) Serial biological concentration of salts – a pilot project. *Australian Journal of Soil and Water Conservation* **9**, 27–31.

Heuperman, A., Mann, L., Greenslade, R. and Heath, J. (2002) *Piloting farm scale serial biological concentration. Final Report for MDBC Project D115*. Victorian Department of Natural Resources, Melbourne, Australia.

Jayawardane, N.S., Biswas, T.K., Blackwell, J. and Cook, F.J. (2001) Management of salinity and sodicity in a land FILTER system, for treating saline wastewater on a saline-sodic soil. *Australian Journal of Soil Research* **39**, 1247–1258.

Jensen, M.C., Middleton, J.E. and Pruitt, W.O. (1961) *Scheduling Irrigation from Pan Evaporation.* Washington Agricultural Experiment Station Circular No. 386.

Jones, A.J. and Bauder, J.W. (1987) Computer-assisted irrigation scheduling: an educational tool. *Applied Agricultural Research* **2**, 260–271.

Lambert, M. and Turner, J. (2000) *Commercial Forest Plantations on Saline Lands.* CSIRO Publishing, Collingwood, Australia.

Lundstrom, D.R., Stegman, E.C. and Werner, H.D. (1981) Irrigation scheduling by the checkbook method. In: *Proceedings of the Irrigation Scheduling Conference*, pp. 187–93. American Society of Agricultural and Biological Engineers.

Maas, E.V. (1990) *Salt, Boron and Chloride Tolerance in Plants.* USDA, ARS www.ars.usda.gov/Services/docs.htm?docid=8908.

Maas, E.V. and Hoffman, G.J. (1977) Crop salt tolerance – Current assessment. *Journal of the Irrigation and Drainage Division* **103**, 115–134.

Mahmood, K., Morris, J., Collopy, J. and Slavich, P. (2001) Groundwater uptake and sustainability of farm plantations on saline sites in Punjab province, Pakistan. *Agricultural Water Management* **48**, 1–20.

Marcar, N.E. and Crawford, D.F. (2004) *Trees for Saline Landscapes.* RIRDC, Canberra, Australia.

Marcar, N.E. and Morris, J. (2005) Plantation productivity in saline landscapes. In: *New Forests: Wood Production and Environmental Services* (eds E.K.S Nambiar and I. Ferguson), pp. 51–74. CSIRO. Collingwood, Australia.

Marcar, N.E. and Termaat, A. (1990) Effects of root-zone solutes on Eucalyptus camaldulensis and Eucalyptus bicostata seedlings: responses to Na+, Mg2+ and Cl−. Plant and Soil **125**, 245–254.

Marschner, H. (1995) *Mineral Nutrition of Higher Plants.* Academic Press, New York.

Munns, R. (2002) Comparative physiology of salt and water stress. *Plant Cell Environment* **25**, 239–250.

Myers, B.J., Benyon, R.G., Theiveyanathan, S., Criddle, R.S, Smith, C.J. and Falkiner, R.A. (1998) Response of effluent-irrigated *Eucalyptus grandis* and *Pinus radiata* to salinity and vapour pressure deficits. *Tree Physiology* **18**, 565–573.

Myers, B.J., Bond, W.J., Benyon, R.G., Falkiner, R.A., Polglase, P.J., Smith, C.J., Snow, V.O. and Theiveyanathan, S. (1999) *Sustainable Effluent-Irrigated Plantations: an Australian Guideline.* CSIRO Forestry and Forest Products, Canberra, Australia.

NRMMC, EPHC (2006) *Australian Guidelines for Water Recycling: Managing Health and Environmental Risks. Phase 1. Natl. Water Quality Management Strategy 21.* Natural Resource Management Ministerial Council. Environmental Protection and Heritage Council Australia Health Ministers' Conference, Canberra, Australia.

Peverill, K.I., Sparrow, L.A. and Reuter, D.J. (1999) *Soil Analysis: An Interpretation Manual.* CSIRO Publ. Collingwood, Australia.

Polglase, P.J., Theiveyanathan S., Benyon, R.G. and Falkiner, R.A. (2002) *Irrigation Management and Groundwater Uptake in Young Tree Plantations Growing Over High Watertables.* Rural Industries Research and Development Corporation RIRDC Publication No. 02/146. RIRDC project No CSF-54A. Canberra, Australia.

Prescod, M.B. (1992) *Wastewater Treatment and Use in Agriculture.* Number 47 in FAO Irrigation and Drainage Paper. FAO, Rome.

QDNR (1997) *Salinity Management Handbook.* Queensland Department of Natural Resources Brisbane, Australia.

Qureshi, R.H. and Barrett-Lennard, E.D. (1998) *Saline Agriculture for Irrigated Land in Pakistan.* Monograph No. 50. Centre for International Agricultural Research (ACIAR), Canberra, Australia.

Rayment, G.E. and Higginson, F.R. (1992) *Australian Laboratory Handbook of Soil and Water Chemical Methods.* Inkata Press, Melbourne, Australia.

Rengasamy, P. (2006) Soil salinity and sodicity. In: *Growing Crops with Reclaimed Wastewater* (ed. D.P. Stevens). CSIRO Pub. Melbourne.

Rhoades, J., Chanduvi, F. and Lesch, S. (1999) *Soil Salinity Assessment: Methods and Interpretation of Electrical Conductivity Measurements.* FAO, Rome.

Rhoades, J.D. and Loveday, L. (1990) Salinity in irrigated agriculture. In: *Irrigation of Agricultural Crops*, vol. 30 (eds A.R. Stewart and D.R. Nielsen), pp. 1089–1142. ASA, CSSA, and SSSA. Madison, WI.

Rhoades, J.D., Kandiah, A. and Mashali, A.M. (1992) *The Use of Saline Waters for Crop Production*. FAO Irrigation and Drainage paper. FAO, Rome.

Shaw, R.J. (1999) Soil salinity – electrical conductivity and chloride. In: *Soil Analysis – an Interpretation Manual* (eds K. Peverill, L. Sparrow and D. Reuter), pp. 129–145. CSIRO Pub. Collingwood, Australia.

Slavich, P. and Petterson, G.H. (1993) Estimating the electrical conductivity of saturated paste extracts from 1:5 soil:water suspensions and texture. *Australian Journal of Soil Research* **31**, 73–81.

Stevens, D. (2006) *Growing Crops with Reclaimed Wastewater*. CSIRO Pub., Melbourne, Australia.

Stevens, D.P., McLaughin, M.J. and Smart, M. (2003) Effects of long-term irrigation with reclaimed water on soils of the Northern Adelaide Plains, South Australia. *Australian Journal of Soil Research* **41**, 933–948.

Theiveyanathan, S., Benyon, R.G., Myers, B.J., Polglase, P.J., Marcar, N.E. and Falkiner, R.A. (2004) An irrigation-scheduling model for sustainable application of saline water to tree plantations. *Forest Ecology and Management* **193**, 97–112.

Yaron, D. (ed). (1981) *Salinity in Irrigation and Water Resources*. Marcel Dekker, New York.

Chapter 9
Physical aspects

Guy J. Levy and Shmuel Assouline

9.1 Introduction

With the increased necessity to use treated wastewater (TWW) for irrigation, especially in arid and semiarid regions, farmers are faced with unique and unfamiliar problems, among which are the possible degradation of soil structural stability and associated physical properties. Probable risks for undesired and adverse changes in the soil and its hydraulic properties, following irrigation with TWW, may stem from the higher load of dissolved organic matter (DOM), suspended solids, sodium adsorption ratio (SAR) and salinity in the TWW compared with its freshwater of origin. The level of suspended solids and DOM in the TWW depends on the quality of the raw sewage water and the degree of its treatment. Conversely, the salt load and composition in the TWW is similar to that of the raw sewage as it is not affected by the treatment and may reach, as for instance was observed in Israel, a level twice that of its freshwater of origin (Feigin et al., 1991).

Processes that may negatively affect soil structural stability and hydraulic properties include clay swelling and dispersion, and aggregate breakdown through slaking or impact of water drops of high kinetic energy (e.g. rain or overhead sprinkler irrigation systems). Irrigation with TWW is expected to enhance the processes of clay swelling and dispersion in comparison to irrigation with freshwater (FW) because of the higher SAR and the presence of DOM in the TWW. Sodic conditions, especially in the absence of electrolytes (Quirk and Schofield, 1955), render soil clays greater sensitivity to swelling and dispersion (Shainberg and Letey, 1984), which in turn adversely affect soil hydraulic properties, namely, the water retention curve and the hydraulic conductivity function (Bresler et al., 1982; Russo, 2005). In addition, there is evidence in the literature that in the presence of dissolved humic substances, as is the case when TWW is used for irrigation, clay dispersivity is enhanced and a higher electrolyte concentration in the soil solution is needed to maintain soil clays in a flocculated state (e.g. Durgin and Chaney, 1984; Frenkel et al., 1992; Gu and Doner, 1993; Tarchitzky et al., 1993, 1999). This effect was attributed to the interaction of negatively charged organic molecules with the positively charged edges of the clay, which prevents the edge-to-face association of clay particles responsible for flocculation (Tarchitzky et al., 1999). The magnitude of the adverse effects of humic

---

Treated Wastewater in Agriculture, First Edition, edited by Guy J. Levy, Pinchas Fine and Asher Bar-Tal

substances on soil properties depends on their concentration, solution pH, and the type of exchangeable cation (Tarchitzky et al., 1993, 1999). It has also been suggested that the presence of DOM in the TWW could have an additional indirect effect on the soil via its effect on the SAR (Halliwell et al., 2001). These authors suggested that the effective SAR of the TWW is greater than that commonly calculated based on the total concentration of cations in the solution (Halliwell et al., 2001). The increase in effective SAR can occur through the formation of complexes between Ca^{2+} and Mg^{2+} and the organic ligands present in the TWW (Metzger et al., 1983). The difference between the commonly calculated SAR and the effective one could reach 48% due to the great affinity of the divalent cations to organic ligands (Fotovat et al., 1996).

During the irrigation season it is expected that the higher electrolyte concentration in the TWW compared with FW will mitigate the above-mentioned possible adverse effects of the higher SAR and DOM levels in the TWW on clay swelling and dispersion. Conversely, during the rainy season when the soil is exposed to rainwater (i.e. water almost free of electrolytes), the sensitivity of the soil clay to swelling and dispersion is high (Shainberg and Letey, 1984), as is the susceptibility of aggregates to slaking (Levy et al., 2003); thus, the aforementioned adverse effects of TWW on soil structural stability and related hydraulic properties could be exacerbated.

The current chapter reviews the state-of-the-art knowledge on the impact of irrigation with TWW on the main soil physical and hydraulic properties from both empirical and modeling perspectives.

9.2 Soil structural stability

The term soil structural stability is commonly used to describe the ability of the soil to retain its arrangement of solid and pore space when exposed to different stresses such as tillage, traffic, and wetting or drying (Kay and Angers, 1999). The term structural stability is often considered synonymous with aggregate stability, especially in cases where the stress to which the soil is exposed arises from wetting under different conditions (Kay and Angers, 1999).

Aggregate breakdown by water may result from a variety of physical and physicochemical mechanisms. Four main mechanisms have been identified: (i) slaking, i.e. breakdown caused by compression of entrapped air during fast wetting (Panabokke and Quirk, 1957); (ii) breakdown by differential swelling during fast wetting (Kheyrabi and Monnier, 1968); (iii) breakdown by impact of raindrops (McIntyre, 1958); and (iv) physicochemical dispersion due to osmotic stress upon wetting with low electrolyte water (Emerson, 1967). These mechanisms differ in the type of energy involved in aggregate disruption. For instance, swelling can overcome attractive pressures in the magnitude of MPa (Rengasamy and Olsson, 1991), whereas slaking and impact of raindrops can overcome attractive pressures in the range of kPa only (Rengasamy and Sumner, 1998). In addition, the various mechanisms may differ in the size distribution of the disrupted products (Farres, 1980; Chan and Mullins, 1994), and in the type of soil properties affecting the mechanism (Le Bissonnais, 1996). There are numerous methods for determining aggregate stability, the most widely used ones being wet sieving (Kemper and Koch, 1966; Kemper and Rosenau, 1986), the drop test technique

(Farres, 1980), and application of ultrasonic energy (North, 1976). Different processes dominate in the breakdown of the aggregates in the various stability tests (Loch, 1994). Thus, unexpectedly, use of different methods for determining aggregate stability has resulted in different rankings of soils studied (Amezketa et al., 1996; Le Bissonnais and Arrouays, 1997).

In general, organic substances have been considered as cementing agents that improve aggregate stability (e.g. Kemper and Koch, 1966; Goldberg et al., 1988; Bronick and Lal, 2005). However, humic substances have been observed to have a dual role as aggregating and disaggregating agents (Emerson, 1983; Oades, 1984). Some studies have indicated that the aforementioned duality is related to the concentration of humic substances; at low concentrations (<0.05–0.10 g/kg) the humic substances stabilize aggregates, whereas at higher concentrations they have a dispersive effect (Piccolo and Mbagwu, 1989, 1994; Mbagwu et al., 1993; Piccolo et al., 1996). Tarchitzky et al. (1999) found that humic substances at concentrations below 0.01 g/kg caused Na-affected clays or soils to disperse, but did not affect Ca-clays.

Levy and Mamedov (2002) studied aggregate resistance to slaking of six soils differing in clay content. Results showed that for coarse- and medium-textured soils (<25% clay), stability ratio (SR), an index representing aggregate stability ($0 < SR < 1$), was relatively low and unaffected by the quality of the irrigation water (Fig. 9.1). Apparently, in soils with unstable aggregates, irrigation with TWW, which induces conditions favoring clay dispersion via increased sodicity, plays a minor role in determining aggregate stability.

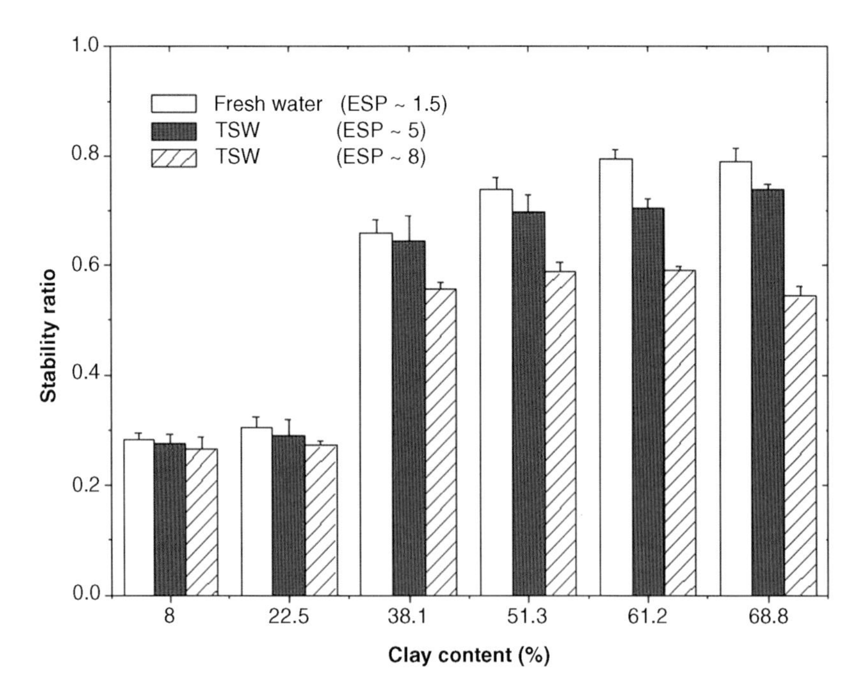

Figure 9.1 Effects of irrigation water quality on aggregate stability (expressed in terms of stability ratio) for six soils represented by their clay content. Reproduced with permission from Levy, G.J, and Mamedov, A.I. (2002) High-energy-moisture-characteristics aggregate stability as a predictor for seal formation. *Soil Science Society of America Journal* **66**, 1603–1609.

Conversely, in the fine-textured soils (>38% clay), the SR was relatively high (>0.5) and related to the fact that the clay acts as a cementing and binding agent in the soil (Kemper and Koch, 1966). Stability of the TWW-irrigated clay soils was lower than that of the FW-irrigated samples. Furthermore, the stability of the aggregates in the TWW-irrigated samples decreased with the increase in soil exchangeable sodium percentage (ESP) (Fig. 9.1). Evidently, in clay soils where the clay stabilizes the aggregates, conditions favoring dispersive behavior of clay (high sodicity and the presence of humic substances in the TWW) adversely affected aggregate resistance to slaking. Levy et al. (2003), who studied the combined impacts of salinity and sodicity on aggregate susceptibility to slaking, arrived at a similar conclusion.

In an additional study, Levy et al. (2006) noted that the greater sensitivity to slaking of aggregates from TWW-irrigated samples is also related to the intensity of cultivation. The greater sensitivity to slaking of the aggregates from the TWW-irrigated samples was noted mainly in aggregates taken from intensively cultivated fields (e.g. field crops). In samples taken from orchards where soil cultivation is minimal and the aggregates are not subjected to frequent disturbance, stability of aggregates from TWW- and FW-irrigated samples was comparable (Levy et al., 2006). Bhardwaj et al. (2007) were also unable to observe differences between the stability of aggregates exposed to irrigation with TWW and those irrigated with FW for a clay soil taken from an orchard.

9.3 Soil hydraulic properties

9.3.1 *Water retention curve*

The water retention curve is a fundamental hydraulic property of the soil that describes the relationship between water content and matric potential (Brooks and Corey, 1964; van Genuchten, 1980; Assouline et al., 1998), with the latter being directly related to surface tension. Furthermore, the hydraulic conductivity function can be derived from the water retention curve (Mualem, 1986; Assouline, 2001). The expression suggested by van Genuchten (1980) for the water retention curve, $S_e(\psi)$, is:

$$S_e(\psi) = 1/\{1 + (\alpha|\psi|^n\}^{(1-1/n)} \quad (9.1)$$

where ψ is the capillary head, S_e is the effective saturation degree, and α and n are empirical parameters. Applying the model of Mualem (1986) to equation 9.1 leads to the following expression for the hydraulic conductivity function, $K(S_e)$:

$$K(S_e) = K_s S_e^{0.5}[1-(1-S_e{}^{(1/n)})^n]^2 \quad (9.2)$$

where K_s is the saturated hydraulic conductivity of the soil.

Lima et al. (1990) studied the effect of salinity and sodicity on the water retention curve. They investigated the combined effects of SAR and salinity on soil–water retention, soil–water diffusivity, Philip's sorptivity (Philip, 1969), and hydraulic conductivity of Yolo loam. These effects were expressed through changes in the parameters α and n of the van Genuchten equation (equation 9.1) that was fitted to the data. As the SAR increased, α

decreased and n increased. The decrease in α corresponds to an increase in the soil air entry value, indicating narrowing of the larger pores. The increase in n could represent narrowing of the pore size distribution itself as higher n values correspond generally to lighter soils. It is interesting to note that whereas the decrease in α could be related to a decrease in K_s, the increase in n would reflect an increase in $K_r = K(S_e)/K_s$ (equation 9.2). Therefore, the net effect on the hydraulic conductivity $K(S_e)$ will depend on the relative importance of the effects on α and n. They also found that the increase in salinity augmented the non-linearity in log(K), especially for higher water contents. They suggested that this "reveals predominance of salt effects on size of larger pores", in accordance with Russo and Bresler (1977). Finally, estimates of hydraulic conductivity based on the water retention curve and the diffusivity function were found to be significantly higher than measured values. This overestimation was related to swelling during the retention measurements and lack of chemical equilibrium during wetting front advance in the diffusivity measurements.

The presence of DOM in the soil solution could affect the wetting properties of the solution such as surface tension and contact angle (Tschapek et al., 1978; Chen and Schnitzer, 1978; Rosen 1989; Anderson et al., 1995; Chenu et al., 2000). Changes in contact angle can also indicate changes in pore surface roughness. Solutes in the wetting liquid, like surfactants and polymers, may coat the solid surfaces with a thin film that will reduce pore surface roughness and affect the contact angle. Consequently, the possible effects of organic matter (OM), present in the TWW, on the water surface tension, surface roughness and contact angle will inevitably reflect on the water retention curve of TWW-irrigated soils.

Coppola et al. (2004) have investigated the impact of TWW on the water retention curve of five representative soils of Sardinia. Comparison with the initial curves that were obtained with FW showed that using TWW led to a change in the pore size distribution, namely, to a translation towards narrower pores.

The expected effects of DOM on the wetting properties of solutions can have an effect not only on the water retention curve but also on its hysteresis. Although this has not been studied directly, we may infer from the results of Ojeda et al. (2006) on the effect of surface applied sewage sludge on the drying and the wetting water retention curves of two soils, assuming that part of the effects of the sewage sludge can be related to the DOM fraction. Their results showed that practically no effect was found on loamy sand. Conversely for a loam, all the treatments affected the retention curves, compared to the untreated soil, by lowering the bulk density and increasing the air entry value. Lowering the soil bulk density being generally related to a decrease in the air entry value, one can assume here that the addition of the solid residues in the sludge increased the porosity and reduced the bulk density, but that the effect of the DOM fraction contributed more in reducing the larger pores so that the final outcome was the increase of the air entry value despite the drop in bulk density. Another interesting result was that all treatments seemed to affect mainly the main wetting branch and practically not the main drying branch of the hysteresis loop, indicating that the sewage sludge application has affected the hysteretic domain of the loam.

9.3.2 Hydraulic conductivity (saturated and unsaturated)

The effects of salinity and sodicity on soil hydraulic properties and flow processes have been studied extensively (e.g. Bresler et al., 1982; Shainberg and Letey, 1984; Jury et al., 1991; Malik et al., 1992; Levy et al., 2005; Russo, 2005). In the early 1950s, it had been generally accepted that ESP 15 separates sodic from non-sodic soils (e.g. US Salinity Laboratory, 1954). In recent years, numerous studies and notably those of Shainberg and Letey (1984) and Mace and Amrhein (2001), have shown that internal swelling and pore clogging can already occur at SARs of 5 and 8, values that characterize the range of SAR levels found in TWW. Mace and Amrhein (2001) have shown that the impact of such SAR levels on the hydraulic conductivity (HC) of a calcareous smectitic soil increases as the electrolyte concentration (C) decreases (Fig. 9.2). These findings indicate that merely from sodicity and salinity considerations, irrigation with TWW could harm soil HC, especially during rainfall.

The quantitative aspects of the impact of salinity and sodicity on soil hydraulic properties were tackled by two types of approaches, empirical and conceptual. The empirical approach, as represented by the work of Suarez and Simunek (1997), used the experimental results of McNeal (1968), to portray the effect of mixed Na/Ca salt solutions on the HC of California soils. The more conceptual, physically based approach, was introduced by Russo and Bresler (1977), and is presented in detail by Russo (1988). The physicochemical interactions between mixed Na/Ca solutions and the soil matrix are derived from theoretical considerations based on the mixed ion diffuse double-layer theory, the structure of clay particles, the pore size distribution of the soil, and

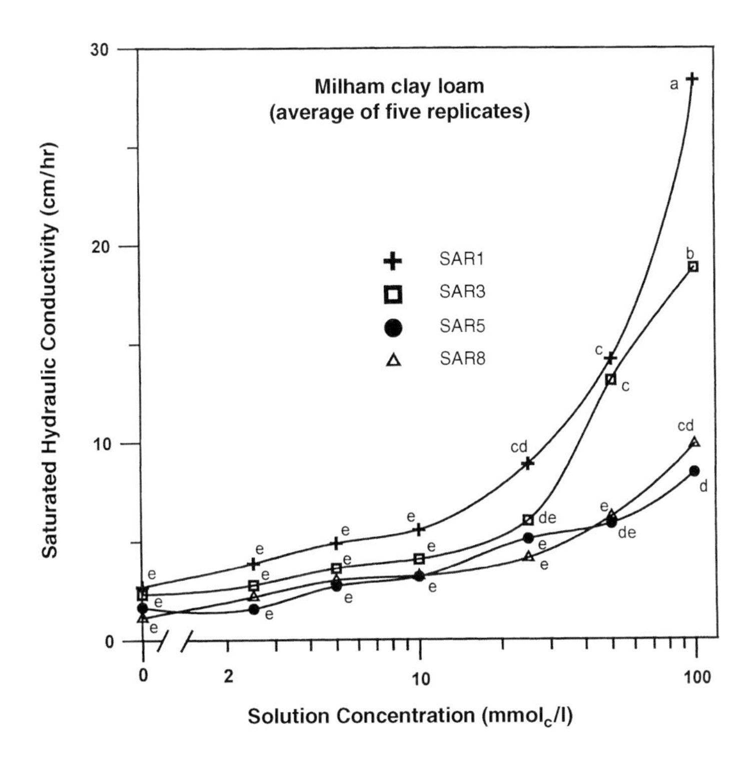

Figure 9.2 Saturated hydraulic conductivity as a function of sodium adsorption ratio (SAR) and electrolyte concentration in solution. Different letters indicate statistical significant difference ($P < 0.05$). Reproduced with permission from Mace, J.E., and Amrhein, C. (2001) Leaching and reclamation of a soil irrigated with moderate SAR waters. *Soil Science Society of America Journal* **65**, 199–204.

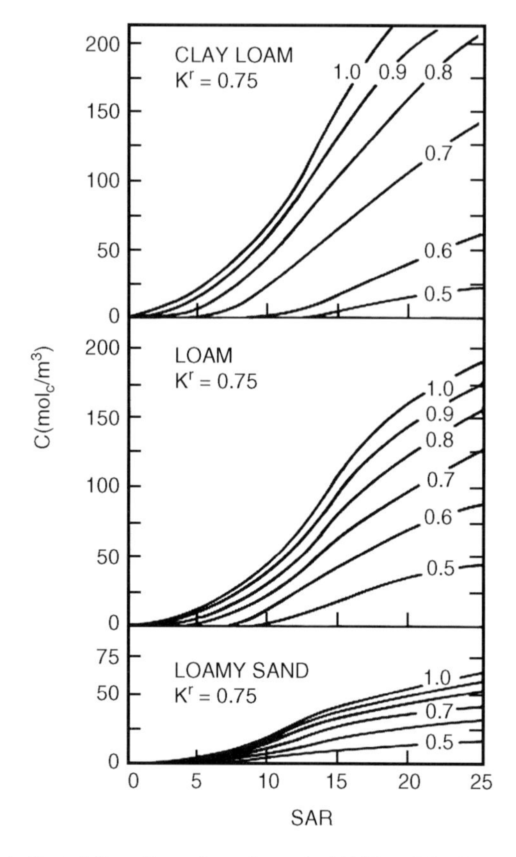

Figure 9.3 Combined effect of C, sodium adsorption ratio (SAR) and the degree of water saturation (the numbers labeling the curves) at which 25% reduction in hydraulic conductivity occurs, for three soils of different texture. Reproduced from Russo, D. (1988) Numerical analysis of nonsteady transport of interacting solutes through unsaturated soil. 1. homogeneous systems. *Water Resources Research* **24**, 271–284; American Geophysical Union.

hydrodynamic principles. The impact of the soil matrix–soil solution interactions on the soil hydraulic properties led to a reduction of the soil HC compared to that of an inert reference soil. This reduction increases as the soil water content, the SAR of the soil solution, and the clay fraction of the soil increase, and its solute concentration decrease, This trend is illustrated in Figure 9.3, where the respective solute concentration, SAR, and water content required to attain a reduction of 25% in the soil HC are depicted for three soil types, the textures of which ranged from loamy sand to clay loam.

The combined effects of sodicity and DOM on soil HC have been the subject of growing interest in recent years. Most studies on the impact of irrigating with TWW on both saturated and unsaturated HC reported decreased conductivity (e.g. de Vries, 1972; Rice, 1974; Lance et al., 1980; Vinten et al., 1983; Cook et al. 1994; Balks et al., 1997; Magesan et al., 2000; Coppola et al., 2004). Notable exceptions are the study of Mathan (1994) that reported an increase in the HC after >10 of irrigation with TWW, and that of Levy et al. (1999) who reported that irrigation with TWW had comparable effects to FW on the HC of three different Israeli soil types.

The effects of irrigation with TWW on soil-saturated HC are expected to take place both during the irrigation season and during the subsequent rainy season. During the irrigation season it is mainly the presence of suspended solids in the TWW (after secondary treatment), especially in poorly TWW, that could adversely affect soil physical and hydraulic properties. Suspended solids present in the TWW may accumulate and physically block water-conducting pores, thereby leading to a sharp decrease in soil HC (de Vries, 1972; Rice, 1974; Vinten et al., 1983; Magesan et al., 2000; Viviani and Iovino, 2004). Blocking of the pores occurs mostly at the upper soil layer (Rice, 1974; Vinten et al., 1983; Siegrist, 1987), and depends on the load of suspended solids in the wastewater (Rice, 1974), and on the environmental conditions prevailing in the soil, that determine the rate of breakdown of the suspended solids (de Vries, 1972). The harmful impact of irrigation with TWW containing high levels of suspended solids on the HC depends on soil texture; the finer the soil texture the greater the decrease in the HC (Levy et al., 1999). Furthermore, these authors also noted that although suspended solids in wastewater are considered, in general, as particles $>1.2\,\mu m$ (Greenberg et al., 1985), suspended solids in the size fraction range of $0.45\,\mu m$ (the size that separates dissolved from particulate OM) to $1.2\,\mu m$ also lead to a significant clogging of conducting pores in soils and to a subsequent decrease in the HC, especially in medium- and fine-textured soils (Levy et al., 1999). Thus, an effective level of suspended solids in TWW could, at times, be much larger than that specified by common analysis, and could, therefore, pose a greater hazard to the soil than anticipated.

In cases where TWW that had been subjected to a higher level of treatment was used, the observed greater reduction in HC following irrigation with TWW compared to leaching with a solution of a salt concentration and composition similar to that found in the TWW was ascribed to the presence of DOM in the TWW. It has been suggested that the DOM enhanced clay swelling and dispersion, even though the electrolyte concentration in the TWW was nearly double that in FW, led to the reduction in HC during leaching with TWW (Tarchitzky et al., 1999). Another possible explanation for the decrease in HC during leaching with TWW of good quality, but of high C-to-N ratio, is the excess growth of microorganisms in the soil due to the presence of nutrients in the TWW, which could also clog, at least partially, water-conducting pores (Magesan et al., 1999).

The impact of irrigation with TWW on soil HC during the subsequent rainy season was evaluated mostly in laboratory studies where soil samples from fields subjected to long term irrigation with TWW were subjected to leaching with deionized water (simulation of rain) (Levy et al., 2005). Results varied and depended on soil texture and conditions prevailing in the soil. For instance, in sandy clay no differences were noted in the relative HC between TWW- and FW-irrigated samples taken from both a cultivated and a non-cultivated field. By comparison, in a clay soil, the relative HC in samples from TWW-irrigated fields was significantly lower than that in samples from FW-irrigated fields. Bhardwaj et al. (2007) studied the HC of intact cores of a clay soil taken from an orchard; the intact soil cores were collected at locations in close proximity to microsprinklers or drippers by inserting stainless steel cylinders into the 0.05–0.15 m surface horizon of the soil. The results showed that the method of irrigation (drip versus microsprinkler), rather than irrigation water quality (TWW versus FW) determined the response of the soil to leaching with deionized water. Moreover, Bhardwaj et al. (2008) showed that replacing saline-sodic irrigation water that had been used for many years with a lesser saline-sodic

TWW, albeit with higher loads of OM and suspended solids, helped the soil regain its structure and improved its hydraulic conductivity.

When the slaking of aggregates was prevented in the HC studies (i.e. the samples in the columns were saturated at a slow rate), the impact of long-term irrigation with TWW on the reduction in HC was of a smaller magnitude compared to conditions where the samples were wetted at a fast rate and aggregate slaking was allowed (Shainberg et al., 2001). The reduction in HC under conditions where slaking was prevented was ascribed to greater clay swelling in the TWW-irrigated samples compared with the FW-irrigated ones, probably due to higher exchangeable sodium percentage in the former samples.

Studies on the impact of irrigation with TWW on the unsaturated HC have also yielded inconclusive results. Magessan et al. (1999) who studied two soils from New Zealand did not detect changes in the unsaturated HC after 5 yrs of irrigation with TWW, albeit for the same soil, Cook et al. (1994) reported a 50% decrease in the saturated HC after 32 months of irrigation with the same TWW. Conversely, Coppola et al. (2004) reported a decrease in both saturated and unsaturated HC of five different soil types from Sardinia following irrigation with TWW, which in turn led to much higher than expected accumulation of solutes (namely heavy metals and boron) in the soil profile.

As can be noted from the above discussion, the combined effects of salinity, sodicity and DOM present in the TWW on soil HC are complex and depend on the quality of the TWW, the soil properties, the conditions prevailing in the soil, and the sequences between irrigation and rainfall. Thus, to date, neither empirically based, nor conceptual models have been developed to assist in predicting the response of soil HC function to the combined effect of salinity, sodicity, and DOM stemming from irrigation with TWW.

9.4 Soil surface sealing, infiltration and runoff

On land, the main components of the hydrological cycle are precipitation, evaporation, infiltration, and runoff. The upper soil layer at the soil surface controls the partitioning of the incoming water (precipitation) into its three outgoing components: infiltration, surface runoff, and evaporation. Infiltration, defined as the flow of water through the soil surface, is an important process in the hydrologic cycle, especially from the agricultural perspective, as it significantly affects the amount of water available for crop production.

When a bare soil is exposed to high-energy rainfall, the formation of a structural seal at the soil surface is the main factor that controls the soil infiltration rate (IR) (McIntyre, 1958; Morin and Benyamini, 1977; Agassi et al., 1981; Ben-Hur et al., 1985; Miller, 1987). The seal formed is a relatively thin layer characterized by greater density and compaction and much lower permeability than the underlying soil (McIntyre, 1958; Chen et al., 1980; Gal et al., 1984; Onofoik and Singer, 1984; Levy et al., 1988; Wakindiki and Ben-Hur, 2002; Assouline, 2004), leading to a drastic reduction in the soil's IR during its formation (Morin and Benyamini, 1977; Levy et al., 1986; Assouline and Mualem, 1997, 2000; Assouline, 2004), and subsequently to the formation of high levels of runoff and erosion (Morin et al., 1981; Assouline and Mualem, 2006; Assouline and Ben-Hur, 2006; Assouline et al., 2007). Two main mechanisms are involved, at different relative intensity depending on the prevailing conditions, in the process of seal formation: (i) physical disintegration of surface soil aggregates, caused by the impact energy of the raindrops (McIntyre, 1958; Betzalel et al., 1995), and (ii) the physicochemical dispersion of soil clays, which migrate into the soil with the infiltrating water and clog the pores immediately beneath the surface to form the "washed-in" zone (Agassi et al., 1981). Consequently, the rate and volume of water that infiltrates into the soil depend mainly on rain properties, the hydraulic properties and conditions at the soil surface and to a lesser degree the hydraulic properties of, and water content distribution, within the soil profile.

The first mechanism is mechanical in nature and is determined by rain intensity, i.e. the wetting rate of the surface aggregates (Loch, 1994; Levy et al., 1997), raindrop kinetic energy (Mamedov et al., 2000a, b) and soil aggregates stability (Loch, 1994). High OM content in the soil is known to improve soil aggregate stability (Kemper and Koch, 1966; Tisdall and Oades, 1982), and consequently reduce soil susceptibility to seal formation. However, recent studies have shown that long-term use of TWW may enhance the decomposition of soil OM, known as the "priming effect" (a more detailed discussion of the "priming effect" appears in Chapter 13). Consequently, OM content in the soil decreases, thus leading to more favorable conditions for soil surface sealing.

The relative importance of the second mechanism depends on the C of the soil solution, and the ESP of the upper layer at the soil surface. As the C decreases and the ESP increases, the reduction in the IR during seal formation is more pronounced (Agassi et al., 1981; Kazman et al., 1983). Consequently, when TWW with a relatively high C level above the flocculation value of the soil clay is used for irrigation, no significant clay dispersion is expected. However, during the rainfall season or if irrigation shifts back to FW, extensive clay dispersion and significant IR reduction would be expected in soils that were previously irrigated with TWW for a period of time sufficient to increase the ESP of the soil surface. This expected result disregards the aforementioned possible favorable effects of OM on soil stability, which could reduce soil susceptibility to slaking and dispersion, and subsequently to seal formation and IR reduction (Lado et al., 2004). However, if depletion in OM content is likely to occur coupled with the presence of humic substances in the TWW, destabilization of aggregates and clay dispersion could be enhanced (e.g. Durgin and Chaney, 1984; Frenkel et al., 1992; Tarchitzky et al., 1999), leading in turn to reduction in the IR of the seal.

Mamedov et al. (2000a) studied the effect of TWW irrigation and rain kinetic energy, in the range of 3.6–15.9 kJ/m, on seal formation and IR of four calcareous Israeli soils under simulated rainfall conditions. For all the soils and kinetic energies, the final IR values of the TWW-irrigated soils were lower than those of the FW-irrigated ones; however, the differences in the final IR were relatively small ($<\sim 2$ mm/h), especially for the rain with high (15.9 kJ/m^3) or low (3.6 kJ/m^3) kinetic energy. Mamedov et al. (2000a) suggested that despite the relatively high ESPs of the TWW-irrigated soils, which enhanced clay dispersion, it is raindrop impact energy, in the case of the high kinetic energy rain, that predominates and controls the process of seal formation at the soil surface. Comparable runoff data for TWW- and FW-irrigated soils subjected to high kinetic energy rain reported by Mamedov et al. (2000b) were in agreement with the above-mentioned final IR trend.

A field study on loam and sandy clay soils from Israel also showed that runoff levels from FW- and TWW-irrigated fields were similar (Agassi et al., 2003). These authors found that the ESP of the upper soil layer in the TWW-irrigated field, which is the layer where seal development occurs, decreased rapidly during the rainy season from 5 to 6 at the end of summer to its natural level of ~ 2, as found in the FW-irrigated field. Agassi et al. (2003) attributed this reclamation of the upper soil layer to the fact that the two soils were calcareous, and further suggested that the rapid decrease in ESP was responsible for the observed similarity in susceptibility to seal formation and runoff production of the soils from the FW- and TWW-irrigated fields. These findings of Agassi et al. (2003) also explain the observed adverse effects of irrigation with TWW on soil sensitivity to sealing in a sandy soil (Mamedov et al., 2001a, b; Lado et al., 2005), as this soil was not calcareous and, therefore, sodicity level at the upper soil level remained almost unchanged during the rainy season, rendering the soil surface a greater susceptibility to seal formation and runoff production.

Mamedov et al. (2001a) studied the effects of aggregate slaking (via manipulation of the wetting rate of the soil) on runoff under simulated rainfall, from five Israeli soils with clay contents ranging from ~8% to ~62% that were subjected to irrigation with FW or TWW for more than 15 yrs. In general, the effects of the various wetting rates on the total runoff were similar in the TWW- and in the FW-irrigated soils. Only for the soil with 8% clay content total runoff was higher in the TWW- than in the FW-irrigated samples, irrespective of rate of wetting. In an additional study, Mamedov and Levy (2001) compared the role of physicochemical clay dispersion to that of aggregate disintegration (slaking) due to fast wetting on seal permeability and runoff in TWW-irrigated soils. The results showed that prevention of either physicochemical dispersion by addition of 5 Mg/ha of phosphogypsum to the soil surface, or aggregate slaking by wetting aggregates slowly, was effective in obtaining high final IR and subsequent low runoff levels (Table 9.1).

Comparison of these two treatments indicated that their efficacy depended on soil clay content. In soils having low-to-medium clay content (<40%), application of phosphogypsum together with fast rate of wetting was more effective than using slow rate of wetting in controlling seal formation, runoff, and erosion. In clay soils (>40% clay), the opposite was true (Mamedov and Levy, 2001).

Table 9.1 Final infiltration rate (FIR), and runoff data for the different treatments

Soil	**Clay** %	**ESP**[a] %	**Treatment**[b]		
			Fast WR	**Fast WR + PG**	**Slow WR**
				FIR (mm/h)	
Loamy sand	9.5	4.6	4.88 g[c]	10.00 e	5.38 fg
Loam	21.2	5.5	3.38 h	10.63 de	6.50 f
Sandy clay	38.6	5.5	3.50 h	11.75 d	6.00 fg
Clay Y	54.6	6.4	3.38 h	11.13 de	18.75 b
Clay E	64.0	3.4	3.63 h	13.25 c	36.00 a
				Runoff (mm)	
Loamy sand	9.5	4.6	39.42 cd[a]	18.60 i	35.05 e
Loam	21.2	5.5	46.30 a	30.40 g	36.66 de
Sandy clay	38.6	5.5	43.35 b	28.56 g	31.99 f
Clay Y	54.6	6.4	43.83 b	29.74 gh	4.09 j
Clay E	64.0	3.4	40.80 c	27.05 h	1.13 k

[a]ESP, exchangeable sodium percentage.
[b]Treatment = fast wetting rate (WR), 64 mm/h; slow WR, 2 mm/h; phosphogypsum (PG).
[c]For FIR and runoff, values followed by same letter are not significantly different at the 0.05 level.
Adapted from Mamedov, A.I. and G.J. Levy. 2001. Clay dispersivity and aggregate stability effects on seal formation and erosion in effluent irrigated soils. Soil Sci. 166:631–639, with permission of Lippincott Williams & Wilkins.

An integrative analysis of numerous factors involved in seal formation in FW- and TWW-irrigated soils, including clay, $CaCO_3$, and OM contents, salinity, wetting rates, and initial conditions, was presented by Lado et al. (2005). The results were in agreement with those described and discussed above, and allowed assessment of their relative importance in relation to the different conditions, within one experimental set-up.

9.5 Soil erosion

Soil erosion by water involves detachment, entrainment, and deposition of soil particles. Therefore, all the interactions between TWW and soil properties that affect stability, infiltration, and runoff are expected to reflect also in a similar manner on the erosion intensity. However, although there is a clear relationship between seal formation and runoff, the relationship between seal formation and soil interrill erosion is not so obvious. Seal formation may affect interrill soil erosion in the following opposite ways: (i) seal development increases the shear strength of the soil surface (Bradford et al., 1987), and thus reduces soil detachment (Moore and Singer, 1990); and (ii) seal formation increases runoff and thus its flow shear force, which in turn increases the transport capacity for entrained material (Moore and Singer, 1990). Initially, rainfall splash is the dominant mechanism for detachment (Hudson, 1971), but due to seal formation and runoff generation, water depth covering the soil surface increases and cushions the impact of raindrops (Ferreira and Singer, 1985). A study of the dynamics of soil erosion during seal formation can be found in Assouline and Ben-Hur (2006). Under soil surface sealing conditions, erosion can be limited either by transport capacity of the shallow flow or by detachment rate at low slopes. At steeper slopes, the earlier stage of erosion is mostly a transport-limited regime and it shifts to a detachment-limited regime as rainfall continues. This shift occurs apparently sooner at higher rainfall intensities and slope gradients. Based on the above, the impact of irrigation with TWW on soil erosion cannot be deduced directly from its impact on soil sealing and runoff.

The susceptibility of soils to erosion is called "erodibility", a terminology introduced in the Universal Soil Loss Equation (Wischmeier and Smith, 1978). A widely used soil erodibility nomograph (Wischmeier et al., 1971) was established based on extensive laboratory and field experiments under simulated and natural rainfall conditions, and indicates to what extent each property of a soil affects its erodibility. The OM content is the second most important indicator of erodibility after the particle-size distribution. The general trend is decreasing erodibility with increasing OM. Similarly, Levy et al. (1994) also observed that soil loss was more closely correlated with soil binding agents (clay content, OM) than soil sodicity and salinity. It is worth noting that although soil erodibility is generally considered as a soil property, at the rainfall event scale, when soil seal formation occurs, it was found to increase at the first stage and then, depending on slope angle, reached a stable value (5% and 9%) or a peak value before declining (15–25%) (Assouline and Ben-Hur, 2006).

Mamedov et al. (2000b) studied the effects of four levels of rain kinetic energy on soil loss from FW- and TWW-irrigated soils. For all the kinetic energy levels studied, soil erosion for the TWW-irrigated samples was higher than for the FW-irrigated ones. Moreover, the difference in soil loss between the TWW- and FW-irrigated samples increased with the increase in rain kinetic energy, suggesting that the increase in the latter magnifies the adverse dispersive effect of soil ESP on soil susceptibility to erosion (Mamedov et al., 2000b). In an additional study, Mamedov et al. (2001b) investigated the combined effects of clay content, water quality, and rate of wetting (a measure to manipulate aggregate slaking) on soil loss. They observed that for soils with medium to fine texture (>20% clay) subjected to non-severe aggregate slaking (wetted rate of <8 mm/hr), comparable soil loss levels were noted for FW- and TWW-irrigated soils (Fig. 9.4).

Lado et al. (2005) investigated the effect of long-term irrigation with TWW on rainfall–runoff–erosion relationships on sandy and clay Israeli soils under simulated rainfall. Two initial soil conditions were considered. In the first, air-dry soil was exposed directly to the rainstorm. This initial condition induces a relatively fast wetting of the aggregates at the soil surface at the beginning of the rainstorm. In the second initial

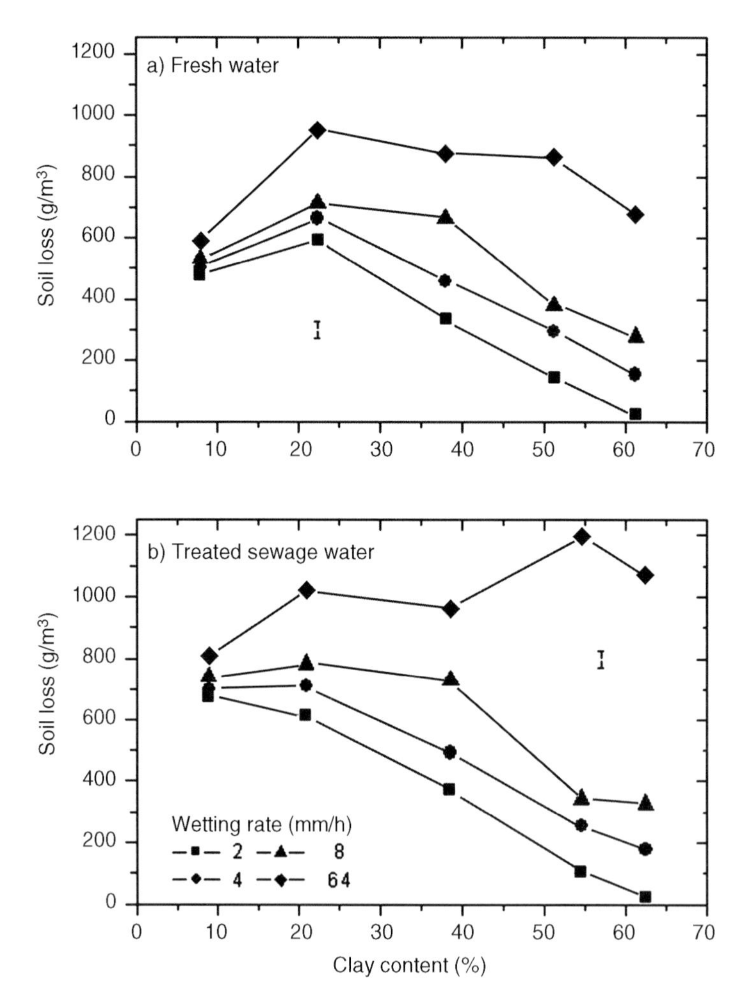

Figure 9.4 Soil loss as a function of clay content and wetting rate for soils irrigated with (a) freshwater and (b) treated sewage water. Bars indicate a single confidence interval value at $\alpha = 0.05$. Reproduced with permission from Mamedov, A.I., Shainberg, I. and Levy, G.J. (2001b) Irrigation with effluents: effects of prewetting rate and clay content on runoff and soil loss. *Journal of Environmental Quality* **30**, 2149–2156.

Table 9.2 Total soil loss for the two freshwater (FW) and treated wastewater (TWW)-irrigated soils, and for the initially dry or prewetted conditions

Irrigation treatment	**Total soil loss (g/m²)**			
	Sandy soil		**Clay soil**	
	Prewetting treatment			
	Dry	**Prewetted**	**Dry**	**Prewetted**
Freshwater	$7.3 \pm 1.6^{\dagger}$	0.0 ± 0	133.6 ± 30.1	18.5 ± 12.4
TWW	28.1 ± 4.2	28.5 ± 3.8	107.2 ± 8.4	23.9 ± 12.9

Mean ± one standard deviation.
Compiled using data from Lado, M., Ben-Hur, M. and Assouline, S. (2005) Effects of effluent irrigation on seal formation, infiltration and soil loss during rainfall. *Soil Science Society of America Journal* **69**, 1432–1439.

condition, the air-dry soil was prewetted by means of a mist of deionized water at a mean rate of 1 mm/h, inducing a much slower wetting of the aggregates during the rainfall simulation. The soil loss data are presented in Table 9.2. The effect of irrigation with TWW on soil loss was significantly more pronounced in the sandy soil than in the clay soil. This was due to the respective processes involved in shaping the rainfall–runoff relationship in each soil. In the sandy soil, clay dispersion and pore clogging was the main process in surface sealing. Irrigation with TWW enhanced clay dispersion due to the increase of the SAR of the soil solution, and a less permeable seal was formed compared to the FW-irrigated soil. Consequently, infiltration was reduced and runoff and soil loss increased in the TWW-irrigated sandy soil, for the two initial conditions. In the dry calcareous clay soil, slaking was the main process in seal formation, and the contribution of clay dispersion due to TWW irrigation was less important. However, when slaking intensity was reduced by initial slow wetting, still no effect of TWW irrigation was observed on soil loss, due to the presence of $CaCO_3$ in this soil. The dissolution of $CaCO_3$ reduced the soil ESP to its natural level by releasing Ca, which could replace the excess of exchangeable Na induced by TWW irrigation. Consequently, the intensity of clay dispersion was similar to that in the FW-irrigated soil, and no effect on soil loss could be measured.

9.6 Water repellency

Water repellency in soils signifies that soils exhibiting hydrophobic properties resist or retard water infiltration when dry (Brandt, 1969). Water-repellent soils have been reported in many countries and may occupy large areas such as the sandy soils of South and Western Australia (De Bano and Letey, 1969). Nevertheless, because severe water-repellency has been considered to be relatively uncommon, the assumption of non-water-repellent behavior is the norm in classical soil physics (Philip, 1969). However, Wallis et al. (1991), who studied 14 New Zealand soils, found the soils to be water-repellent at field moisture conditions. This finding led Wallis et al. (1991) to suggest that some degree of water-repellency is the norm rather than the exception. It should be pointed out that Wallis et al. (1991) employed the sensitive technique developed by Tilman et al. (1989) for measuring water-repellency, in which the repellency index was based on the ratio of soil sorptivity to ethanol to soil sorptivity to water, rather than the commonly used water drop penetration time (WDPT) method (Letey, 1969).

The main effect of water-repellency is the reduced infiltration rate at the initial phase of the infiltration, which leads to overland flow and soil erosion; this can at times be followed by a possible increase in the infiltration rate with time (Wallis et al., 1991). Water-repellency can also lead to fingered flow and preferential flow paths resulting in severely irregular wetting fronts (e.g. Dekker and Ritsema, 1994; Ritsema and Dekker, 1998), variations in water content in the soil profile and poor seed germination (Wallis and Horne, 1992). Furthermore, the water-repellent induced preferential flow can accelerate leaching of contaminants to groundwater (e.g. Blackwell, 2000).

Attempts to determine the causes for water-repellency have yielded inconclusive findings, which suggests that there are a number of causes and that our understanding of the repellency phenomenon is incomplete (see extensive reviews of Wallis and Horne, 1992; De Bano, 2000; Doerr et al., 2000). It is agreed though that coating of soil surfaces by OM imparts water-repellency (Wallis and Horne, 1992). This notion is supported by scanning electron micrographs that show the presence of OM coating in repellent sand grains and the absence of the organic coating in non-repellent sand grains (Rankin and Ross, 1982). More recent studies have shown that water-repellency can also be induced by fire (Shakesby et al., 1993), the presence of various types of plant species (Doerr, 1998), bacterial activity (Chan, 1992) and the presence of specific organic compounds such as aliphatic hydrocarbons (Roy and McGill, 2000).

Based on the understanding that water-repellency is associated with the OM in the soil coupled with the suggestion of Wallis and Horne (1992) that the characteristics of the OM are more important in imparting water-repellency in soils than its quantity, one would expect that irrigation with TWW would be accompanied by studies and monitoring of water-repellency in the soils. However, despite the wide interest in water-repellency over the years (e.g. see reviews by Wallis and Horne, 1992; De Bano, 2000), little attention has been paid to the possible impact of irrigation with TWW on water-repellency. In Israel, observations made by farmers and extension officers who noted that irrigation with TWW was associated with a narrower wetted zone below the soil surface and the formation of a smaller wetting-diameter at the soil surface in drip-irrigated orchards, leading to the undesired condition of a non-continuous wet strip along the tree row, prompted research in this direction (e.g. Chen et al., 2003; Lerner et al., 2003, Wallach et al., 2005; Tarchitzky et al., 2007). Chen et al. (2003) tested the hypothesis that hydrophobic organic materials that penetrate the soil with the TWW affect the degree of soil repellency and noted that diameter of the saturated area (on the soil surface) and the volumetric water content (at a depth of 0–10 cm) were smaller with TWW irrigation than with FW irrigation. Furthermore, results of a WDPT test indicated no water-repellency in soils irrigated with FW, whereas soils irrigated with TWW exhibited various degrees of water-repellency (Chen et al., 2003). In a continuation of the former study, Tarchitzky et al. (2007) studied water-repellency in banana plantations and orchards irrigated with TWW in four different soil types. It was observed that the TWW-irrigated soils exhibited water-repellency, whereas the FW-irrigated soils were hydrophilic. Fourier transform infrared spectroscopy (FTIR) and ^{13}C-NMR analyses of organic components extracted from the soils indicated that the water-repellency in the TWW-irrigated soils could be associated with changes in the OM characteristics of 0–2 cm of the upper soil layer (Tarchitzky et al., 2007). Wallach et al. (2005) studied water-repellency with the WDPT method and its recurrence following the rainy season in a long-term TWW-irrigated orchard. Extreme to severe water-repellency was noted in the TWW-irrigated orchard, whereas in an adjacent FW-irrigated orchard no repellency or mild repellency was observed. Water-repellency was very variable spatially and with depth. The non-uniform distribution of soil moisture content and fingered flow that were observed in the soil profile after both the irrigation and the rainy season led Wallach et al. (2005) to the conclusion that water-repellency in TWW-irrigated soils has a persistent effect on water flow and distribution in the soil profile. A similar conclusion can be drawn from the study of Taumer et al. (2005). In their study, the water-repellency from a former dumping site for poorly treated sewage water was studied on soil samples taken nearly 20 yrs after dumping of the sewage water had been stopped. The results indicated a severe degree of water-repellency that had a major influence on the water flow even during wintertime, led to preferential flow paths, and reduced the amount of plant available water.

The reviewed studies raise the concern that irrigation with TWW may impart water-repellency and its undesired consequences on the irrigated soils. Furthermore, Graber et al. (2006) have shown that the use of disturbed samples for the WDPT test in the evaluation of water-repellency may distort the results in comparison to non-disturbed samples, towards the direction of less repellency. As, in most water-repellency studies, use of disturbed samples is the norm, the danger of underestimating the degree of water-repellency in TWW-irrigated soils and thus its accompanying adverse effects should be recognized and taken into consideration in planning of possible counter measures.

9.7 Concluding comments

The demand for FW by the urban and industrial sectors and the subsequent production of wastewater are steadily increasing worldwide. Consequently, TWW is increasingly being considered as a significant source of water for irrigated agriculture, especially in arid and semiarid regions where FW is scarce.

Treated wastewater is characterized by a higher load of DOM, suspended solids, SAR, and salinity compared with its FW of origin. These characteristics of the TWW could affect, at different levels, soil properties such as aggregate stability, water retention and hydraulic conductivity, water-repellency and, consequently, the rainfall–infiltration–runoff–erosion relationships. Thus, long-term intensive irrigation with TWW could possibly lead to a substantial degradation of soil structural stability and associated physical properties, which in turn could harm soil productivity.

Review of numerous studies that evaluated the impact of irrigation with TWW on different parameters representing soil structural stability indicates that effects of using TWW for irrigation were inconsistent. The observed inconsistency may stem from (i) the fact that different mechanisms control the response of the different parameters tested; (ii) the impact of specific soil properties; (iii) differences in conditions prevailing in the soil; and (iv) variation in the quality of the TWW, i.e. the load and characteristics of the OM in the water. Furthermore, despite the large volume of research dedicated to the use of TWW in agriculture, quantitative aspects of its impact on soil structural stability are sparse. In addition, long-term effects of irrigation with TWW are still scarce and history of plots actually being irrigated with TWW is not always retractable, which precludes in-depth analyses and the identification of possible trends. The existing knowledge is also insufficient to support essential modeling for the prediction of soil response to the use of TWW. All of the above strongly suggest that further research is needed before definite conclusions can be drawn regarding the conditions under which deterioration in the physical properties of the soil, following the use of TWW for irrigation, can be expected.

Our soils are a valuable resource that, once degraded, is extremely difficult and costly to reclaim. Thus, despite the expected constant improvement in the quality of the treatment to which TWW is subjected, irrigation with TWW should be accompanied by close monitoring of different indicators of soil physical properties in order to maintain the soils in a satisfactory state and to ensure their sustainable productivity for future generations.

References

Agassi, M., Shainberg, I. and Morin, J. (1981) Effect of electrolyte concentration and soil sodicity on the infiltration rate and crust formation. *Soil Science Society of America Journal* **45**, 848–851.

Agassi, M., Tarchitzky, J., Keren, R., Chen, Y., Goldstein, D. and Fizik, E. (2003) Effects of prolonged irrigation with treated municipal effluent on runoff rate. *Journal of Environmental Quality* **32**, 1053–1057.

Amezketa, E., Singer, M.J. and Le Bissonnais, Y. (1996) Testing a new procedure for measuring water-stable aggregates. *Soil Science Society of America Journal* **60**, 888–894.

Anderson, M.A., Hung, A.Y.C., Mills, D. and Scott, M.S. (1995) Factors affecting the surface tension of soil solutions and solutions of humic acids. *Soil Science* **160**, 111–116.

Assouline, S. (2001) A model of soil relative hydraulic conductivity based on water retention curve characteristics. *Water Resources Research* **37**, 265–271.

Assouline, S. (2004) Rainfall-induced soil surface sealing: a critical review of observations, conceptual models and solutions. *Vadose Zone Journal* **3**, 570–591.

Assouline, S. and Ben-Hur, M. (2006) Effects of rainfall intensity and slope gradient on the dynamics of interrill erosion during soil surface sealing. *Catena* **66**, 211–220.

Assouline, S. and Mualem, Y. (1997) Modeling the dynamics of seal formation and its effect on infiltration as related to soil and rainfall characteristics. *Water Resources Research* **33**, 1527–1536.

Assouline, S. and Mualem, Y. (2000) Modeling the dynamics of soil seal formation: Analysis of the effect of soil and rainfall properties. *Water Resources Research* **36**, 2341–2349.

Assouline, S. and Mualem, Y. (2006) Runoff from heterogeneous small bare catchments during soil surface sealing. *Water Resources Research* **42**, W12405, doi: 10.1029/2005WR004592.

Assouline, S., Selker, J. and Parlange, J-Y. (2007) A simple accurate method to predict time of ponding under variable intensity rainfall. *Water Resources Research* **43**, W03426, doi: 10.1029/2006WR005138.

Assouline, S., Tessier, D. and Bruand, A. (1998) A conceptual model of the soil water retention curve. *Water Resources Research* **34**, 223–231.

Balks, M.R., Mclay, C.D.A. and Harfoot, C.G. (1997) Determination of the progression in soil microbial response, and changes in soil permeability, following application of meat processing effluent to soil. *Applied Soil Ecology* **6**, 109–116.

Ben-Hur, M., Shainberg, I., Bakker, D. and Keren, R. (1985) Effect of soil texture and $CaCO_3$ content on water infiltration in crusted soils as related to water salinity. *Irrigation Science* **6**, 281–284.

Betzalel, I., Morin, J., Benyamini, Y., Agassi, M. and Shainberg, I. (1995) Water drop energy and soil seal properties. *Soil Science* **159**, 13–22.

Bhardwaj, A.K., Goldstein, D., Azenkot, A. and Levy, G.J. (2007) Irrigation with treated wastewater under two different irrigation methods: effects on hydraulic conductivity of a clay soil. *Geoderma* **140**, 199–206.

Bhardwaj, A.K., Mandal, U.K., Bar-Tal, A., Gilboa, A. and Levy, G.J. (2008) Replacing saline-sodic irrigation water with treated wastewater: Effects on saturated hydraulic conductivity, slaking and swelling. *Irrigation Science* **26**, 139–146.

Blackwell, P.S. (2000) Management of water repellency in Australia: The risks associated with preferential flow, pesticide concentration and leaching. *Journal of Hydrology* (Amsterdam) 231–232, 384–395.

Bradford, J.B., Ferris, J.E. and Remley, P.A. (1987) Interrill soil erosion processes: I. Effect of surface sealing on infiltration, runoff and soil splash detachment. *Soil Science Society of America Journal* **51**, 1566–1577.

Bresler, E., McNeal, B.L. and Carter, D.L. (1982) *Saline and Sodic Soils: Principles-Dynamics-Modeling*. Springer-Verlag, Berlin.

Bronick, C.J. and Lal, R. (2005) Soil structure and management: a review. *Geoderma* **124**, 3–22.

Brooks, R.H., and Corey, A.T. (1964) Hydraulic properties of porous media. *Hydrology Papers* **3**, 27 pp. Colorado State University, Fort Collins.

Chan, K.Y. (1992) Development of seasonal water repellence under direct drilling. *Soil Science Society of America Journal* **56**, 326–329.

Chan, K.Y., and Mullins, C.E. (1994) Slaking characteristics of some Australian and British soils. *European Journal of Soil Science* **4**, 273–283.

Chen, Y. and Schnitzer, M. 1978. Surface-tension of aqueous-solutions of soil humic substances. *Soil Science* **1251**, 7–15.

Chen, Y., Lerner, O. and Tarchitzky, J. (2003) Hydraulic conductivity and soil hydrophobicity: effect of irrigation with reclaimed wastewater. In: *9th Nordic IHSS Symposium on Abundance and Functions of Natural Organic Matter Species in Soil and Water* (ed. U. Lundstrom), pp. 19. Mid-Sweden University, Sundsvall, Sweden.

Chen, J., Tarchitzky, J., Morin, J. and Banin, A. (1980) Scanning electron microscope observations on soil crust and their formation. *Soil Science* **130**, 49–55.

Chenu, C., Le Bissonnais, Y. and Arrouays, D. (2000) Organic matter influence on clay wettability and soil aggregate stability. *Soil Science Society of America Journal* **64**, 1479–1486.

Cook, F.J., Kelliher, F.M. and McMahon, S.D. (1994) Changes in infiltration during wastewater irrigation of a highly permeable soil. *Journal of Environmental Quality* **23**, 476–482.

Coppola, A., Santini, A., Botti, P., Vacca, S., Comegna, V. and Severino, G. (2004) Methodological approach for evaluating the response of soil hydrological behaviour to irrigation with treated municipal wastewater. *Journal of Hydrology* **292**, 114–134.

De Bano, L.F. (2000) Water repellency in soil: a historical overview. *Journal of Hydrology* 231–232, 4–32.

De Bano, L.F. and Letey, J. (1969) *Proceedings of the Symposium on Water Repellent Soils*. University California Riverside, CA, USA. 6–10 May 1968. pp. 1–321.

Dekker, L.W., and Ritsema, C.J. (1994) How water moves in a water repellent sandy soil. 1. Potential and actual water repellency. *Water Resources Research* **30**, 2507–2517.

Doerr, S.H. (1998) On standardizing the "Water Drop Penetration Time" and the "Molarity of an Ethanol Droplet" techniques to classify soil hydrophobicity. A case study using medium textured soils. *Earth Surface Processes and Landforms* **23**, 663–668.

Doerr, S.H., Shakesby, R.A. and Walsh, R.P.D. (2000) Soil water repellency: its causes, characteristics and hydro-geomorphological significance. *Earth-Science Reviews* **51**, 33–65.

Durgin, P.B. and Chaney, J.G. (1984) Dispersion of kaolinite by dissolved organic matter from Douglas-fir roots. *Canadian Journal of Soil Science* **64**, 445–455.

Emerson, W.W. (1967) A classification of soil aggregates based on their coherence in water. *Australian Journal of Soil Research* **5**, 47–57.

Emerson, W.W. (1983) Interparticle bonding. In: *Soils: An Australian viewpoint*, p. 447–498. Academic Press, London, UK.

Farres, P.J. (1980) Some observations on the stability of soil aggregates to raindrop impact. *Catena* **7**, 223–231.

Feigin, A., Ravina, I. and Shalhevet, J. (1991) Irrigation with sewage effluent. *Advanced Series in Agricultural Science* Vol. 17, Springer-Verlag, Berlin.

Ferreira, A.G. and Singer, M.J. (1985) Energy-dissipation for water drop impact into shallow pools. *Soil Science Society of America Journal* **49**, 1537–1542.

Fotovat, A., Smith, L., Naidu, R.J. and Oades, J. (1996) Analysis of indigenous zinc in alkaline sodic soil solutions by graphite furnace atomic absorption spectrometry. *Communications in Soil Science and Plant Analysis* **27**, 2997–3012.

Frenkel, H., Fey, M.V. and Levy, G.J. (1992) Organic and inorganic anion effects on reference and soil clay critical flocculation concentration. *Soil Science Society of America Journal* **56**, 1762–1766.

Gal, M., Arcon, L., Shainberg, I. and Keren, R. (1984) The effect of exchangeable Na and phosphogypsum on the structure of soil crust – SEM observation. *Soil Science Society of America Journal* **48**, 872–878.
van Genuchten, M.Th. (1980) A closed-form equation for predicting the hydraulic conductivity of unsaturated soils. *Soil Science Society of America Journal* **44**, 892–898.
Goldberg, S., Suarez, D.L. and Glaubig, R.A. 1988. Factors affecting clay dispersion and aggregate stability of arid-zone soils. *Soil Science* **146**, 317–325.
Graber, E.R., Ben-Arie, O. and Wallach, R. (2006) Effect of sample disturbance on soil water repellency determination in sandy soils. *Geoderma* **136**, 11–19.
Greenberg, A.E., Trussell, R.R. and Clesceri, L.S. (eds) (1985) *Standard Methods for Examination of Water and Wastewater*. American Public Health Association, Washington, DC.
Gu, B. and Doner, H.E. (1993) Dispersion and aggregation of soils as influenced by organic and inorganic polymers. *Soil Science Society of America Journal* **57**, 709–716.
Hudson, N. (1971) *Soil Conservation*. Cornell Univ. Press, Ithaca, NY.
Halliwell, D.J., Barlow, K.M. and Nash, D.M. (2001) A review of the effects of wastewater sodium on soil physical properties and their implications for irrigation systems. *Australian Journal of Soil Research* **39**, 1259–1267.
Jury, W.A., Gardner, W.R. and Gardner, W.H. (1991) *Soil Physics*. 5th ed. John Wiley & Sons, Inc., New York.
Kay, B.D. and Angers, D.A. (1999) Soil Structure. In: *Handbook of Soil Science* (ed. M.E. Sumner), pp. A229–A276. CRC Press, Boca Raton, FL.
Kazman, Z., Shainberg, I. and Gal, M. (1983) Effect of low levels of exchangeable Na and applied phosphogypsum on the infiltration rate of various soils. *Soil Science* **35**, 184–192.
Kemper, W.D. and Koch, E.J. (1966) Aggregate stability of soils from western U.S. and Canada. Tech. Bull. No. 1355, U.S. Dept. Agric. Arid zone soils. *Soil Science* **146**, 317–325.
Kemper, W.D. and Rosenau, R. (1986) Aggregate stability and size distribution. In: *Methods of Soil Analysis* Part 1, 2nd ed (ed. A. Klute), pp. 425–442. Agron. Monogr. 9. ASA and SSSA, Madison WI.
Kheyrabi, D. and Monnier, G. (1968) Etude experimentale de l'influence de la composition granulometrique des terres leur stabilite structurale. *Annales Agronomiques* **19**, 129–152.
Lado, M., Paz, A. and Ben-Hur, M. (2004) Organic matter and aggregate size interactions in infiltration, seal formation and soil loss. *Soil Science Society of America Journal* **68**, 935–942.
Lado, M., Ben-Hur, M. and Assouline, S. (2005) Effects of effluent irrigation on seal formation, infiltration and soil loss during rainfall. *Soil Science Society of America Journal* **69**, 1432–1439.
Lance, J.C., Rice, R.C. and Gilbert, R.G. (1980) Renovation of wastewater by soil columns flooded with primary effluent. *Journal of the Water Pollution Control Federation* **52**, 381–388.
Le Bissonnais, Y. (1996) Aggregate stability and assessment of soil crustability and erodibility. I. Theory and methodology. *European Journal of Soil Science* **47**, 425–437.
Le Bissonnais, Y. and Arrouays, D. (1997) Aggregate stability and assessment of soil crustability and erodibility. II. Application to humic loamy soils with various organic carbon contents. *European Journal of Soil Science* **48**, 39–48.
Lerner, O., Brener, A., Chen, Y., Shani, U., Gilboa, A. and Tarchitzky, J. (2003) Changes in soil water distribution in soils irrigated with effluents. *Water and Irrigation* **437**, 22–28 (in Hebrew).
Letey, J. (1969) Measurement of contact angle, water drop penetration time, and critical surface tension. In: *Proceedings of the Symposium on Water Repellent Soils* (eds L.S. Debano and J. Letey), pp. 43–47. University of California, Riverside.
Levy, G.J, and Mamedov, A.I. (2002) High-energy-moisture-characteristics aggregate stability as a predictor for seal formation. *Soil Science Society of America Journal* **66**, 1603–1609.
Levy, G.J., Berliner, P.R., du Plessis, H.M. and van der Watt, H.v.H. (1988) The microtopographical characteristics of artificially formed crusts. *Soil Science Society of America Journal* **52**, 784–791.
Levy, G.J., Goldstein, D. and Mamedov, A.I. (2005) Saturated hydraulic conductivity of semi arid soils: Combined effects of salinity, sodicity and rate of wetting. *Soil Science Society of America Journal* **69**, 653–662.
Levy, G.J., Goldstein, D., Trachitzky, J. and Chen, Y. (2006) *Effects of Irrigation with Treated Waste Water and of Tillage on the Structure and Stability of Soils*. Final report submitted to the Chief Scientist, Ministry of Agriculture, State of Israel (in Hebrew).

Levy, G.J., Levin, J. and Shainberg I. (1994) Seal formation and interrill erosion. *Soil Science Society of America Journal* **58**, 205–209.
Levy, G.J., Levin, J. and Shainberg, I. (1997) Prewetting Rate and aging effect on seal formation and interrill soil erosion. *Soil Science* **162**, 131–139.
Levy, G.J., Mamedov, A.I. and Goldstein, D. (2003) Sodicity and water quality effects on slaking of aggregates from semi-arid soils. *Soil Science* **168**, 552–562.
Levy, G.J., Rosenthal, A., Shainberg, I., Tarchitzky, J. and Chen, Y. (1999) Soil hydraulic conductivity changes caused by irrigation with reclaimed waste water. *Journal of Environmental Quality* **28**, 1658–1664.
Levy, G.J., Shainberg, I. and Morin, J. (1986) Factors affecting the stability of soil crusts in subsequent storms. *Soil Science Society of America Journal* **50**, 196–201.
Lima, L.A., Grismer, M.E. and Nielsen, D.R. (1990) Salinity effects on Yolo loam hydraulic properties. *Soil Science* **150**, 451–458.
Loch, R.J. (1994) Structure breakdown on wetting. In: *Sealing Crusting and Hardsetting Soils* (ed. H.B. So), pp. 113–132. Australian Society of Soil Science.
Mace, J.E., and Amrhein, C. (2001) Leaching and reclamation of a soil irrigated with moderate SAR waters. *Soil Science Society of America Journal* **65**, 199–204.
Magesan, G.N., Williamson, J.C., Sparling, G.P., Schipper, L.A. and Lloyd-Jones, A.R.H. (1999) Hydraulic conductivity in soils irrigated with wastewaters of differing strengths: field and laboratory studies. *Australian Journal of Soil Research* **37**, 391–402.
Magesan, G.N., Williamson, J.C., Yeates, G.W. and Lloyd-Jones, A.R.H. (2000) Wastewater C:N ratio effects on soil hydraulic conductivity and potential mechanisms for recovery. *Bioresource Technology* **71**, 21–27.
Malik, M., Mustafa, M.A. and Letey, J. (1992) Effect of mixed Na/Ca solutions on swelling, dispersion and transient water flow in unsaturated montmorillonitic soils. *Geoderma* **52**, 17–28.
Mamedov, A.I. and Levy, G.J. (2001) Clay dispersivity and aggregate stability effects on seal formation and erosion in effluent irrigated soils. *Soil Science* **166**, 631–639.
Mamedov, A.I., Shainberg, I. and Levy, G.J. (2000a) Irrigation with effluent water: Effect of rainfall energy on soil infiltration. *Soil Science Society of America Journal* **64**, 732–737.
Mamedov, A.I., Shainberg, I. and Levy, G.J. (2000b) Rainfall energy effects on runoff and interrill erosion in effluent irrigated soils. *Soil Science* **165**, 535–544.
Mamedov, A.I., Levy, G.J., Shainberg, I. and Letey, J. (2001a) Wetting rate, sodicity and soil texture effects on infiltration rate and runoff. *Australian Journal of Soil Research* **39**, 1293–1303.
Mamedov, A.I., Shainberg, I. and Levy, G.J. (2001b) Irrigation with effluents: effects of prewetting rate and clay content on runoff and soil loss. *Journal of Environmental Quality* **30**, 2149–2156.
Mathan, K.K. (1994). Studies on the influence of long-term municipal sewage-effluent irrigation on soil physical properties. *Bioresource Technology* **48**, 275–276.
Mbagwu, J.S.C., Piccolo, A. and Mbila, M.O. (1993) Water stability of aggregates of some tropical soils treated with humic substances. *Pedologie* **43**, 269–284.
McIntyre, D.S. (1958) Permeability measurements of soil crust formed by raindrop impact. *Soil Science* **85**, 158–189.
McNeal, B.L. (1968) Prediction of the effects of mixed–salt solutions on soil hydraulic conductivity. *Soil Science Society of America Proceedings* **32**, 190–193.
Metzger, L., Yaron, B. and Mingelgrin, U. (1983) Soil hydraulic conductivity as affected by physical and chemical-properties of effluents. *Agronomie (Paris)* **3**, 771–777.
Miller, W.P. (1987) Infiltration and soil loss of three gypsum-amended Ultisols under simulated rainfall. *Soil Science Society of America Journal* **51**, 1314–1320.
Moore, D.C. and Singer, M.J. (1990) Crust formation effects on soil-erosion processes. *Soil Science Society of America Journal* **54**, 1117–1123.
Morin, J. and Benyamini, Y. (1977) Rainfall infiltration into bare soils. *Water Resources Research* **14**, 813–837.
Morin, J., Benyamini, Y. and Michaeli, A. (1981) The effect of raindrop impact on the dynamics of soil surface crusting and water movement in the profile. *Journal of Hydrology* **52**, 321–335.
Mualem, Y. (1986) Hydraulic conductivity of unsaturated soils, predictions and formulas. In: *Methods of Soil Analysis*, Part 1 2nd ed (ed. A. Klute), pp. 799–823. Agronomy Monograph 9, American Society of Agronomy and Soil Science Society of America, Madison, WI.
North, P.F. (1976) Towards an absolute measurement of soil structural stability using ultrasound. *Journal of Soil Science* **27**, 451–459.

Oades, J.M. (1984) Soil organic matter and structural stability: Mechanisms and implications for management. *Plant Soil* **76**, 319–337.
Ojeda, G., Perfect, E., Alcaniz, J.M. and Ortiz, O. (2006) Fractal analysis of soil water hysteresis as influenced by sewage sludge application. *Geoderma* **131**, 386–401.
Onofoik, O. and Singer, M.J. (1984) Scanning electron microscope studies of soil surface crusts formed by simulated rainfall. *Soil Science Society of America Journal* **48**, 1137–1143.
Panabokke, C.R. and Quirk, J.P. (1957) Effect of initial water content on stability of soil aggregates in water. *Soil Science* **83**, 185–195.
Philip, J.R. (1969) Theory of infiltration. In: *Advances in Hydroscience* Vol. 5 (ed. V.T. Chow), pp. 215–296. Academic Press, New York.
Piccolo, A. and Mbagwu, J.S.C. (1989) Effects of humic substances and surfactants on the stability of soil aggregates. *Soil Science* **147**, 47–54.
Piccolo, A. and Mbagwu, J.S.C. (1994) Humic substances and surfactants effects on the stability of two tropical soils. *Soil Science Society of America Journal* **58**, 950–955.
Piccolo, A., Pietramellara, G. and Mbagwu, J.S.C. (1996) Effects of coal derived humic substances on water retention and structural stability of Mediterranean soils. *Soil Use Management* **12**, 209–213.
Quirk, J.P., and Schofield, R.K. (1955) The effect of electrolyte concentration on soil permeability. *Journal of Soil Science* **6**, 163–178.
Rankin, P. and Ross, C. (1982) Dry patch – so what is new? *Sports Turf Review* **139**, 61–66. N.Z. Turf Culture Institute.
Rengasamy, P. and Olsson, K.A. (1991) Sodicity and soil structure. *Australian Journal of Soil Research* **29**, 65–76.
Rengasamy, P. and Sumner, M.E. (1998) Processes involved in sodic behavior In: *Sodic Soils* (eds M.E. Sumner and R. Naidu), pp. 35–50. Oxford University Press, New York.
Rice, R.C. (1974) Soil clogging during infiltration of secondary effluent. *Journal of Water Pollution Control* **46**, 708–716.
Ritsema, C.J., and Dekker, L.W. (1994) Three-dimensional patterns of moisture, water repellency, bromide and pH in a sandy soil. *Journal of Contaminant Hydrology* **31**, 295–313.
Rosen, M.J. (1989) *Surfactants and interfacial phenomenon.* 2nd Ed. John Wiley & Sons, New York.
Roy, J.L. and McGill, W.B. (2000) Flexible conformation in organic matter coatings: An hypothesis about soil water repellency. *Canadian Journal of Soil Science* **80**, 143–152.
Russo, D. (1988) Numerical analysis of nonsteady transport of interacting solutes through unsaturated soil. 1. homogeneous systems. *Water Resources Research* **24**, 271–284.
Russo, D. (2005) Physical aspects of soil salinity. In: *Encyclopedia of Soils in the Environment* Vol. 3, (ed. D. Hillel), pp. 442–453. Elsevier Ltd, Oxford, U.K.
Russo, D. and Bresler, E. (1977) Analysis of saturated-unsaturated hydraulic conductivity in a mixed Na-Ca soil system. *Soil Science Society of America Journal* **41**, 706–710.
Siegrist, R.L. (1987) Soil clogging during subsurface waste water infiltration as affected by effluent composition and loading rate. *Journal of Environmental Quality* **16**, 181–187.
Shainberg, I. and Letey, J. (1984) Response of soils to sodic and saline conditions. *Hilgardia* **52**, 1–57.
Shainberg, I., Levy, G.J., Goldstein, D., Mamedov, A.I. and Letey, J. (2001) Prewetting rate and sodicity effects on the hydraulic conductivity of soils. *Australian Journal of Soil Research* **39**, 1279–1291.
Shakesby, R.A., Coelho, C.O.A., Ferreira, A.J.D., Terry, J.P. and Walsh, R.P.D. (1993) Wildfire impacts on soil erosion and hydrology in wet Mediterranean forest. *International Journal of Wildland Fire* **9**, 95–110.
Suarez D.L. and Simunek, J. (1997) UNSATCHEM: Unsaturated water and solute transport model with equilibrium and kinetic chemistry. *Soil Science Society of America Journal* **61**, 1633–1646.
Sumner, M.E. (1993) Sodic soils: New perspectives. *Australian Journal of Soil Research* **31**, 683–750.
Tilman, R.W., Scotter, D.R., Wallis, M.G. and Clothier, B.E. (1989) Water repellency and its measurement using intrinsic sorptivity. *Australian Journal of Soil Research* **27**, 637–644.
Tarchitzky, J., Chen, Y. and Banin, A. (1993) Humic substances and pH effects on sodium and calcium montmorillonite flocculation and dispersion. *Soil Science Society of America Journal* **52**, 1449–1452.
Tarchitzky, J., Golobati, Y., Keren, R. and Chen, Y. (1999) Wastewater effects on montmorillonite suspensions and hydraulic properties of sandy soil. *Soil Science Society of America Journal* **63**, 554–560.

Tarchitzky, J., Lerner, O., Shani, U., Arye, G., Lowengart-Aycicegi, A., Brener, A. and Chen, Y. (2007) Water distribution pattern in treated wastewater irrigated soils: hydrophobicity effect. *European Journal of Soil Science* **58**, 573–588.

Taumer, K., Stoffregen, H. and Wessolek, G. (2005) Determination of repellency distribution using soil organic matter and water content. *Geoderma* **125**, 107–115.

Tisdall, J.M. and Oades, J.M. (1982) Organic matter and water-stable aggregates in soils. *Journal of Soil Science* **33**, 141–163.

Tschapek, M., Scoppa, C.O. and Wasowski, C.J. (1978) The surface tension of soil water. *J Journal of Soil Science* **29**, 17–21.

US Salinity Laboratory Staff (1954) *Diagnosis and Improvement of Saline and Alkali Soils*. USDA Handbook 60. Washington, D.C.

Vinten, A.J.A., Mingelgrin, U. and Yaron, B. (1983) The effect of suspended solids in wastewater on soil hydraulic conductivity: II. Vertical distribution of suspended solids. *Soil Science Society of America Journal* **47**, 408–412.

Viviani, G. and Iovino, M. (2004) Wastewater reuse effects on soil hydraulic conductivity. *Journal of Irrigation and Drainage Engineering* **130**, 476–484.

de Vries, J. (1972) Soil filtration of wastewater effluent and the mechanism of soil clogging. *Journal of the Water Pollution Control Federation* **44**, 565–573.

Wakindiki, I.I.C. and Ben-Hur, M. (2002) Soil mineralogy and texture effects on crust micromorphology, infiltration and erosion. *Soil Science Society of America Journal* **66**, 897–905.

Wallach, R., Ben-Arie, O. and Graber, E.R. (2005) Soil water repellency induced by long term irrigation with treated sewage effluent. *Journal of Environmental Quality* **34**, 1910–1920.

Wallis, M.G. and Horne, D.J. (1992) Soil water repellency. *Advances in Soil Science* **20**, 91–146.

Wallis, M.G., Scotter, D.R. and Horne, D.J. (1991) An evaluation of the intrinsic sorptivity water repellency index on a range of New Zealand soils. *Australian Journal of Soil Research* **29**, 353–362.

Wischmeier, W.H. and Smith, D.D. (1978) *Predicting rainfall erosion losses – A guide to conservation planning*. Agriculture Handbook No. 537. USDA, Washington D.C.

Wischmeier, W.H., Johnson, C.B. and Cross, B.V. (1971) A soil erodibility monograph for farmland and construction sites. *Journal Soil Water Conservation* **26**, 189–193.

Chapter 10
Fouling in microirrigation systems applying treated wastewater effluents

Carlos G. Dosoretz, Jorge Tarchitzky, Ilan Katz, Elisha Kenig and Yona Chen

10.1 Introduction

A major agrotechnical constraint concerning the application of wastewater effluents in irrigation systems is the potential for fouling of pipes and distribution devices, and especially clogging of emitters in microirrigation. Clogging has been observed and described in the literature since the early 1980s (Bucks et al., 1979; Nakayama and Bucks, 1981, 1991; Adin, 1986, 1987; Adin and Sacks, 1991). Emitter clogging in microirrigation is directly related to the quality of the irrigation water. Four interlinked mechanism types of clogging of irrigation devices can be discerned (Bucks et al., 1979; Nakayama and Bucks, 1981; Nakayama et al., 2007): (i) particulate matter clogging narrow flow paths, due to presence of suspended solids in the feedwater; (ii) scaling, due to formation/chemical precipitation of soluble salt at a concentration above the saturation product (e.g. carbonate, phosphate, sulfate); (iii) adsorption, due to hydrophobic interaction of soluble or colloidal organic macromolecules (e.g. humic substances, soluble microbial products, cell debris); and (iv) biological, due to biofilm formation and algal growth. The final formation of a fouling layer in distribution lines or clogging of emitters is usually the concerted action of more than one of this individual type of fouling. As suspended materials can be avoided and algae growth can be controlled, the more acute form of fouling is the in situ formation of particulate material by supersaturation (scaling), hydrophobic interaction (adsorption), and biofouling (biofilm).

This chapter aims to describe qualitative and quantitative aspects of clogging-related phenomena, characterize the properties of the fouling layer (the organic and inorganic components), and elaborate the formation mechanism. A distinction between clogging, which is the final consequence of the fouling process, and fouling will be established. Due to their higher water use efficiency, microirrigation devices (microsprinklers and drippers) are becoming increasingly popular both in agriculture and landscape irrigation. As fouling

Treated Wastewater in Agriculture, First Edition, edited by Guy J. Levy, Pinchas Fine and Asher Bar-Tal

is more critical in microirrigation, and especially drip-irrigation, and largely a major cause of emitter malfunction, emphasis will be put on drip-irrigation. Furthermore, as this irrigation method is considered to minimize risks (e.g. less contact with plants and operators, does not form aerosols, less percolation), drip-irrigation has been considered since its implementation as particularly suitable for treated wastewater (TWW) effluent reuse (Capra and Scicolone, 1998, 2004, 2005).

10.2 Quality of treated effluents as a source of irrigation water

The hydraulic loading regime of microirrigation systems is conceived based on even supply distribution, i.e. the same amount of water per dosing cycle. Emitter clogging is the main source of uneven irrigation, and even a few clogged emitters can greatly reduce the uniformity of water application (Bralts et al., 1981; Nakayama and Bucks, 1981). As a result of low water velocities, microirrigation devices are prone to the development of biofilms and scaling depositions (Ravina et al., 1992). Uneven water distribution in the field means areas with excess or insufficient water. Excess water means leaching of water and nutrients below root-zone and downwards to the underground water. Insufficient water means that there is not enough water for the plants to develop optimally and a potential salt accumulation.

Regarding the flow rate changes in drip-irrigation, two types of drippers can be distinguished:

(i) non-regulated drippers in which the flow path is made of a rigid material, and therefore only flow rate reduction can take place when dirt particles accumulate in their pathway. Flow rate decrease is relatively easy to distinguish and monitor by means of an irrigation controller;
(ii) regulated drippers in which in addition to the rigid pathway there is a flexible diaphragm. Accumulation of particles in the pathway can cause flow rate reduction, but if the accumulation occurred within the regulation cell, flow rate increase can occur.

As both phenomena may take place simultaneously in the same field, it will be impossible to distinguish flow rate changes from an irrigation controller output. Therefore, follow-up of individual drippers' discharge (at least 25), should be performed in the field in order to evaluate performance.

Emitter clogging and, consequently, irrigation efficiency in microirrigation are directly related to water quality and emitter design. Nevertheless, no foolproof quantitative correlation between irrigation water quality and clogging has been established to date. Emitters with long pathways and without a self-cleaning mechanism were found to be more susceptible to clogging (Ravina et al., 1997; Cararo et al., 2006). Applying effluents of moderate to very low quality and comparing different techniques of filtration, Capra and Scicolone (2004, 2005) concluded that vortex emitters were more sensitive to clogging than labyrinth emitters. A series of criteria for potential clogging or hazard rating, classified as minor, moderate, and severe and based on a few water quality parameters, including total suspended solids (TSS), total dissolved solids (TDS), pH, bacterial counts and ion composition, were published elsewhere (Nakayama and Bucks, 1981, 1991; Taylor et al., 1995; Capra and Scicolone, 1998; Nakayama et al., 2007).

Nakayama and Bucks (1991) classified the clogging criteria for drippers with discharge rates of 2–4 l/hr at an operating pressure of 101.2 kPa. Capra and Scicolone (1998) classified the hazard for large-size drippers with discharge rates of 8–16 l/hr and sprayers. Mainly based on the removal of suspended solids and applying moderate- to very low-quality effluents, Capra and Scicolone (2004) reported that these clogging risk classifications proposed for clean water can only be considered reliable for wastewater when labyrinth emitters and gravel or good-quality disk filters are used; they are not adequate for vortex emitters or screen filters.

As a general rule, the higher the water quality, the lower the expected rate and extent of emitter clogging. Thus, as the current technological feasibility for upgrading effluent quality is almost unlimited, the optimal compromise for irrigation management applying TWW rests upon economical and public health considerations, as well as on environmental regulations.

Treated wastewater contains macro- and micronutrients, which often contribute to the fouling of irrigation and distribution equipment. The composition and quality of TWW is closely related to the degree of treatment applied. Nitrogen and phosphorous species, as well as residual dissolved organic carbon (DOC), are those present at the most significant levels depending on the type of treatment employed (Table 10.1).

The main technologies generating the major quantities of effluents of quality complying with existing regulations in developed countries comprise conventional activated sludge (CAS); enhanced secondary treatment: CAS with biological nutrients removal (BNR); tertiary treatment: CAS with BNR, depth filtration and disinfection; membrane bioreactor (MBR) equipped with either microfiltration (MF) or ultrafiltration (UF) membranes;

Table 10.1 Typical ranges of main nutrients in treated wastewater according to the type of treatment applied

Type of treatment[a]	Parameter range (mg/l)							
	TSS[b]	Colloids	COD	TOC	Total N	Ammonia-N	Nitrate-N	Total P
Secondary	5–25	5–25	40–80	10–40	15–35	10–35[d]	1–5[d]	4–10
Enhanced secondary	5–20	5–10	20–40	10–20	3–10	1–3	1–10	1–2
Tertiary	1–3	1–5	20–30	5–10	2–10	1–2	1–10	<2
Membrane bioreactor[c]	<1	≤1	<10–30	5–10	5–10	<1–5	5–10	0.3–5
Quaternary	<1	≤1	<2–10	0.1–1.0	<1	<0.1	<1	<0.05

[a]Secondary treatment: conventional activated sludge (CAS); enhanced secondary treatment: CAS with biological nutrients removal (BNR); tertiary treatment: CAS with BNR, depth filtration and disinfection; quaternary treatment: CAS with BNR, microfiltration (MF), reverse osmosis (RO) and disinfection.
[b]TSS: total suspended solids.
[c]Without BNR.
[d]These values may change depending on whether nitrification is applied.
Source: Amy and Drewes, 2006; Asano et al., 2007; Katz and Dosoretz, 2008; Lazarova, 2004; Tchobanoglous et al., 2003.

quaternary treatment: CAS with BNR, MF, reverse osmosis (RO) and disinfection; and soil-aquifer treatment (SAT) of CAS effluents. Typical TDS remain almost unchanged during the conventional wastewater treatment (500–1500 mg/l), and only desalination, e.g. RO, reduces TDS values ($\leq$5–50 mg/l). For high-quality TWW, TDS $\approx EC_{25}$ (dS/cm) $\times$ 640, where EC_{25}: electrical conductivity at 25 °C (Tchobanoglous et al., 2003). Soil-aquifer treatment is a procedure applied on large scale in Israel for unrestricted irrigation (Dan Region Project) and in Tucson and Mesa, AZ, in the USA among others, where TWW of improved quality with DOC < 5 mg/l is produced (Amy and Drewes, 2006).

Advanced effluent treatment, using membrane technologies can significantly enhance the final effluent quality. A complete array of pressure-driven membrane separation systems of commercial availability is presented in Figure 10.1, displaying the range of molecular size of particles, colloids, and molecules removed and pressure required for appropriate operation of these systems. Membrane technologies are capable of generating effluents free of pathogens, colloids, dissolved solids, scaling salts, organic matter, and nutrients, as well as emerging contaminants, as required.

The relative size distribution of potential soluble and colloidal materials present in TWW is shown in Figure 10.1. Besides desalination, i.e. RO, levels of micronutrients and TDS remain almost invariable throughout all the other treatments (Tchobanoglous et al., 2003; Lazarova, 2004; Lazarova et al., 2004). Nevertheless, levels of calcium, magnesium, and sulfate can be highly variable according to the source of the freshwater, the sewage origin (domestic, industrial, or agricultural) and the treatment applied.

Organic matter (OM) originating from TWW consists of a variety of macromolecules ranging over a wide range of molecular weights. Chemically, the OM contains polysaccharides, proteins, aminosugars, nucleic acids, humic and fulvic acids, and cell debris. The origin of these materials is attributed to four different sources: (i) refractory natural organic matter (NOM) present in the water sources; (ii) synthetic organic compounds (SOC) introduced as a result of household and personal care products; (iii) soluble

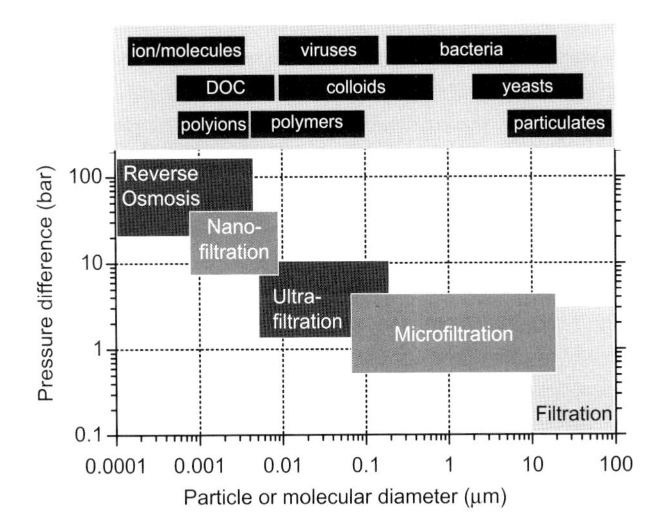

Figure 10.1 Classification of pressure-driven membrane processes. Modified from IVT, Aachen (www.ivt.rwth-aachen.de/index.php?id=617&L=2). DOC, dissolved organic carbon.

microbial products (SMP) produced during biological processes of wastewater treatment; and (iv) disinfection by-products (DBPs) formed during disinfection processes (e.g. chlorination) (Jarusutthirak and Amy, 2001; Jarusutthirak et al., 2002; Her et al., 2003). Most of the OM in secondary effluents is found in the soluble fraction (86% of the chemical oxidation demand, COD) (Shon et al., 2006). The presence of nutrients and residual carbon also affect malodor generation and algal growth during storage and distribution as well as biofilm formation.

Speciation of inorganic nitrogen depends on the source of TWW (see Table 10.1). In reclaimed effluents from CAS treatment without nitrification, NH_4^+ is the primary species (80–95% of total N). Residual nitrogen in effluents of CAS with either nitrification or BNR treatment and MBR plants is predominantly in its inorganic oxidized form, NO_3^-. Phosphate is a regular component of secondary and tertiary effluents. Average phosphate concentration in TWW is in the range of 0.05–10 mg/l, depending on the type of treatment, and it is predominantly in inorganic forms, mainly orthophosphate. In addition to the agronomic constraints of the presence of high level of phosphate in the irrigation water, phosphate along with carbonates is a major contributor to scaling and thus clogging of irrigation devices (Lazarova and Asano, 2004; Asano et al., 2007).

10.3 Emitter clogging in relation to irrigation water quality

Water quality factors affecting fouling and clogging of irrigation devices (pipes, distribution lines, and emitters) can be classified into four different categories as listed in Table 10.2. The factors involved are interrelated and their intensity depends on the

Table 10.2 Main mechanisms of clogging of microirrigation system, denoting factor/consequence relationship

Suspended material (particles and colloids)	Chemical precipitate (scaling)[c]	Organic adhesion (adsorption)[c]	Biological precipitate (biofouling)[d]
Inorganic particles[a]	*Salts*	*Coating*	*Biofilms*
Sand	Ca/Mg carbonate	SMP	Bacteria
Clay	Ca/Mg bicarbonate	NOM	Slime
Silt	Ca/Mg/Fe phosphate	Phenols	*Suspended biomass*[b]
	Ca/Fe sulfate	Tannins	Algae
	Ca silicates		Sloughing biomass
Organic particles[b]	*Hydroxides*	*Sequestering*	*Microbial deposit*
Phytoplankton/algae	Ca	Fe	S oxidation
Zooplankton	Mg	Mn	Fe oxidation
Biosolids			Mn oxidation
Planktonic bacteria			

[a]Depends on the TWW distribution system applied, e.g. open channels.
[b]Cell debris is included in these categories.
[c]SMP: soluble microbial products; NOM: natural organic matter.
[d]In cases in which fertilizer are introduced into the irrigation lines, they may contribute to scaling and biofilm formation.
Source: Asano et al., 2007; Bucks et al., 1979; Nakayama and Bucks, 1991; Nakayama et al., 2007.

conditions. For example colloids and particulates, including biofilms, increase the rate of crystallization by enhancing nucleation; solubility of certain chemical species (e.g. hardness) decreases as temperature increases, whereas microbial activity increases (Ford, 1979); humic acids sequester Fe^{2+} and insoluble Fe^{3+} thereby facilitating its bacterial uptake (Nakayama et al., 2007). Fertilizers added into microirrigation systems (e.g. fertigation) contribute directly to chemical fouling and indirectly to biofouling and algal development but they will not be considered in this chapter.

10.3.1 Suspended material

A suspended solids concentration of 50 mg/l and above is a characteristic of low-quality TWW and includes both organic and inorganic particles of broad particle size distribution (Adin and Sacks, 1991; Nakayama and Bucks, 1991; Capra and Scicolone, 1998). Most perturbing suspended material in TWW of acceptable quality for unrestricted irrigation is of biological origin, and especially microbial aggregates and algae developed in reservoirs. Particulate clogging, which was extensively treated in the literature in the last three decades, is nowadays the most feasible to manage and the least prejudicial when high-quality TWW is applied.

10.3.2 Chemical deposition

Salt, oxides, and hydroxides of bivalent alkaline cations and transition cations are the major mechanism of chemical deposition and clogging in microirrigation in general and in reclaimed water in particular (see Table 10.2). Salinity, which is an important environmental and agronomical water quality parameter, does not contribute directly to chemical deposition (Nakayama et al., 2007; Ravina et al., 1997). The most prevalent reactions leading to precipitate formation involve soluble cationic species such as Ca^{+2}, Mg^{+2}, and Fe^{+2}, and soluble anionic species such as carbonate/bicarbonate, phosphate, and sulfate, and to a minor extent, silicate (Fig. 10.2). The scale forming reaction in irrigation systems depends on pH, temperature, supersaturation concentration, contact time and other factors such as common ion effect, presence of particulates, evaporation rate, flow rate and irrigation regime. Once precipitates are formed, they do not redissolve under natural conditions, and therefore reclamation of clogged emitters requires drastic procedures such as acid treatment.

Compounds carried by the irrigation water as soluble constituents may precipitate and form scale as a result of pressure drop, temperature change, flow velocity alterations, pH or alkalinity change, or other water conditions. Scale is a hard, adherent mineral deposit that usually precipitates from solution and grows in place. Unlike other types of deposition, it is a complex crystallization process. The time it takes for an initial scale layer to form and its subsequent rate of growth are determined by the interaction of several rate-determining processes (e.g. supersaturation, nucleation, diffusion, chemical reaction and molecular arrangement of the scale crystal lattices). Solubility of most calcium and magnesium salts decreases with an increase in temperature (Eaton et al., 1995). Therefore, evaporation in between irrigation periods along with elevated temperatures (45 °C and above) developing in the irrigation lines (most are black to suppress phototrophic activity) lower the equilibrium solubility and promote precipitation.

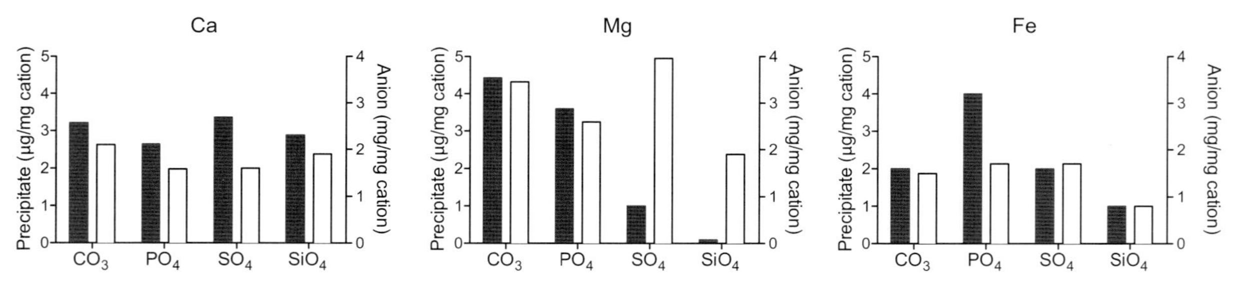

Figure 10.2 Relative precipitation potential rates (scaling) for main complex ions on irrigation systems. (■): precipitate/cation ratio; (□): anion/cation ratio. Theoretical calculations were made on a basis of 100 mg/l for Ca and Mg and 1 mg/l for Fe at STP conditions. Anion concentrations represent the amount required for complete displacement of the reaction. Adapted from Nakayama et al., copyright 2007 with permission of Elsevier.

Changes in the pH or alkalinity of the irrigation water have a major effect on the solubility of scaling ions. An increase in alkalinity reduces solubility of calcium carbonate and also affects the solubility of calcium and iron phosphates. An increase in pH also decreases the solubility of most calcium salts, with the exception of calcium sulfate. Addition of ammonium salts as a fertilizer into the irrigation line raises pH and encourages precipitation of calcium and magnesium. The presence of other complex anions such as sulfate, sulfide, and silicates may encourage precipitation (Asano et al., 2007). Sulfate and silicate scaling is not a common problem in reuse of TWW.

Ways to estimate chemical deposition consist of chemical tests and calculation of saturation index at equilibrium for each individual species based on solubility product, pH, Ca/Mg/Fe concentration, alkalinity and anion concentration, such as the Langelier saturation index (LSI) for carbonates. Application of commercially available chemical equilibrium software, such as MINEQL+ (www.mineql.com/) allows a relatively accurate prediction of chemical precipitation of multiple species expected under defined conditions. Nevertheless, considering the complexity of the conditions (chemical and biological deposition, biofouling, evaporation rates, temperature shifts, etc), predictions should be validated by tests under field conditions.

Precipitation of calcium carbonate ($CaCO_3$) and hardness at supersaturation is the most common scale problem in most water applications and also in irrigation water. Most natural waters, including TWW, contain bicarbonate alkalinity and are at a $pH < 8.2$. When bicarbonate concentration is higher than 2 meq/l and $pH > 7.5$, Ca, Mg, and Fe can be precipitated, the problem is especially acute in high hardness waters. Precipitation may occur in between irrigation periods, when the water remaining in the line evaporates and thus minerals reach supersaturation. Whereas calcium bicarbonate is soluble, calcium bicarbonate will be converted to calcium carbonate if the pH and/or the water temperature are raised, according to reaction 10.1:

$$Ca^{+2} + 2HCO_3^- \rightarrow CaCO_3(s) + H_2O + CO_2 \tag{10.1}$$

Calcium carbonate appears in three polymorphs named calcite (trigonal), aragonite (orthorhombic), and vaterite (hexagonal). A supersaturated solution will begin precipitating as an unstable or metastable form, i.e. amorphous calcium carbonate or aragonite, which will progressively crystallize to the most stable form. Although the final crystal form can change with temperature, calcite is the most thermodynamic stable polymorph at ambient temperature and pressure, and therefore the most common polymorph found in natural systems. High temperatures developed in irrigation devices during summer (main irrigation season) favor formation of calcite.

Precipitation of calcium phosphate would also occur in a similar fashion to carbonate. Phosphate scales occur when conditions are favorable for tricalcium phosphate or whitlockite, $Ca_3(PO_4)_2$, which has a very low-solubility product (2.07×10^{-33}). Temperature, pH, calcium, and orthophosphate concentrations affect the formation of calcium phosphate. The solubility product of the other combinations of calcium phosphate is strongly dependent on pH and most common forms are sparingly soluble. These include hypophosphate, $Ca_2P_2O_6$; metaphosphate, $(Ca(PO_3)_2)_n$; pyrophosphate, $Ca_2P_2O_7$; and the orthophosphate brushite, $CaHPO_4.2H_2O$ (Nakayama et al., 2007). Precipitation of calcium phosphate from unsaturated solutions at common pH values of TWW, is quick and starts initially with the formation of the amorphous precursor, which undergoes a slow transformation, upon nucleation, into the thermodynamically stable form hydroxyapatite, $Ca_5(PO_4)_3(OH)$. Although not specifically described for irrigation systems, during utilization of low-quality TWW, the formation of struvite, magnesium ammonium phosphate hexahydrate ($MgNH_4PO_4.6H_2O$), is also possible. Accumulation of struvite on surfaces that are in contact with TWW is a widely reported problem in the wastewater treatment industry. Struvite precipitation is possible in cases in which the activities of the Mg^{+2}, NH_4^+, and PO_4^{3-} ions exceed the respective solubility product. The extent of struvite precipitation and the characteristics of the precipitating solid depend on the solution pH, supersaturation, temperature, and the presence of additional ions (Kofina and Koutsoukos, 2005; Ohlinger et al., 1998).

10.3.3 Biological deposition

Several lithotrophic filamentous and planktonic slime-forming microorganisms are capable of oxidizing soluble Fe^{+2} (and H_2S when present) into the insoluble Fe^{+3} congener, i.e. $Fe(OH)_3$, which can become a serious clogging problem at an iron concentration above 0.2 mg/l and pH above 8. For this reason, irrigation water containing iron concentrations above 0.5 mg/l is impractical for use in microirrigation systems unless the water is treated. Typical Fe^{+2}-oxidizing microorganisms include algae, fungi, and actinomycetes commonly present in all surface waters. Examples of iron-oxidizing microorganisms reported in irrigation systems are *Thiothrix*, *Leptothrix*, *Gallionela*, *Crenothrix*, and *Sphaerotilus* (Ford, 1979; Gilbert et al., 1982; Brigmon et al., 1997; Asano et al., 2007; Nakayama et al., 2007). The presence of natural iron-sequestering materials, even with low complex stability constants, such as humic acids, phenols, and tannins, can chelate reduced iron species, increasing its concentration and thus increasing its rate of deposition. Furthermore, adsorption of these complexes onto deposited materials or directly onto the irrigation devices can increase the rate of biological scaling (Ford and Tucker, 1975; Nakayama and Bucks, 1991). As reduced iron and sulfur species are at low levels in most TWW, especially those of high quality, iron-induced biological deposition is a minor obstacle during utilization of reclaimed water for microirrigation. However, only short irrigation periods seem to be required for sufficient iron deposit buildup to levels sufficient to support biofilm formation. Reported typical concentrations for total soluble iron in TWW range from 0.2 mg/l after secondary treatment to less than 0.05 mg/l after quaternary treatment (Asano et al., 2007).

10.3.4 Biofouling

Development of biofilms and algae is the most difficult and critical problem to control using TWW. It affects not only emitters but also pipes, storage, and distribution systems (Bucks et al., 1979; Adin and Sacks, 1991; Taylor et al., 1995; Rowan, 2004). Major contaminating bacteria include filamentous and slime-forming heterotrophs, and to a lower extent lithotrophic iron- and sulfur-oxidizers mentioned above. A common practice is to maintain an excess of residual chlorine and perform periodic flushing of distribution lines to reduce microbial growth (Asano et al., 2007; Nakayama et al., 2007).

Algae can be a very severe clogging factor in terms of in situ generation of suspended solids. A primary measure aiming to limit algal growth is restricting nutrient supply and avoiding exposure of the TWW to sunlight to suppress phototrophic activity. However, the concerted action with other algaecides is needed as they can reproduce on the outside orifice of the emitters exposed to sunlight, especially microsprinklers. Microbial predators, and especially protozoa, are also common constituents of biofilms developed on micro-irrigation systems, and in particular when TWW is applied (Dehghanisanij et al., 2004, 2005; Ivnitsky et al., 2005; Dosoretz, 2006; Dosoretz et al., 2006; Asano et al., 2007).

Biofouling is referred to as the undesired development of biofilms on surfaces. Biofilm is a microbial community developing in aggregates that occur in the solid–liquid interface of any flowing system. Organic matter is the main source of biofilm formation in TWW recycling systems. In irrigation systems the film is formed on the surface of the pipes and emitters, and results in the stabilization of conditions favoring hydrophobic interactions, or alternatively on an electrostatically charged surface that will facilitate sorption or attachment of bacteria and other colloids. Following the initial colonization of the site by a small number of bacteria, their population rapidly increases if the environmental conditions support their growth, while exopolymeric substances (EPS) are released and act as a further support for the colony. The EPS consist of polysaccharides, polyuronic acid, proteins, nucleic acids, and lipids; the relative fraction of each of these components varies with the OM composition, the type of dominant bacteria, the physicochemical, and the hydrodynamic conditions. The proportion of EPS can vary between 50% and 80% of the OM and is the main structural component of biofilms (Characklis and Marshall, 1990; Flemming, 1996; Bachmann and Edyvean, 2005).

The physical properties of the biofilm are largely determined by the EPS, whereas the physiological properties are determined by the bacterial cells. Thus, the originally dissolved nutrients are now locally immobilized and converted from a soluble state into a semisolid state. Hence, biofouling is the result of interaction between the surface, dissolved substances, fluid flow parameters, and microorganisms. Biofilms usually exhibit gel-type EPS textures, which induce their physical elasticity, allowing the whole matrix to withstand high water flow rates and the related convection and shear stress forces. Biofilms have, in general, a life-cycle and span similar to that of bacteria in suspension, and their regrowth and/or re-establishment involves a release of cells and cell clusters into the flowing water, thereby facilitating the formation of new colonies/biofilm-supporting sites (de Beer and Stoodley, 2000; Stoodley et al., 2002; Wuertz et al., 2004). A schematic representation of the life-cycle span and organization of a biofilm is depicted in Figure 10.3. Often, along the pipes, these biofilms can act as clogging sites for the pipe lines, taps, valves, or emitters. The physical structure and physiology of the colony and bacterial groups within the biofilm greatly increase their resistance to disinfecting chemicals, such as chlorine and bactericides, and therefore the control of biofilm growth and intensity is greatly enhanced in comparison with the individual small bacterial colonies (de Beer et al., 1994; Flemming, 1997).

The operation of either continuous or intermittent elements involving high surface areas, coupled with the tendency of cells to be transported towards the irrigation head, increases the probability of contact between the microorganism and the head surface. Biofilm formation in systems like irrigation devices can be maintained below a trouble-some level through the application of a series of concerted actions, such as continuous

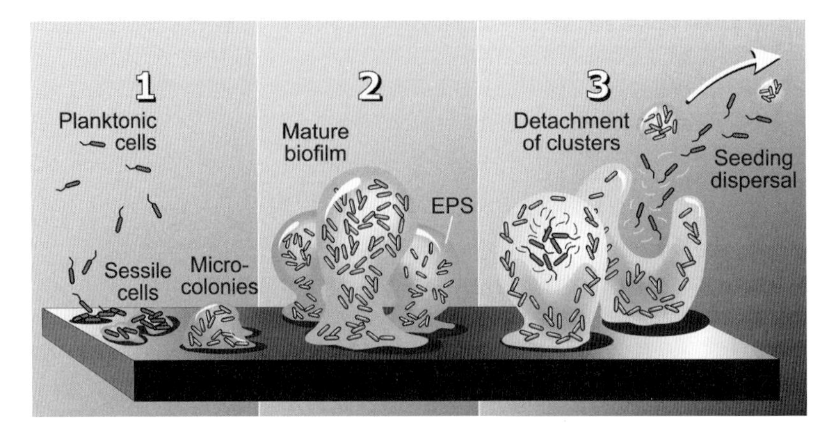

Figure 10.3 Conceptual illustration of biofilm life-cycle and organization. EPS, extracellular polymeric substances. Source: MSU Center for Biofilm Engineering, P. Dirckx.

dosing of sufficiently high biocide concentration, reduction of nutrients, and regular inspection and monitoring (Flemming, 2002). Application of biocides, however, cannot be viewed as an ultimate solution, considering costs, growing environmental concerns, and strict legislative regulations related to the release of oxidizing material into the environment, especially when chlorination is practiced. A more specific concern is the negative effect that the oxidizing material may have on root cells.

Biofilms usually develop in a subtle manner due to the gradual accumulation of the biological deposit. At present, we have no information about the actual growth rates of biofilms in irrigation systems. The concentration of assimilable organic carbon is a key parameter controlling the level of a steady biofilm. Adhesion seems to be the strategy of bacteria to survive low nutrient concentrations. Forming biofilms enhances their capability to scavenge nutrients even at a very low concentration in the feedwater. Early detection of biofilm development and the optimization of countermeasures depend on effective monitoring systems (Flemming, 2003; Lewandowski and Beyenal, 2003).

Biofilms can increase pressure loss due to friction along the length of irrigation lines due to: (i) reduction of the cross-sectional flow area; (ii) oscillation of filaments attached to the biofilms; and (iii) increased roughness (Bishop, 2007; Characklis and Marshall, 1990). Biofilm formed on the surface of the pipeline generates increased roughness, which, in turn, increases friction along the irrigation line, leading to increased resistance to flow and therefore to increased pressure loss along the length of the irrigation line with a consequent decrease of emission uniformity. In addition, biofilm can augment roughness by adsorption of colloidal particles other than the bacterial cells such as clays, $CaCO_3$, and $Ca_3(PO_4)_2$, metal hydroxides, and humic substances. In drip-irrigation, due to the changing flow conditions around and inside the emitter with the formation of a deposition layer, the flow in and out of the dripper varies, increasing or decreasing according to the dripper design. Energy loss in addition to uneven distribution of irrigation water is often observed. Icekson-Tal et al. (2003) reported on pressure losses of more than 35% due to biofilm formation in the pipeline carrying TWW from the Tel-Aviv wastewater treatment plant to their SAT site. Treatment with chlorine reduced the pressure loss but did not eradicate it.

Biofilm contamination of water carriers in large and long-distance pipes is usually the default rather than the exception. Dissolved organic matter (DOM), macro- and micronutrients at increased concentrations, typical of secondary or even tertiary TWW, seem to form the basis for microbial growth in general, and of bacteria in particular. Biofilms formed may also provide the inoculum for contamination of smaller water pipes, namely the irrigation systems in the field.

Although biofilm research has advanced and drawn attention in areas other than irrigation, little is known about water pipes in general and irrigation systems in particular. Wingender and Flemming (2004) reported that, in spite of massive inoculation of pipes with biofilms in drinking water systems, the population did not exceed 10^6–10^7 bacteria/cm^2, and that usually pathogenic bacteria were not present in this population. Biofilms studied in pipes carrying secondary effluents treated with nanofiltration have shown population density was 10^6–10^7 colony forming units (CFU)/cm^2, the same as formerly reported for freshwater systems (Ivnitsky et al., 2005, 2007). This similarity between the bacteria concentration in fresh and treated effluents has been explained by the delicate balance between the input of nutrients and the physical flow stress on the biofilm (Ivnitsky et al., 2005, 2007).

The composition of the microbial populations colonizing the deposition layer in irrigation systems employing TWW has been sparsely reported. However, based on numerous data analyzed in water distribution systems and membrane separation devices, it seems that in spite of a high diversity of microbial species found on water systems biofilms, generic patterns of bacterial colonization can be established (Ivnitsky et al., 2007). According to this pattern, proteobacteria are the predominant heterotrophic species with minor contribution of Gram-positives and other Gram-negatives. This profile seems to be related to a broad ability of surface colonization, adaptability to changing nutrients and capability to proliferate under oligotrophic conditions of these organisms (Ridgway et al., 1983; Baker and Dudley, 1998; Schmeisser et al., 2003; Chen et al., 2004; Horsch et al., 2005). Dominant heterotrophs reported on surface water irrigation biofilms comprise aerobic/facultative species of *Pseudomonas*, *Flavobacterium*, *Cytophaga*, *Micrococcus*, *Acinetobacter*, *Brevibacterium*, *Vibrio*, and *Bacillus* (Gilbert et al., 1982). Development of methane-oxidizing bacteria, as well as hyphae of the fungus *Trichoderma*, clogging drippers in greenhouses were also reported (de Kreij et al., 2003).

Algae are commonly found in biofilms on irrigation systems; however, normally they do not reproduce on it and their presence is a function of the prefiltration efficiency, quality of irrigation water, and storage conditions (e.g. covered or uncovered reservoirs). Common species include among others *Chlorella* (green *Chlorophyta*), diatoms (yellow-green *Chrysophyta*) such as *Navicula*, *Diatomea*, and *Fragilaria* (de Kreij et al., 2003; Dosoretz, 2006; Dosoretz et al., 2006; Nakayama et al., 2007). An example of the deposition layer developed on a drip-irrigation pipe, depicting the presence of different deposition materials, bacteria, bacterial predators, and green algae is presented in Figure 10.4.

10.4 Management of emitter clogging

Emitter clogging should be managed through the control of TWW quality, filtration, emitter design, and appropriate operation and maintenance (Nakayama and Bucks, 1991;

Figure 10.4 Scanning electron microscopy micrographs of the deposition layer developed on drip irrigation pipes, sampled from a commercial field irrigated with low-quality effluents in central Israel. Pipes were cross-sectioned before fixation. Source: Dosoretz et al., 2006.

Tajrishy et al., 1994; Taylor et al., 1995; Ravina et al., 1997; Ould Ahmed et al., 2007). The manufacturers of microirrigation systems, and in particular products employed in drip-irrigation systems applying TWW, have made numerous modifications to emitter design and other system components aiming to reduce clogging, and especially biofouling, and ensure an even water supply. Measures to prevent clogging comprise quality monitoring, filtration, selection of appropriate equipment, control of flow rates, flushing, and disinfection (Asano et al., 2007; Trooien and Hills, 2007). Partial clogging by biofilm buildup and particle accumulation in the corners of tortuous pathways within the dripper is the main mechanism of emitter clogging in TWW irrigation; short pathways and efficient self-cleaning membranes are desired emitter characteristics for wastewater irrigation (Cararo et al., 2006).

Proper management of emitter clogging includes the concerted action of several factors, comprising filtration to remove particulate material, dosing of biocides to restrict microbes, acidification and/or sequestering agents to reduce scale formation and flushing to empty the irrigation line. During application of treated effluents as sources of irrigation water, special attention should be paid to their nutrient content to ensure long-term function without disturbances.

10.4.1 Filtration of particulate material

As microirrigation is sensitive to clogging by suspended material, its removal is a prerequisite for long-term performance. Sediment deposition begins at the ends of irrigation lines, where flow velocities are reduced to a value that no longer keeps particulate material in suspension (Shanon et al., 1982; Ravina et al., 1992; Trooien et al., 2000; Trooien and Hills, 2007). This trend varies according to the irrigation regime applied, type and mode of operation of emitter, and line length.

The most typical types of filters applied for removal of suspended matter in micro-irrigation are sand media (1-mm particle), disk, cartridge, and screen filters, and their combinations. When using secondary-treated quality effluents, filtration with at least

80 mesh opening (equivalent to 180 μm) and bimonthly chlorination are required (Ravina et al., 1992; Nakayama et al., 2007; Trooien and Hills, 2007). Filter operation requires back-wash and disinfection (chlorination), especially in media filters, which are more prone to biofilm development and algae accumulation. Algae development due to water exposed to sunlight will increase the need for chemical oxidants and back-washing and therefore should be avoided. According to the supplier and type, filters are equipped with automatic self-cleaning control based on elapsed time or pressure drop limit. However, it should be noted that some filters require manual care. Filtration of irrigation water has been extensively described and discussed in the literature, and therefore will not be further described here (for most recent reviews consult Asano et al., 2007; Nakayama et al., 2007).

Filtration alone is not enough to ensure long-term performance of microirrigation emitters. Conventional filtration methods applied in water reclamation for irrigation are unable to remove supracolloidal particles (1–10 μm) and colloids (<1 μm) (see Fig. 10.1). Although control of biofouling and scaling requires further treatment, removal of small particles and colloids will certainly reduce clogging and improve irrigation efficiency. Self-cleaning emitters are not clogged by large particles (400 μm and above); however, proliferation of algae (~150 μm), bacterial aggregates (10–100 μm), biofilms and scaling deposition would decrease emitter efficiency, reduce irrigation uniformity, and finally clog them (Adin and Sacks, 1991; Feigin et al., 1991; Ravina et al., 1992; Tajrishy et al., 1994; Ravina et al., 1997; Trooien and Hills, 2007). Reduction of supracolloids and colloids is possible by application of membrane filtration (see Fig. 10.1). It should be noted that upgraded effluents upon either tertiary treatment with membrane filtration or MBR or RO are free of small supracolloids and colloids (see Table 10.1).

10.4.2 Biofouling control

Biofouling is an operational definition, referring to the amount of biofilm development that interferes with technical or economic requirements (Flemming et al., 1997; Flemming, 2002). To date, biocides are the most common answer to biofouling. However, as microorganisms are ubiquitous and disinfection does not lead to sterility in technical systems, they will always be present and reinfect the system as soon as the biocide application is suspended. The efficacy of biocides against biofilm organisms is lower than that against the same organisms suspended in the water phase, and therefore high doses should be maintained (Flemming, 1997; Davey and O'Toole, 2000). As already mentioned, adhesion seems to be the strategy of bacteria to survive in flowing systems, forming biofilms capable of scavenging nutrients at concentrations higher than the feed water. Therefore, a series of concerted actions should be maintained in order to control biofilm formation and keep biofuling below economic detrimental levels (Flemming, 2003). For practical irrigation purposes, these actions include reduction of nutrients content and dosing of biocides.

10.4.2.1 Reduction of nutrient content

As it is virtually impossible to keep a field system completely sterile, microorganisms on surfaces will always be present. All biotransformable substances, including inorganic sources in the case of lithotrophic organisms, must be considered as potential biomass. Basically, the extent of biofilm growth is controlled generally by the availability of nutrients and the shear forces. Although no single action is effective in controlling biofilm development, due to the low shear forces developed in microirrigation systems, reduction of nutrients is a basic action to be taken for biofouling control (for nutrient levels in reclaimed water refer to Table 10.1). Furthermore, reduction of nutrient levels of TWW will also reduce soil and groundwater pollution and help to maintain sustainable agriculture.

10.4.2.2 Dosing of biocides

Dosing of oxidizing biocides, and particularly chlorination, is the most common method practiced to prevent clogging by biofilm formation and algal proliferation when TWW is applied for microirrigation (Ravina et al., 1992, 1997; Tajrishy et al., 1994; Rav-Acha et al., 1995; Asano et al., 2007). Due to the increased resistance of biofilms to biocides, biocides dosing, in order to be effective, should be performed as a preventive treatment before clogging of the emitters becomes extensive. As an example, the active chlorine concentration within a biofilm decreases to 20% or less of that in the bulk liquid, to almost none close to the substratum surface, because of the protective and shielding nature of the biofilm (Characklis and Marshall, 1990; Flemming, 1996; de Beer and Stoodley, 2000; Bachmann and Edyvean, 2005).

Chlorination to a free chlorine residual concentration of 0.5–1 mg HOCl/l at the end of each irrigation cycle, in combination with periodic flushing of distribution lines, is an accepted preventive protocol employed to control biological clogging to reduce bacterial growth (Tajrishy et al., 1994; Cararo et al., 2006). Chlorination is managed by either continuous or intermittent ("shock") application, based on a target residual chlorine concentration. In continuous chlorination an acceptable level of residual chlorine is 0.5 mg/l measured at the most distant emitter. Higher levels can be deleterious to plants (Ford, 1976; Ravina et al., 1992, 1997; Tajrishy et al., 1994; Lazarova, 2004). Intermittent chlorination at concentrations of 2 mg/l residual chlorine was reported to be as satisfactory as 0.4–0.5 mg/l in continuous chlorination (Capra and Scicolone, 1998, 2004, 2005, 2007; Trooien and Hills, 2007). It should be noted that the efficiency of chlorination measured in terms of emitter performance is strongly dependent on the quality of the effluents, extent of filtration, flow velocity, type of emitter, and mode of operation and regime of irrigation. An example of the effect of chlorination on biofouling control is depicted in Figure 10.5, summarizing a pilot plant experiment conducted under controlled conditions in which drip lines were exposed to irrigation with high-quality secondary effluent in the presence and absence of NaOCl (Dosoretz et al., 2006). The drippers treated with chlorine were visually clean with minor salt deposition, whereas those fed with secondary effluent were covered by a slimy organic fouling layer.

The efficiency of chlorination strongly depends on the presence of reducing species in the irrigation water. Several components in TWW, such as DOM, ammonium and amines, consume active chlorine. Therefore, the required concentration of active chlorine at the discharge point should be calculated or measured by trial and error according to the effluent composition. As active chlorine can react with suspended matter, chlorination is commonly applied after filtration. Chlorine can be consumed to form insoluble oxidizing oxides and hydroxides by reacting with dissolved reduced species such as Fe^{+2} and Mn^{+2}

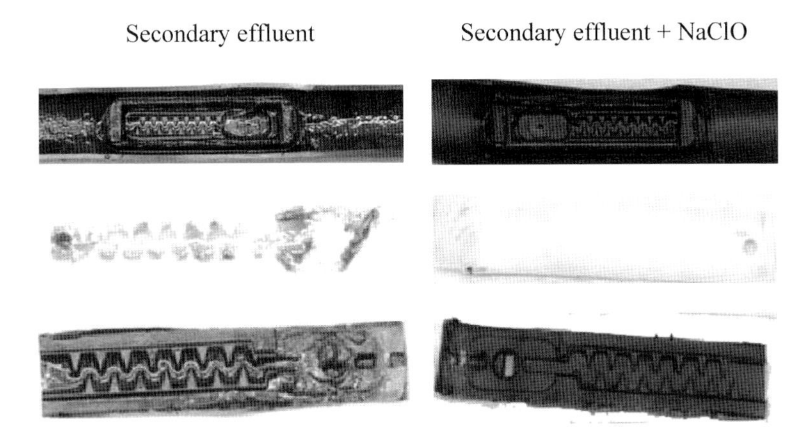

Figure 10.5 Photographs of the cross-section of irrigation line samples showing a regulated dripper (upper panel), and disassembled diaphragm (mid panel) and labyrinth (bottom panel), at the end of an experimental irrigation season with high-quality secondary effluents. Note the slimy (bright) material coating the pipe without cleaning treatment (secondary effluent) and deposit accumulated on the dripper diaphragm (yellowish) and labyrinth when no chlorination was applied. Source: Dosoretz et al., 2006.

forming particulate material. Therefore, special attention should be paid when the concentration of these species is high. Appropriate management of the contact time may result in the avoidance of coagulation of the oxides and formation of agglomerates (Asano et al., 2007; Nakayama et al., 2007). The same is true for sulfides (S^{-2}, HS^-). Active chlorine tends to react with free ammonia and amine groups to produce chloramines. Even though these compounds exhibit longer residual activity, they have approximately 80 times less biocidal power (Feigin et al., 1991; de Beer et al., 1994; Tchobanoglous et al., 2003). These, added to the increased resistance of mature biofilms to penetration of biocides, considerably decrease chlorination efficiency. The final pH of the chlorinated effluents changes according to the source of chlorine and buffer capacity of the effluents and should also be checked to ensure effective control of biofilm formation. When Cl_2 gas is the source of hypochlorous acid, there is a release of H^+, which may consume alkalinity and decrease pH according to reaction 10.2:

$$Cl_2 + H_2O \leftrightarrows H^+ + Cl^- + HOCl \qquad (10.2)$$

When sodium or calcium hypochlorite are the source of HOCl there is a release of hydroxyls which may increase pH according to reactions 10.3 and 10.4:

$$NaClO + H_2O \leftrightarrows Na^+ + OH^- + HOCl \qquad (10.3)$$

$$Ca(ClO)_2 + 2H_2O \leftrightarrows Ca^{+2} + 2OH^- + 2HOCl \qquad (10.4)$$

Other oxidizing biocidal procedures were studied to prevent biofilm formation during irrigation, including ClO_2, peracetic acid, ozone, photocatalytic TiO_2 and ultraviolet irradiation (Rav-Acha et al., 1995; Hills et al., 2000; Lazarova, 2004; Lubello et al., 2004;

Asano et al., 2007; Trooien and Hills, 2007). Although most have not been applied in field scale, these procedures were focused on public health-related issues of restricted irrigation, and especially eradication of viruses, rather than biofilm control. Although very effective in controlling algal growth, the use of copper sulfate, which was widely applied in the past, is decreasing due to Cu toxicity.

The use of non-oxidizing biocides in irrigation as dosing disinfectants has not been reported, most probably because of their high price. Nonetheless, quaternary ammonium salts and other non-oxidizing biocides are commonly used for disinfection of filters applied in irrigation lines (Nakayama et al., 2007). Biocides containing heavy metals (their use is nowadays restricted in most developed countries), were incorporated as an impregnating material into irrigation equipment by some manufactures to prevent slime formation. Examples of these biocides are the organoarsenic compound 10,10-oxybisphenoxyarsine, Vinyzene™ and the organotin compound tributyltin maleate (Ultra Fresh DM50™) (Rowan, 2004).

10.4.3 Control of chemical and biological deposition

Scale prevention due to precipitation and crystallization of calcium phosphate and carbonate principally, is common in application of TWW (Nakayama et al., 2007; Trooien and Hills, 2007). This phenomenon depends on the quality of the effluents applied, presence of particulate material and biofilm formation, roughness and flow velocity, and mode of operation and regime of irrigation. As pH increases the rate of scaling increases, therefore acidification to pH 6.2–6.5 by continuous dilute acid treatment can help to reduce mineral deposit to almost completion (most natural waters contain bicarbonate alkalinity and are at a pH 8.2). Calcium carbonate scale control involves the release of gaseous CO_2 according to reaction 10.5:

$$Ca^{+2} + 2HCO_3^- + 2H+ \rightarrow Ca^{+2} + 2CO_2(g) + 2H_2O \qquad (10.5)$$

Reduction of phosphate scale by pH control can be achieved by a shift in the equilibrium of $Ca_3(PO_4)_2$ precipitation, which has a very low solubility product, and therefore reduction of pH can be effective at relatively low dissolved orthophosphate concentrations (up to 2–4 mg/l). Higher concentration will require pretreatment of the effluents prior to use aiming to avoid scaling formation (Katz and Dosoretz, 2008). Iron control by pH reduction is ineffective. Although this is not a common scaling factor in treated effluent irrigation, use of antiscalants or stabilizers (e.g. chelation with sequestering agents) or pretreatment should be considered if concentrations are above 2 mg/l.

The required amount of acid to be added can be estimated from chemical equilibrium calculations based on the initial pH, buffer capacity, and precipitating species concentration. According to Nakayama et al. (2007), for water with an initial pH of 8 an addition of 0.5 meq/l acid will cause a drop in pH of approximately one unit, whereas dosing of 1.0 meq/l acid would decrease the pH to 6.0–6.5. These values can differ significantly according to the buffer capacity and alkalinity of the irrigation water, and for the case of treated effluents according to the source and degree of treatment applied (see Table 10.1). Although phosphoric or sulfuric acid are preferable based on soil structure considerations, the use of hydrochloric acid may be a more beneficial, cheaper and less dangerous practice used to reduce total concentration of precipitating species and control alkalinity. The contribution of acid dosing, as well as other chemicals, to TDS and salinity should be taken into account depending on the sensibility of the crop at the growth stage at the treatment time and on the soil buffer properties. Inert growth media is a particular case where special caution has to be taken.

Controlling the pH of the water is a key factor for both chemical and biocidal effects when chlorination is applied. Chlorination is relatively inefficient at $pH > 7.5$. Calcium and magnesium carbonates precipitates at higher rates at $pH > 7.2$ (depending of the temperature). At $pH \leq 6.5$, stability of complexes or iron with tannins and humic substances increases where phosphate precipitation rate decreases.

10.4.4 Flushing of the irrigation line

Due to economical considerations, removal of particles $\leq 10\%$ of the emitter orifice, i.e. small particles, supracolloids, and colloids, from irrigation water is not practiced (Nakayama et al., 2007). As described above, although these particles are small enough to be discharged, they form an integral part of the clogging or fouling layer, they tend to agglomerate, become entrapped in the biofilm EPS and can be occluded in the scaling layer. These small particles and colloid are transported at high flow velocities, but they start settling and depositing as the flow velocity decreases along the irrigation line. At the very end of the irrigation line and laterals, the flow rate is reduced to that of the emitter discharge, where clogging is more intense.

Flushing of the irrigation lines within and prior to the irrigation period is a routine procedure applied to remove particulate material out from the irrigation line. The frequency of flushing depends on the type of emitter, particle cut-off and efficiency of filtration, and combination with other maintenance procedures. Flushing of the irrigation lines, performed at flow velocities of 0.3–0.6 m/s, much higher than irrigation velocities, can to some extent erode the fouling layer deposited at much lower shear conditions and washout sloughed agglomerates (Adin and Sacks, 1991; Ravina et al., 1997; Hills and Brenes, 2001; Nakayama et al., 2007). Flushing frequencies of once every 2 weeks were reported for dripper irrigation (Ravina et al., 1997). Although, flushing alone cannot solve emitter clogging, periodical application in conjunction with other maintenance procedures can alleviate the problem and extend the period uniform efficiency.

10.5 Recovery of clogged emitters

Restoration of emitter efficiency is based on the nature of the most permanent deposit, namely removal of scaling and biofilm layer. Therefore, procedures for recovery of clogged emitters comprise use of acid to pH 2 to remove the reversible scale and alkaline hydrolysis and/or chemical oxidant to remove biological slime material. Recovery procedures should be employed in accordance with the chemical resistance of the different components of the emitters and tubing. Recovery procedures are always accompanied by flushing of irrigation water in order to remove the dissolved salts and the residual acid.

Chemical weakening of the biofilm matrix, which is the most complex fouling layer to remove, is mostly attained by use of oxidants (chlorine, ozone, hydrogen peroxide, peracetic acid, etc.), alkaline treatment (NaOH, pH 10), enzymes, or by dispersants. Dispersant are based on polyethylene glycol and are supposed to weaken the interactions in the EPS matrix, as well as the interaction between biofilm and support material. If an oxidizing biocide is used, it will weaken the biofilm matrix as a side effect. A combination of various agents may increase the treatment efficiency (Flemming, 2002).

It is important to establish some testing system for quantifying the success of irrigation system recovery. To date, the efficacy of a cleaning measure is assessed by the recovery measured as close as possible to the initial emitter discharge, which in most cases is never recovered to its initial level and does not account for pressure losses due to increase of roughness of the tubing surface. However, this is an indirect method and it is not very well suited to reliable assessment of success. Sporadic autopsy of irrigation devices may give an overall estimation of clogging tendency and recovery efficiency in addition to agronomic tests.

10.6 Concluding remarks and future prospects

Emitter fouling is the main cause of uneven water discharge in microirrigation, and even a few clogged emitters can greatly reduce the uniformity of water application and affect crop yield and productivity. Even though no foolproof quantitative correlation between irrigation water quality and clogging can be established, as a general rule, the expected rate and extent of emitter malfunction is inversely proportional to the irrigation water quality. As existing technological capabilities for upgrading effluent quality are almost unlimited, the optimal compromise for irrigation management applying TWW rests on economic and public health considerations, and environmental regulations. Upgrading TWW quality will alleviate soil and groundwater pollution, reduce public health risks, reduce irrigation costs, and help to maintain sustainable agriculture. Upgraded TWW are characterized by low nutrient and OM concentrations with a concomitant decrease in chlorine demand, easier control of biofilm formation, and reduced scaling deposition.

Proper management of emitter clogging includes the integrated action of several factors, comprising filtration to remove particulate material, dosing of biocides to restrict microbes, acidification and/or antiscalant dosing to reduce scale formation and flushing to empty the irrigation lines. It is important to collect periodic flow rate measurements and to check discharge of individual drippers to determine system performance. Measurement of initial and final flow rates provides only a snapshot analysis of the irrigation system that is not necessarily representative of its overall performance. Chemical analyses of the irrigation water amended by chemical equilibrium calculations can provide a fair estimate of the expected chemical precipitation/deposition under defined evaporation rates and temperature shift conditions.

During application of TWW as a source of irrigation water, special attention should be paid to the OM and nutrient contents to ensure long-term functioning without disturbances. Development of biofilms is the most difficult and critical problem to control using TWW. It affects not only emitters, but also pipes, storage and distribution systems. Organic matter is the main source of biofilm formation in TWW recycling systems. Major contaminating bacteria include filamentous and slime-forming heterotrophs, and to a lower extent lithotrophic iron- and sulfur-oxidizers.

Additional research is still needed to confirm the pattern of biofilm formation during microirrigation and its quantitative correlation with emitter discharge applying TWW as irrigation water. Determining the composition of the biofilms is also recommended. Understanding the microbial community structure growing in the driplines and emitters could improve the rational application of cleaning and disinfecting protocols and treatment effectiveness. Whether or not pathogenic bacteria can develop in biofilms on irrigation lines also needs to be assessed, as this may be a limitation for sustainable water reuse in agriculture.

References

Adin, A. (1986) Problems associated with particulate matter in water reuse for agricultural irrigation and their prevention. *Water Science and Technology* **18**, 185–95.

Adin, A. (1987) Clogging in irrigation systems reusing pond effluents and its prevention. *Water Science and Technology* **19**, 323–328.

Adin, A., and Sacks, M. (1991) Dripper-clogging factors in wastewater irrigation. *Journal of Irrigation and Drainage Engineering* **117**, 813–826.

Amy, G., and Drewes, J. (2006) Soil Aquifer Treatment (SAT) as a natural and sustainable wastewater reclamation/reuse technology: fate of wastewater effluent organic matter (Efom) and trace organic compounds. *Environmental Monitoring and Assessment* **12**, 181–190.

Asano, T., Burton, F.L., Leverenz, H.L., Tsuchihashi, R. and Tchobanoglous, G. (2007) *Water Reuse: Issues, Technologies and Applications*. Metcalf & Eddy|AECOM. McGraw Hill Inc., New York, NY.

Bachmann, R.T., and Edyvean, R.G.J. (2005) Biofouling: an historic and contemporary review of its causes, consequences and control in drinking water distribution systems. *Biofilms* **2**, 197–227.

Baker, J.S. and Dudley, L.Y. (1998) Biofouling in membrane system-a review. *Desalination* **118**, 81–90.

Bishop, P.L. (2007) The role of biofilms in water reclamation and reuse. *Water Science and Technology* **55**, 19–26.

Bralts, V.F., Wu, I.P. and Gitlin, H.M. (1981) Drip irrigation uniformity: considering emitter plugging. Transactions of the ASAE **24**, 1234–1240.

Brigmon, R.L., Martin, H.W. and Aldrich, H.C. (1997) Biofouling of groundwater systems by Thiothrix species. *Current Microbiology* **35**, 169–174.

Bucks, D.A., Nakayama, F.S. and Gilbert, R.G. (1979) Trickle irrigation water quality and preventive maintenance. *Agricultural Water Management* **2**, 149–162.

Capra, A. and Scicolone, B. (1998) Water quality and distribution uniformity in drip/trickle irrigation systems. *Journal of Agricultural Engineering Research* **70**, 355–365.

Capra, A. and Scicolone, B. (2004) Emitter and filter tests for wastewater reuse by drip irrigation. *Agricultural Water Management* **68**, 135–149.

Capra, A. and Scicolone, B. (2005) Assessing dripper clogging and filtering performance using municipal wastewater. *Irrigation and Drainage* **54**, S71–S79.

Capra, A. and Scicolone, B. (2007) Recycling of poor quality urban wastewater by drip irrigation systems. *Journal of Cleaner Production* **15**, 1529–1534.

Cararo, D.C., Botrel, T.A., Hills, D.J. and Leverenz, H.L. (2006) Analysis of clogging in drip emitters during wastewater irrigation. *Applied Engineering in Agriculture* **22**, 251–257.

Characklis, W.G., and Marshall, K.C. (1990) Biofilms. John Wiley & Sons, New York, NY.

Chen, C.L., Liu, W.T., Chong, M.L., Wong, M.T., Ong, S.L., Seah, H. and Ng, W.J. (2004) Community structure of microbial biofilms associated with membrane-based water purification processes as revealed using a polyphasic approach. *Applied Microbiology and Biotechnology* **63**, 466–473.

Davey, M.E., and O'Toole, G.A. (2000) Microbial biofilms: from ecology to molecular genetics. *Microbiology and Molecular Biology Reviews* **64**, 847–867.

de Beer, D. and Stoodley, P. (2000) *Microbial Biofilms*, http://et.springer-ny.com:8080/prokPUB/chaprender/jsp/showchap.jsp?chapnum=267 The Prokaryotes. Springer, New York, NY.

de Beer, D., Srinivasan, R. and Stewart, P.S. (1994) Direct measurement of chlorine penetration in biofiloms during disinfection. *Applied Environmental Microbiology* **60**, 4339–4344.

de Kreij, C., van-der Burg, A.M.M. and Runia, W.T. (2003) Drip irrigation emitter clogging in Dutch greenhouses as affected by methane and organic acids. *Agricultural Water Management* **60**, 73–85.

Dehghanisanij, H., Yamamoto, T., Rasiah, V., Utsunomiya, J. and Inoue, M. (2004) Impact of biological clogging agents on filter and emitter discharge characteristics of microirrigation systems. *Irrigation and Drainage* **53**, 363–373.

Dehghanisanij, H., Yamamoto, T., Ould Ahmad, B., Fujiyama, H. and Miyamoto, K. (2005) The effect of chlorine on emitter clogging induced by algae and protozoa and the performance of drip irrigation. *Transactions of the ASAE* **48**, 519–527.

Dosoretz, C.G. (2006) *Biofouling Build-Up on Dense Membranes in Pressure-Driven Separation Processes for Wastewater Treatment*. U.S.–Israeli Workshop on Nanotechnology for Water Purification, Arlington, VA. INNI-Water Campus-NSF. www.nanoisrael.org/download/nanowater1/US-IL%20NanoWater%20Workshop%20Program%20Final%2009-MAR.pdf.

Dosoretz, C.G., Katz, I., Tarchitzky, J., Kenig, E., Gips, A. and Chen, Y. (2006) *Study of Biofilm Formation in Irrigation and Treated Effluents Supply Systems*. Project No. 838-0527-05, Chief Scientist Ministry of Agriculture, Bet Dagan, Israel.

Eaton, A.D., Clesceri, L.S. and Greenberg, A.E. (1995) *Standard Methods for the Examination of Water and Wastewater*, 19th ed. American Public Health Association, Washington, DC.

Feigin, A., Ravina, I. and Shalhevet, J. (1991) *Irrigation with Treated Sewage Effluent: Management for Environmental Protection*. Springer-Verlag, Heidelberg, Germany.

Flemming, H.C. (1996) Antifouling strategies in technical systems - a short review. *Water Science and Technology* **34**, 517–524.

Flemming, H.C. (1997) Reverse osmosis membrane biofouling. *Experimental Thermal and Fluid Science* **14**, 1382–1391.

Flemming, H.C. (2002) Biofouling in water systems - cases, causes and countermeasures. *Applied Microbiology and Biotechnology* **59**, 629–640.

Flemming, H.C. (2003) Role and levels of real-time monitoring for successful anti-fouling strategies - an overview. *Water Science and Technology* **47**, 1–8.

Flemming, H.C., Scaule, G., Griebe, T., Schmitt, J. and Tamachkiarowa, A. (1997) Biofouling-the Achilles heel of membrane processes. *Desalination* **113**, 215–225.

Ford, H.W. (1976) The use of chlorine in drip irrigation systems. *Proceedings of the Florida State Horticultural Society* **88**, 1–4.

Ford, H.W. (1979) Characteristics of slime and ochre in drainage and irrigation systems. *Transactions of the ASAE* **22**, 1093–6.

Ford, H.W. and Tucker, D.P.H. (1975) Blockage of drip irrigation filters and emitters by iron-sulfur-bacterial products. *HortScience* **10**, 62–64.

Gilbert, R.G., Nakayama, F.S., Bucks, D.A., French, O.F., Adamson, K.C. and Johnson, R.M. (1982) Trickle irrigation: predominant bacteria in treated Colorado river water and biologically clogged emitters. *Irrigation Science* **3**, 123–132.

Her, N., Amy, G., McKnight, D., Sohn, J. and Yoon, Y. (2003) Characterization of DOM as a function of MW by fluorescence EEM and HPLC-SEC using UVA, DOC, and fluorescence detection. *Water Research* **37**, 4295–303.

Hills, D.J., and Brenes, M.J. (2001) Microirrigation of wastewater effluent using drip tape. *Applied Engineering in Agriculture* **17**, 303–308.

Hills, D.J., Tajrishy, M.A. and Tchobanoglous, G. (2000) The influence of filtration on ultraviolet disinfection of secondary effluent for microirrigation. *Transactions of the ASAE* **43**, 1499–1505.

Horsch, P., Gorenflo, A., Fuder, C., Deleage, A. and Frimmel, F.H. (2005) Biofouling of ultra- and nanofiltration membranes for drinking water treatment characterized by fluorescence in situ hybridization (FISH). *Desalination* **172**, 41–52.

Icekson-Tal, N., Avraham, O., Sack, J. and Cikurel, H. (2003) Water reuse in Israel - the Dan Region Project: evaluation of water quality and reliability of plant's operation. *Water Science and Technology* **3**, 231–237.

Ivnitsky, H., Katz, I., Minz, D., Shimoni, E., Chen, Y., Tarchitzky, J., Semiat, R. and Dosoretz, C.G. (2005) Characterization of membrane biofouling in nanofiltration processes of wastewater treatment. *Desalination* **185**, 255–268.

Ivnitsky, H., Katz, I., Minz, D., Volvovic, G., Shimoni, E., Kesselman, E., Semiat, R. and Dosoretz, C.G. (2007) Bacterial community composition and structure of biofilms developing on nanofiltration membranes applied to wastewater treatment. *Water Research* **41**, 3924–3935.

Jarusutthirak, C., and Amy, G. (2001) Membrane filtration of wastewater effluents for reuse: effluent organic matter rejection and fouling. *Water Science and Technology* **43**, 225–32.

Jarusutthirak, C., Amy, G. and Croué, J.-P. (2002) Fouling characteristics of wastewater effluent organic matter (EfOM) isolates on NF and UF membranes. *Desalination* **145**, 247–255.

Katz, I., and Dosoretz, C.G. (2008) Desalination of domestic wastewater effluents: phosphate removal as pretreatment. *Desalination* **222**, 230–242.

Kofina, A.N., and Koutsoukos, P.G. (2005) Spontaneous precipitation of struvite from synthetic wastewater solutions. *Crystal Growth and Design* **5**, 489–496.

Lazarova, V. (2004) Wastewater treatment for water recycle. In: *Water Reuse for Irrigation: Agriculture, Landscape and Turf Grass* (eds V. Lazarova and A. Bahri), pp. 163–234. CRC Press, Boca Raton, FL.

Lazarova, V., and Asano, T. (2004) Challenges of sustainable irrigation with recycled water. In: *Water Reuse for Irrigation: Agriculture, Landscape and Turf Grass* (eds V. Lazarova and A. Bahri), pp. 1–30. CRC Press, Boca Raton, FL.

Lazarova, V., Bower, H. and Bahri, A. (2004) Water quality considerations. In: *Water Reuse for Irrigation: Agriculture, Landscape and Turf Grass* (eds V. Lazarova and A. Bahri), pp. 31–60. CRC Press, Boca Raton, FL.

Lewandowski, Z. and Beyenal, H. (2003) Biofilm monitoring: a perfect solution in search of a problem. *Water Science and Technology* **47**, 9–18.

Lubello, C., Gori, R., Nicese, F.P. and Ferrini, F. (2004) Municipal-treated wastewater reuse for plant nurseries irrigation. *Water Research* **38**, 2939–2947.

Nakayama, F.S. and Bucks, D.A. (1981) Emitter clogging effects on trickle irrigation uniformity. *Transactions of the ASAE* **24**, 77–80.

Nakayama, F.S., and Bucks, D.A. (1991) Water quality in drip/trickle irrigation. *Irrigation Science* **12**, 187–192.

Nakayama, F.S., Boman, B.J. and Pitts, D.J. (2007) Maintenance. In: *Microirrigation for Crop Production. Design, Operation and Management* (eds F. R. Lamm, et al.), pp. 389–430. Elsevier, Amsterdam, The Netherlands.

Ohlinger, K.N., Young, T.M. and Schroeder, E.D. (1998) Predicting struvite formation in digestion. *Water Research* **32**, 3607–3614.

Ould Ahmed, B.A., Yamamoto, T., Fujiyama, H. and Miyamoto, K. (2007) Assessment of emitter discharge in microirrigation system as affected by polluted water. *Irrigation and Drainage Systems* **21**, 97–107.

Rav-Acha, C., Kummel, M., Salamon, I. and Adin, A. (1995) The effect of chemical oxidants on effluent constituents for drip irrigation. *Water Research* **29**, 119–129.

Ravina, I., Paz, E., Sofer, Z., Marcu, A., Shisha, A. and Sagi, G. (1992) Control of emitter clogging in drip irrigation with reclaimed wastewater. *Irrigation Science* **13**, 129–139.

Ravina, I., Paz, E., Sofer, Z., Marcu, A., Schischa, A., Sagi, G., Yechialy, Z. and Lev, Y. (1997) Control of clogging in drip irrigation with stored treated municipal sewage effluent. *Agricultural Water Management* **33**, 127–137.

Ridgway, H.F., Kelly, A., Justice, C. and Olson, B.H. (1983) Microbial fouling of reverse-osmosis membranes used in advanced wastewater treatment technology: chemical, bacteriological, and ultrastructural analyses. *Applied Environmental Microbiology* **45**, 1066–1084.

Rowan, M.A. (2004) *The Utility of Drip Irrigation for the Distribution of On-Site Wastewater Effluent*. PhD dissertation, Food, Agricultural, and Biological Engineering, Ohio State University, Columbus, OH.

Schmeisser, C., Stöckigt, C., Raasch, C., Wingender, J., Timmis, K.N., Wenderoth, D.F., Flemming, H.-C., Liesegang, H., Schmitz, R.A., Jaeger, K.-E. and Streit, W.R. (2003) Metagenome Survey of Biofilms in Drinking-Water Networks. *Applied Environmental Microbiology* **69**, 7298–7309.

Shanon, W.M., James, L.G., Basset, D.L. and Mih, W.C. (1982) Sediment transport and depositon in trickle irrigation laterals. *Transactions of the ASAE* **25**, 160–164.

Shon, H.K., Vigneswaran, S. and Snyder, S.A. 2006. Effluent organic matter (EfOM) in wastewater: constituents, effects, and treatment. *Critical Reviews in Environmental Science and Technology* **36**, 327–374.

Stoodley, P., Sauer, K., Davies, D.G. and Costerton, J.W. (2002) Biofilms as complex differentiated communities. *Annual Review of Microbiology* **56**, 187–209.

Tajrishy, M.A., Hills, D.J. and Tchobanoglous, G. (1994) Pretreatment of secondary effluent for drip irrigation. *Journal of Irrigation Drainage and Engineering* **120**, 716–731.

Taylor, H.D., Bastos, R.K.X., Pearson, H.W. and Mara, D.D. (1995) Drip irrigation with waste stabilization pond effluents: solving the problem of emitter fouling. *Water Science and Technology* **31**, 417–424.

Tchobanoglous, G., Burton, F.L. and Stensel, H.D. (2003) *Wastewater Engineering: Treatment, Disposal and Reuse*. 4rd ed. Metcalf & Eddy. McGraw Hill, New York, NY.

Trooien, T.P., and Hills, D.J. (2007) Application of biological effluents. In: *Microirrigation for Crop Production. Design, Operation and Management* (eds F.R. Lamm et al.), pp. 329–356. Elsevier, Amsterdam, The Netherlands.

Trooien, T.P., Lamm, F.R., Stone, L.R., Alam, M., Rogers, D.H., Clark, G.A. and Schlegel, A.J. (2000) Subsurface drip irrigation using livestock wastewater: dripline flow rates. *Applied Engineering in Agriculture* **16**, 505–508.

Wingender, J., and Flemming, H.-C. (2004) Contamination potential of drinking water distribution network biofilms. *Water Science and Technology* **49**, 277–286.

Wuertz, S., Okabe, S. and Hausner, M. (2004) Microbial communities and their interactions in biofilm systems: an overview. *Water Science and Technology* **49**, 327–336.

Chapter 11

Effects of treated municipal wastewater irrigation on soil microbiology

Dror Minz, Rachel Karyo and Zev Gerstl

11.1 Introduction

Soil is a highly complex and heterogeneous environment, containing an immense microbial community with as many as 10^9 bacterial cells per gram of soil. It is highly diverse both chemically and physically, as well as biologically, and contains a wide range of physicochemical factors that affect growth, activity and survival of indigenous microorganisms (Torsvik et al., 1990; Stotzky, 1997; van Elsas and Rutgers, 2005). Microbial communities play essential roles in many soil processes by regulating major biogeochemical nutrient cycling, such as carbon, nitrogen, and sulfur turnover.

In addition to their role in affecting soil chemical processes, microbial communities have a critical role in organic matter decomposition, in shaping the physical characteristics of the soil and the formation and stabilization of soil structure and hydrological properties and energy flow (Oades, 1993; Jastrow et al., 1998; Amezketa, 1999; Doran and Zeiss, 2000; Šantrůčková et al., 2004). Hence, changes in microbial diversity and in species composition of soil microbial community can influence microbial processes in the soil and may result in changes in soil function, both qualitatively and quantitatively (Lawton, 1994; Schimel, 1995; Cavigelli and Robertson, 2000; Buckley and Schmidt, 2003).

There is a growing interest in the effect of environmental factors and human activity, including agricultural practices such as irrigation with wastewater, on microbial diversity, especially in light of concerns about the preservation of biological diversity in terrestrial ecosystems (Buckley and Schmidt, 2003; Clegg et al., 2003; Grayston et al., 2004). Limited knowledge is available on the structure of soil microbial communities, their response to changes in their environment or the consequences that alterations in microbial community structure have on soil ecosystem function. This lack of information is a consequence of the enormous complexity and genetic diversity of soil microbial communities (Torsvik et al., 1990; Torsvik and Øvreås, 2002). It is also a result of the fact that

Treated Wastewater in Agriculture, First Edition, edited by Guy J. Levy, Pinchas Fine and Asher Bar-Tal

the microorganisms that can be isolated from soil, cultivated and studied in the laboratory represent only a small portion of the microorganisms present in situ (Amann et al., 1995; Hugenholtz et al., 1998). It has been estimated that 95–99% of the total amount of viable soil bacteria are unculturable (Torsvik et al., 1990; Amann et al., 1995). However, the ability to characterize microbial diversity and population shifts in environmental systems has recently increased with the introduction of novel and effective molecular methods (for reviews see Kozdroj and van Elsas, 2001; Torsvik and Øvreås, 2002; Nannipieri et al., 2003; Kirk et al., 2004; Collins et al., 2006; Schmidt, 2006).

Due to the increase in water demand and the growing use of treated wastewater (TWW) for irrigation, there is a need to identify effects that such irrigation has on soil microbial community composition and function. Wastewater irrigation provides water, minerals, and nutrients such as nitrogen and phosphorus, as well as organic matter to the soil (Pettygrove et al., 1985; Siebe, 1998; World Health Organization (WHO), 2006). All of these factors may have beneficial or detrimental effects on the soil biota. High levels of minerals (i.e. sodium), dissolved organic carbon, detergents, pharmaceuticals and other organic chemicals (Chang and Page, 1985; Bouwer, 2000; Oppel et al., 2004; Harrison et al., 2006), pollutants such as pesticides (Tomson et al., 1981; Muszkat et al., 1993a,b; Graber et al., 1995), and toxic metals (Siebe and Fischer, 1996; Ramirez-Fuentes et al., 2002; Gans et al., 2005) introduced via wastewater irrigation may accumulate in soil (Pettygrove et al., 1985; Siebe and Cifuentes, 1995) (for a review, see WHO Guidelines for the Safe Use of Wastewater, Excreta and Greywater, 2006). This may affect the structure and function of microbial communities, which, in turn, influence soil fertility (Ross et al., 1978; Goyal et al., 1995; Monnett et al., 1995; Meli et al., 2002; Ramirez-Fuentes et al., 2002; Speir, 2002; Gans et al., 2005). Furthermore, effluents may contain large amounts of microorganisms, including bacteria, protozoa and viruses, some of which can directly affect soil microbial diversity and activity and even impact human health. However, these negative effects of irrigation with TWW on soil microbiology may be minimized by applying proper agricultural practices (Estrada et al., 2004; Karyo et al., unpublished data).

Changes in soil biological characteristics have been proposed as sensible indicators of changes in soil quality, as they are more dynamic and often more sensitive than physical or chemical soil properties (Elliott et al., 1996; Anderson, 2003; Bending et al., 2004).

11.2 Soil microbial ecology and activities

Soil enzymes, primarily of microbial origin, are major participants in all biological and biogeochemical processes in soils (Coleman and Crossley, 1996; Nannipieri and Badalucco, 2003) and offer a useful assessment of soil "function". Microbial activities are thus considered as some of the best indicators of soil health and quality (Powlson et al., 1987; Kennedy and Papendick, 1995; Dick, 1997; Lalande et al., 2000; Killham and Staddon, 2002; Anderson, 2003).

Several microbial activities are considered indicators for soil health, measuring complex metabolic processes in soil. These include, for example, the evolution of CO_2 that reflects the aerobic catabolic degradation, substrate-induced respiration (SIR) (Torstensson et al., 1998; Stenberg, 1999), and potential ammonia oxidation and nitrification (Belser and Mays, 1980; Torstensson, 1993, ISO, 2004).

It has been hypothesized that diversity of soil organisms provides resistance to stress and that a decrease in diversity will cause declines in the resistance of soils to stress or disturbance (Elliott and Lynch, 1994; Brussaard et al., 1997; Giller et al., 1997; Degens et al., 2001). Hence, soil microbial diversity and community composition may be another indicator of soil health and quality.

Community fingerprinting based on rRNA and its coding genes (Head et al., 1998), or phospholipid fatty acid (PLFA) (Bååth et al., 1992; Ibekwe and Kennedy, 1998; Debosz et al., 2002) are valuable tools allowing estimation of community composition of the microorganisms in soils. However, they provide only limited information on soil functional diversity. Hence, although these techniques make it possible to describe microbial communities in the soil, the link between soil function and biodiversity is not yet well established (for a review see Nannipieri et al., 2003). Studies on metabolically active microorganisms in the environment usually use ribosome content as a measure of activity, as active cells will usually have higher ribosome content and therefore more rRNA to be detected (Davis et al., 1986; Givskov et al., 1994). Several methods have been employed to target microorganisms possessing the potential for specific activity, by targeting for specific functional genes or total mRNA. Thus, in some cases, research has succeeded in linking community composition with corresponding functions (Waldrop et al., 2000; Nogales et al., 2002; Burgman et al., 2003). Using stable isotope probing (SIP) (Radajewski et al., 2000; Rangel-Castro et al., 2005), it is possible to identify the organisms that respond directly to the addition of nutrients. One should bear in mind the enormous complexity of the soil microbial community and its metabolic activities and functions, which limits the detection to the few most active organisms. Due to the large diversity of soil microbiota (Gans et al., 2005), functional redundancy is often found, as many functions can be performed by several taxonomically different groups of microorganisms (Andren and Balandreau, 1999; Bardgett and Shine, 1999; Loreau et al., 2001). Thus, functional stability in soil does not necessarily imply a stable community composition (Loreau et al., 2001).

As changes in microbial activities and community composition can be useful for detecting changes in soil systems, they may be used to detect environmental effects of wastewater irrigation. Soil microbial biomass has been used to compare natural (Ross et al., 1982) and disturbed (Sparling et al., 1981) ecosystems. Soil biomass can be measured as the biomass C or N content. Biomass C can serve as a sensitive indicator of changes in soil organic matter and it is one of the general indices to soil microbial activities (Wick et al., 1998). Although one can expect effects of wastewater irrigation on microbial biomass in soil, there are conflicting reports in the literature as to the nature of these effects. For example, Schipper et al. (1996) and Meli et al. (2002) have shown no difference in microbial biomass between wastewater- and freshwater-irrigated soils. Conversely, other studies show an increase in microbial biomass following wastewater-irrigation. Ramirez-Fuentes et al. (2002) showed that the application of wastewater for periods up to 90 yrs increased the organic C and total N and microbial biomass C and N. Goyal et al. (1995) and Barkle et al. (2000), studying soil irrigated with distillery and dairy farm wastewater (respectively), have also reported an increase in microbial biomass in soil.

In addition to bulk soil biomass, several enzymatic activities (such as dehydrogenase, urease, and proteases) are often used as indicators for soil health. Both positive and adverse effects on soil microbial enzymatic activities have been observed to result from wastewater irrigation (Monnett et al., 1995; Niewolak and Tucholski, 1999; Barkle et al., 2000; Ramirez- Fuentes et al., 2002; Speir, 2002). Several examples for such effects are described below. Degens et al. (2000) studied microbial biomass and activity, and the relative use of substrates in an allophanic soil after 22 yrs of irrigation with a dairy factory effluent. Microbial biomass C and basal respiration activity increased by 4- and 1.6-fold, respectively, in the surface (0–10 cm) layer of the irrigated soil. Measurements of relative use of substrates indicated that the increased microbial biomass found in the effluent-irrigated soil was supported by the inputs of available C in the effluent, rather than by greater decomposition of the organic C in the soil. In related field studies, Sparling et al. (2001) assessed the effects of irrigation with dairy factory effluent on the properties of three soils. Two soils had been irrigated for 22 yrs and one for 2 yrs with the dairy effluent, and matched, non-irrigated areas were sampled for comparison. Whereas irrigation for 22 yrs in two of the soils caused no change, or a slight decrease in total C and N in the topsoil, microbial C and mineralizable N contents were more than doubled, and N cycling activity much increased.

While evaluating microbial and biochemical activities as indicators of soil quality, Filip et al. (1999) measured the effect of long-term (up to 100 yrs) wastewater-irrigation on development of oligotrophic and copiotrophic bacteria and fungi. In addition, they examined numerous enzymatic activities such as β-glucosidase, β-acetylglucosaminidase, proteinase, and phosphatase. In this, and in a subsequent study (Filip et al., 2000), they observed an increase in microbial biomass as a result of long-term wastewater-irrigation. Although the soil microflora originally consisted of mainly oligotrophic microorganisms, the application of wastewater stimulated the development of copiotrophic microorganisms. Furthermore, an enhancement in the above-mentioned enzymatic activities was observed, implying increased substrate concentration and high biochemical capacity to hydrolyze different organic compounds. The highest numbers of bacteria, actinomycetes, and fungi were found in the organic fraction of the long-term wastewater-irrigated soil. The ATP content, a measure of active microbial biomass, and the microbial counts were higher in the long-term irrigated soil than in the soil that was never irrigated.

Schipper et al. (1996), monitoring as many as 14 biochemical properties of soil, reported that as little as 49 mm of effluent irrigation per week was enough to increase pH, denitrification, mineralizable N, nitrate, and invertase activity. Similarly, Kannan and Oblisami (1990) found an increase in different groups of microorganisms and an increase in cellulase, amylase, dehydrogenase, and phosphatase activities in soil after 15 yrs of paper-mill effluent irrigation. Tam (1998) compared the microbial responses in mangrove soils receiving wastewater of different salinities. The results showed that addition of wastewater to soils, irrespective of the salinities of the wastewater, increased population sizes of heterotrophic bacteria, ammonium- and nitrite-oxidizers, and denitrifiers. However, in their study, ATP content, dehydrogenase, and alkaline phosphatase activities were not affected by the wastewater.

Meli et al. (2002) examined a number of chemical and microbial parameters in soils irrigated with lagooned wastewater. After 15 yrs of irrigation they observed increases of hydrolase and phosphatase activities, as well as higher metabolic efficiency of the microflora.

Gelsomino et al. (2006) estimated the changes in chemical and biological properties in an agricultural soil subjected repeatedly to flooding with wastewater containing alluvial sediments and potentially hazardous compounds. The authors used, among other approaches, a community-level redox technique (Biolog®) based on C-source utilization, for estimating functional microbial community diversity and community-level physiological profile approach. Their results showed an increase in microbial biomass, SIR, and soil basal respiration (SBR) in the wastewater-flooded soil, which indicates an increase in microbial biomass and metabolic activity. The measured SIR and microbial biomass were positively correlated with total organic carbon (TOC) and total N, whereas SBR showed a positive correlation only with TOC. Additionally, functional and compositional changes in soil bacterial communities as a response to repeated flooding were evident, resulting in microbial populations that were functionally more uniform. They also noticed reduced microbial diversity in certain microbial groups in soil, resulting from the wastewater.

Dehydrogenase is a group of enzymes, which oxidize organic compounds, leading to a loss of protons and reflecting the total oxidative activity of the soil microflora. Dehydrogenase is usually correlated with soil respiration (Brookes, 1995) and has been used as an index of microbial activity and as a sensitive indicator of metal toxicity (Doelman and Haanstra, 1979; Rossel and Tarradellas, 1991). Tam (1998) reported that dehydrogenase activity was not affected by the addition of wastewater containing heavy metals. Brzezinska et al. (2001) determined dehydrogenase and catalase activities and redox potential (an index of soil aeration status, useful especially under flood conditions) in fields irrigated with two levels of municipal wastewaters. Dehydrogenase activity increased following wastewater irrigation, although differently under different plantation types. Irrigation with wastewater resulted in a decrease of redox potential, as well as of soil catalase activity. In a more recent study (Brzezinska et al., 2006) involving irrigation of a variety of crops for 4 yrs, they described contradicting results. In their study they compared two levels of wastewater irrigation with two levels of tap water (of undescribed characteristics). Their results, when significant, mostly suggested that tapwater-irrigated soils had higher dehydrogenase, acid, and alkaline phosphatase and CO_2 evolution activities, whereas urease activity was higher in soils irrigated with municipal wastewater.

Friedel et al. (2000) studied the correlation between compounds present in wastewater (i.e. minerals such as heavy metals and organic matter) and found no negative effect of 80 yrs of wastewater irrigation on dehydrogenase and nitrification activity. However, denitrification capacities decreased with increasing time of irrigation, suggesting that denitrification potentials can indicate the effect of pollutants in wastewater on microbial activity.

In addition to broad biochemical activities in soil, important specific groups of organisms may be affected by nutrients introduced to soil via wastewater irrigation and may be used for environmental monitoring. Among these are the ammonia-oxidizing prokaryotes, nitrite-oxidizing and carbon- and nitrogen-fixing bacteria and actinomycetes.

Microbial chemolithoautotrophic nitrification, the conversion of ammonia to nitrate via nitrite is a key process in the global nitrogen cycle. Autotrophic nitrification is carried out by two distinct groups of microorganisms, the ammonia-oxidizers, which convert ammonia to nitrite and nitrite-oxidizers, which convert it to nitrate. It is generally assumed that the first step is the rate-limiting one (Paul and Clark, 1989; Prosser, 1989; Jetten et al., 1997). The ammonia-oxidizing bacteria (AOB) have previously been shown to be affected by a variety of chemical conditions including, but not limited to, levels of ammonium, organic matter, toxic metals, and salinity (Paul and Clark, 1989; Koops and Moller, 1992; Prinčič et al., 1998; Schramm et al., 1998; Stephen et al., 1999; Bruns et al., 1999; Mendum et al., 1999), and are therefore often used as indicator organisms to study different kinds of stress in soil (Hastings et al., 1997; Stephen et al., 1999; Phillips et al., 2000; Chang et al., 2001; Oved et al., 2001; Avrahami et al., 2002, 2003). In many soils, *Nitrosospira*-like AOB are the most dominant (Hiorns et al., 1995; Mendum et al., 1998; Stephen et al., 1999; Hastings et al., 2000; Kowalchuk et al., 2000; Avrahami et al., 2002).

In our studies, irrigation with TWW was shown to result in an increase in the activity of several microbial processes, including nitrification, during the irrigation season. In the winter, when no irrigation was applied, the soils previously irrigated with freshwater were more active (Minz et al., unpublished). It is easier to explain an increase of activities related to the input of organic matter in soil. However, increased nitrification potential is harder to explain as the AOB are autotrophs (i.e. fix CO_2 to organic matter using the energy obtained from oxidizing ammonia). The effect of effluent irrigation on community composition and function of ammonia-oxidizing bacteria (AOB) in soil was evaluated by Oved et al. (2001). A significant and consistent shift in the population composition was detected in soil irrigated with effluent. This shift was absent in soils irrigated with fertilizer-amended freshwater. At the end of the irrigation period, *Nitrosospira*-like populations were dominant in soils irrigated with fertilizer-amended freshwater, whereas *Nitrosomonas*-like populations were dominant in effluent-irrigated soils. Indeed, *Nitrosomonas* strains are the most common type of ammonia-oxidizers found in wastewaters (Watson et al., 1989; Wagner et al., 1995).

Several components of wastewater are expected to be involved in determining which AOB species will dominate an environmental niche. Among them are: ammonium concentration (Suwa et al., 1994; Schramm et al., 1998; Prinčič et al., 1998), organic content (Wagner et al., 1995; Hastings et al., 1997; Rotthauwe et al., 1997), and pH (Stephen et al., 1996; Prinčič et al., 1998). Oved et al. (2001) suggested that in effluent-irrigated soil, as in other organic-rich environments, environmental conditions may favor the *Nitrosomonas* species.

Ammonia concentration in soil was shown to have a major effect on nitrification (Schuster and Conrad, 1992; Müller et al., 1998; Avrahami et al., 2002). Webster et al. (2005) demonstrated high inputs of ammonia, followed by lags in nitrification were correlated with relative abundance of particular AOB phylotypes (*Nitrosospira* clusters 3a and 3b).

Ammonia oxidation, in environments rich in ammonium (such as wastewater-irrigated soils), increases the oxygen uptake and lowers the pH. Such modifications of environmental conditions affect not only nitrification, but can also select for specific groups of microorganisms specialized for these new conditions. Prinčič et al. (1998) found that ammonium at a very high concentration selected for novel nitrifier populations and that pH extremes (pH 6.0 and 8.2), modified the community. However, this shift in population composition did not affect the nitrification rates as the ammonium consumption rates were rapid and equal before and after the structural change. In addition, they found that the pH-mediated structural changes took longer to develop than those caused by increased ammonia concentration and the nitrification rate in the low-pH environment was retarded relatively to the control. An important finding of their study is that the community that exhibited the greatest structural shift was able to reacquire its original structure in a relatively short time after the selecting conditions were eliminated.

11.2.1 Nitrogen fixation

Microbial nitrogen fixation is an essential component of the nitrogen cycle in natural and agricultural systems. Nitrogen-fixing microorganisms (diazotrophs) are more diverse than AOB, having representatives in several major phyla, including most subclasses of proteobacteria, cyanobacteria, firmicutes, and euryarchaea (Raymond et al., 2004).

Very little attention has been given to describing the effects of wastewater irrigation on nitrogen-fixing bacteria in soils and in symbiosis with plants. However, soil salinity was shown to have a strong effect on the rhizobial-legume symbiotic performance (Cordovilla et al., 1999; Maatallah, et al., 2002; Bolaños 2003; Bouhmouch et al., 2005), suggesting that wastewater-irrigation has a large potential to affect nitrogen-fixing bacteria.

We have found (Farkash, Jurkevitch and Minz, unpublished data) that irrigation of chickpea (*Cicer arietinum*) with recycled wastewater affected plant parameters: reducing dry weight, and increasing nitrogen content and nitrogen fixation potential in comparison to control plants irrigated with freshwater. Although a single operative taxonomic unit closely related to *Mesorhizobium mediterraneum* was found in all cases, the dominant *M. mediterraneum* genotype was influenced by irrigation type and irrigation history. Some genotypes, undetected in nodules of plants irrigated with freshwater, formed dominant groups in soils irrigated with TWW in the past or in the described study.

11.2.2 Denitrification

Denitrification is a respiratory process in which oxidized nitrogen compounds are used as alternative electron acceptors for energy production when oxygen is limited and carbon and nitrate are available (Tiedje, 1988). Bacteria capable of denitrification are very diverse and widely distributed in the environment (Tiedje, 1988).

The maximum denitrification rate is determined by the size of the denitrifying population (Tiedje, 1982), which is influenced by soil and environmental factors. Barton et al. (1999), studying a forested land-based wastewater treatment system during a 12-month field study, found that denitrification rates were small, despite weekly irrigation with tertiary TWW containing nitrate and that there was a decrease in annual denitrification in wastewater-irrigated soils, similarly to that in unirrigated soils. However, denitrification activity increased in wastewater-irrigated soils when oxygen was limited (by flooding). These results were reconfirmed in a later study (Barton et al., 2000). However, additions of carbon and nitrate to anaerobic soils did not increase the denitrifying population in the wastewater-irrigated soils compared to the control soils (Barton et al., 2000). Furthermore, the size of the denitrifying population in the irrigated soils was small and not significantly different from the unirrigated soils. Although wastewater-irrigation did not increase the size of the denitrifying population, Barton et al. (2000) found that it altered the short-term response of denitrifiers to oxygen concentrations in soil. Under low-oxygen conditions, denitrifiers in the wastewater-irrigated soils produced enzymes sooner and at a greater rate than soils without a history of wastewater-irrigation.

It was proposed that the size of the denitrifying population cannot be expected to be large in free-draining, coarsely textured soils, even when provided with additional nitrogen and water inputs. It is likely that soils are rarely anaerobic enough to promote denitrification activity because of their free-draining nature (Cook et al. 1994).

In a subsequent study, Barton and Schipper (2001) determined the effect of dairy farm effluent (DFE) irrigation on N_2O emissions from a surface-drained peat soil and a freely drained mineral soil. Nitrous oxide emissions increased immediately following DFE irrigation to both soils, and were generally greater than emissions following the application of inorganic fertilizer with water. Increased N_2O emissions following DFE irrigation coincided with increased soil water contents and mineral N and CO_2 emissions. It was suggested that DFE application increased N_2O emissions more than inorganic N fertilizer by enhancing denitrification either by increasing C availability and/or decreasing soil aeration following increased respiration.

11.2.3 Organic matter

Soil organic matter (SOM) plays a major role in terrestrial ecosystem development and function (Brady and Weil, 2002). Even though microorganisms are estimated to comprise only 1–3% of the total soil carbon and approximately 5% of the total soil N (Jenkinson, 1988; Smith and Paul, 1990), they are critical for both soil structure and fertility. Soil organic matter is often suggested to be the most important parameter in describing soil quality (Ruark and Zarnoch, 1992; McConnell et al., 1993; Giusquiani et al., 1995). In addition to supplying plant nutrients after being mineralized, the type and amount of SOM influences several soil properties, particularly those related to physical characteristics (Stevenson, 1994; Van-Camp et al., 2004; Clapp et al., 2005).

Generally, the addition of organic matter is beneficial to soil fertility, improving soil structure, increasing water-holding capacity, and stimulating microbial activity (Ritz et al., 1997, Simek et al., 1999; Marinari et al., 2000). Increased microbial activity will, in turn, lead to a further release of substantial amounts of plant nutrients from both the newly added and indigenous SOM. It has been reported that wastewater irrigation leads to an increase in soil total organic carbon (TOC) contents (Siebe and Cifuentes, 1995; Friedel et al., 2000). As SOM accumulates, ecosystem productivity becomes increasingly dependent on mineralization by the microbial community (Parfitt et al., 2005). The rate-limiting step in this mineralization is thought to be degradation by microbial enzymes (Sinsabaugh et al., 1993; Waldrop and Firestone, 2004).

Application of wastewater containing organic matter and nutrients has been found to increase total soil biomass and activity in field experiments (Saber, 1986; Goyal et al., 1995; Friedel et al. 2000). As the C:N ratio is known to be important in regulating microbial community composition (Swift et al., 1979; Myrold, 1998; Högberg, 2007), organic matter inputs via wastewater irrigation should have an effect on microbial community in soil. Magesana et al. (2000), studying the changes in soil biological properties and hydraulic conductivity in soil receiving wastewater with three C:N ratios, found that an increase in the wastewater C:N ratio increased microbial biomass.

The potential of the labile fraction of SOM as a soil quality indicator has been widely suggested, as it provides the substrate for microbial biomass turnover and can respond rapidly to changes in C supply (Bol et al., 2003; Chantigny, 2003). Several studies found that olive oil wastewater application on soils increased SOM content, leading to a change in microbial community composition and an increase in biological activities, which enhanced the mineralization of the labile organic matter introduced in the wastewater (Mekki et al., 2006; Sierra et al., 2007). Still, oil mill wastewater is a special case of wastewater as it contains compounds that may be toxic to plants and some soil microorganisms (Yesilada and Sam, 1998; Mekki et al., 2007). The degradation of wastes containing high amounts of soluble organic carbon, in forms such as amino acids or carbohydrates, immediately leads to an increase in CO_2 evolution. This may cause high CO_2 and low oxygen concentrations, which in turn may lead to oxygen deficiency in the rhizosphere, and reducing conditions in the soil (Bernal et al., 1998), thus inhibiting the nitrification process (see discussion on microbial activity above).

Several xenobiotics are found in large amounts in wastewater. These include pesticides, and pharmaceuticals (including veterinary and human antibiotics, hormones, and sterols) (for a review see Nikolaou et al., 2007). Their fate and effects in wastewater-irrigated soils are described below.

11.2.4 Pesticides

Degradation of pesticide in soils can occur by either chemical (abiotic) or microbial (biotic) pathways (Golovleva et al., 1990), both of which may be influenced by wastewater irrigation. Chemical processes such as hydrolysis, oxidation–reduction, substitution, elimination, dehalogenation, and photolysis are influenced by the soil's organic matter, clay content, and pH, as well as the temperature. In soils with high amounts of organic carbon or clay content, photolytic degradation of organic chemicals may occur at the soil surface by sunlight. Factors affecting chemical degradation show that effluent irrigation may affect chemical degradation by altering the soil pH and increasing dissolved organic matter (DOM) content; however, no studies in which the impact of wastewater irrigation on these interactions have been reported (for a review see Müller et al., 2007).

Wastewater-irrigation can modify microbial degradation of pesticides through a number of processes including changes in the size and composition of microbial populations, enhancing nutrient and moisture status of the soil and providing excessive carbon as an energy source.

The effect of wastewater-irrigation on atrazine (an s-triazine-ring herbicide) degradation was reported in a series of papers. An external organism capable of mineralizing atrazine was used as a bacterial probe to determine the effect of TWW application on the potential for pesticide degradation in soils that otherwise did not degrade atrazine rapidly, or in soils where intrinsic atrazine degradation ability had been modified (Masaphy and Mandelbaum, 1997). Irrigation with TWW for an extended period of time modified the ability of soils to degrade atrazine. Only one of the tested soils exhibited intrinsic ability to mineralize atrazine, and in the corresponding soil irrigated with TWW no atrazine mineralization could be detected. In soils augmented with *Pseudomonas sp.* strain ADP (P. ADP), atrazine mineralization depended on TWW irrigation history and on preincubation of atrazine in the soil before inoculation. In soils irrigated with TWW and inoculated 1 d after atrazine application, rapid mineralization occurred. However, in the corresponding soils, not irrigated with TWW, a longer lag phase was observed. A different degradation pattern was observed when atrazine was preincubated for 12 d in the soils before inoculation. In soils irrigated with TWW, atrazine mineralization was retarded and less than 20% of it was mineralized within 20 d, whereas in the corresponding soils not irrigated with TWW, atrazine mineralization reached 60–80%. From these results it was concluded that aging of the herbicide in the soil has a pronounced effect on the mineralization results. Adsorption of atrazine was positively correlated with organic matter content in the TWW-irrigated soils and may be partially responsible for the inhibition of mineralization after long preincubation of atrazine in soil. This observation suggests that if atrazine-degrading bacteria are present in the soil, long-term TWW irrigation could enhance mineralization of newly applied atrazine in the first days after application, but would adversely affect mineralization of atrazine that had been aged in the soil for several days. As the atrazine half life in soils is in the order of weeks, it is to be expected that the overall effect of TWW irrigation on atrazine degradation would be toward retardation of mineralization rates.

One of the obstacles to biological treatment of wastewaters is their high salt content. To enable biological treatment, bacteria capable of atrazine mineralization (P. ADP) isolated from contaminated soils were adapted to biodegradation of atrazine at salt concentrations relevant to atrazine manufacturing wastewater (Shapir et al., 1998). Their results suggest that salt-adapted *Pseudomonas sp.* strain ADP can be used for atrazine degradation in salt-containing wastewater. They later demonstrated (Shapir et al., 2000) the ability of indigenous soil bacteria to mineralize atrazine in a sandy soil. They observed that even though wastewater irrigation increased the competition between indigenous populations and the introduced bacteria, *Pseudomonas sp.* strain ADP was able to continue mineralizing atrazine.

Despite the above works showing an effect of wastewater on atrazine degradation in soils, Huang et al. (2000) have shown that animal-derived lagoon effluents have, at most, a minor effect on chlorpyrifos (an organophosphate insecticide) degradation in soils. They showed that the use of animal-derived effluents in low pH soils induced increases in the soil pH and increased hydrolysis by no more than a few percent. Based on their findings they concluded that soil properties, and not effluent properties, control chlorpyrifos degradation. However, long-term studies are needed to assess the impact of long-term effluent irrigation on changes to soil properties and microbial ecology related to herbicide degradation.

11.2.5 Pharmaceuticals

The accumulation of pharmaceuticals in soils, sediments, and aquatic environments is of great concern as they contain relatively recalcitrant, non-biodegradable compounds that persist longer than previously thought (for data see Kümmerer, 2008). High levels of consumed pharmaceuticals are secreted from the body, as they are only partly metabolized (McArthur and Tuckfield, 2000), thus arriving at wastewater treatment plants where even the best tertiary treatment is not designed to fully degrade them or effectively disable their activity. As a result, pharmaceuticals are likely to be released into the soil via TWW irrigation and accumulate (Koplin et al. 2002; Kinney et al., 2006).

Due to their increasing ubiquity, it is essential to understand the fate of these pollutants in soils, sediments, and water bodies. Highly mobile pollutants have the potential to leach into groundwater, whereas strongly sorbing compounds can accumulate in the top layer of soils. These compounds may subsequently be taken up by plants (thus integrated into the food chain) and also affect the soil microbial community (Sören, 2003 and references therein). Furthermore, existence of antibiotics in the environment increases the risk of bacterial resistance development and transfer. Many excellent reviews of the current state of knowledge are available (Witte, 2001; Teuber, 2001; White et al., 2002; McDermott et al., 2003), but the importance of environmental pollution pathways to the proliferation of resistant bacteria has only recently received attention (for a review see Rooklidge, 2004). The extent of bacterial resistance caused by trace antimicrobials in the environment is not yet clearly defined, and contributions from agriculture and especially wastewater-irrigation are contradictory (Witte, 1998; Aarestrup and Wegener, 1999; Piddock et al. 2000; Schwarz et al., 2001; Kinney et al., 2006).

Kinney et al. (2006) showed that TWW irrigation results in soil pharmaceutical concentrations that vary through the irrigation season and that some compounds (the most commonly detected pharmaceuticals were diphenhydramine, fluoxetine, carbamazepine, and the antibiotic erythromycin) persist for months after irrigation. Kong et al. (2006) showed that addition of the antibiotic oxytetracycline decreased the functional diversity of the soil microbial community by 63% at 43 mM and decreased functional evenness by 41% at 109 mM.

11.2.6 Heavy metals

Prolonged wastewater irrigation may lead to increases in concentrations of some heavy metals in soils (Ramirez- Fuentes et al., 2002; Siebe and Cifuentes, 1995). Heavy metals at high concentrations severely affect the growth, morphology, and metabolism of microorganisms in bulk soils (Kandeler et al., 1996) through functional disturbance, protein denaturation, or the destruction of the integrity of cell membranes (Leita et al. 1995). Hence, high levels of heavy metals are toxic and can reduce soil microbial biomass and cause a change in community composition (Ohya et al., 1985; Naidu and Reddy, 1988; Aoyama et al., 1993; Kozdroj, 1995; Leita et al., 1995; Speir et al., 1995; Kandeler et al., 1996; Knight et al., 1997; Giller et al., 1998: Gans et al., 2005), as well as reduce soil microbial activities including respiration, ammonification, nitrification, and enzyme activities (Bååth 1989; Doelman and Haanstra, 1989; Tyler et al., 1989; Kannan and Oblisami 1990; Aoyama and Nagumo, 1996). Valsecchi et al. (1995) reported that heavy metals appeared to cause an alteration in the soil C cycle, and modified the energy metabolism of soil microorganisms, leading to decreased net SOM mineralization. However, other research has shown that heavy metals at low concentrations, or inputs of heavy metals with organic matter, stimulate bacterial growth and population size and stimulate mineralization and nitrification processes (Dusek, 1995).

Although most studies on the effects of heavy metals on soil microbial community structure and function deal with concentrations that are much higher than typically found in municipal treated wastewater, a few contradictory reports on the effect and toxicity of low concentrations exist. These disparities are a result of differences in bioavailability (McGrath, 1994) and differences in sensitivity of microorganisms or microbial processes to different types of metals, which can vary even between strains of the same species. Differences in methodology, soil type, and acidity, and the microbial biomass and community composition all have the potential to affect the results of such experiments (for a review see Giller et al., 1998). Furthermore, microorganisms respond mainly to the soluble metal fraction and the proportion of total metals present in this fraction may differ with soil type, environmental conditions, and time of measurement relative to metal inputs (Dar and Mishra, 1994).

Kong et al. (2006) showed that the addition of Cu decreased the functional diversity of the soil microbial community by 35% at 100 mM. Other studies (Barnhart and Vestal, 1983; Capone et al., 1983; Jonas et al., 1984; Vives-Rego et al., 1986; Said and Lewis, 1991) also demonstrated that short-term exposure to toxic heavy metals affects a variety of microbial processes. In addition to the effects heavy metals have on microbial activities, molecular techniques for studying community composition have demonstrated an impact of metal exposure on the entire indigenous soil community (Bååth et al., 1998; Konopka et al., 1999; Konstantinidis et al., 2003; Moffett et al., 2003).

Tam (1998) found that the heavy metals present in wastewater, and their quantities retained in soils, did not show any significant harmful effect on the bacterial population sizes and activities in mangrove soils. Soils with high organic matter content have cation exchange capacity, which has an important role in reducing the toxicity of the metals in soil environments (Babich and Stotzky, 1985). In addition, the adsorption of heavy metals to particulate and soluble organic matter reduces its biological availability and toxicity in soil. Thus, organic amendments such as composts or peat, containing a high proportion of humifed organic matter, may decrease the bioavailability of heavy metals in soil by adsorption and by forming stable complexes of the heavy metals with humic substances (Shuman, 1999).

There is a lack of studies dealing with the combined effect of metals and organic carbon on microbial community composition and activity. It has been shown that when soil is co-contaminated with organic matter and metals, although the effect of both contaminations was reduced a shift in microbial community was observed (Baath et al., 1998). Nakatsu et al. (2005) suggested that the shift is mostly due to the organic carbon amendment, rather than the metal stress. It has been reported that heavy metals could be mobilized from soil particulate matter and become dissolved complexes when exposed to an increase concentration of Na, K, Ca, Mg, and chloride (Comans and Van Dijk, 1988; Paalman et al., 1994). Similarly, Gambrell et al. (1991) reported that the mobilization/immobilization of trace and toxic metals in brackish marsh soils was affected by salinity; increasing salinity seemed to increase soluble Cd, Cr and Cu levels, but not Pb, Ni and Zn. Such a change in metal availability levels will then influence the microbial response.

Heavy metal contamination due to wastewater irrigation was found to decrease microbial carbon/TOC ratios significantly (Valscchi et al., 1995). Furthermore, Nakatsu et al. (2005) showed that the microbial community structure in a long-term mixed waste contaminated site might reflect both metal and aromatic hydrocarbon concentrations in soil. For individual microbes to persist under complex conditions, they must tolerate both local metal and hydrocarbon contaminants.

Kandeler et al. (2000) performed particle-size fractionation of a heavy metal polluted soil. In their study, they found that heavy metal pollution influenced the structure and function of the microbial community in the bulk soil and in particle-size fractions. There was an increase in the fungal:bacteria ratio in polluted soils. Furthermore, microbial biomass decreased significantly in the silt and clay fraction, which was enriched with heavy metals, and the microbial community produced fewer enzymes (urease, phosphatase, and arylsulfatase). Consequently, a reduction of enzyme activity due to heavy metal pollution was noticed in all fractions. Dehydrogenase activity and denitrification rates were also shown to be sensitive to heavy metal pollution (Bardgett et al., 1994; Kandeler et al., 1996).

11.2.7 pH

Soil pH is a major factor affecting the activity and survival of microorganisms in soil and determining microbial community composition (Alexander, 1977; Pennanen, 2001; Bååth and Anderson, 2003; Högberg et al., 2007). A decrease in the fungi:bacteria ratio was found when soil pH increased across a gradient in boreal forest (Högberg et al., 2003). Wastewater-irrigation may have a minor effect on soil pH. The magnitude and direction of any change in pH is determined by both wastewater and soil characteristics. Irrigation with wastewater was shown to result in either a slight increase (Schipper et al., 1996; Qian and Mecham, 2005) or decrease in soil pH (Mohammad and Mazahreh, 2003; Angin et al., 2005). It has been shown that the survival of heterotrophic bacteria in acidic soils (pH 3–5) is commonly lower than in alkaline soil, as the low pH could act to negatively affect the viability of bacteria and the availability of nutrients (Tate, 1978). Furthermore, changes in soil pH may affect the net microbial activity by affecting enzymatic function. Total microbial biomass and respiration rates are lower in acidic than in neutral soils (Anderson and Domsch, 1993). A decrease in pH has both direct and indirect effects on the biomass size, composition, and activity of microorganism in soil (Prescott and Parkinson, 1985; Tabatabai, 1985; Berg et al., 1991; Myrold and Nason, 1992).

Different reports exist in the literature regarding pH effects on fungi in soil. Fritze et al. (1992) and Esher et al. (1992) have shown a decrease in fungal biomass due to acidification, whereas Bååth et al. (1979) showed no change and Ruess et al. (1996) reported an increase in fungal biomass as a response to acidification. Killham et al. (1983) exposed a soil to different pH levels in order to examine the effect of acid rain on soil microbial activity. They found an increase or decrease in activities of soil enzymes such as urease, phosphatase, dehydrogenase, and arylsulfatase, according to the amount of the acid load used in the experiments. Furthermore, only the pH 2.0 input caused inhibition of both respiration and enzyme activities. They concluded that individual microbial processes will have different sensitivities to acid and changes in the supply of N were evaluated as the major mechanism through which simulated acid rain affects soil microbial activity. Batra and Manna (1997) found that a reduction in pH from 10.6 to 8.5 increased dehydrogenase activity in an alkali soil.

Prinčič et al. (1998) studied shifts in nitrifying community structure and function in response to different pH values (pH 6.0, 7.0, and 8.2) in experimental reactors inoculated with nitrifying bacteria from a wastewater treatment plant. The authors found that, at the two pH extremes tested, the community structure of the nitrifying community was altered and the changes were irreversible.

Soil pH and heavy metal concentrations are bound together as pH is one of the major factors affecting metal solubility. Under acidic conditions, heavy metals are usually more available and toxic than in similar soil of higher pH (Collins and Stotzky, 1989).

11.2.8 Salinity

Wastewater usually contains increased salt concentration. In general, changes in salinity affect microbial activity and function (Zahran, 1997). The direct effect of salinity and especially salinity increase due to wastewater-irrigation on soil microbial biomass, community composition, and function, has not been described in detail so far. In the absence of these data, the effect of salinity per se is described. Both increases and decreases in mineralization of C and N with increasing salinity have been observed (Singh et al., 1969; Laura, 1974; McClung and Frankenberger, 1987; Nelson et al., 1996; Pathak and Rao, 1998). Omar et al. (1994) found that 5% w/w NaCl significantly decreased fungal, bacterial, and actinomycete population sizes in soil. Similarly, Polonenko et al. (1981) found a decrease in bacterial levels in highly saline soil. By contrast, Matsuguchi and Sakai (1995) found an increase of fluorescent pseudomonads population in three different soils with increasing levels of salinity. However, in rhizosphere soil, their levels decreased whereas numbers of total bacteria (including Gram negative bacteria) increased.

Yuan et al. (2007) studied the effects of salinity on the size, activity, and community structure of soil microorganisms in salt-affected arid soils. They found that the active and total microbial communities were adversely affected by soil salinity. Their results confirm a pattern found in naturally occurring saline soils where a negative correlation is usually found between microbial biomass and total soluble salt content (Ragab, 1993; Rietz and Haynes, 2003; Sardinha et al., 2003); however, a positively correlation with soil organic C contents (Zahran et al., 1992). Low percentages of organic C present as microbial biomass C show that the high salinity was detrimental to soil microorganisms.

Friedel et al. (2000) found a negative effect of salinization on soil respiration. However, this effect was masked by a predominant TOC effect on respiratory activity.

Yuan et al. (2007) described a negative relationship between the fluorescein diacetate (FDA) hydrolysis rate, arginine ammonification rate, and high salinity, indicating the inhibitory effects that salinity inflict on soil microbial activity.

It has been shown before in pure culture that NaCl can reduce the activity of nitrogenase in some nitrogen fixers (Hassouna et al., 1995). Nitrification is also inhibited by 50% and 70% in soils at 5 and 10 dS/m^2, respectively (Kumar and Wagenet, 1985).

Sarig and Steinberger (1994) studied the relationships between seasonal changes in soil salinity and the microbial biomass and activity in arid soils. According to their findings, microbial biomass sizes were not directly affected by the degree of soil salinity.

Batra and Manna (1997) studied the dehydrogenase activity and microbial biomass in typical saline and saline–alkali sandy soils. They showed that dehydrogenase activity increased in saline–alkali sandy soil of an arid region. Dehydrogenase activity was negatively correlated with pH and salinity (measured as electrical conductance), whereas it was positively correlated with organic C.

11.2.9 Boron

Boron is a plant micronutrient at low concentrations. Boron is not an essential growth element for bacteria or fungi with the exception of some cyanobacteria (Bonilla et al., 1990). It accumulates in soils when irrigated by boron-rich water, especially from detergent degradation in wastewater. As boric acid is an inorganic substance it does not degrade in sewage treatment plants (HERA, 2002), and reaches the TWW. Above a minimal crucial concentration, boron becomes toxic to plants. However, the effect of soil contamination with boron on microbial community structure or function in situ is largely unknown. Nelson and Mele (2007) examined the impact of B and NaCl on soil microbial community structure in wheat rhizosphere. They supplemented the soil with 0, 12, and 24 μg/g B as boric acid. A significant decrease was found in functional diversity and richness of the rhizosphere community when exposed to high B concentrations. However, no significant change was observed in specific bacterial community structure such as the ammonia-oxidizing bacteria, and the authors suggest that boron addition to soil is more likely to affect rhizosphere microbial community structure indirectly through root exudates rather than directly through microbial toxicity.

Several studies using pure cultures have shown boron can inhibit growth of bacteria (Bringmann and Kuhn, 1980; Butterwick et al., 1989; Guhl, 1996), fungi (Bowen and Gauch, 1966), and algae (Guhl, 1996), mainly at relatively high concentrations.

11.3 Human pathogens

Large numbers of bacteria, including human pathogens, are added to soil by wastewater effluent-irrigation (Frankenberger, 1985; Abdulraheem, 1989; Pepper and Gerba, 1989; Vasseur et al., 1996; Gibbs et al., 1997; Santamaria and Toranzos, 2003; Gerba and Smith, 2005). Common pathogens reaching soils as a result of wastewater-irrigation include pathogenic bacteria such as *Escherichia coli* 0157:H7, *Salmonella* spp., *Listeria monocytogenes*, *Campylobacter* spp., *Clostridium botulinum*, *Clostridium perfringens*, *Shigella* spp. and *Streptococcus fecalis*; parasites such as *Giardia* sp. and *Cryptosporidium parvum* and *Entamoeba histolytica*; and several human pathogenic viruses (West, 1991; Santamaria and Toranzos, 2003; Gerba and Smith, 2005). These pathogens can contaminate the environment and are a potential health hazard to animals grazing wastewater-irrigated pastures, agricultural workers, residents living in proximity to agricultural fields and consumers of the agricultural products (Fattal et al., 1986; Shuval et al., 1986; Abdulraheem, 1989; Hespanol, 1990; Strauss, 1994; Bouwer, 2000; Solomon et al., 2002; Wachtel et al., 2002; Warriner et al., 2003; Steele and Odumeru, 2004).

Fecal bacteria are known to survive in soil for long periods of time (Lau and Ingham, 2001; Jiang et al., 2002, Nicholson et al., 2005). The survival of bacteria in soil is influenced by biotic and abiotic factors. Some of the abiotic factors found to affect survival are temperature (Zibilske and Weaver, 1978; Jiang et al., 2002), soil type (van Elsas et al., 1986, 1991; Lau and Ingham, 2001; Nicholson et al., 2005), soil management types (Franz et al., 2005), origin and type of wastewater and soil moisture content (Zibilske and Weaver, 1978; Chandler and Craven, 1980), sunlight, pH, and the availability of organic matter (Gerba and Goyal, 1985; Franz et al., 2005) (for reviews see Gerba et al., 1975; England et al., 1993; Gerba and Smith 2005). Biotic factors suggested to regulate population size of bacteria introduced to soil environments include competition with natural microorganisms for nutrients (van Veen et al., 1997) and for biological space (Nannipieri et al., 1983), inhibition by antimicrobial components secreted by soil biota, and predation by indigenous microorganisms (Heijnen et al., 1988; Kuikman et al., 1991; Recorbet et al., 1992; Ashelford et al., 2000) and nematodes (Bardgett and Griffiths, 1997).

It has been shown that wastewater-irrigation leads to an increase in pathogenic microorganisms in the irrigated soils (Abdulraheem, 1989; Pepper and Gerba, 1989; Santamaria and Toranzos, 2003; Gerba and Smith, 2005). However, many studies reported a decline of pathogens in soil with time. The reported survival period of fecal coliforms in manure-amended soils vary considerably from several days (Malkawi and Mohammad, 2003), to weeks (Nicholson et al., 2005), to months (Kearny et al., 1993; Nicholson et al., 2000; Jiang et al., 2002).

Sun et al. (2006) studied the effect of sewage sludge application on changes in fecal coliform counts in soils in order to evaluate the hygienic risks of sludge application. They found that although sludge application increased the hygiene risk, coliform counts decreased with time and reached a very low level after 56 d. We have also noticed a decrease in coliform counts in soils treated with sewage sludge within 30 d of incubation. This decrease was affected by temperature and disappeared in sterile soils, suggesting it may be due to biological processes.

Little is known about the possible accumulation of pathogenic microorganisms at different depths in the soil following irrigation of fields with effluents (Gerba and Goyal, 1985). Malkawi and Mohammad (2003) studied the effect of wastewater-irrigation on soil microbiology and on the survival and accumulation of bacteria of fecal origin at different depths in the soil. They examined a field of forage crops irrigated with secondary-TWW during 2 yrs. No significant difference was detected between the total aerobic bacterial counts in soils irrigated with secondary-TWW and with potable water. However, they reported an increase of about 10-fold in total coliform bacteria counts in soils irrigated with TWW compared to soils irrigated with potable water but this decreased with time. Malkawi and Mohammad (2003) reported no threat as a result of wastewater-irrigation in regard to nematodes and helminth eggs.

Most of the studies cited deal with pathogenic bacteria, mainly enteric. Only a few studies focus on viruses. Enteric viruses have been found in drinking water, surface water, wastewater and groundwater (Bales et al., 1995; Jin et al., 1997; Nicosia et al., 2001). Conventional wastewater disinfection cannot completely remove or inactivate enteric viruses (Straub et al., 1992; Jin et al., 1997). Virus persistence and mobility in soils is variable and dependent on the complex interactions of the soil environment; soil texture, temperature, moisture content, soil microbial activity, pH, certain inorganic salts, organic compounds, virus type, and virus physical state influence virus persistence and movement (Gerba et al., 1975; Meschke and Sobsey, 1998). Virus inactivation periods (loss of infectivity) are reported to range from 24 h to 6 months in soils (Straub et al., 1992; Nicosia et al., 2001; for a review see Frankenberger, 1985). Hence, wastewater-irrigation poses a health risk. However, wastewater application to soil appears to be very effective in pathogen immobilization and inactivation (Frankenberger, 1985). Seeded polioviruses were shown by Tierney et al. (1977) to survive in soil after irrigation with effluent for 11 d during the summer and 96 d during the winter.

Hurst et al. (1980) examined the effects of soil and sewage effluent microflora upon enterovirus survival under different abiotic conditions in laboratory experiments. Their results indicated that temperature was a significant predictor of virus survival in soil. Virus survival was not significantly affected by the concentration of sewage effluent.

The protozoan pathogen *Cryptosporidium* spp. oocysts and *Giardia* spp. cysts have been found to be the cause of a number of water-borne and food-borne outbreaks (Mackkenzie et al., 1994) caused by wastewater contamination. *Giardia* spp. cysts and *Cryptosporidium* spp. oocysts are highly resistant to chlorination (Li et al., 2001) and may persist for prolonged periods of time under various environmental conditions.

Although a reduction of up to one order of magnitude in the *Cryptosporidium* oocyst concentration during conventional wastewater treatment processes is common, only a relatively low amount of oocysts is needed for it to be infectious (100 oocysts for healthy individuals) (DuPont et al. 1995). Nasser et al. (2003) studied the die-off of *Giardia* spp. cysts and *Cryptosporidium* spp. oocysts in loamy soil saturated with secondary effluent. They reported a prolonged persistence of these parasites in saturated soil. Only one log reduction in *Cryptosporidium* spp. oocyst number was observed after 20 d incubation and a two log die-off for *Giardia* spp. cysts. In a later study, Nasser et al. (2007) reported a one log reduction in oocyst infectivity at 30 °C in saturated soil and a three log reduction when incubated for 10 d in a dry loamy soil at 32 °C, whereas no change of oocyst viability was recorded in either cases. Mawdsley et al. (1996) studied the potential for transfer of *Cryptosporidium parvum* through soil to land drains and, subsequently, water courses, following application of livestock waste to land in laboratory experiments using various types of soils. Following irrigation over a period of 21 d, the majority of *C. parvum* oocysts were found in the top 2 cm of soil, with numbers decreasing with depth.

Jenkins et al. (2002) performed a laboratory pot study to determine the factors (soil type, temperature, and soil water potential) affecting *Cryptosporidium* oocyst survival in soil. They reported that survival was significantly greater in the silt loam compared to silty clay loam and loamy sand soils. In this study it was clear that temperature had the greatest effect on oocyst inactivation, with greatest survival at 4 °C. Furthermore, under constant laboratory conditions, oocyst survival was significantly greater in all three soil types than in distilled water (Jenkins et al., 2002). Kato et al. (2004) studied the environmental parameters that affect inactivation of *C. parvum* oocysts in soil under field conditions at a dairy farm and found that after 120 d there were still viable and potentially infective oocysts in the soil. Although studies on the survival of these pathogens in field soils are still needed, the results currently available emphasize the risk of *Cryptosporidium* infection as a result of wastewater-irrigation.

11.4 Summary and conclusions

The effects of several compounds found in wastewater on soil microbiology are well described. However, effects of wastewater as a whole on soil microbiology have not been studied very well. Of note, data are missing on the effects of these compounds, separately or in combination, on microbial community composition in soils. With the increasing significance of TWW in irrigation and the introduction of new methods to soil microbiology studies, we expect an increase in the near future in the number and impact of studies that will describe the effects of wastewater irrigation on community composition and function in soil.

References

Aarestrup, F. and Wegener, H. (1999) The effects of antibiotic usage in food animals on the development of antimicrobial resistance of importance for humans in *Campylobactor* and *Escherichia coli*. *Microbes and Infection* **1**, 639–644.

Abdulraheem, M.Y. (1989) Health Considerations in Using Treated Industrial and Municipal Effluents for Irrigation. *Desalination* **72**, 81–113.

Alexander, M. (1977) *Introduction to Soil Microbiology*, 2nd ed. John Wiley & Sons, New York.

Amann, R.I., Ludwig, W. and Schleifer, K.H. (1995) Phylogenetic identification and in-situ detection of individual microbial-cells without cultivation. *Microbial Reviews* **59**, 143–169.

Amezketa, E. (1999) Soil aggregate stability: a review. *Journal of Sustainable Agriculture* **14**, 83–151.

Anderson, T.H. (2003) Microbial eco-physiological indicators to assess soil quality. *Agriculture, Ecosystems and Environment* **98**, 285–293.

Anderson, T.H. and Domsch, K.H. (1993) The metabolic quotient for CO_2 (qCO_2) as a specific activity parameter to assess the effects of environmental conditions, such as pH, on the microbial biomass of forest soils. *Soil Biology and Biochemistry* **25**, 393–395.

Andren, O. and Balandreau, J. (1999) Biodiversity and soil functioning - from black box to can of worms? *Applied Soil Ecology* **13**, 105–108.

Angin, I., Yaganoglu, A.V. and Turan, M. (2005) Effects of long-term wastewater irrigation on soil properties. *Sustainable Agriculture* **26**, 31–42.

Ashelford, K.E., Norris, S.J., Fry, J.C., Bailey, M.J. and Day, J.M. (2000) Seasonal Population Dynamics and Interactions of Competing Bacteriophages and Their Host in the Rhizosphere. *Applied Environmental Microbiology* **66**, 4193–4199.

Aoyama, M., Itaya, S. and Otowa, M. (1993) Effect of copper on the decomposition of plant residus, microbial biomass and beta-glucosidase activity in soils. *Soil Science and Plant Nutrition (Tokyo)* **39**, 557–566.

Aoyama, M. and Nagumo, T. (1996) Factors affecting microbial biomass and dehydrogenase activity in apple orchard soils with heavy metal accumulation. *Soil Science and Plant Nutrition (Tokyo)* **42**, 821–831.

Avrahami, S., Conrad, R. and Braker, G. (2002) Effect of soil ammonium concentration on N2O release and on the community structure of ammonia oxidisers and denitrifiers. *Applied Environmental Microbiology* **68**, 5685–5692.

Avrahami, S., Liesack, W. and Conrad, R. (2003) Effects of temperature and fertilizer on activity and community structure of soil ammonia oxidizers. *Environmental Microbiology* **5**, 691–705.

Bååth, E., Lundgren, B. and Söderström. B. (1979) Effects of artificial acid rain on microbial activity and biomass. *Bulletin of Environmental Contamination and Toxicology* **23**, 737–740.

Bååth, E. (1989) Effects of heavy metals in soil on microbial processes and population, a review. *Water Air Soil Pollution* **47**, 335–379.

Bååth, E., Frostegård, A. and Fritze, H. (1992) Soil bacterial biomass, activity, phospholipid fatty acid pattern, and pH tolerance in an area polluted with alkaline dust deposition. *Applied Environmental Microbiology* **58**, 4026–4031.

Bååth, E., Diaz-Ravina, M., Frostegard, A. and Campbell, C.D. (1998) Effect of metal-rich sludge amendments on the soil microbial community. *Applied Environmental Microbiology* **64**, 238–245.

Bååth, E. and Anderson, T.H. (2003) Comparison of soil fungal/bacterial ratios in a pH gradient using physiological and PLFA-based techniques. *Soil Biology and Biochemistry* **35**, 955–963.

Babich, H. and Stotzky, G. (1985) Heavy metal toxicity to microbe-mediated ecologic processes: a review and potential application to regulatory policies. *Environmental Research* **36**, 111–137.

Bales, R.C., Li, S., Maguire, K.M., Yahya, M.T., Gerba, C.P. and Harvey, R.W. (1995) Virus and bacteria transport in a sandy aquifer, Cape Cod, MA. *Ground Water* **33**, 653–661.

Bardgett, R.D., Speir, T.W., Ross, D.J., Yeates, G.W. and Kettles, H.A. (1994) Impact of pasture contamination by copper, chromium and arsenic timber preservatives on soil microbial properties and nematodes. *Biology and Fertility of Soils* **18**, 71–79.

Bardgett, R.D. and Griffiths, B.S. (1997) Ecology and biology of soil protozoa, nematodes and micro-arthropods. In: *Modern Soil Microbiology* (eds Elasas, J.D., van Elsas, J.D., Trevors, J.T. and Wellington, E.M.H.), pp. 129–163. Marcel Dekker, New York.

Bardgett, R.D. and Shine, A. (1999) Linkages between plant litter diversity, soil microbial biomass and ecosystem function in temperate grasslands. *Soil Biology and Biochemistry* **31**, 317–321.
Barkle, G.F., Stenger, R., Singleton, P.L. and Painter, D.J. (2000) Effect of regular irrigation with dairy farm effluent on soil organic matter and soil microbial biomass. *Australian Journal of Soil Research* **38**, 1087–1097.
Barnhart, C.L. and Vestal, R. (1983) Effect of environmental toxicant on metabolic activity of natural microbial communities. *Applied Environmental Microbiology* **46**, 970–977.
Barton, L., McLay, C.D.A., Schipper, L.A. and Smith, C.T. (1999) Denitrification rates in a wastewater-irrigated forest soil in New Zealand. *Journal of Environmental Quality* **28**, 2008–2014.
Barton, L., Schipper, L.A., Smith, C.T. and McLay, C.D.A. (2000) Denitrification enzyme activity is limited by soil aeration in a wastewater-irrigated forest soil. *Biology and Fertility of Soils* **32**, 385–389.
Barton, L. and Schipper, L.A. (2001) Regulation of Nitrous Oxide Emissions from Soils Irrigated with Dairy Farm Effluent. *Journal of Environmental Quality* **30**, 1881–1887.
Batra, L. and Manna, M.C. (1997) Dehydrogenase activity and microbial biomass carbon in salt affected soils of semiarid and arid regions. *Arid Soil Research and Rehabilitation* **11**, 295–303.
Belser, L.W. and Mays, E.L. (1980) Specific Inhibition of Nitrite Oxidation by Chlorate and Its Use in Assessing Nitrification in Soils and Sediments. *Applied Environmental Microbiology* **39**, 505–510.
Bending, G.D., Turner, M.K., Rayns, F., Marx, M.C. and Wood, M. (2004) Microbial and biochemical soil quality indicators and their potential for differentiating areas under contrasting agricultural management regimes. *Soil Biology and Biochemistry* **36**, 1785–1792.
Berg, B., Ekbohm, G., Söderström, B. and Staaf, H. (1991) Reduction of decomposition rates of Scots pine needle litter due to heavy-metal pollution. *Water Air Soil Pollution* **59**, 165–177.
Bernal, M.P., Sánchez-Monedero, M.A., Paredes, C. and Roig, A. (1998) Carbon mineralization from organic wastes at different composting stages during their incubation with soil. *Agriculture, Ecosystems and Environment* **69**, 175–189.
Brady, N.C. and Weil, R.R. (2002) Soil organic matter. In: The Nature and Properties of Soils, 13th ed (eds Brady, N.C. and Weil, R.R.), pp. 498–591. Prentice Hall, Upper Saddle River, NJ, USA.
Bringmann, G. and Kuhn, R. (1980) Comparison of the toxicity thresholds of water pollutants to bacteria, algae and protozoa in the cell multiplication inhibition test. *Water Research* **14**, 231–241.
Brookes, P.C. (1995) The use of microbial parameters in monitoring soil pollution by heavy metals. *Biology and Fertility of Soils* **19**, 269–279.
Bruns, M.A., Stephen, J.R., Kowalchuk, G.A., Prosser, J.I. and Paul, E.A. (1999) Comparative diversity of ammonia oxidizer 16S rRNA gene sequences in native, tilled and successional soils. *Applied Environmental Microbiology* **65**, 2994–3000.
Brussaard, L., Behanpelletier, V.M., Bignell, D.E., Brown, V.K., Didden, W., Folgarait, P., Fragoso, C., Freckman, D.W., Gupta, V.V.S.R., Hattori, T., Hawksworth, D.L., Klopatek, C., Lavelle, P., Malloch, D.W., Rusek, J., Soderstrom, B., Tiedje, J.M. and Virginia, R.A. (1997) Biodiversity and ecosystem functioning in soil. *Ambio* **26**, 563–570.
Brzezinska, M., Stepniewska, Z. and Stepniewski, W. (2001) Dehydrogenase and catalase activity of soil irrigated with municipal wastewater. *Polish Journal of Environmental Studies* **10**, 307–311.
Brzezinska, M., Tiwari, S.C., Stepniewska, Z., Nosalewicz, M., Bennicelli, R.P. and Samborska, A. (2006) Variation of enzyme activities, CO_2 evolution and redox potential in an Eutric Histosol irrigated with wastewater and tap water. *Biology and Fertility of Soils* **43**, 131–135.
Bol, R., Kandeler, E., Amelung, W., Glaser, B., Marx, M.C. and Preedy, N. (2003) Short-term effects of dairy slurry amendment on carbon sequestration and enzyme activities in a temperate grassland. *Soil Biology and Biochemistry* **35**, 1411–1421.
Bolaños, L., El-Hamdaoui, A. and Bonilla, I. (2003) Recovery of development and functionality of nodules and plant growth in salt-stressed *Pisum sativum-Rhizobium leguminosarum* symbiosis by boron and calcium. *Journal of Plant Physiology* **160**, 1493–1497.
Bonilla, I., Garcia-Gonzalez, M. and Mateo, P. (1990) Boron requirement in cyanobacteria: its possible role in the early evolution of photosynthetic organisms. *Plant Physiology* **94**, 1554–1560.
Bouhmouch, I., Souad-Mouhsine, B., Brhada, F. and Aurag, J. (2005) Influence of host cultivars and Rhizobium species on the growth and symbiotic performance of *Phaseolus vulgaris* under salt stress. *Journal of Plant Physiology* **162**, 1103–1113.
Bouwer, H. (2000) Groundwater problems caused by irrigation with sewage effluent. *Environmental Health* **10**, 17–20.

Bowen, J.E. and Gauch, H.G. (1966) Nonessentiality of boron in fungi and the nature of its toxicity. *Plant Physiology* **41**, 319–324.
Buckley, D.H. and Schmidt, T.M. (2003) Diversity and dynamics of microbial communities in soils from agro-ecosystems. *Environmental Microbiology* **5**, 441–452.
Burgman, H., Widmer, F., Sigler, W.V. and Zeyer, J. (2003) mRNA extraction and reverse transcription-PCR protocol for detection of *nifH* gene expression by *Azotobacter vinelandii* in soil. *Applied Environmental Microbiology* **69**, 1928–1935.
Butterwick, L., De Oude, N. and Raymond, K. (1989) Safety assessment of boron in aquatic and terrestrial environments. *Ecotoxicology and Environmental Safety* **17**, 339–371.
Capone, D.G., Reese, D. and Kiene, R.P. (1983) Effects of metals on methanogenesis, sulfate reduction, carbon dioxide evolution, and microbial biomass in anoxic salt marsh sediments. *Applied Environmental Microbiology* **45**, 1586–1591.
Cavigelli, M.A. and Robertson, G.P. (2000) The functional significance of denitrifier community composition in a terrestrial ecosystem. *Ecology* **81**, 1402–1414.
Chandler, D.S. and Craven, J.A. (1980) Relationship of soil moisture to survival of *Escherichia coli* and *Salmonella typhimurium* in soils. *Australian Journal of Agricultural Research* **31**, 547–555.
Chang, A.C. and Page, A.L. (1985) Fate of wastewater constituents in soil and groundwater: trace organics. In: Irrigation with Reclaimed Wastewater: A Guidance Manual (eds G.S. Pettygrove and T. Asano), pp. 1–20. Lewis Publishers Inc., Chelsea, MI.
Chang, Y.J., Anwar Hussain, A.K.M., Stephen, J.R., Mullens, M.D., White, D.C. and Peacock, A. (2001) Impact of herbicides on the abundance and structure of indigenous β-subgroup ammonia-oxidizer communities in soil microcosms. *Environmental Toxicology and Chemistry* **20**, 2462–2468.
Chantigny, M.H. (2003) Dissolved and water-extractable organic matter in soils: a review on the influence of land use and management practices. *Geoderma* **113**, 357–380.
Clapp, C.E., Hayes, M.H.B., Simpson, A.J. and Kingery, W.L. (2005) Chemistry of soil organic matter. In: *Chemical Processes in Soils* (eds M.A. Tabatabai and D.L. Sparks), pp. 1–150. Soil Science Society of America, Madison, WI.
Clegg, C.D., Lovell, R.D.L. and Hobbs, P.J. (2003) The impact of grassland management regime on the community structure of selected bacterial groups in soils. *FEMS Microbiology and Ecology* **43**, 263–270.
Coleman, D.C. and Crossley, D.A. (1996) *Fundamentals of Soil Ecology*. Academic Press, London.
Collins, Y.E. and Stotzky, G. (1989) Factors affecting the toxocity of heavy metals to microbes. In: *Metal Ions and Bacteria* (eds T.J Beweridge and R.J. Doyle), pp. 31–90. John Wiley & Sons, New York, USA.
Collins, G., Kavanagh, S., McHugh, S., Connaughton, S., Kearney, A., Rice, O., Carrigg, C., Scully, C., Bhreathnach, N., Mahony, T., Madden, P., Enright, A.M. and O'Flaherty, V. (2006) Accessing the Black Box of Microbial Diversity and Ecophysiology: Recent Advances Through Polyphasic Experiments. *Journal of Environmental Science and Health* **41**, 897–922.
Comans, R.N.J. and Van Dijk, C.P.J. (1988) Role of complexation processes in cadmium mobilization during estuarine mixing. *Nature* **336**, 151–154.
Cordovilla, M.P., Ligero, F. and Lluch, C. (1999) Effect of salinity on growth, nodulation and nitrogen assimilation in nodules of faba bean (*Vicia faba* L.). *Applied Soil Ecology* **11**, 1–7.
Cook, F.J., Kelliher, F.M. and McMahon, S.D. (1994) Changes in infiltration and drainage during wastewater irrigation of a highly permeable soil. *Journal of Environmental Quality* **23**, 476–482.
Dar, G.H. and Mishra, M.M. (1994) Influence of cadmium on carbon and nitrogen mineralization in sewage-sludge amended soils. *Environmental Pollution* **84**, 285–290.
Davis, B.D., Luger, S.M. and Tai, P.C. (1986) Role of ribosome degradation in the death of starved *Escherichia coli* cells. *Journal of Bacteriology* **166**, 439–445.
Debosz, K., Petersen, S.O., Kure, L.K. and Ambus, P. (2002) Evaluating effects of sewage sludge and household compost on soil physical and microbial properties. *Applied Soil Ecology* **19**, 237–248.
Degens, B.P., Schipper, L.A., Sparling, G.P. and Vojvodic-Vukovic, M. (2000) Decreases in organic C reserves in soils can reduce the catabolic diversity of soil microbial communities. *Soil Biology and Biochemistry* **32**, 189–196.
Degens, B.P., Schipper, L.A., Sparling, G.P. and Duncan, L.C. (2001) Is the microbial community in a soil with reduced catabolic diversity less resistant to stress or disturbance? *Soil Biology and Biochemistry* **33**, 1143–1153.

Dick, R.P. (1997) Soil enzyme activities as integrative indicators of soil health. In: *Biological Indicators of Soil Health* (eds C.E. Pankhurst, B.M. Doube and V.V.S.R. Gupta), pp. 121–156. CABI Publishers, Wallingford, UK.
Doelman, P. and Haanstra, L. (1979) Effect of lead on soil respiration and dehydrogenase activity. *Soil Biology and Biochemistry* **11**, 475–479.
Doelman, P. and Haanstra, L. (1989) Short- and long-term effects of heavy metals on phosphatase activity in soils: an ecological dose- response model approach. *Biology and Fertility of Soils* **8**, 235–241.
Doran, J.W. and Zeiss, M.R. (2000) Soil health and sustainability: managing the biotic component of soil quality. *Applied Soil Ecology* **15**, 3–11.
DuPont, H.L., Chappell, C.L., Sterling, C.R., Okhuysen, P.C., Rose, J.B. and Jakubowski, W.I. (1995) The infectivity of *Cryptosporidium* in healthy volunteers. *New England Journal of Medicine* **332**, 855–859.
Dusek, L. (1995) The effect of cadmium on the activity of nitrifying populations in two different grassland soils. *Plant Soil* **177**, 43–53.
Elliott, L.F. and Lynch, J.M. (1994) Biodiversity and soil resilience. In: *Soil Resilience and Sustainable Land Use* (eds D.J. Greenland and I. Szabolcs), pp. 353–364. CAB International, Wallingford, UK.
Elliott, L.F., Lynch, J.M. and Papendick, R.I. (1996) The microbial component of soil quality. In: *Soil Biochemistry*, vol. 9 (eds G. Stotzky and J.M. Bollag), pp. 1–21. Marcel Dekker Inc., New York, USA.
van Elsas, J.D., Dijkstra, A.F., Govaert, J.M. and van Veen, J.A. (1986) Survival of *Pseudomonas fluorescens* and *Bacillus subtilis* introduced into two soils of different texture in field microplots. *FEMS Microbial Letters* **38**, 151–160.
van Elsas, J.D., Heijnen, C.E. and van Veen, J.A. (1991) The fate of introduced genetically engineered microorganisms in soil, in microcosms and the field: impact of soil textural aspects. In: *Biological Monitoring of Genetically Engineered Plants and Microbes* (eds D.R. MacKenzie and S.C. Henry), pp. 67–79. Agricultural Research Institute, Bethesda, MD, USA.
van Elsas, J.D. and Rutgers, M. (2005) Estimating soil microbial diversity and community composition. In: *Microbiological Methods for Assessing Soil Quality* (eds J. Bloem, D. W. Hopkins and A. Benedetti), pp. 183–186. Cabi Publishing, Cambridge, MA, USA.
England, L.S., Lee, H. and Trevors, J.T. (1993) Bacterial survival in soil: Effect of clays and protozoa. *Soil Biology and Biochemistry* **25**, 525–531.
Esher, R.J., Marx, D.H., Ursic, S.J., Baker, R.L., Brown, L.R. and Coleman, D.C. (1992) Simulated acid rain effects on fine roots, ectomycorrhizae, microorganisms, and invertebrates in pine forests of the southern United States. *Water Air Soil Pollution* **61**, 269–278.
Estrada, I.B., Aller, A., Aller, F., Gomez, X. and Moran, A. (2004) The survival of *Escheria coli,* faecal coliforms and enterobacteriaceae in general in soil treated with sludge from wastewater treatment plants. *Bioresource Technology* **93**, 191–198.
Fattal, B., Wax, Y., Davies, M. and Shuval, H.I. (1986) Health risks associated with wastewater irrigation: an epidemiological study. *American Journal of Public Health* **76**, 977–979.
Filip, Z., Kanazawa, S. and Berthelin, J. (1999) Characterization of effects of a long-term wastewater irrigation on soil quality by microbiological and biochemical parameters. *Journal of Plant Nutrition and Soil Science* **162**, 409–413.
Filip, Z., Kanazawa, S. and Berthelin, J. (2000) Distribution of microorganisms, biomass ATP, and enzyme activities in organic and mineral particles of a long-term wastewater irrigated soil. *Journal of Plant Nutrition and Soil Science* **163**, 143.
Frankenberger, W.T. (1985) Fate of Wastewater Constituents in Soil and Groundwater: Pathogens. In: *Irrigation with Reclaimed Municipal Wastewater – A Guidance Manual* (eds G.S. Pettygrove and T. Asano), pp. 14-1–14-25. Lewis Publishers, Inc., Chelsea MI USA.
Franz, E., van Diepeningen, A.D., de Vos, O.J. and van Bruggen, A.H. (2005) Effects of cattle feeding regimen and soil management type on the fate of *Escherichia coli* O157:H7 and *Salmonella enterica Serovar Typhimurium* in Manure, Manure-Amended Soil, and Lettuce. *Applied Environmental Microbiology* **71**, 6165–6174.
Friedel, J.K., Langer, T., Siebe, C. and Stahr, K. (2000) Effects of long-term waste water irrigation on soil organic matter, soil microbial biomass and its activities in central Mexico. *Biology and Fertility of Soils* **31**, 414–421.
Fritze, H., Kiikkilä, O., Pasanen, J. and Pietikäinen, J. (1992) Reaction of forest soil microflora to environmental stress along a moderate pollution gradient next to an oil refinery. *Plant Soil* **140**, 175–182.

Gambrell, R.P., Wiesepage, J.B., Patrick, Jr, W.H. and Duff, M.C. (1991) The effects of pH, redox, and salinity on metal release from a contaminated sediments. *Water Air Soil Pollution* **57–58**, 359–367.
Gans, J., Wolinsky, M. and Dunbar, J. (2005) Computational improvements reveal great bacterial diversity and high metal toxicity in soil. *Science (Washington D.C.)* **309**, 1387–1390.
Gelsomino, A., Badalucco, L., Ambrosoli, R., Crecchio, C., Puglisi, E. and Meli, S.M. (2006) Changes in chemical and biological soil properties as induced by anthropogenic disturbance: A case study of an agricultural soil under recurrent flooding by wastewaters. *Soil Biology and Biochemistry* **38**, 2069–2080.
Gerba, C.P., Wallis, C. and Melnick, J.L. (1975) Fate of wastewater bacteria and viruses in soil. *Journal of Irrigation and Drainage* American Society Civil Engineering **101**, 157–174.
Gerba, C.P. and Goyal, S.M. (1985) Pathogen removal from wastewater during groundwater recharge. In: Artificial Recharge of Groundwater (ed. T. Asano), pp. 283–317. Butterworths Publishers, Boston, MA, USA.
Gerba C.P. and Smith, Jr., J.E. (2005) Source of pathogenic microorganisms and their fate during land application of wastes. *Journal of Environmental Quality* **34**, 42–48.
Gibbs, R.A., Hu, C.J., Ho, G.E. and Unkovich, I. (1997) Regrowth of faecal coliforms and salmonella in stored biosolid and soil amended with biosolids. *Water Science and Technology* **35**, 269–275.
Giller, K.E., Beare, M.H., Lavelle, P., Izac, A.M.N. and Swift, M.J. (1997) Agricultural intensification, soil biodiversity and agroecosystem function. *Applied Soil Ecology* **6**, 3–16.
Giller, K.E., Witter, E. and McGrath, S.P. (1998) Toxicity of heavy metals to microorganisms and microbial processes in agricultural soils: a review. *Soil Biology and Biochemistry* **30**, 1389–1414.
Giusquiani, P.L., Pagliai, M., Gigliotti, G., Businelli, D. and Benetti, A. (1995) Urban waste compost: effects on physical, chemical and biochemical soil properties. *Journal of Environmental Quality* **24**, 175–182.
Givskov, M.I., Eberl, L., Moller, S., Poulsen, L.K. and Molin, S. (1994) Responses to nutrient starvation in *Pseudomonas putida* KT2442: Analysis of general cross-protection, cell shape and macromolecular content. *Journal of Bacteriology* **176**, 7–14.
Graber, E.R., Gerstl, Z., Fischer, E. and Mingelgrin, U. (1995) Enhanced transport of atrazine under irrigation with effluent. *Soil Science Society of America Journal* **59**, 1513–1519.
Grayston, S.J., Campbell, C.D., Bardgett, R.D., Mawdsley, J.L., Clegg, C.D., Ritz, K., Griffiths, B.S., Rodwell, J.S., Edwards, S.J. and Davies, W.J. (2004) Assessing shifts in microbial community structure across a range of grasslands of differing management intensity using CLPP, PLFA and community DNA techniques. *Applied Soil Ecology* **25**, 63–84.
Golovleva, L.A., Aharonson, N., Greenhalgh, R., Sethunathan, N. and Vonk, J.W. (1990) The role and limitations of microorganisms in the conversion of xenobiotics. *Pure Applied Chemistry* **62**, 351–364.
Goyal, S., Chander, K. and Kaboor, K. (1995) Effect of distillery wastewater application on soil microbiological properties and plant growth. *Environmental Ecology* **13**, 1–89.
Guhl, W. 1996. Ecological aspects of boron. *SÖFW-Journal* **118**, 1159–1168 (in German).
Harrison, E.Z., Oakes, S.R., Hysell, M. and Hay, A. (2006) Organic chemicals in sewage sludges. *Science of the Total Environment* **31** (367), 481–97.
Hassouna, M.S., Madkour, M.A., Helmi, S.H.E. and Yacout, S.I. (1995) Salt tolerance of some free-living nitrogen fixers. *Alexandria Journal of Agricultural Research* **40**, 389–413.
Hastings, R.C., Ceccherini, M.T., Miclaus, N., Saunders, J.R., Bazzicalupo, M. and McArthy, A.J. (1997) Direct molecular biological analyses of ammonia oxidising bacteria populations in cultivated soil plots treated with swine manure. *FEMS Microbiology and Ecology* **23**, 45–54.
Hastings, R., Butler, C., Singleton, I., Saunders, J. and McCarthy, A. (2000) Analysis of ammonia-oxidising bacteria populations in acid forest soil during conditions of moisture limitation. *Letters in Applied Microbiology* **30**, 14–18.
Head, I.M., Saunders, J.R. and Pickup, R.W. (1998) Microbial evolution, diversity, and ecology: a decade of ribosomal RNA analysis of uncultivated microorganisms. *Microbial Ecology* **35**, 1–21.
Heijnen, C.E., van Elsas, J.D., Kuikman, P.J. and van Veen, J.A. (1988) Dynamics of Rhizobium leguminosarum biovar trifolii introduced in soil: the effect of bentonite clay on predation by protozoa. *Soil Biology and Biochemistry* **20**, 483–488.
HERA (2002) Human and Environmental Risk Assessment on ingredients of European household cleaning products. *Risk Assessment of Sodium Perborate Mono- and Tetrahydrate*. Hera publication, Brussels, Belgium.

Hespanol, I. (1990) Guidelines and integrated measures for public health protection and agricultural reuse systems. *Journal of Water Supply: Research and Technology* **39**, 237–249.

Hiorns, W., Hastings, R., Head, I., McCarthy, A., Saunders, J., Pickup, R. and Hall, G. (1995) Amplication of 16s ribosomal RNA genes of autotrophic ammonia-oxidising bacteria demonstrates the ubiquity of *Nitrosospiras* in the environment. *Microbiology* **141**, 2793–2800.

Högberg, M.N., Bååth, E., Nordgren, A., Arnebrant, K. and Högberg, P. (2003) Contrasting effects of nitrogen availability on plant carbon supply to mycorrhizal fungi and saprotrophs—a hypothesis based on field observations in boreal forest. *New Phytologist* **160**, 225–238.

Högberg, M.N., Högberg, P. and Myrold, D.D. (2007) Is microbial community composition in boreal forest soils determined by pH, C-to-N ratio, the trees, or all three? *Oecologia* **150**, 590–601.

Huang, X., Lee, L.S. and Nakatsu, C. (2000) Impact of animal waste lagoon effluents on chlorpyrifos degradation in soils. *Environmental Toxicology and Chemistry* **19**, 2864–2870.

Hugenholtz, P., Goebel, B.M. and Pace, N.R. (1998) Impact of culture-independent studies on the emerging phylogenetic view of bacterial diversity. *Journal of Bacteriology* **180**, 4765–4774.

Hurst, C.J., Gerba, C.P. and Cech, I. (1980) Effects of Environmental Variables and Soil Characteristics on Virus Survival in Soil. *Applied Environmental Microbiology* **40**, 1067–1079.

Ibekwe, A.M. and Kennedy, A.C. (1998) Phospholipid fatty acid profiles and carbon utilization patterns for analysis of microbial community structure under field and greenhouse conditions. *FEMS Microbiology and Ecology* **26**, 151–163.

ISO 15685 (2004) *Soil Quality – Determination Of Potential Nitrification – Rapid Test By Ammonium Oxidation*. International Organization for Standardization (ISO), Geneva.

Jastrow, J.D., Miller, R.M. and Lussenhop, J. (1998) Contributions of interacting biological mechanisms to soil aggregate stabilization in restored prairie. *Soil Biology and Biochemistry* **30**, 905–916.

Jenkins, M.B., Bowman, D.D, Fogarty, E.A. and Ghiorse, W.C. (2002) *Cryptosporidium parvum* oocyst inactivation in three soil types at various temperatures and water potentials. *Soil Biology and Biochemistry* **34**, 1101–1109.

Jenkinson, D.S. (1988) Determination of microbial biomass carbon and nitrogen in soil. In: *Advances in Nitrogen Cycling in Agricultural Ecosystems* (ed. J.R. Wilson), pp. 368–386. C.A.B. International Wallingford, Brisbane, Australia.

Jetten, M.S., Logemann, S., Muyzer, G., Robertson, L.A., de Vries, S., van Loosdrecht, M.C. and Kuenen, J. G. (1997) Novel principles in the microbial conversion of nitrogen compounds. *Antonie Leeuwenhoek International Journal* **71**, 75–93.

Jiang, X., Morgan, J. and Doyle, M.P. (2002) Fate of *Escherichia coli* O157:H7 in manure-amended soil. *Applied Environmental Microbiology* **68**, 2605–2609.

Jin, Y., Yates, M.V., Thompson, S.S. and Jury, W.A. (1997) Sorption of viruses during flow through saturated sand columns. *Environmental Science and Technology* **31**, 548–555.

Jonas, R.B., Gilmour, C.G., Stoner, D.L., Weir, M.M. and Tuttle, J.H. (1984) Comparison of methods to measure acute metal and organometal toxicity to natural aquatic microbial communities. *Applied Environmental Microbiology* **47**, 1005–1011.

Kandeler, E., Kampichter, C. and Horak, O. (1996) Influence of heavy metals on the functional diversity of soil microbial communities. *Biology and Fertility of Soils* **23**, 299–306.

Kandeler, E., Tscherko, D., Bruce, K.D., Stemmer, M., Hobbs, P.J., Bardgett, R.D. and Amelung, W. (2000) Structure and function of the soil microbial community in microhabitats of a heavy metal polluted soil. *Biology and Fertility of Soils* **32**, 390–400.

Kannan, K. and Oblisami, G. (1990) Influence of paper mill effluent irrigation on soil enzyme activities. *Soil Biology and Biochemistry* **22**, 923–926.

Kato, S., Jenkins, M., Fogarty, E. and Bowman, D. (2004) *Cryptosporidium parvum* oocyst inactivation in field soil and its relation to soil characteristics: analyses using the geographic information systems. *Science of the Total Environment* **321**, 47–58.

Kearny, T.E., Larkin, M.J. and Levett, P.N. (1993) The effects of slurry storage and anaerobic digestion on survival of pathogenic bacteria. *Journal of Applied Microbiology* **74**, 86–93.

Kennedy, A.C. and Papendick, R.I. (1995) Microbial characteristics of soil quality. *Journal of Soil and Water Conservation* **50**, 243–248.

Killham, K., Firestone, M.K. and McColl, J.G. (1983) Acid rain and soil microbial activity: effects and their mechanisms. *Journal of Environmental Quality* **12**, 133–137.

Killham, K. and Staddon, W.J. (2002) Bioindicators and sensors of soil health and the application of geostatistics. In: *Enzymes in the Environment: Activity, Ecology and Applications* (eds R.G. Burns and R.P. Dick), pp. 391–394. Marcel Dekker, New York, USA,.

Kinney, C.A., Furlong, E.T., Werner, S.L. and Cahill, J.D. (2006) Presence and distribution of wastewater-derived pharmaceuticals in soil irrigated with reclaimed water: *Environmental Toxicology and Chemistry* **25**, 317–326.

Kirk, J.L., Beaudette, L.A., Hart, M., Moutoglis, P., Klironomos, J.N., Lee, H. and Trevors, J.T. (2004) Methods of studying soil microbial diversity. *Journal of Microbiological Methods* **58**, 169–188.

Knight, B., McGrath, S.P. and Chaudri, A.M. (1997) Biomass carbon measurements and substrate utilization patterns of microbial population from soils amended with cadium, copper or zinc. *Applied Environmental Microbiology* **63**, 39–43.

Kong, W.D., Zhu, Y.G., Fu, B.J., Marschner, P. and He, J.Z. (2006) The veterinary antibiotic oxytetracycline and Cu influence functional diversity of the soil microbial community. *Environmental Pollution* **143**, 129–137.

Konopka, A., Zakharova, T., Bischoff, M., Oliver, L., Nakatsu, C. and Turco, R.F. (1999) Microbial biomass and activity in lead-contaminated soil. *Applied Environmental Microbiology* **65**, 2256–2259.

Konstantinidis, K.T., Isaacs, N., Fett, J., Simpson, S., Long, D.T. and Marsh, T.L. (2003) Microbial diversity and resistance to copper in metal-contaminated lake sediment. *Microbial Ecology* **45**, 191–202.

Koplin, D.W., Furlong, E.T., Meyer, M.T., Thurman, E.M., Zaugg, S.D., Barber, L.B. and Buxton, H.T. (2002) Pharmaceuticals, Hormones, and Other Organic Wastewater Contaminants in U.S. Streams, 1999–2000: A National Reconnaissance. *Environmental Science and Technology* **36**, 1202–1211.

Kozdroj, J. (1995) Microbial responses to single or successive soil contamination with Cd or Cu. *Soil Biology and Biochemistry* **27**, 1459–1465.

Kozdroj, J. and van Elsas, J. D. (2001) Structural diversity of microorganisms in chemically perturbed soil assessed by molecular and cytochemical approaches. *Journal of Microbiology Methods* **43**, 197–212.

Koops, H.P. and Moller, U.C. (1992) The lithotrophic ammonia-oxidizing bacteria. In: The Prokaryotes, 2nd ed (eds A. Balows, H. G. Trüper, M. Dworkin, W. Harder, and K. H. Schleifer), pp. 2625–2637. Springer-Verlag, New York, USA.

Kowalchuk, G.A., Stienstra, A.W., Hellig, G.H.J., Stephen, J.R. and Woldendorp, J.W. (2000) Changes in the community structure of ammonia-oxidising bacteria during secondary succession of calcareous grasslands. *Environmental Microbiology* **2**, 99–110.

Kuikman, P.J., Jansen, A.G. and van Veen, J.A. (1991) 15N-nitrogen mineralization from bacteria by protozoan grazing at different soil moisture regimes. *Soil Biology and Biochemistry* **23**, 193–206.

Kümmerer, K. (ed.) (2008) *Pharmaceuticals in the Environment - Sources, Fate, Effects and Risks* - 3rd Edition. Springer, Berlin, Germany. 521 pp.

Kumar, V. and Wagenet, R.J. (1985) Salt effects on urea hydrolysis and nitrification during leaching through laboratory soil columns. *Plant Soil* **85**, 219–227.

Lalande, R., Gagnon, B., Simard, R.R. and Côte, D. (2000) Soil microbial biomass and enzyme activity following liquid hog manure application in a long-term field trial. *Canadian Journal of Soil Science* **80**, 263–269.

Lau, M.M. and Ingham, S.C. (2001) Survival of faecal indicator bacteria in bovine manure incorporated into soil. *Letters in Applied Microbiology* **33**, 131–136.

Laura, R.D. (1974) Effects of neutral salts on carbon and nitrogen mineralization of organic matter in soil. *Plant Soil* **41**, 113–127.

Lawton, J.H. (1994) What Do species do in ecosystems? *Oikos* **71**, 367–374.

Leita, L., Denobili, M., Muhlbachova, G., Mondini, C., Marchiol, L. and Zerbi, G. (1995) Bioavailability and effects of heavy metals on soil microbial biomass survival during laboratory incubation. *Biology and Fertility of Soils* **19**, 103–108.

Li, H., Finch, G.R., Smith, D.W. and Belosevic, M. (2001) Sequential inactivation of Cryptosporidium parvum using ozone and chlorine. *Water Research* **35**, 4339–4348.

Loreau, M., Naeem, S., Inchausti, P., Bengtsson, J., Grime, J.P. and Hector, A. (2001) Biodiversity and ecosystem functioning: current knowledge and future challenges. *Science (Washington D.C.)* **294**, 804–808.

Maatallah, J., Berraho, E.B., Munoz, S., Sanjuan, J. and Lluch, C. (2002) Phenotypic and molecular characterization of chickpea rhizobia isolated from different areas of Morocco. *Journal of Applied Microbiology* **93**, 531–540.

McArthur, J.V. and Tuckfield, R.C. (2000) Spatial patterns in antibiotic resistance among stream bacteria: effects of industrial pollution. *Applied Environmental Microbiology* **66**, 3722–26.

McClung, G. and Frankenberger, W.T. (1987) Nitrogen mineralization rates in saline vs. salt amended soils. *Plant Soils* **104**, 13–21.

McConnell, D.B., Shiralipour, A. and Smith, W.H. (1993) Compost application improves soil properties. *Biocycle* **3**, 61–63.

McDermott, P., Walker, R. and White, D. (2003) Antimicrobials: modes of action and mechanisms of resistance. *International Journal of Toxicology* **22**, 135–143.

McGrath, S.P. (1994) Effect of heavy metals from sewage-sludge on soil microbes in agriculture ecosystems. In: *Toxic Metals in Soil Plant Systems* (ed. S.M. Ross), pp. 247–273. John Wiley & Sons, Chichester.

Mackkenzie, W.R., Hoxie, N.J., Proctor, M.E., Gradus, M.S., Blair, K.A., Peterson, D.E., Kazmierczak, J.J., Addiss, D.G., Fox, K.R., Rose, J.B. and Davis, J.P. (1994) A massive outbreak in Milwaukee of *Cryptosporidium* infection transmitted through the public water supply. *New England Journal of Medicine* **331**, 161–167.

Magesana, G.N., Williamsona, J.C., Yeatesb, G.W. and Lloyd-Jonesa, A.Rh. (2000) Wastewater C:N ratio effects on soil hydraulic conductivity and potential mechanisms for recovery. *Bioresources Technology* **71**, 21–27.

Malkawi, H.I. and Mohammad, M.J. (2003) Survival and accumulation of microorganisms in soils irrigated with secondary treated wastewater. *Journal of Basic Microbiology* **43**, 47–55.

Marinari, S., Masciandaro, G., Ceccanti, B. and Grego, S. (2000) Influence of organic and mineral fertilisers on soil biological and physical properties. *Bioresources Technology* **72**, 9–17.

Masaphy, S. and Mandelbaum, R.T. (1997) Atrazine mineralization in slurries from soils irrigated with treated wastewater. *Applied Soil Ecology* **6**, 283–291.

Mawdsley, J.L., Brooks, A.E. and Merry, R.J. (1996) Movement of the protozoan pathogen *Cryptosporidium parvum* through three contrasting soil types. *Biology and Fertility of Soils* **21**, 30–36.

Matsuguchi, T. and Sakai, M. (1995) Influence of soil salinity on the populations and composition of fluorescent pseudomonads in plant rhizosphere. *Soil Science and Plant Nutrition* **41**, 497–504.

Mekki, A., Dhouib, A. and Sayadi, S. (2006) Changes in microbial and soil properties following amendment with treated and untreated olive mill wastewater. *Microbiology Research* **161**, 93–101.

Mekki, A., Dhouib, A., Feki, F. and Sayadi, S. (2007) Assessment of toxicity of the untreated and treated olive mill wastewaters and soil irrigated by using microbiotests. *Ecotoxicology and Environmental Safety* **69**, 488–495.

Meli, S., Porto, M., Belligno, A., Bufo, S.A., Mazzatura, A. and Scopa, A. (2002) Influence of irrigation with lagooned urban wastewater on chemical and microbiological soil parameters in a citrus orchard under Mediterranean conditions. *Science of the Total Environment* **285**, 69–77.

Mendum, T.A., Sockett, R.E. and Hirsch, P.R. (1998) The detection of Gram negative bacterial mRNA from soil by RT-PCR. *FEMS Microbiology Letters* **164**, 369–373.

Mendum, T.A., Sockett, R.E. and Hirsch, P.R. (1999) Use of molecular and isotopic techniques to monitor the response of autotrophic ammonia-oxidizing populations of the beta subdivision of the class *Proteobacteria* in arable soils to nitrogen fertilizer. *Applied Environmental Microbiology* **65**, 4155–4162.

Meschke, J.S. and Sobsey, M.D. (1998) Comparative adsorption of Norwalk virus, poliovirus 1, and F + RNA coliphages MS2 to soils suspended in treated wastewater. *Water Science and Technology* **38**, 187–189.

Moffett, B.F., Nicholson, F.A., Uwakwe, N.C., Chambers, B.J., Harris, J.A. and Hill, T.C.J. (2003) Zinc contamination decreases the bacterial diversity of agricultural soil. *FEMS Microbiology and Ecology* **43**, 13–19.

Mohammad, M.J. and Mazahreh, N. (2003) Changes in soil fertility parameters in response to irrigation of forage crops with secondary treated wastewater. *Communications in Soil and Plant Analysis* **34**, 1281–1294.

Monnett, G.T., Reneau, R.B. and Hagedon, C. (1995) Effects of domestic wastewater spray irrigation on denitrification rates. *Journal of Environmental Quality* **24**, 940–946.

Müller, C., Sherlock, R.R. and Williams, P.H. (1998) Field method to determine N_2O emission from nitrification and denitrification. *Biology and Fertility of Soils* **28**, 51–55.

Müller, K., Magesan, G.N. and Bolan, N.S. (2007) A critical review of the influence of effluent irrigation on the fate of pesticides in soil. *Agriculture, Ecosystems and Environment* **120**, 93–116.

Muszkat, L., Lahav, D., Ronen, D. and Magaritz, M. (1993a) Penetration of pesticides and industrial organics deep into soil and into groundwater. *Archives of Insect Biochemistry and Physiology* **22**, 487–499.

Muszkat, L., Raucher, D., Magaritz, M., Ronen, D. and Amiel, A.J. (1993b) Unsaturated zone and groundwater contamination by organic pollutants in a sewage-effluent-irrigated site. *Ground Water* **31**, 556–565.

Myrold, D.D. (1998) Transformations of nitrogen. In: *Principles and Applications of Soil Microbiology* (eds D.M. Sylvia, J.J. Fuhrmann, P.G. Hartel and D.S. Zuberer), pp. 259–294. Prentice Hall Inc., Upper Saddle River, NJ, USA.

Myrold, D.D. and Nason, G.E. (1992) Effect of acid rain on soil microbial processes. In: *Environmental Microbiology* (ed. R. Mitchell), pp. 59–81. Wiley-Liss, New York.

Naidu, C.K. and Reddy, T.K.R. (1988) Effect of cadmium on microorganisms and microbe-mediated mineralization process in soil. *Bulletin of Environmental Contamination and Toxicology* **41**, 657–663.

Nakatsu, C.H., Carmosini, N., Baldwin, B., Beasley, F., Kourtev, P. and Konopka, A. (2005) Soil microbial community responses to additions of organic carbon substrates and heavy metals (Pb and Cr). *Applied Environmental Microbiology* **71**, 7679–7689.

Nannipieri, P., Muccini, L. and Ciardi, C. (1983) Microbial biomass and enzyme activities: Production and persistence. *Soil Biology and Biochemistry* **15**, 679–685.

Nannipieri, P. and Badalucco, L. (2003) Biological processes. In: *Handbook of Processes and Modeling in the Soil–Plant System* (eds D.K. Bembi and R. Nieder), pp. 57–76. The Haworth Press, Binghamton, NY, USA.

Nannipieri, P., Ascher, J., Ceccherini, M.T., Landi, L., Pietramellara, G. and Renella, G. (2003) Microbial diversity and soil functions. *European Journal of Soil Science* **54**, 655–670.

Nasser, A.M., Huberman, Z., Zilberman, A. and Greenfeld, S. (2003) Die-off and retardation of *Cryptosporidium* spp. oocyst in loamy soil saturated with secondary effluent. *Water and Science Technology: Water Supply* **3**, 253–259.

Nasser, A.M., Tweto, E. and Nitzan, Y. (2007) Die-off of *Cryptosporidium parvum* in soil and wastewater effluents. *Journal of Applied Microbiology* **102**, 169–176.

Nelson, P.N., Ladd, J.N. and Oades, J.M. (1996) Decomposition of 14C labelled plant material in a salt affected soil. *Soil Biology and Biochemistry* **28**, 433–441.

Nelson, D.R. and Mele, P.M. (2007) Subtle changes in rhizosphere microbial community structure in response to increased boron and sodium chloride concentrations *Soil Biology and Biochemistry* **39**, 340–351.

Nicholson, F.A., Hutchison, M.C., Smith, K.A., Keevil, C.W., Chambers, B.J. and Moore, A. (2000) *A Study on Farm Manure Application to Agricultural Land and an Assessment of the Risks of Pathogen Transfer Into The Food Chain.* Project number FS2526, London: Final Report to the Ministry of Agriculture, Fisheries and Food.

Nicholson, F.A., Groves, S.J. and Chambers, B.J. (2005) Pathogen survival during livestock manure storage and following land application. *Bioresources Technology* **96**, 135–143.

Nicosia, L.A., Rose, J.B., Stark, L. and Stewart, M.T. (2001) A field study of virus removal in septic tank drain fields. *Journal of Environmental Quality* **30**, 1933–1939.

Niewolak, S. and Tucholski, S. (1999) The effect of meadow irrigation with biologically treated sewage on the occurrence of microorganisms indicatory of pollution and sanitary state and of potentially pathogenic bacteria in grass. *Polish Journal of Environmental Studies* **8**, 39–46.

Nikolaou, A., Meric, S. and Fatta, D. (2007) Occurrence patterns of pharmaceuticals in water and wastewater environments. *Analytical and Bioanalytical Chemistry* **387**, 1225–1234.

Nogales, B., Timmis, K.N., Nedwell, D.B. and Osborn, A.M. (2002) Detection and Diversity of Expressed Denitrification Genes in Estuarine Sediments after Reverse Transcription-PCR Amplification from mRNA. *Applied Environmental Microbiology* **68**, 5017–5025.

Oades, J.M. (1993) The role of biology in the formation, stabilization and degradation of soil structure. *Geoderma* **56**, 377–400.

Ohya, H., Komai, Y and Yamaguchi, M. (1985) Zinc effect on soil microflora and glucose metabolites in soil amended with ^{14}C-glucose. *Biology and Fertility of Soils* **1**, 17–122.

Omar, S.A., Abdel-Sater, M.A., Khallil, A.M. and Abd-Alla, M.H. (1994) Growth and enzyme activities of fungi and bacteria in soil salinized with sodium chloride. *Folia Microbiology (Prague)* **39**, 23–28.

Oppel, J., Broll, G., Loffler, D., Meller, M., Rombke, J. and Ternes, T. (2004) Leaching behavior of pharmaceuticals in soil-testing-systems: a part of an environmental risk assessment for groundwater protection. *Science of the Total Environment* **328**, 265–273.
Oved, T., Shaviv, A., Goldrath, T., Mandelbaum, R. and Minz, D. (2001) Influence of effluent irrigation on community composition and function of ammonia oxidizing bacteria in soil. *Applied Environmental Microbiology* **67**, 3426–3433.
Paalman, M.A.A., Van Der Weijden, C.H. and Loch, J.P.G. (1994) Sorption of cadmium on suspended matter under estuarine conditions; competition and complexation with major sea-water ions. *Water Air Soil Pollution* **73**, 49–60.
Parfitt, R.L., Ross, D.J., Coomes, D.A., Richardson, S.J., Smale, M.C. and Dahlgren, R.A. (2005) N and P in New Zealand soil chronosequences and relationships with foliar N and P. *Biogeochemistry* **75**, 305–328.
Pathak, H. and Rao, D.L.N. (1998) Carbon and nitrogen mineralization from added organic matter in saline and alkali soils. *Soil Biology and Biochemistry* **30**, 695–702.
Paul, E.A. and Clark, F.E. (1989) *Soil Microbiology and Biochemistry*. Academic Press, Inc., San Diego, CA.
Pennanen, T. (2001) Microbial communities in boreal coniferous forest humus exposed to heavy metals and changes in soil pH - a summary of the use of phospholipid fatty acids, Biolog® and ^{3}H-thymidine incorporation methods in field studies. *Geoderma* **100**, 91–126.
Pepper, I.L. and Gerba, C.P. (1989) Pathogens. In: Agricultural Sludge Reclamation Study, Part 6, pp. 94–146. Pima County Wastewater Management Dept. Tucson, AZ, USA.
Pettygrove, G.S., Davenport, D.C. and Asano, T. (1985) Introduction: California's reclaimed municipal wastewater resource. In: *Irrigation with Reclaimed Municipal Wastewater–A Guidance Manual*, 2nd ed (eds G. S. Pettygrove and T. Asano), pp. 1–14. Lewis Publishers, Chelsea, MI, USA.
Phillips, C.J., Harris, D., Dollhopf, S.L., Gross, K.L., Prosser, J.I. and Eldor, A.P. (2000) Effects of agronomic treatments on structure and function of ammonia-oxidizing communities. *Applied Environmental Microbiology* **66**, 5410–5418.
Piddock, L., Ricci, V., Stanley, K. and Jones, K. (2000) Activity of antibiotics used in human medicine for *Campylobacter jejuni* isolated from farm animals and their environment in Lancashire, UK. *Journal of Antimicrobial Chemotherapy* **46**, 303–306.
Polonenko, D.R., Dumbroff, E.B. and Mayfield, C.I. (1981) Microbial responses to salt-induced osmotic stress II. Population changes in the rhizoplane and rhizosphere. *Plant Soil* **63**, 415–426.
Powlson, D.S, Brookes, P.C. and Christensen, B.T. (1987) Measurement of soil microbial biomass provides an early indication of changes in total soil organic matter due to straw incorporation. *Soil Biology and Biochemistry* **19**, 159–164.
Prescott, C.E. and Parkinson, D. (1985) Effects of sulphur pollution on rates of litter decomposition in a pine forest. *Canadian Journal of Botany* **63**, 1436–1443.
Prinčič, A., Mahne, I., Megusar, F., Paul, E.A. and Tiedje, J.M. (1998) Effects of pH and oxygen and ammonium concentrations on the community structure of nitrifying bacteria from wastewater. *Applied Environmental Microbiology* **64**, 3584–3590.
Prosser, J.I. (1989) Autotrophic nitrification in bacteria. *Advanced Microbiology and Physiology* **30**, 125–181.
Qian, Y.L. and Mecham, B. (2005) Long-Term Effects of Recycled Wastewater Irrigation on Soil Chemical Properties on Golf Course Fairways. *Agronomy Journal* **97**, 717–721.
Radajewski, S., Ineson, P., Parekh, N.R. and Murrell, J.C. (2000) Stable-isotope probing as a tool in microbial ecology. *Nature* **403**, 646–649.
Ragab, M. (1993) Distribution pattern of soil microbial population in salt affected soils. In: *Towards Rational Use of High Salinity Tolerant Plants, Deliberations about High Salinity Tolerant Plants and Ecosystems*, vol. 1 (eds H. Lieth and A.A. Al-Masoom), pp. 467–472. Kluwer Academic Publishers, Dordrecht, Netherlands.
Ramirez-Fuentes, E., Lucho-Constantino, C., Escamilla-Silva, E. and Dendooven, L. (2002) Characteristics, and carbon and nitrogen dynamics in soil irrigated with wastewater for different lengths of time. *Bioresources Technology* **85**, 179–187.
Rangel-Castro, J.I, Killham, K., Ostle, N., Nicol, G.W., Anderson, I.C., Scrimgeour, C.M., Ineson, P., Meharg, A. and Prosser, J.I. (2005) Stable isotope probing analysis of the influence of liming on root exudate utilization by soil microorganisms. *Environmental Microbiology* **7**, 828–838.

Raymond, J., Siefert, J.L., Staples, C.R. and Blankenship, R.E. (2004) The natural history of nitrogen fixation. *Molecular Biology and Evolution* **21**, 541–554.
Recorbet, G., Steinberg, C. and Faurie, G. (1992) Survival in soil of genetically engineered *Escherichia coli* as related to inoculum density, predation and competition. *FEMS Microbial Ecology* **10**, 251–260.
Rietz, D.N. and Haynes, R.J. (2003) Effects of irrigation induced salinity and sodicity on soil microbial activity. *Soil Biology and Biochemistry* **35**, 845–854.
Ritz, K., Wheatley, R.E. and Griffiths, B.S. (1997) Effects of animal manure application and crop plants upon size and activity of soil microbial biomass under organically grown spring barley. *Biology and Fertility of Soils* **24**, 372–377.
Ross, D.J., Cainrs, A. and Speir, T.W. (1978) Effect of irrigation with municipal water or sewage effluent on the biology of soil cores. *New Zealand Journal of Agricultural Research* **21**, 411–417.
Ross, D.J., Speir, T.W., Tate, K.R., Cairns, A., Meyrick, K.F. and Pansier, E.A. (1982) Restoration of pasture after topsoil removal: effects on soil carbon and nitrogen mineralization, microbial biomass and enzyme activities. *Soil Biology and Biochemistry* **14**, 575–581.
Rossel, D. and Tarradellas, J. (1991) Dehydrogenase activity of soil microflora: significance in ecotoxicological tests. *Environmental Toxicology and Water Quality* **6**, 17–33.
Rooklidge, S.J. (2004) Environmental antimicrobial contamination from terraccumulation and diffuse pollution pathways. *Science of the Total Environment* **325**, 1–13.
Rotthauwe, J.H., Witzel, K.P. and Liesack, W. (1997) The ammonia monooxygenase structural gene *amoA* as a functional marker: molecular fine-scale analysis of natural ammonia-oxidizing populations. *Applied Environmental Microbiology* **63**, 4704–4712.
Ruark, G.A. and Zarnoch, S.J. (1992) Soil carbon, nitrogen and fine root biomass sampling in a pine stand. *Soil Science Society of America Journal* **56**, 1945–1950.
Ruess, L., Sandbach, P., Cudlin, P., Dighton, J. and Crossley, A. (1996) Acid deposition in a spruce forest soil: effects on nematodes, mycorrhizas and fungal biomass. *Pedobiologia* **40**, 51–66.
Saber, M.S.M. (1986) Prolonged effect of land disposal of human waste on soil conditions. *Water Science and Technology* **18**, 371–374.
Said, W.A. and Lewis, D.L. (1991) Quantitative assessment of the effects of metals on microbial degradation of organic chemicals. *Applied Environmental Microbiology* **57**, 1498–1503.
Santamaria, J. and Toranzos, G.A. (2003) Enteric pathogens and soil: a short review. *International Microbiology* **6**, 5–9.
Šantrůčková, H., Vrba, J., Picek, T., and Kopáček, J. (2004) Soil biochemical activity and phosphorus transformations and losses from acidified forest soils. *Soil Biology and Biochemistry* **36**, 1569–1576.
Sardinha, M., Muller, T., Schmeisky, H. and Joergensen, R.G. (2003) Microbial performance in soils along a salinity gradient under acidic conditions. *Applied Soil Ecology* **23**, 237–244.
Sarig, S. and Steinberger, Y. (1994) Microbial biomass response to seasonal fluctuation in soil salinity under the canopy of desert halophytes. *Soil Biology and Biochemistry* **26**, 1405–1408.
Schimel, J. (1995) Ecosystem consequences of microbial diversity and community structure. In: *Arctic and Alpine Diversity: Patterns, Causes, and Ecosystem Consequences* (eds F.S. Chapin and C. Körner), pp. 239–254. Springer-Verlag, Berlin.
Schipper, L.A., Williamson, J.C., Kettles, H.A. and Speir, T.W. (1996) Impact of land-applied tertiary-treated effluent on soil biochemical properties. Soil denitrification dynamics: spatial and temporal variations of enzyme activity, populations and nitrogen gas loss. *Journal of Environmental Quality* **25**, 1073–1077.
Schmidt, T.M. (2006) The maturing of microbial ecology. *International Microbiology* **9**, 217–23.
Schramm, A., De Beer, D., Wagner, M. and Amann, R. (1998) Identification and activities in situ of *Nitrosospira* and *Nitrospira* spp. as dominant populations in a nitrifying fluidized bed reactor. *Applied Environmental Microbiology* **64**, 3480–3485.
Schuster, M. and Conrad, R. (1992) Metabolism of nitric-oxide and nitrousoxide during nitrification and denitrification in soil at different incubation conditions. *FEMS Microbiology and Ecology* **101**, 133–143.
Schwarz, S., Kehrenberg, C. and Walsh, T. (2001) Use of antimicrobial agents in veterinary medicine and food animal production. *International Journal of Antimicrobial Agents* **17**, 431–437.
Shapir, N., Mandelbaum, R.T. and Gottlieb, H. (1998) Atrazine degradation in saline wastewater by *Pseudomonas* sp strain ADP. *Journal of Industrial Microbiology and Biotechnology* **20**, 153–159.

Shapir, N., Mandelbaum, R.T. and Fine, P. (2000) Atrazine mineralization by indigenous and introduced *Pseudomonas* sp. strain ADP in sand irrigated with municipal wastewater and amended with composted sludge. *Soil Biology and Biochemistry* **32**, 887–897.
Shuman, L.M. (1999) Organic waste amendments effect on zinc fractions of two soils. *Journal of Environmental Quality* **28**, 1442–1447.
Shuval, H., Adnin, A., Fattal, B., Rawtzr, E. and Yekutiel, P. (1986) *Wastewater Irrigation in Developing Countries; Health Effects and Technical Solutions*. Technical paper No. 51. World Bank, Washington D.C.
Siebe, C. and Cifuentes, E. (1995) Environmental impact of wastewater irrigation in central Mexico: an overview. *International Journal of Environmental Health Research* **5**, 161–173.
Siebe, C. and Fischer, W.R. (1996) Adsorption of Pb, Cd, Cu and Zn by two soils of volcanic origin under long-term irrigation with untreated sewage effluent in Central Mexico. *Journal of Plant Nutrition and Soil Science* **159**, 357–364.
Siebe, C. (1998) Nutrient inputs to soils and their uptake by alfalfa through long-term irrigation with untreated sewage effluent in Mexico. *Soil Use Management* **13**, 1–5.
Sierra, J., Martí, E., Garau, M.A. and Cruañas, R. (2007) Effects of the agronomic use of olive oil mill wastewater: Field experiment. *Science of the Total Environment* **378**, 90–94.
Simek, M., Hopkins, D.W., Kalcík, J., Picek, T., Santrucková, H., Stana, J. and Trávník, K. (1999) Biological and chemical properties of arable soils affected by long-term organic and inorganic fertilizer applications. *Biology and Fertility of Soils* **29**, 300–308.
Singh, B.R., Agarwal, A.S. and Kanehiro, Y. (1969) Effect of chloride salts on ammonium nitrogen release in two Hawaiian soils. *Soil Science Society of America Proceedings* **33**, 557–568.
Sinsabaugh, R.L., Antibus, R.K., Linkins, A.E., McClaugherty, C.A., Rayburn, L., Repert, D. and Weiland, T. (1993) Wood decomposition: nitrogen and phosphorus dynamics in relation to extracellular enzyme activity. *Ecology* **74**, 1586–1593.
Smith, J.L. and Paul, E.A. (1990) The significance of soil microbial biomass estimations. In: *Soil Biochemistry*, vol. 6 (eds Bollag J-M and G. Stotzky), pp. 357–396. Marcel Dekker, New York, USA.
Sparling, G.P., Ord, B.G. and Vangham, D. (1981) Microbial biomass and activity in soils amended with glucose. *Soil Biology and Biochemistry* **13**, 99–104.
Sparling, G.P., Schipper, L.A. and Russell, J.M. (2001) Changes in soil properties after application of dairy factory effluent to New Zealand volcanic ash and pumice soils. *Australian Journal of Soil Research* **39**, 505–518.
Speir, T.W. (2002) Soil biochemical properties as indices of performance and sustainability of effluent irrigation systems in New Zealand—a review. *Journal of the Royal Society of New Zealand* **32**, 535–553.
Speir, T.W., Kettles, H.A., Parshotam, A., Searle, P.L. and Vlaar, L.N.C. (1995) A simple kinetic approach to derive the ecological dose value. ED(50), for the assessment of Cr(VI) toxicity to soil biological properties. *Soil Biology and Biochemistry* **27**, 801–810.
Steele, M. and Odumeru, J. (2004) Irrigation water as source of food borne pathogens on fruit and vegetables. *Journal of Food Protection* **67**, 2839–2849.
Stenberg, B. (1999) Monitoring Soil Quality of Arable Land: Microbiological Indicators. *Acta Agriculturae Scandinavica* **49**, 1–24.
Stephen, J.R., Chang, Y.J., Macnaughton, S.J., Kowalchuk, G.A., Leung, K.T., Flemming, C.A. and White, D.C. (1999) Effect of toxic metals on indigenous soil beta-subgroup proteobacterium ammonia oxidizer community structure and protection against toxicity by inoculated metal-resistant bacteria. *Applied Environmental Microbiology* **65**, 95–101.
Stephen, J.R., McCaig, A.E., Smith, Z., Prosser, J.I. and Embley, T.M. (1996) Molecular diversity of soil and marine 16S rRNA gene sequences related to beta-subgroup ammonia-oxidizing bacteria. *Applied Environmental Microbiology* **62**, 4147–4154.
Stevenson, F.J. (1994) *Humus Chemistry: Genesis, Composition, Reactions* (2nd ed.), John Wiley & Sons, NY.
Stotzky, G. (1997) Soil as an environment for microbial life. In: *Modern Soil Microbiology* (eds J.D. van Elsas, J.T. Trevors, E.M.H. Wellington), pp. 1–20. Marcel Dekker Inc., New York, USA.
Straub, T.M., Pepper, I.L. and Gerba, C.P. (1992) Persistence of viruses in desert soils amended with anaerobically digested sewage sludge. *Applied Environmental Microbiology* **58**, 636–641.
Strauss, M. (1994) *Health Implications of Excreta and Wastewater Use*. Hubei Environmental Sanitation Study, 2nd workshop. Hubei, Wuhan.

Solomon, E.B., Yaron, S. and Matthews, K.R. (2002) Transmission of *Escherichia coli* O157:H7 from Contaminated Manure and Irrigation Water to Lettuce Plant Tissue and Its Subsequent Internalization. *Applied Environmental Microbiology* **68**, 397–400.

Sören, T.B. (2003) Pharmaceutical antibiotic compounds in soils - a review. *Journal of Plant Nutrition and Soil Science* **166**, 145–167.

Sun, Y.H., Luo, Y.M., Wu, L.H., Li, Z.G., Song, J. and Christie, P. (2006) Survival of faecal coliforms and hygiene risks in soils treated with municipal sewage sluges. *Environmental Geochemistry and Health* **28**, 97–101.

Suwa, Y., Imamura, Y., Suzuki, T., Tashiro, T. and Urushigawa, Y. (1994) Ammonium-oxidizing bacteria with different sensitivities to $(NH_4)_2SO_4$ in activated sludges. *Water Research* **28**, 1523–1532.

Swift, M.J., Heal, O.W. and Anderson, J.M. (1979) *Decomposition in Terrestrial Ecosystems. Studies in Ecology*. Vol 5. Blackwell Publishing Ltd., Oxford.

Tabatabai, M.A. (1985) Effect of acid rain in soils. *CRC Critical Reviews in Environmental Control* **15**, 65–110.

Tam, N.F.Y. (1998) Effects of wastewater discharge on microbial populations and enzyme activities in mangrove soils. *Environmental Pollution* **102**, 233–242.

Tate, R.L. (1978) Cultural and environmental factors affecting the longevity of Escherichia coli in histosols. *Applied Environmental Microbiology* **35**, 25–29.

Teuber, M. (2001) Veterinary use and antibiotic resistance. *Current Opinions in Microbiology* **4**, 493–499.

Tiedje, J.M. (1982) Denitrification. In: *Methods of Soil Analysis: Chemical and Microbiological Properties*, part 2, 2nd ed. (Agronomy monograph no 9) (eds A.L. Page, R.H. Miller and D.R. Keeney), pp. 1011–1026. ASA–SSSA, Madison, WI, USA.

Tiedje, J.M. (1988) Ecology of denitrification and dissimilatory nitrate reduction to ammonium. In: *Biology of Anaerobic Microorganisms* (ed. A. J. B. Zehnder), pp. 179–244. John Wiley & Sons, Inc., New York, N.Y, USA.

Tierney, J.T., Sullivan, R. and Larkin, E.P. (1977) Persistence of poliovirus 1 in soil and on vegetables grown in soil previously flooded with inoculated sewage sludge or effluent. *Applied Environmental Microbiology* **33**, 109–113.

Tomson, M.B., Dauchy, J., Hutchins, S.R., Curran, C., Cook, C.F. and Ward, C.H. (1981) Groundwater contamination by trace level organics from a rapid infiltration site. *Water Resources* **15**, 1109–1116.

Torstensson, L. (1993) *Guidelines–Soil Biological Variables in Environmental Hazard Assessment*. Swedish Environmental Protection Agency, Report No. 4262.

Torstensson, L., Pell, M. and Stenberg, B. (1998) Need of a strategy for evaluation of arable soil quality. *Ambio* **27**, 4–8.

Torsvik, V., Goksoyr, J. and Daae, L. (1990) High diversity in DNA of soil bacteria. *Applied Environmental Microbiology* **56**, 782–787.

Torsvik, V. and Øvreås, L. (2002) Microbial diversity and function in soil: from genes to ecosystems. *Current Opinions in Microbiology* **5**, 240–245.

Tyler, G., Balsberg-Pahlsson, A., Bengtsson, G., Baath, E. and Tranvik, L. (1989) Heavy metal ecology of terrestrial plants, microorganisms and invertebrates: a review. *Water, Air Soil Pollution* **47**, 189–215.

Valscchi, G., Gighotti, C. and Farini, A. (1995) Microbial biomass, activity and organic matter accumulation in soils contaminated with heavy metals. *Biology and Fertility of Soils* **20**, 253–259.

Van-Camp, L., Bujarrabal, B., Gentile, A.R., Jones, R.J.A., Montanarella, L., Olazabal, C. and Selvar-adjou, S.K. (2004) *Reports of the Technical Working Groups Established under the Thematic Strategy for Soil Protection*, EUR 21319 EN/3, Office for Official Publications of the European Communities, Luxembourg.

Vasseur, L., Cloutier, C., Labelle, A., Duff, J.N., Beaulieu, C. and Ansseau, C. (1996) Responses of indicator bacteria to forest soil amended with municipal sewage sludge from aerated and non- aerated ponds. *Environmental Pollution* **92**, 67–72.

van Veen, J.A., van Overbeek, L.S. and van Elsas, J.D. (1997) Fate and activity of microorganisms introduced into soil. *Microbiology and Molecular Biology Reviews* **61**, 121–135.

Vives-Rego, J., Vaque, D. and Martinez, J. (1986) Effect of heavy metals and surfactants on glucose metabolism, thymidine incorporation and exoproteolytic activity in sea water. *Water Research* **20**, 1411–1415.

Wachtel, M.R., Whitehand, L.C. and Mandrell, R.E. (2002) Association of *Escherichia coli* O157:H7 with preharvest leaf lettuce upon exposure to contaminated irrigation water. *Journal of Food Protection* **65**, 18–25.

Wagner, M., Rath, G., Amann, R., Koops, H.P. and Schleifer, K.H. (1995) *In situ* identification of ammonia oxidizing bacteria. *Systems in Applied Microbiology* **18**, 251–264.
Waldrop, M.P., Balser, T.C. and Firestone, M.K. (2000) Linking microbial composition to function in a tropical soil. *Soil Biology and Biochemistry* **32**, 1837–1846.
Waldrop, M.P., and Firestone, M.K. (2004) Microbial community utilization of recalcitrant and simple carbon compounds: impact of oak-woodland plant communities. *Oecologia (Berlin)* **138**, 275–284.
Warriner, K., Ibrahim, F., Dickinson, M., Wright, C. and Waites, W.M. (2003) Interaction of *Escherichia coli* with growing spinach plants. *Journal of Food Protection* **66**, 1790–1797.
Watson, S.W., Bock, E., Harms, H., Koops, H.P. and Hooper, A.B. (1989) Nitrifying bacteria. In: *Bergey's Manual of Systematic Bacteriology*, vol. 3 (eds J. T. Staley, M. P. Bryant, N. Pfennig and J. G. Holt), pp. 1808–1834. The Williams and Wilkins Co., Baltimore, MD, USA.
Webster, G., Embley, T.M., Freitag, T.E., Smith, Z. and Prosser, J.I. (2005) Links between ammonia oxidiser species composition, functional diversity and nitrification kinetics in grassland soils. *Environmental Microbiology* **7**, 676–684.
West, P.A. (1991) Human pathogenic viruses and parasites: emerging pathogens in the water cycle. *Journal of Applied Bacteriology Symposium Supplement* **70**, 107S–114S.
White D., Zhao, S., Simjee, S., Wagner, D. and McDermott, P. (2002) Antimicrobial resistance of foodborne pathogens. *Microbes and Infection* **4**, 405–412.
WHO. (2006) *Guidelines for the Safe Use of Wastewater, Excreta and Greywater*. Vol. 2. Wastewater use in agriculture. World Health Organization, Geneva.
Wick, B., Kühne, R.F. and Vlek, P.L.G. (1998) Soil microbiological parameters as indicators of soil quality under improved fallow management systems in South Western Nigeria. *Plant Soil* **202**, 97–107.
Witte, W. (1998) Medical consequences of antibiotic use in agriculture. *Science (Washington D.C.)* **279**, 996–997.
Witte, W. (2001) Selective pressure by antibiotic use in livestock. *International Journal of Antimicrobial Agents* **16**, 19–24.
Yesilada, Ö. and Sam, M. (1998) Toxic effects of biodegraded and detoxifie olive oil mill wastewater on the growth of Pseudomonas aeruginosa. *Toxicology and Environmental Chemistry* **65**, 87–94.
Yuan, B.C., Li, Z.Z., Liu, H., Gao, M. and Zhang, Y.Y. (2007) Microbial biomass and activity in salt affected soils under arid condition *Applied Soil Ecology* **35**, 319–328.
Zahran, H.H., Moharram, A.M. and Mohammad, H.A. (1992) Some ecological and physiological studies on bacteria isolated from salt affected soils of Egypt. *Journal of Basic Microbiology* **35**, 269–275.
Zahran, Z. (1997) Diversity, adaptation and activity of the bacterial flora in saline environments. *Biology and Fertility of Soils* **25**, 211–223.
Zibilske, L.M. and Weaver, R.W. (1978) Effect of environmental factors on survival of *Salmonella typhimurium* in soil. *Journal of Environmental Quality* **7**, 593–597.

Chapter 12

Impact of irrigation with treated wastewater on pesticides and other organic microcontaminants in soils

Zev Gerstl and Ellen R. Graber

12.1 Introduction

The reuse of treated effluents in agriculture for irrigation has become the mainstay in Israel and other parts of the world due to increasing population and the limited amount of high-quality water. The sustainable use of effluent for irrigation requires consideration not only of the short-term benefits such as mitigating water shortages and providing plant nutrients, but also assessment of its disadvantages and potential risks. Among the long-term potential risks associated with effluent irrigation are elevated concentrations of anthropogenic chemicals (e.g. hormones, pharmaceuticals, pesticides), as well as enhanced transport of these chemicals due to their interaction with dissolved organic matter (DOM).

Herein, we review the long-term risks posed by the presence of DOM and anthropogenic chemicals in treated effluents used for irrigation. The reader is referred to Muller et al. (2007) for a comprehensive review of the fate of pesticides in soils under effluent irrigation, and to Shon et al. (2006) for a recent study on the composition of effluent organic matter and a review of various wastewater treatment methods. The effect of irrigation with effluents on the microbial activity and pesticide degradation is presented in Chapter 11.

12.2 The effect of DOM on the chemical behavior of organic xenobiotics

12.2.1 DOM binding with xenobiotic organic chemicals (XOCs)

The transport, stability, and bioavailabilty of organic pollutants in the environment are strongly affected by their interactions with organic macrospecies. The term macrospecies is defined here as all entities that are larger than the more common organic monomers and

Treated Wastewater in Agriculture, First Edition, edited by Guy J. Levy, Pinchas Fine and Asher Bar-Tal

includes soluble macromolecules, colloidal particles, and immobile solid components of the system of interest.

Sorption of small organic compounds on immobile surfaces retards their transport, diminishes their availability for uptake by plants, and, more often than not, decreases their susceptibility to biotic degradation. Interaction with soluble macromolecules or mobile colloidal particles, on the other hand, can increase the apparent solubility of hydrophobic molecules in aqueous solution and hence their mobility. The effect of sorption of organic compounds by suspended particles on their transport depends on the relation between particle size and soil pore size distribution. Factors that affect this relationship (such as ionic strength, which controls the tendency of colloids to flocculate) will also determine whether sorption on suspended particles will retard or accelerate transport.

Despite the contrasting effects that XOC sorption on immobile surfaces and their complexation with soluble or suspended macrospecies may have on their transport, the nature of the interaction between small XOC molecules and the immobile and mobile organic matter (OM) macrospecies is similar in many respects. This is so because these interactions are governed by the same physical principles and parameters, regardless of the identity or size of the specific OM macrospecies.

The overall interaction of a dissolved organic chemical in a system containing both dissolved macrospecies and the solid phase consists of three sets of interactions as can be seen from Figure 12.1, where k_d is the distribution constant of the XOC between the aqueous (C_{eq}) and soil (S) phase, k_b is the binding constant between the aqueous XOC (C_{eq}) and the DOM (C_{DOM}), and k_{DOM} is the distribution constant of the complex XOC–DOM (C^b) between the aqueous phase and the soil (S_{DOM}).

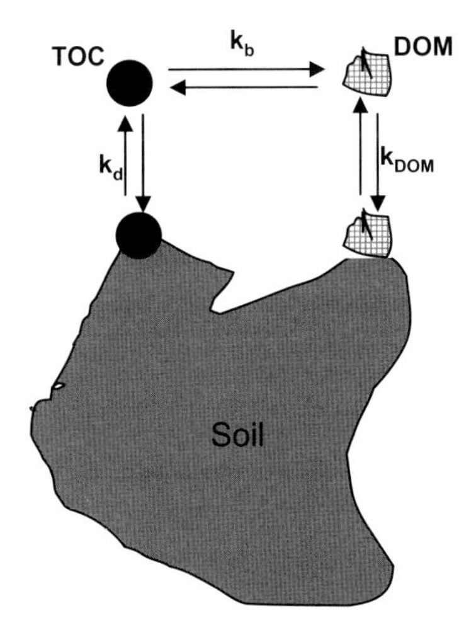

Figure 12.1 Partitioning of a xenobiotic organic compound (XOC) and dissolved organic matter (DOM) with soil.

As a simplification, sorption of the XOC by the solid soil phase is assumed to follow a linear isotherm such that

$$S = k_d C_{eq} \tag{12.1}$$

Binding of the XOC by DOM is represented by the following equilibrium:

$$C_{DOM} + C_{eq} \overset{k_b}{\longleftrightarrow} C^b \tag{12.2}$$

Thus, the binding constant of the XOC to the DOM is:

$$k_b = \frac{C^b}{C_{eq} C_{DOM}} \tag{12.3}$$

Following the approach of Rav-Acha and Rebhun (1992), we assume that DOM-bound XOC sorbs to the solid soil surface independently of free XOC (represented in equation 12.1), and also follows a linear sorption isotherm. Thus, in addition to equation 12.1 we have:

$$S_{DOM} = k_{DOM} C^b \tag{12.4}$$

The overall sorption coefficient for the XOC, K_p, can be given as:

$$K_p = \frac{S + S_{DOM}}{C_{eq} + C^b} \tag{12.5}$$

Substituting for S and S_{DOM} from equations 12.1 and 12.4 into equation 12.5 we obtain:

$$K_p = \frac{k_{DOM} C^b + k_d C_{eq}}{C^b + C_{eq}} \tag{12.6}$$

and rearranging equation 12.3 to express C^b as a function of C_{eq} and substituting in equation 12.6 yields:

$$K_p = \frac{k_{DOM} k_b\, C_{DOM} + k_d}{k_b\, C_{DOM} + 1} \tag{12.7}$$

For simplicity, it may be assumed that the sorption of the XOC–DOM complex is identical to the sorption of DOM itself.

When $k_d \gg k_{DOM}$, equation 12.7 reduces to:

$$K_p = \frac{k_d}{k_b C_{DOM} + 1} \tag{12.8}$$

indicating that the sorption of the XOC in the presence of DOM is always less than that in a DOM-free system. The fraction of a XOC remaining in solution as a function of k_d and k_b is presented in Figure. 12.2 and Figure12.3. In Figure 12.2 we can see that as k_d increases, the fraction of XOC in solution decreases. At a DOM concentration of 10 mg/l, no significant

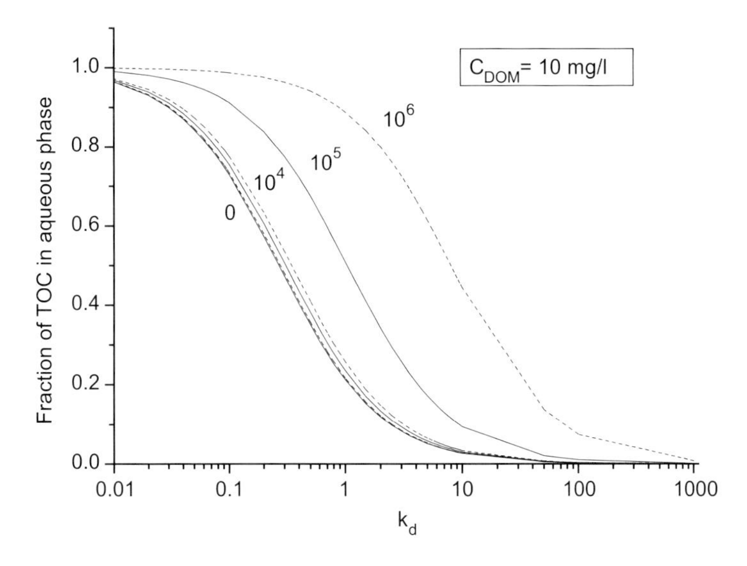

Figure 12.2 Effect of binding constant (k_b) on the fraction of xenobiotic organic compound (XOC) remaining in solution in the presence of dissolved organic matter (DOM) at a concentration of 10 mg/l. TOC, total organic carbon.

change in the fraction of XOC in solution occurs for binding constants, k_b, up to 1000 ml/g (Fig. 12.2). For higher binding constants ($>10^4$) we can observe significant solubilization of the XOC. A plot of the effect of DOM concentration on solubilization of an XOC with a binding constant of 10^5 ml/g, demonstrates that even 10 mg/l of DOM is enough to result in a strong increase the apparent solubilization of the XOC (Fig. 12.3).

The importance of equation 12.7 for movement of XOCs in soils can be seen by substituting K_p for k_d in the retardation factor, R, which is a measure of the degree to which

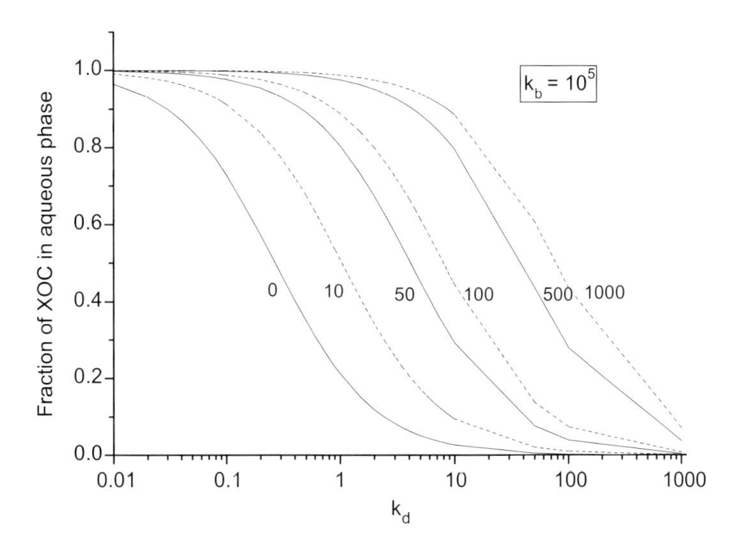

Figure 12.3 Effect of dissolved organic matter (DOM) concentration C_{DOM} on solubilization of an xenobiotic organic compound (XOC) with a binding constant of 10^5 ml/g.

the movement of a sorbing compound is slowed down in comparison to a non-sorbing compound (Bouwer, 1991). The retardation factor is generally presented as:

$$R = 1 + \frac{\varrho k_d}{\theta} \qquad (12.9)$$

where ϱ is the bulk density of the soil and θ the volumetric soil moisture content. After substitution and rearrangement, equation 12.9 becomes:

$$R^* = 1 + \frac{\varrho k_D}{\theta(k_b C_{DOM} + 1)} + \frac{\varrho k_b C_{DOM} k_{DOM}}{\theta(k_b C_{DOM} + 1)} \qquad (12.10)$$

When both k_d and k_{DOM} are 0 (e.g. no sorption of the XOC or of the DOM), the retardation factor is 1 and the compound moves freely through the soil. When $k_d > 0$ and k_{DOM} is 0 then the retardation factor is reduced as a result of binding of the XOC to the DOM. If, however, $k_{DOM} > 0$, then the effect of the DOM on the transport of a XOC depends on the values of k_D and k_{DOM}. If XOC sorption is greater than that of the DOM ($k_d > k_{DOM}$), then the effect of the DOM will still be enhanced transport due to solubilization. If, however, DOM sorption is greater than that of the XOC ($k_{DOM} > k_d$), we will observe an decrease in the overall transport of the XOC as shown in Figure 12.4 where equation 12.10 is plotted for various values of k_{DOM} for a XOC with a k_d of 25 ml/g. For a more detailed analysis of the many different possible scenarios the reader is referred to Totsche and Kogel-Knabner (2004).

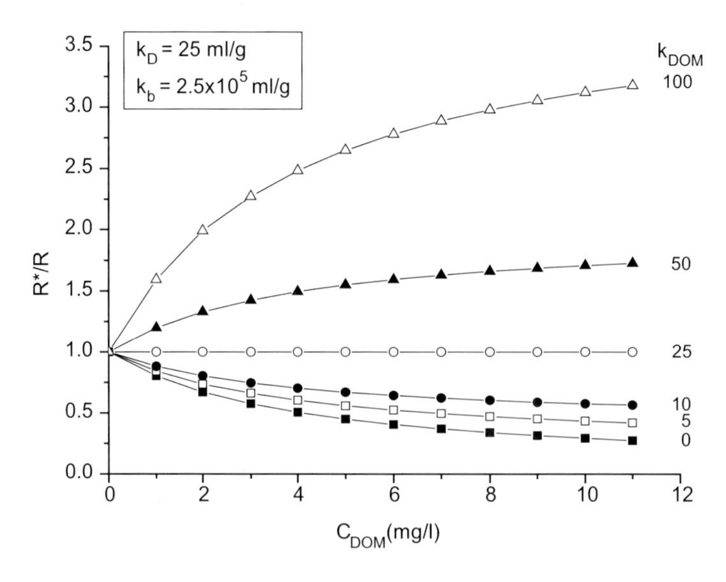

Figure 12.4 Retardation factor as a function of the dissolved organic matter (DOM) concentration for different values of k_{DOM}.

12.2.2 Binding constant – DOM relationships

The most common approach for describing the binding of XOCs with DOM is by the linear free-energy relationship (LFER):

$$\log k_b = a \log K_{ow} + b \qquad (12.11)$$

that correlates two partitioning-like processes for an XOC: (i) between water and the DOM; and (ii) between K_{ow}, the water, and n-octanol, respectively. In this equation, a and b are empirical constants. However, it has been noticed in several studies that the values of these empirical constants can be class-specific, as well as DOM source-specific (Gauthier et al., 1986; Rebhun et al., 1992, 1996); that is, the values of a and b vary significantly with the class of XOC under consideration or with DOM from different sources but the same chemical class of XOC.

The source of DOM has been shown to have a profound effect on k_b values for a series of 20 phenols and chlorinated phenols (Ohlenbusch and Frimmel, 2001), where values of 0.314 and 1.342 were obtained for a (and values of 1.595 and -0.496 for b) on Aldrich humic acid and bovine serum albumin (BSA), respectively. Perminova et al. (1999) in a study of 26 different dissolved humic materials reported ranges of k_b values of over an order of magnitude for pyrene, fluoranthene, and anthracene. Similarly, Evans (1998), looking at the DOC from 12 lakes and streams observed that the k_b varied by over an order of magnitude for PCB 52 and PCB 153. Tanaka et al. (2005) found that log k_b values for a chlorinated dioxin on humic acids from tropical peat, brown forest, and ando soils were in the range of 7.3–7.6; however, the log k_b values on fulvic acids and peat-derived humic acids were 0.5–1 log units lower. Several explanations have been advanced for these differences. Ohlenbusch et al. (2000) attributed the higher binding by BSA to specific interactions, namely, the presence of basic amino groups in the DOM. Perminova et al. (1999) and Tanaka et al. (2005) proposed the greater binding in the DOMs they studied as being due to their greater aromaticity or polarity. Raber et al. (1998) ascribed the greater binding constants for polycyclic aromatic hydrocarbons (PAHs) in DOM from acid forest soils to the higher proportion of hydrophobic components compared to DOM from mineral soils. Evans (1988), on the other hand, found that there were no significant correlations between the k_b values and various chemical parameters in the lake and stream DOMs.

Burkhard (2000) conducted a literature search for dissolved organic carbon/water partition coefficients for non-ionic organic chemicals (k_b). Data were collected from more than 70 references and included DOM from sediment porewaters, soil porewater and groundwater, surface waters, soil humic and fulvic acids, and Aldrich humic acid. The k_b values were evaluated as a function of the octanol/water partition coefficients (K_{ow}). It was observed that the variability in the structure and composition of the DOC from sediments, soils, and ambient waters, and the analytical difficulties in determining k_b resulted in substantial variability in measured k_b values. The upshot of these findings is that predictive relationships based solely on the hydrophobicity (K_{ow}) of the chemical will have large uncertainties despite the underlying importance of hydrophobicity of the chemical. It was suggested that any model for predicting the association of a wide range of XOCs to DOM will require, in addition to the K_{ow}, the inclusion of a parameter for describing how the different components of the DOM interact with XOCs. The relatively large variability observed among investigations using Aldrich humic acid as the DOM source suggests that refinements and improvements in measurement techniques will also be required in order to reduce uncertainties. A step in this direction has recently been taken by Poerschmann and Kopinke (2001) and Georgi and Kopinke (2002), who observed that binding of different classes of XOCs resulted in class-specific k_b–K_{ow} relationships on a series of different DOMs. Using a modified Flory-Huggins approach and Hildebrand solubility parameters specific for each DOM and XOC, an equation was developed that allows accurate prediction of binding of different XOCs with DOM.

12.2.3 Effect of DOM on sorption and transport of XOCs

Numerous soil-applied pesticides are used in agriculture (and many crop-applied pesticides reach the soil via leaf-washoff). A major concern in using effluents for irrigation is that the downward transport of the pesticides may be affected by the DOM in the effluent.

Several cases of enhanced sorption of XOCs by soils in the presence of DOM or decreased transport of XOCs by DOM have been reported in the literature (Totsche et al., 1997; Huang and Lee, 2001; Gonzalez-Pradas et al., 2005; Ling et al., 2005; Flores-Cespedes et al., 2006; Ling et al., 2006). Flores-Cespedes et al. (2006) and Gonzalez-Pradas et al. (2005) found that natural DOM increased the sorption of 3,4-dichloroaniline and 4-bromoaniline over a concentration range of 15–100 mg/l. Similarly, Ling et al. (2005, 2006) reported that DOM at relatively lower concentrations significantly enhanced the sorption of atrazine by soil, whereas it inhibited atrazine sorption at higher concentrations. These findings show that the assumption $k_d \gg k_{DOM}$ is not always true. The critical concentration of DOM, below which DOM would enhance atrazine sorption, was negatively correlated with SOC but was generally in the range of 10–15 mg/l. The influence of DOM on the mobility of PAHs was studied in medium-scale soil column experiments under unsaturated flow conditions (Totsche et al., 1997). The addition of DOM resulted in reduced mobility of PAHs and reduced PAH effluent concentrations. This was attributed primarily by sorption of DOM-associated PAHs to the bulk soil, although the continuous sorption of DOM by the bulk phase resulting in an increased sorption capacity for free PAHs cannot be disregarded.

Although reduced transport due to DOM sorption has been noted, solubilization and enhanced transport of XOCs complexed by DOM is a much more commonly reported phenomenon (Graber et al., 1995; Maxin and Kogelknabner, 1995; Baskaran et al., 1996; Kalbitz et al., 1997; Celis et al., 1998; Fang et al., 1998; Nelson et al., 1998; Mitra and Dickhut, 1999; Nelson et al., 2000; Williams et al., 2000; Graber et al., 2001; Flores-Cespedes et al., 2002; Ben-Hur et al., 2003; Persson et al., 2003; Sabbah et al., 2004; Amiri et al., 2005). For example, Sabbah et al. [2004] have shown that concentrations as low as 8 mg/l were able to significantly enhance the transport of fluoranthene ($\log K_{ow} = 5.22$) in sand columns, but 30 mg/l was necessary to have an effect on phenanthrene ($\log K_{ow} = 4.57$) transport (Fig. 12.5).

Muszkat et al. (1993a,b) compared the distribution of various organic contaminants in soil profiles to a depth of 20 m in two citrus groves irrigated over a period of at least 20 yrs, one with sewage effluents and the other with groundwater. In the effluent-irrigated grove,

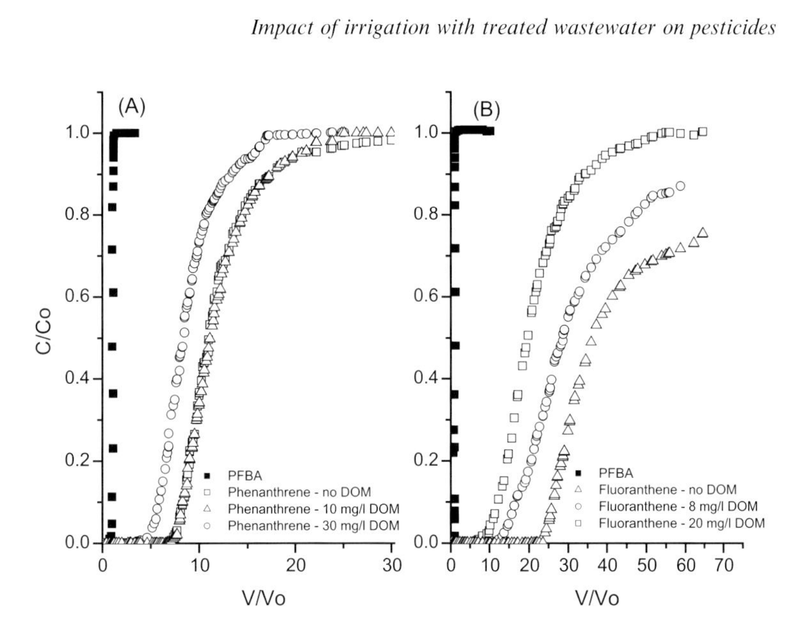

Figure 12.5 Experimental breakthrough curves (BTCs) for pentafluorobenzoic acid (PFBA – a non-sorbed tracer), phenanthrene and fluoranthene through a sand column in the presence of different concentrations of humic acids. DOM, dissolved organic matter. Reproduced from Sabbah et al., copyright 2004 with permission of Elsevier.

pollutants were detected in all of the soil layers and the extent of contamination increased with depth, e.g. terbumeton was detected to a depth of 20 m. In the citrus grove irrigated with high-quality groundwater, on the other hand, contaminants were confined to the upper soil layers with only a few reaching a depth of 5 m. Muszkat et al. (1993a) attributed the enhanced transport of these contaminants to the presence of surface active compounds in the effluents, such as detergents and solvents. A different explanation was put forth by Muszkat et al. (1993b); asserting that pollutants formed complexes with colloidal particles, which facilitated their transport through the soil.

Graber et al. (1995) studied the effect of irrigation water quality on the transport of atrazine in a fine clayey soil. Cores to 4 m were obtained from 10 effluent-irrigated and 10 high-quality water-irrigated plots after only two growing seasons and two winter rainy seasons. In most of the effluent-irrigated cores, atrazine was widely distributed with depth, with an average center of atrazine mass at 115 ± 39 cm (Fig. 12.6). In the high-quality water-irrigated cores, atrazine was concentrated in the upper soil horizons with an average center of mass at 63 ± 64 cm.

In a series of studies on the complexation of herbicides by DOM, Ben-Hur and coworkers (Nelson et al., 1998; Williams et al., 2000; Nelson et al., 2000; Ben-Hur et al., 2003) showed that DOM can greatly enhance the transport of herbicides. The amount of napropamide found in the initial leachate from columns amended with sewage sludge was twice that of non-amended soils and was shown to be associated with DOM (Nelson et al., 1998). In a second study (Williams et al., 2000) in which water was applied to a series of soil columns to which napropamide was previously surface-applied, it was found that both DOM and napropamide concentrations were highest in the first 0.22 cm of leachate and that 17–56% of the napropamide was associated with DOM. Nelson et al. (2000) concluded that increased napropamide movement through soils was the result of stable

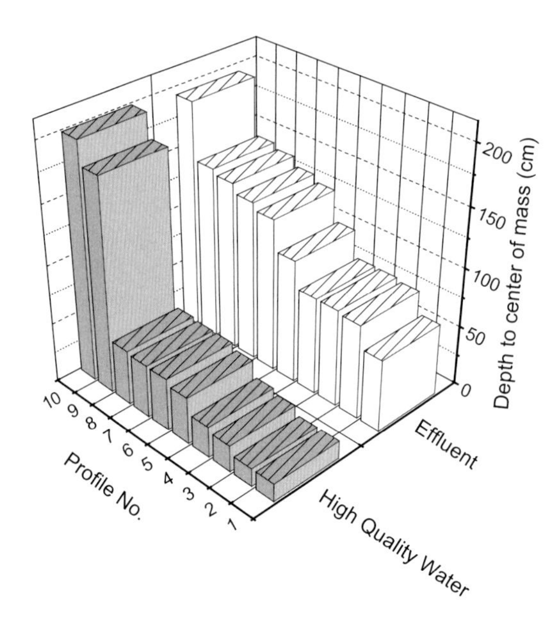

Figure 12.6 Centers of mass for 10 high-quality water and 10 effluent-irrigated profiles. Reproduced from Graber et al., 1995, with permission.

DOM complexes that did not exhibit any propensity to sorb to the soil. Although this amounted to less than 6% of the herbicide applied, facilitated transport to this extent can pose serious risks to groundwater quality. Ben-Hur et al. (2003) found that atrazine forms complexes with DOM in soils and the extent of sorption of these complexes depends on the soil OM content. If the DOM concentration in the soil solution exceeded an equilibrium value (e.g. the DOM concentration in solution from a DOM-sorption isotherm), the sorption of a DOM–atrazine complex was increased. The implication of these findings is that as atrazine–DOM complexes move through a soil profile in which the soil OM content decreases with depth, sorption of the complex can occur, thus reducing the leaching of atrazine.

Seol and Lee (2000) studied the effect of DOM from several sources (secondary effluents, swine-derived waste, and Aldrich humic acid) on the sorption of two triazine herbicides by soils in laboratory experiments. Dissolved organic matter of up to 150 mg OC/l did not significantly decrease herbicide sorption. These findings agreed with predictions from a model they developed, using independently measured binary distribution coefficients and assuming linear sorption of the herbicides and the DOM by the soil. A similar study was undertaken by Huang and Lee (2001) with the organophosphate insecticide, chlorpyrifos. They found that for all soil–DOM combinations, chlorpyrifos sorption by soils decreased in the presence of DOM; however, the predicted reduction in sorption due to chlorpyrifos–DOM complex formation was consistently lower than the measured values. They pointed out that effluent DOM is a mixture of molecules ranging in size and polarity, and showed that k_b, the binding constant of a XOC with DOM, and k_{DOM}, the sorption constant of DOM by soils, increased with decreasing DOM polarity. Thus, the more non-polar DOM molecules were suggested to sorb to the soil preferentially leaving the less hydrophobic DOM molecules in solution to form chlorpyrifos-DOM complexes, the net result being a reduced effect of DOM on chlorpyrifos sorption than would otherwise be expected.

Environmental conditions can influence the behavior of XOC–DOM complexes in soils. Nelson et al. (2000) showed that drying of the soil induced enhanced herbicide movement upon wetting due to the formation of stable herbicide–DOM complexes that were not adsorbed by the soil. Seol and Lee (2001) also observed that drying enhanced dissolution of soil OM; however, they also found that drying could result in upward movement of pesticide-containing pore water, thus offsetting to a certain degree the enhanced movement effect of the dissolved DOM.

The above review indicates that effluent irrigation affects pesticide fate in soil by altering sorption as a result of DOM in the soil solution. Laboratory studies have shown that DOM can also facilitate the movement of soil-borne pesticides by forming soluble pesticide complexes. Field studies have borne this out showing that long- and short-term effluent irrigation is likely to result in an enhanced downward pesticide transport. The concentration and composition of DOM in the soil solution, as well as the hydrophobicity of the XOC are important factors in determining the extent of the above phenomena. As of 2001, 22% of the effluents in Israel had biological oxygen demand (BOD) values greater than 60 mg O_2/l (Gabbay, 2002), which is equivalent to a total organic carbon (TOC) of 84 mg/l based on a BOD–TOC ratio of 1.4 (Fine et al., 2002). Use of such effluents for irrigation without further treatment may result in enhanced leaching of the more hydrophobic pesticides in use today.

In addition to the effects of DOM on the behavior of soil-applied pesticides in effluent-irrigated soil, the presence of effluent-borne XOCs is also of concern and is addressed in the next section.

12.3 Effluent-borne organic contaminants

Although the majority of the OM load in wastewater is incorporated into the sludge phase during wastewater treatment, there are organic compounds that remain in the effluent, making up the dissolved and suspended OM load. Much of the effluent DOM consists of proteins, polysaccharides, low molecular weight carboxylic acids, lipids, and polycarboxylic acids (Shon et al., 2006). Specific contaminants that are identifiable by chromatographic techniques are present usually at low, sub-mg/l levels. For example, non-ionic surfactants (between 1 and 30 μg/l), antioxidants, caffeine, fecal steroids and cholesterol have all been identified in treated effluents (Brown et al., 1999). Other compounds that have been identified in treated wastewater include pharmaceuticals, fire retardants, plasticizers, industrial solvents, disinfectants, PAHs, and high-use domestic pesticides (lindane, dieldrin, diazinon, chlorpyrifos, etc.) (Daughton and Ternes, 1999). Estrogens such as 17-estradiol, 17-ethinyl estradiol, and estrone have also been found in various wastewaters at levels of low ng/l up to several tens of ng/l (Falconer et al., 2006). Almost 25% of a total of 126 wastewater samples from diverse locations around the USA were found to contain organic micropollutants (Brown et al., 1999). Many of the myriad of organic pollutants identified in effluents are endocrine-disrupting compounds (EDCs). Because of public concern, coupled with a lack of scientific knowledge involving EDCs and the risks they pose in the environment, they are of particular concern when reusing wastewater in agriculture (Falconer et al., 2006). Irrigation of crops with treated effluent and the use of sewage sludge or biosolids as soil amendments can introduce these pollutants and others into surface runoff (Pedersen et al., 2002) and from there into surface water bodies, or into the soil, and ultimately, into groundwater. For many reasons, then, there is considerable interest in evaluating the fate of such compounds, commonly termed "emerging compounds" in the environment.

The presence of potentially toxic dissolved and particle-associated organic chemicals in surface runoff from agricultural fields irrigated with disinfected tertiary recycled water or wastewater effluent-dominated streamwater was examined by Pedersen et al. (2003, 2005). They found that certain traditionally targeted pollutants (e.g. regulated USEPA priority pollutants, specific pesticides) were sometimes present at concentrations commensurate with acute and chronic water quality criteria and 96-h LC50 values for sensitive invertebrate species (Pedersen et al., 2003) (Table 12.1). As a number of organophosphorus insecticides were usually present at the same time, and due to their potential additive toxicity, it was suggested that even when concentrations of individual compounds were below levels of concern, their combined effect could result in toxicity to non-target organisms (Pedersen et al., 2003). Concentrations of organophosphorus insecticides and other compounds exhibited significant variability within runoff events (varying as much as several orders of magnitude), as well as great variability between events and between sites (Pedersen et al., 2003, 2005). Examples of highly variable concentrations of two synthetic polycyclic musks and an alkyl phosphate flame retardant over the course of a single runoff event can be seen in Figure 12.7, reproduced from Pedersen et al. (2005).

Kinney et al. (2006) assessed the presence and distribution of pharmaceuticals in the upper 30 cm of soil irrigated with treated effluent derived from urban wastewater at three sites in Colorado. Differences in behavior (accumulation versus leaching) were reported at the different sites. For example, cimetidine, warfain, gemfibrozil, and codeine did not accumulate in soils, and were usually present as very low percentages of the applied mass. On the other hand, compounds such as acetaminophen, fluoxetine, caffeine, erythromycin, and carbamazepine accumulated at much higher masses than the estimated applied mass, suggesting their strong retention and long-term persistence in the upper 30 cm soil profile. Other measured pharmaceutical compounds displayed inconsistent behavior that varied

Table 12.1 Concentrations of dissolved organophosphorous insecticides in surface runoff from agricultural fields

Compound	**Dissolved concentration (ng/l)**			
	Irrigation runoff	**Storm runoff**	**Water quality criteria, acute/chronic**	**96-h LC_{50}**
Chlorpyrifos	<3–981	<3–409	20/14	38
Diazinon	<3–231	<3–105000	80/50	200
Dichlorivos	<3–89	<3–1340	–	100
Dimethoate	<13–1550	<13–381000	–	43 000
Disulfoton	<5–442	<5–106	–	3900
Malathion	<5–187	<5–216	–/100	500
Methyl parathion	<13–9150	<13	80	3800

Adapted from Pedersen et al., copyright 2003 with permission of the American Chemical Society.

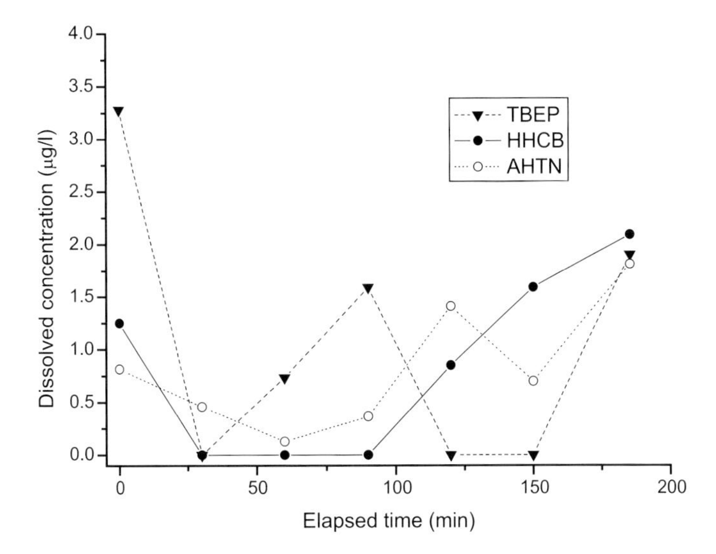

Figure 12.7 Variation in concentrations over the course of a runoff event from an agricultural field irrigated with effluent-dominated streamwater. HHCB and AHTN are synthetic polycyclic musks, and TBEP is an alkyl phosphate flame-retardant chemical. Reproduced from Pedersen et al., copyright 2005 with permission of Elsevier.

from site to site. When concentrations were normalized to soil organic carbon content, the accumulated amounts at each site were more similar to each other, suggesting that soil OM played an important role in the retention of the studied compounds. Supporting this, it was found that the most frequently detected pharmaceutical compounds (erythromycin, carbamazepine, fluoxetine, and diphenhydramine) had lower water solubility and higher log K_{ow} values than the other studied compounds. Those compounds with water solubility lower than 100 mg/l were most consistently retained in the soil cores, indicating that water solubility can serve as an initial indicator of pharmaceutical accumulation and retention. Strongly retained compounds were also readily identified 6 months after the cessation of irrigation and leaching by about 215 mm precipitation. Depth distribution of pharmaceuticals throughout the 30-cm long cores demonstrated that some of the studied compounds were relatively mobile (Fig. 12.8), raising the possibility that such compounds could leach eventually to groundwater.

A study conducted in China surveyed the upper 10 cm soil layer from farmlands irrigated by effluents from biological treatment plants for PAHs and organochlorine pesticides (OCPs) (Chen et al., 2005). The results showed that wastewater-irrigation could cause accumulation of PAHs in soils close to the pollution discharge. In contrast to trends that could be expected on the basis of the high affinity of such pollutants for soil OM, they did not find a significant relationship between the accumulation of persistent organic pollutants and the soil OM content. Contamination was characterized by the accumulation of mainly high molecular weight PAHs, the concentrations of some of which exceeded the soil quality standard for biodegraded soils. Soil genotoxicity was found to be affected by the PAH accumulation (Chen et al., 2004). Polycyclic aromatic hydrocarbon residues in effluents and emission from a nearby coal/coke plant were both considered to be the sources responsible for the contamination observed.

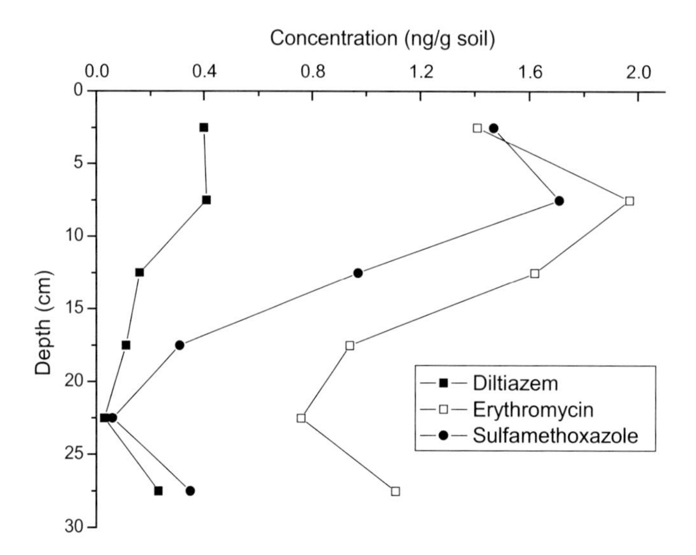

Figure 12.8 Concentration of wastewater-derived pharmaceuticals with depth in soil at a reclaimed-water facility (from Kinney et al., 2006).

Recently, the effect of more than 45 yrs of irrigation of agricultural lands with secondary treated sewage effluent on the vertical distribution of 52 pharmaceuticals and personal care products (PPCPs) in soil and groundwater was reported by Ternes et al. (2007). In the study area (Braunschweig, Germany), the sandy topsoil (0.55 m) is underlain by fine and medium sand, with groundwater occurring at only 1.6 m below ground surface. To improve water holding capacity of the soil, digested sludge had been routinely mixed into the irrigation water at levels of 0.3 g/l. They reported that out of 52 analyzed PPCPs, only four were detected in the groundwater, and six in soil water. Diatrizoate (an iodinated contrast medium) was present at the highest concentrations, up to 10 μg/l, whereas the others were present at less than 1 μg/l. The study concluded that even in a soil with a relatively thin Ap horizon (0.55 m) characterized by a very low clay and OM content (~3% clay and 0.5% OM), there is a very high removal efficiency for many PPCPs, either via sorption or degradation. Due to the relative high polarity and absence of specific sorbing functionalities (e.g. positively charged amino groups) under the ambient neutral pH conditions, sorption was considered to be of minor significance for most of the compounds investigated. However, wastewater-irrigation did lead to contamination of the shallow groundwater with selected PPCPs such as diatrizoate, carbamazepine (antiepileptic), and sulfamethoxazole (antibiotic), albeit at very low concentrations.

The accumulation and leaching potential of organic compounds in the soil and subsurface will depend on the properties of the porous media (OM and clay contents; isoelectric point, pH etc.), chemical properties of the compound (pK_a, log K_{ow}, aqueous solubility), and microbial degradation potential of the compounds (Lorphensri et al., 2006). Although porous media are predominantly negatively charged at neutral pH (e.g. silica sands, clays), alumina and some iron oxides that have a net positive charge at neutral pH will often be present in small amounts or as surface coatings on framework grains (Lorphensri et al., 2006). At pH values below the isoelectric point (point of zero charge; PZC; equal number of negatively and positively charged sites), the net surface charge is positive and anion adsorption dominates. At pH values higher than the PZC, the net charge is negative and cation adsorption dominates. Pharmaceutical compounds and pesticides frequently contain both non-polar and ionizable functional groups, and sometimes have multiple functional groups such that they can both accept and donate protons at a given pH (Lorphensri et al., 2006). The result is that these compounds can have cationic, neutral, zwitterionic, and anionic forms as a function of pH. The complex chemistry of many effluent-borne contaminants means that their retention and transport behavior in soils will be difficult to predict a priori using simple linear free energy relationships (LFERS) such as the K_{ow}–K_{oc} relationships used widely to predict sorption and transport behavior of non-polar organic pollutants. Such effects will strongly complicate the task of understanding the behavior of effluent-borne pharmaceuticals in irrigated soils.

An example of the complicated sorption behavior that can be evinced by pharmaceuticals is observed in Figure 12.9, reproduced from the study by Lorphensri et al. (2006). There, strong pH-dependent changes in the sorption coefficient (K_d) of norfloxacin (an antibiotic) to alumina and silica can be seen, as well as a pH-dependent octanol-water distribution coefficient (log K_{ow}). Fractions of the compound in anionic, zwitterionic, neutral, and cationic forms are also shown by the dotted lines. From the results presented, it is clear that translating such data to complex, heterogeneous and changeable soil environments is a daunting task.

In addition to variable sorption processes, variable and environmentally affected degradation processes will operate on effluent-borne chemicals. For example, the estrogenic steroids (estrone, 17β-estradiol, estriol, and 12α-ethynylestradiol) and testosterone are known to degrade rapidly in aerobic soil environments, mediated by microbial activity (Colucci et al., 2001; Colucci and Topp, 2001; Lorenzen et al., 2005; Ying and Kookana, 2005). However, in anaerobic soil, degradation of estrogenic steroids slows considerably, even when the microbial activity is robust (Ying and Kookana, 2005). Thus,

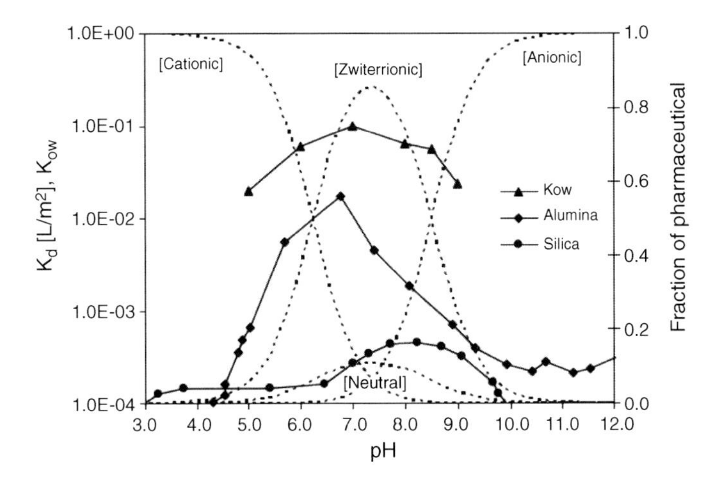

Figure 12.9 pH-sorption profiles of nalidixic acid to alumina and silica are shown along with the fraction of neutral and anionic forms of nalidixic acid and the pH-dependent octanol–water partition coefficient. Reproduced from Lorphensri et al., copyright 2006 with permission of Elsevier.

in soils with aeration problems (more commonly developed in effluent-irrigated soils), there is a greater possibility that estrogenic steroids will accumulate.

12.4 Summary and conclusions

The limited quantity of unpolluted water available for future use for food production and drinking water supply is a major challenge facing the world today. Treated wastewater can increase the water supply in areas in which the water demand by the urbanized population has exceeded the available natural water resources, yet reuse of these waters has certain associated risks.

The introduction of soluble or DOM is of considerable concern because of its ability to interact with organic pollutants, and, thus, affect the fate of these pollutants in soil or aquatic systems. Dissolved organic matter may reduce the sorption of organic pollutants through stable DOM–pollutant interactions in solution or by competing with the pollutant molecules for the sorption sites on the soil surface. The extent and nature of the interactions of DOM with organic pollutants depend on factors such as the nature of the organic compound, (e.g. its molecular weight and polarity), and, also, the size, polarity and molecular configuration of the DOM. Therefore, the impact of DOM on sorption and subsequent transport is dependent on the intrinsic nature of the solute, soil, and DOM, and the competition between solute–soil, solute–DOM, and DOM–soil interactions.

In addition to the potential risks associated with effluent-borne DOM, contaminants present in treated wastewater may leach into groundwater used as a drinking water source or enter aquatic ecosystems through irrigation runoff. Studies examining the long-term effects of effluent irrigation on groundwater quality have focused primarily on dissolved salts, nutrients, and heavy metals; however, concerns about effluent-derived organic microcontaminants entering aquifers used as drinking water sources need to be addressed. Attention should also be paid to the introduction of effluent-derived organic xenobiotics in runoff from fields irrigated with treated wastewater into aquatic ecosystems. A large number of potentially problematic organic micropollutants have been identified in treated wastewater and in streams receiving treated effluent. Such contaminants may be present in runoff from effluent irrigated fields. Classes of pollutants identified or potentially present in treated wastewater include pharmaceuticals, personal care product ingredients, nutraceuticals/herbal remedies, flame retardant chemicals, plasticizers, and disinfection byproducts. Uptake of these compounds by crop plants may also warrant concern. In regions using treated wastewater for irrigation, application of advanced treatment technologies may be required to further reduce the introduction of these compounds into agricultural and aquatic ecosystems.

References

Amiri, F., Bornick, H. and Worch, E. (2005) Sorption of phenols onto sandy aquifer material: the effect of dissolved organic matter (DOM). *Water Research* **39**, 933–941.

Baskaran, S., Bolan, N.S., Rahman, A. and Tillman, R.W. (1996) Effect of exogenous carbon on the sorption and movement of atrazine and 2,4-D by soils. *Australian Journal of Soil Research* **34**, 609–622.

Ben-Hur, M., Letey, J., Farmer, W.J., Williams, C.F. and Nelson, S.D. (2003) Soluble and solid organic matter effects on atrazine adsorption in cultivated soils. *Soil Science Society of America Journal* **67**, 1140–1146.

Bouwer, H. (1991) Simple derivation of the retardation equation and application to preferential flow and macrodispersion. *Ground Water* **29**, 41–46.

Brown, G.K., Zatig, S.D. and Barber, L.B. (1999) *Wastewater Analysis by Gas Chromatography/Mass Spectrometry*. 99-4018B, U.S. Geological Survey, West Trenton, NJ.

Burkhard, L.P. (2000) Estimating dissolved organic carbon partition coefficients for nonionic organic chemicals. *Environmental Science and Technology* **34**, 4663–4668.

Celis, R., Barriuso, E. and Houot, S. (1998) Sorption and desorption of atrazine by sludge-amended soil: Dissolved organic matter effects. *Journal of Environmental Quality* **27**, 1348–1356.

Chen, Y., Wang, C.X. and Wang, Z.J. (2005) Residues and source identification of persistent organic pollutants in farmland soils irrigated by effluents from biological treatment plants. *Environment International* **31**, 778–783.

Chen, Y., Wang, C.X., Wang, Z.J. and Huang, S.B. (2004) Assessment of the contamination and genotoxicity of soil irrigated with wastewater. *Plant Soil* **261**, 189–196.

Colucci, M.S., Bork, H. and Topp, E. (2001) Persistence of estrogenic hormones in agricultural soils: I. 17β-estradiol and estrone. *Journal of Environmental Quality* **30**, 2070–2076.

Colucci, M.S. and Topp, E. (2001) Persistence of estrogenic hormones in agricultural soils: II. 17α-ethynylestradiol. *Journal of Environmental Quality* **30**, 2077–2080.

Daughton, C.G. and Ternes, T.A. (1999) Pharmaceuticals and personal care products in the environment: Agents of subtle change? *Environmental Health Perspectives* **107**, 907–938.

Evans, H.E. (1988) The binding of three PCB congeners to dissolved organic carbon in freshwaters. *Chemosphere* **17**, 2325–2338.

Falconer, I.R., Chapman, H.F., Moore, M.R. and Ranmuthugala, G. (2006) Endocrine-disrupting compounds: A review of their challenge to sustainable and safe water supply and water reuse. *Environmental Toxicology* **21**, 181–191.

Fang, F., Kanan, S., Patterson, H.H. and Cronan, C.S. (1998) A spectrofluorimetric study of the binding of carbofuran, carbaryl, and aldicarb with dissolved organic matter. *Analytica Chimica Acta* **373**, 139–151.

Flores-Cespedes, F., Fernandez-Perez, M., Villafranca-Sanchez, M. and Gonzalez-Pradas, E. (2006) Cosorption study of organic pollutants and dissolved organic matter in a soil. *Environmental Pollution* **142**, 449–56.

Flores-Cespedes, F., Gonzalez-Pradas, E., Fernandez-Perez, M., Villafranca-Sanchez, M., Socias-Viciana, M. and Urena-Amate, M.D. (2002) Effects of dissolved organic carbon on sorption and mobility of imidacloprid in soil. *Journal of Environmental Quality* **31**, 880–888.

Gauthier, T.D., Shane, E.C., Guerin, W.F., Seitz, W.R. and Grant, C.L. (1986) Fluorescence quenching method for determining equilibrium-constants for polycyclic aromatic-hydrocarbons binding to dissolved humic materials. *Environmental Science and Technology* **20**, 1162–1166.

Georgi, A. and Kopinke, F.D. (2002) Validation of a modified Flory-Huggins concept for description of hydrophobic organic compound sorption on dissolved humic substances. *Environmental Toxicology and Chemistry* **21**, 1766–1774.

Gonzalez-Pradas, E., Fernandez-Perez, M., Flores-Cespedes, F., Villafranca-Sanchez, M., Urena-Amate, M.D., Socias-Viciana, M. and Garrido-Herrera, F. (2005) Effects of dissolved organic carbon on sorption of 3,4-dichloroaniline and 4-bromoaniline in a calcareous soil. *Chemosphere* **59**, 721–728.

Graber, E.R., Dror, I., Bercovich, F.C. and Rosner, M. (2001) Enhanced transport of pesticides in a field trial with treated sewage sludge. *Chemosphere* **44**, 805–811.

Graber, E.R., Gerstl, Z., Fischer, E. and Mingelgrin, U. (1995) Enhanced transport of atrazine under irrigation with effluent. *Soil Science Society of America Journal* **59**, 1513–1519.

Huang, X.J. and Lee, L.S. (2001) Effects of dissolved organic matter from animal waste effluent on chlorpyrifos sorption by soils. *Journal of Environmental Quality* **30**, 1258–1265.

Kalbitz, K., Popp, P., Geyer, W. and Hanschmann, G. (1997) beta-HCH mobilization in polluted wetland soils as influenced by dissolved organic matter. *Science of the Total Environment* **204**, 37–48.

Kinney, C.A., Furlong, E.T., Werner, S.L. and Cahill, J.D. (2006) Presence and distribution of wastewater-derived pharmaceuticals in soil irrigated with reclaimed water. *Environmental Toxicology and Chemistry* **25**, 317–326.

Ling, W.T., Wang, H.Z, Xu, J.M. and Gao, Y.Z. (2005) Sorption of dissolved organic matter and its effects on the atrazine sorption on soils. *Journal of Environmental Sciences-China* **17**, 478–482.
Ling, W.T., Xu, J.M. and Gao, Y.Z. (2006) Dissolved organic matter enhances the sorption of atrazine by soil. *Biology and Fertility of Soils* **42**, 418–425.
Lorenzen, A., Chapman, R., Hendel, J.G. and Topp, E. (2005) Persistence and pathways of testosterone dissipation in agricultural soil. *Journal of Environmental Quality* **34**, 854–860.
Lorphensri, O., Intravijit, J., Sabatini, D.A., Kibbey, T.C.G., Osathaphan, K. and Saiwan, C. (2006) Sorption of acetaminophen, 17 alpha-ethynyl estradiol, nalidixic acid, and norfloxacin to silica, alumina, and a hydrophobic medium. *Water Research* **40**, 1481–1491.
Maxin, C.R. and Kogel-Knabner, I. (1995) Partitioning of polycyclic aromatic hydrocarbons (PAH) to water- soluble soil organic matter. *European Journal of Soil Science* **46**, 193–204.
Mitra, S. and Dickhut, R.M. (1999) Three-phase modeling of polycyclic aromatic hydrocarbon association with pore-water-dissolved organic carbon. *Environmental Toxicology and Chemistry* **18**, 1144–1148.
Muller, K., Magesan, G.N. and Bolan, N.S. (2007) A critical review of the influence of effluent irrigation on the fate of pesticides in soil. *Agricultural Ecosystems and Environment* **120**, 93–116.
Muszkat, L., Lahav, D., Ronen, D. and Magaritz, M. (1993a) Penetration of pesticides and industrial organics deep into soil and into groundwater. *Archives of Insect Biochemistry and Physiology* **22**, 487–499.
Muszkat, L., Raucher, D., Magaritz, M., Ronen, D. and Amiel, A.J. (1993b) Unsaturated zone and groundwater contamination by organic pollutants in a sewage-effluent-irrigated site. *Ground Water* **31**, 556–565.
Nelson, S.D., Letey, J., Farmer, W.J., Williams, C.F. and Ben-Hur, M. 1998. Facilitated transport of napropamide by dissolved organic matter in sewage sludge-amended soil. *Journal of Environmental Quality* **27**, 1194–1200.
Nelson, S.D., Letey, J., Farmer, W.J., Williams, C.F. and Ben-Hur, M. 2000. Herbicide application method effects on napropamide complexation with dissolved organic matter. *Journal of Environmental Quality* **29**, 987–994.
Ohlenbusch, G. and Frimmel, F.H. (2001) Investigations on the sorption of phenols to dissolved organic matter by a QSAR study. *Chemosphere* **45**, 323–327.
Ohlenbusch, G., Kumke, M.U. and Frimmel, F.H. (2000) Sorption of phenols to dissolved organic matter investigated by solid phase microextraction. *Science of the Total Environment* **253**, 63–74.
Pedersen, J.A., Soliman, M. and Suffet, I.H. (2005) Human pharmaceuticals, hormones, and personal care product ingredients in runoff from agricultural fields irrigated with treated wastewater. *Journal of Agricultural and Food Chemistry* **53**, 1625–1632.
Pedersen, J.A., Yeager, M.A. and Suffet, I.H. (2002) Characterization and mass load estimates of organic, compounds in agricultural irrigation runoff. *Water Science and Technology* **45**, 103–110.
Pedersen, J.A., Yeager, M.A. and Suffet, I.H. (2003) Xenobiotic organic compounds in runoff from fields irrigated with treated wastewater. *Journal of Agricultural and Food Chemistry* **51**, 1360–1372.
Perminova, I.V., Grechishcheva, N.Y. and Petrosyan, V.S. (1999) Relationships between structure and binding affinity of humic substances for polycyclic aromatic hydrocarbons: Relevance of molecular descriptors. *Environmental Science and Technology* **33**, 3781–3787.
Persson, L., Alsberg, T., Odham, G. and Ledin, A. (2003) Measuring the pollutant transport capacity of dissolved organic matter in complex matrixes. *International Journal of Environmental and Analytical Chemistry* **83**, 971–986.
Poerschmann, J. and Kopinke, F.D. (2001) Sorption of very hydrophobic organic Compounds (VHOCs) on dissolved humic organic matter (DOM). 2. Measurement of sorption and application of a Flory-Huggins concept to interpret the data. *Environmental Science and Technology* **35**, 1142–1148.
Raber, B., Kogel-Knabner, I., Stein, C. and Klem, D. (1998) Partitioning of polycyclic aromatic hydrocarbons to dissolved organic matter from different soils. *Chemosphere* **36**, 79–97.
Rav-Acha, Ch. and Rebhun, M. (1992) Binding of organic solutes to dissolved humic substances and its effect on adsorption and transport in the aquatic environment. *Water Research* **26**, 1645–1654.
Rebhun, M., Kalabo, R., Grossman, L., Manka, J. and Rav-Acha, Ch. (1992) Sorption of Organics on Clay and Synthetic Humic Clay Complexes Simulating Aquifer Processes. *Water Research* **26**, 79–84.
Rebhun, M., Desmedt, F., and Rwetabula, J. (1996) Dissolved humic substances for remediation of sites contaminated by organic pollutants. Binding-desorption model predictions. *Water Research* **30**, 2027–2038.

Sabbah, I., Rebhun, M. and Gerstl, Z. (2004) An independent prediction of the effect of dissolved organic matter on the transport of polycyclic aromatic hydrocarbons. *Journal of Contaminant Hydrology* **75**, 55–70.

Seol, Y. and Lee, L.S. (2000) Effect of dissolved organic matter in treated effluents on sorption of atrazine and prometryn by soils. *Soil Science Society of America Journal* **64**, 1976–1983.

Seol, Y. and Lee, L.S. (2001) Coupled effects of treated effluent irrigation and wetting- drying cycles on transport of triazines through unsaturated soil columns. *Journal of Environmental Quality* **30**, 1644–1652.

Shon, H.K., Vigneswaran, S. and Snyder, S.A. (2006) Effluent organic matter (EfOM) in wastewater: Constituents, effects, and treatment. *CRC Critical Reviews of Environmental Science and Technology* **36** (4), 327–374.

Tanaka, F., Fukushima, M., Kikuchi, A., Yabuta, H., Ichikawa, H. and Tatsumi, K. (2005) Influence of chemical characteristics of humic substances on the partition coefficient of a chlorinated dioxin. *Chemosphere* **58**, 1319–1326.

Ternes, T.A., Bonerz, M., Herrmann, N., Teiser, B. and Andersen, H.R. (2007) Irrigation of treated wastewater in Braunschweig, Germany: An option to remove pharmaceuticals and musk fragrances. *Chemosphere* **66**, 894.

Totsche, K.U., Danzer, J. and Kogel-Knabner, I. 1997. Dissolved organic matter-enhanced retention of polycyclic aromatic hydrocarbons in soil miscible displacement experiments. *Journal of Environmental Quality* **26**, 1090–1100.

Totsche, K.U. and Kogel-Knabner, I. (2004) Mobile organic sorbent affected contaminant transport in soil: Numerical case studies for enhanced and reduced mobility. *Vadose Zone Journal* **3**, 352–367.

Williams, C.F., Agassi, M., Letey, J., Farmer, W.J., Nelson, S.D. and Ben-Hur, M. (2000) Facilitated transport of napropamide by dissolved organic matter through soil columns. *Soil Science Society of America Journal* **64**, 590–594.

Ying, G.G. and Kookana, R.S. (2005) Sorption and degradation of estrogen-like-endocrine disrupting chemicals in soil. *Environmental Toxicology and Chemistry* **24**, 2640–2645.

Chapter 13

Organic matter in wastewater and treated wastewater-irrigated soils: properties and effects

Yona Chen, Carlos G. Dosoretz, Ilan Katz, Elizabeth Jüeschke, Bernd Marschner and Jorge Tarchitzky

13.1 Introduction

The use of treated wastewater (TWW) for crop irrigation as an alternative to wastewater disposal is common in water-scarce countries worldwide. In arid and semiarid regions, the use of TWW for agriculture means that a larger amount of freshwater (FW) is available for domestic use. The combination of severe water shortages, contamination of water resources, densely populated urban areas, and highly intensively irrigated agriculture, makes it essential that arid and semiarid countries, such as Israel, place wastewater treatment and reuse high on their list of national priorities. In Israel, for example, national policy calls for the gradual replacement of FW allocations to agriculture by TWW, such that in 2020, TWW will constitute about 50% of this sector's consumption. A similar trend has been observed in other countries around the world (Feigin et al., 1991; EPA, 1992; Asano, 1998).

13.2 Organic matter in wastewater

One of the major concerns related to the use of TWW is its organic matter (OM) load. Wastewater originating from municipal, industrial, and agricultural sources contains OM at different levels, which is gradually removed during sewage water treatment. The treatment path includes physical, chemical, and biological methods integrated into four major treatment groups (Shon et al., 2006):

(i) preliminary, aimed at removing coarse and readily settleable inorganic solids;
(ii) primary, designed to remove most (70–90%) of the suspended solids and 30–40% of the biodegradable organics;

Treated Wastewater in Agriculture, First Edition, edited by Guy J. Levy, Pinchas Fine and Asher Bar-Tal

(iii) secondary, aimed at reducing biodegradable dissolved organic matter (DOM) and nutrients; and
(iv) tertiary–specifically designed to treat the remaining organic solids and pathogenic microorganisms.

The content and class of the OM in raw sewage and TWW can be expressed by several parameters (Table 13.1). The chemical oxygen demand (COD) level in raw sewage and in TWW after the primary and secondary treatments has been reported to be 250–1000, 150–750 and 30–160 mg/l, respectively (Feigin et al., 1991). After the purification effect of a soil-aquifer system, the COD concentration was 6 mg/l (Icekson-Tal et al., 2007). However, the COD value is a comprehensive and non-specific parameter that provides no characterization of the OM in the water. More specific classification based on the physical and chemical characteristics would contribute to improving wastewater treatment and to studies on the potential effects of OM on irrigation systems, soils, and crops.

13.2.1 Particle-size distribution

In their review, Sophonsiri and Morgenroth (2004) classified the OM in raw wastewater, primary, and secondary effluents into four particle-size fractions: soluble (<0.08 μm), colloidal (0.08–1 μm), supracolloidal (1–100 μm), and settleable (>100 μm). According to these investigators, particles in untreated and TWW that are more than 1 μm in size constitute a significant fraction of the overall OM and require extracellular microbial hydrolysis before they can be metabolized by bacteria. Rickert and Hunter (1971) found that 53% of the total organic carbon was in the soluble and colloidal fractions of the sewage water, increasing to 75% of the total organic carbon in the secondary effluent.

According to Levine et al. (1991), the particle-size distribution in untreated wastewater is site-specific and a significant fraction of the OM is in the colloidal to supracolloidal size range. Treatment operations and processes provide physicochemical removal of certain

Table 13.1 Typical range of untreated and treated wastewater after secondary treatment (modified from Asano et al., 2007)

Treatment / Parameter	Untreated	Conventional activated sludge	Activated sludge with biological nutrient removal	Membrane bioreactor
	mg/l			
Biochemical oxygen demand (BOD)	110–350	5–25	5–15	<1–5
Chemical oxygen demand (COD)	250–800	40–80	20–40	<10–30
Total organic carbon (TOC)	80–260	10–40	8–20	0.5–5
Volatile organic compounds (VOC)	< 100– > 400	10–40	10–20	10–20

organic size fractions, as well as solubilization and biological conversion of the OM. Thus, size distribution shifts according to the specific treatment process. Following biological treatment, the organic content is reduced and a significant fraction of the remaining OM is in the submicron size ranges. Differences in molecular weight (MW) distribution between secondary and tertiary wastewater treatment processes, and as a result of the season, have been reported by Amy et al. (1987). These authors found a shift in MW distribution due to a slight removal of high-MW organics (>100 000) and an increase in the <500 MW range for the tertiary effluent. Shon et al. (2006) distinguished between particulate organic carbon, with a size above 0.45 μm, and dissolved organic carbon (DOC), consisting of particles smaller than 0.45 μm – mainly cell fragments and macromolecules in the $10–10^9$ Da MW range. Painter (1973) found 60% and 40% of the organic carbon in secondary effluent to be >10 000 Da and <10 000 Da, respectively. The MW of humic substances (HS) in secondary effluents was reported to range from 500 to 50 000 Da (Rebhun and Manka, 1971; Manka et al., 1974), and their distribution is shown in Table 13.2 (Manka et al., 1974). Sachedev et al. (1976) reported that 58–62% of the DOC present in wastewater has a MW of <700 Da, whereas the >5000 Da fraction amounts to between 20% and 29% only. Humic acid (HA) isolated from refinery effluent exhibited a lower MW than natural HS (Lingbo et al., 2005). The number-average MW (M_n) and the weight-average MW (M_w) were 1069 Da and 2934 Da, respectively, and those of fulvic acid (FA) were 679 Da and 1212 Da, respectively (Lingbo et al., 2005). It should be noted, however, that the determination of the MW in polydispersed systems such as TWW is complicated and often related to measurement procedures, as evidenced by the variability in the reported results.

13.2.2 Chemical composition

The molecular properties of OM present in TWW have been described in several studies. In the municipal raw wastewater stream of four treatment plants, total proteins, carbohydrates, and lipids were found to be 28%, 18% and 31% of the total COD, respectively (Raunkjær et al., 1994). Sophonsiri and Morgenroth (2004) reported that in municipal wastewater, most of the OM (measured as COD) is accounted for by the sum of proteins, carbohydrates, and lipids. Dignac et al. (2000) identified 40% of the DOC in raw sewage, and attributed it to sugars (12%), proteins (23%), and fatty acids (5%). However, in the TWW, only 22% of the DOC was identified, including sugars (9%), proteins (13%), and lipids (less than 1%). Proteins and lipids accounted for most of the loss of OM during the sewage treatment. Dignac et al. (2000), using spectroscopic analysis, found that the

Table 13.2 Molecular-weight distribution of humic substances isolated from secondary effluents (modified from manka et al., 1974)

	Percentage in the fraction (%)			
Molecular weight (kDa)	<500	500–5000	5000–50 000	>50 000
Humic acid (HA)	17.9	35.6	44.5	2.0
Fulvic acid (FA)	27.5	43.5	24.7	4.3

predominance of aliphatic carbon in long polymethylenic chains in the untreated wastewater disappears in the TWW.

Rebhun and Manka (1971) identified 42% of the DOC in secondary effluents as HS: 25% (FA), 11% HA, and 6% hymathomelanic acid. According to Manka and Rebhun (1982), HS were the largest organic group in a municipal sewage treatment plant after different treatment steps, with proteins next in order. The total HS were 35.9% of the soluble COD after a secondary treatment by oxidation ponds (Table 13.3), and they decreased to 21.6% after a lime clarifier and to 16.7% after a polishing pond (Manka and Rebhun, 1982). The highest removal during treatment was of the HS and carbohydrates. In another study, the HS content in TWW was found to be 45% of the DOC, and it increased by 40% during a further degradation process (Namour and Müller, 1998). Amy et al. (1987) reported HS content of 54–58% after an activated sludge treatment. Painter (1973) identified 80% of the smaller fraction of ultrafiltered effluent by dialysis membrane (cut-off 10 000) and classified it into various known organic compounds. However, he identified only 10% of the larger fraction, assuming that some of the unidentified components in the high-MW fraction could be FA and HA. Fluorescence excitation-emission matrix (EEM) spectroscopy was used to identify the chemical properties of the DOM in TWW (Her et al., 2003). These researchers found a mixture of polysaccharide-like/protein-like substances and minimum fulvic-like substances in the high-MW fraction (>10 000 Da), but the medium MW range showed higher aromaticity (FA).

A comparison of the HS extracted from streamwater, wastewater, and soil indicated that the water-originated compounds were very similar to each other and that both were easily distinguishable from the soil HS (Peschel and Wildt, 1988). The hydrophilic fractions (Leenheer, 1981; Stevenson, 1994) (such as low-MW polysaccharides, carbohydrates, proteins, peptides, and amino acids, which are widespread in sewage water) were degraded during the aerobic wastewater treatment, resulting in a relative increase in aromatic and aliphatic compounds (Chefetz et al., 1998). In effluents treated with activated carbon and reverse osmosis, over 50% of the DOC was classified as hydrophilic OM, whereas

Table 13.3 Group composition in wastewater after successive treatments (modified from Manka and Rebhun, 1982)

Treatment / Organic group	Oxidation pond	Lime clarifier	Polishing pond
	Percent of soluble COD		
Proteins	26.1	22.9	17.7
Anionic detergents	14.8	15.4	12.9
Carbohydrates	7.0	4.7	4.5
Tannins and lignins	7.0	4.7	4.5
Ether extract pH 2	10.8	9.3	8.2
Ether extract pH 10	4.1	4.0	1.7
Total ether extract	14.9	13.3	9.9
Fulvic acid	21.1	12.4	12.4
Humic acid	8.5	3.1	4.3
Hymathomelanic acid	6.3	6.9	
Total humic substances	35.9	21.6	16.7
Total identifiable COD	103.3	82.2	67.1
COD (mg/l)	116	84	55

COD, chemical oxygen demand.

33–34% was classified as FA (Fujita et al., 1996). In groundwater that had been partially recharged by these effluents, only 16% of the DOC was hydrophilic, 50% of the DOC was FA, and 6.5–10% was recovered as HA. Similarly, Ma et al. (2001) found that the effluent of a wastewater-treatment plant was almost devoid of HA, the FA was 42.5% of the total organic carbon and the hydrophilic fraction was 55.8% of the total organic carbon. The differences between the data of Chefetz et al. (1998) and those of Ma et al. (2001) stem in part from differences in the separation procedures: the first group used separation techniques based on protocols of the International Humic Substances Society (IHSS), whereas the second (Ma et al., 2001) used reverse-osmosis techniques, which reject large molecules such as HA. In contrast, natural water samples from two rivers, a lake, and a swamp contained high concentrations of FA (53.7–68.4%), lower levels of the hydrophilic fraction and significant concentrations of HA (13.5–28.7%) (Ma et al., 2001). The hydrophilic fraction and the aquatic HS represented 32–74% and 3–28% of the DOC, respectively, in effluent with a low biochemical oxygen demand (BOD) concentration from sewage and human-waste treatment plants (Imai et al., 2002).

Shon et al. (2006) distinguish between (Fig. 13.1):

- particulate organic carbon – characterized by particle sizes >0.45 μm, which includes zooplankton, algae, bacteria, and debris OM from soil and plants;
- dissolved organic carbon – particle sizes <0.45 μm, mainly cell fragments and macromolecules exhibiting a wide range of MW (10–10^9 Da). Chemically, these materials

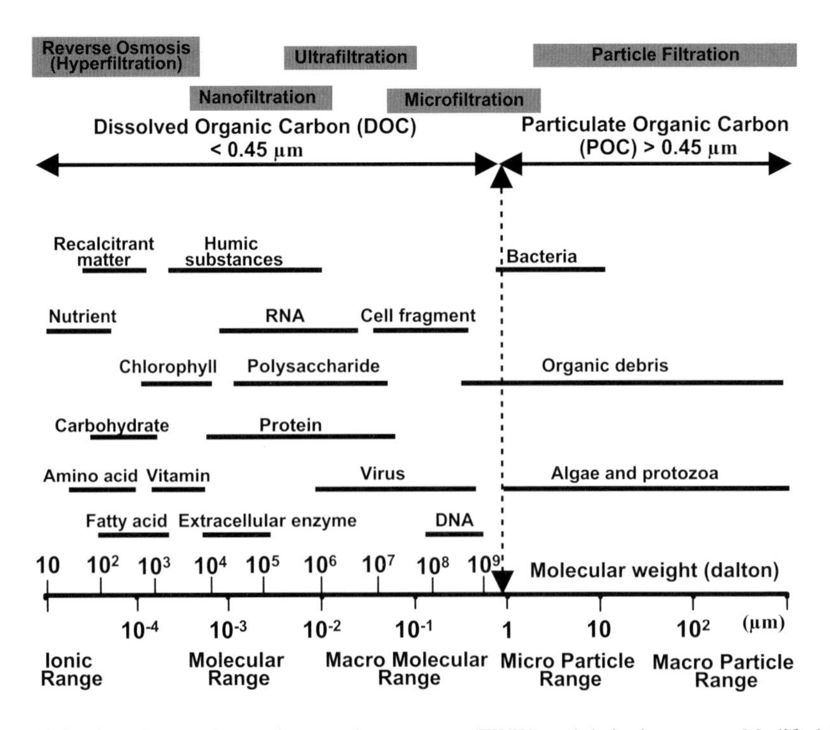

Figure 13.1 Organic constituents in treated wastewater (TWW) and their size ranges. Modified from Shon et al., 2006.

consist of a vast array of chemicals: recalcitrant material, HS, nutrients, chlorophyll, polysaccharides, carbohydrates, proteins, amino acids, vitamins, fatty acids, RNA, DNA, and extracellular enzymes.

In refinery effluent, which is obviously different from regular municipal effluent, the concentration of DOC was 9.9 mg/l, of which HA and FA accounted for 2.3% and 34.6%, respectively (Lingbo et al., 2005). Elemental analysis, the specific ultraviolet absorbance at 254 nm ($SUVA_{254}$) and ^{13}C-nuclear magnetic resonance (NMR) analyses of refinery effluent-derived HS indicated the presence of four kinds of carbon structures: aliphatic, aromatic hetero-aliphatic, and carboxyl carbon. The results support the interpretation that these HS are rich in aliphatic and poor in aromatic carbon.

Most (>90%) of the OM accumulated in a gravel-bed constructed wetland receiving dairy farm wastewater over a 5-yr period consisted of stable (recalcitrant) OM fractions, whereas the water-soluble carbon fractions accounted for <10% of the sediment OM. Humic acid, FA and humin were the predominant stable OM fractions, accounting for 63% to 96% of the OM (Nguyen, 2000).

13.3 Soil organic matter (SOM)

Most arable soils contain 0.1–5% OM by weight. The extremely low figures represent sandy soils of arid zones, whereas the higher values are typical of clayey soils in temperate zones. Most of these soils are physically and chemically influenced by the OM they contain. In addition to its nutritional value, OM plays a critical role in the formation and stabilization of soil structure, which in turn produces desirable drainage and resistance to erosion. Although HS have already been discussed in relation to their concentration in TWW, they are further defined here as we are now dealing with fractions of SOM rather than water.

As organic materials in the soil decay, humus consisting of macromolecules of a mixed aliphatic and aromatic nature is formed. The term humus is considered by many scientists as synonymous to SOM (e.g. Stevenson, 1994). Others distinguish between total SOM and humus. Some scientists refer to soil HS as natural OM (NOM, or DOM, the latter referring to water-borne OM). Humus consists of the total NOM in soil excluding identifiable plant and animal tissues and living biomass. Humic substances, including HA, FA and humin, comprise soil humus. Defined biochemicals, even though soluble, are not considered to be a component of the various HS fractions.

The chemical and colloidal properties of NOM can only be properly studied in the free state, i.e. when freed of inorganic components. Thus, one must first separate NOM, and in particular, SOM from the inorganic matrix. Alkali, usually 0.1–0.5 M NaOH, as been widely used to extract SOM. Various researchers have also used other methods of extraction. However, this subject is beyond the scope of this chapter, and extraction procedures will not be described. A detailed discussion of the extraction procedures can be found elsewhere (Stevenson, 1994; Swift, 1996). Humic substances are major components of the NOM in soil and water, as well as of geological organic deposits such as sediments, peats, brown coals, and shales. They make up much of the characteristic brown color of decaying plant debris and contribute to the brown or black color in surface soils. They can also impart color to surface waters, especially in brown FW ponds, lakes, and streams. In leaf litter or composts, the color may be yellowish-brown to black, depending on the degree of decay.

Humic substances are the most important components of soil material, which affect the soil's physical and chemical properties and improve its fertility. In aqueous systems, such as rivers, about 50% of the DOM is from HS, which affect the pH and alkalinity of the water. In both terrestrial and aquatic systems they affect the chemistry, cycling, and bioavailability of chemical elements, as well as the transport and degradation of organic chemicals and xenobiotics. They also affect biological productivity in blackwater ecosystems, as well as the formation of disinfection byproducts during water treatment. Humic substances are complex, heterogeneous, colloidal mixtures formed by biochemical and chemical reactions during the transformation of plant and microbial remains (humification). Plant lignin, as well as polyphenols, polysaccharides, melanin, cutin, proteins, fatty acids from lipids. and other polymeric components, are important structural components taking part in this process.

As stated earlier, HS can be divided into three main fractions: HA, FA, and humin. Humic acid and FA are extracted from soil material using strong bases (NaOH or KOH). Humic acid, in contrast to FA, is insoluble at low pH, and thus the HA is separated from the FA by adding strong acid (adjusting pH to 1 with HCl). Humins make up the fraction that is not extracted at either low or high pH. As a result, water HS contain only HA and FA. For scientific purposes, these components are generally removed from the water by lowering the pH to 2 and adsorbing both components on a suitable insoluble resin. Humic acid and FA are then extracted from the resin with a strong base followed by lowering the pH to 1 to precipitate out the HA. Analogous processes can be used to obtain HA and FA from any source.

Humic substances are highly chemically reactive, yet recalcitrant with respect to biodegradation. Most data on HA, FA, and humin refer to average properties and structures of a large ensemble of diverse macromolecules. The precise properties and structure of a given HS sample depend on the water or soil source and the specific conditions of extraction. Nevertheless, the average properties of all HA, FA and humins are remarkably similar.

13.4 The influence of treated-wastewater-borne organic matter on soil organic matter

The content and accumulation of OM in soils differ according to climatic conditions and vegetation type. The use of TWW can also produce changes in SOM content or in the ratios between its different components and their properties. Organic matter can also cause changes in soil properties (Jüeschke et al., 2008). Very little information is available on the changes in SOM due to long-term use of TWW. More studies have been published, however, on the effects of sewage sludge on soil properties, with particular attention to SOM.

In general, the input of particulate organic carbon, DOC and/or colloidal OM increases SOM content in the topsoil (Friedel et al., 2000; An et al., 2004; Jüeschke et al., 2004), although much of the TWW-borne OM is easily degradable (Fine et al., 2002; Meli et al., 2002; An et al., 2004; Saadi et al., 2006). Some studies have also shown that some of the added DOC is translocated to deeper soil layers (Fine et al., 2002; An et al., 2004). Meli et al. (2002) determined several chemical and microbiological soil parameters, in a citrus orchard irrigated with lagoon-treated urban wastewater for 15 yrs. Mean values of OM as soluble carbon during the irrigation season were always higher in the TWW-irrigated soil than in the FW-irrigated one. The values decreased for both water classes in the spring at the end of the irrigation season. Ramirez-Fuentes et al. (2002) described a field site irrigated since 1886 with domestic and industrial untreated wastewater. A significant increase in organic carbon (1.4-fold) was found in the upper 20 cm of the soil profile. Irrigation with distillery effluent with a high organic load (high BOD) increased the total organic carbon and total Kjeldahl nitrogen two- to threefold compared to a control, but a decrease (up to 50%) in the bacterial population was observed (Kaushik et al., 2005). Similar results were found in soils irrigated with tannery wastewater (Alvarez-Bernal et al., 2006).

However, the incorporation of easily degradable substances can cause an increase in microbial population and activity, and, consequently, enhanced mineralization of the incorporated and native OM (i.e. a priming effect) (Kuzyakov et al., 2000; De Nobili et al., 2001; Hamer and Marschner, 2005). As previously described, some of the easily degradable substances are present in the TWW at different concentrations and relative ratios. Various mechanisms may play a role in the activation of microorganisms for priming effects, but the most important ones activated once TWW is applied are:

(i) co-metabolism of easily available organic substances incorporated during TWW-irrigation;
(ii) osmotic stress of microorganisms due to high salinity, and the subsequent release of carbon and nitrogen upon their death and lysis; and
(iii) acceleration of SOM mineralization through improved aeration and destruction of aggregates (Kuzyakov et al., 2000).

Gloaguen et al. (2007) studied the soil-solution chemistry of a Brazilian oxisol irrigated with effluent water. Enhanced mineralization of the DOC could be detected in the soil solution, causing a long-term decrease in DOC. This was explained by a reduction in SOM due to intensification of microbial activity.

In Israel, a decrease in OM was observed in the subsoil of some fields with a long history of TWW-irrigation (Jüeschke et al., 2004). This observation can be attributed to inputs of easily degradable DOC compounds, which can stimulate microbial activity and thus induce priming effects. Such stimulatory effects of TWW-irrigation on microbial or enzyme activity are commonly observed in topsoils (Filip et al., 1999; Friedel et al., 2000; Meli et al., 2002; Criquet et al., 2007; Gloaguen et al., 2007) and changes in microbial composition have been reported (Oved et al., 2001; Gelsomino et al., 2006).

Little information is available regarding possible modifications of the chemical and physicochemical characteristics of SOM as a result of OM application during TWW-irrigation in comparison with these properties in FW-irrigated soils. In contrast, several studies have evaluated the effects of sewage sludge on SOM properties. Data presented by Boyd et al. (1980) indicated that sludge OM is recovered in HA fractions of the sludge-amended soil. Proteinaceous and aliphatic materials were the primary sludge constituents associated with HA extracted from sludge and sludge-amended soil. These constituents were indicated by an increased intensity of amide and C-H infrared (IR) absorption bands, increased nitrogen content, and decreased C/H ratios. Similarly, the IR spectra of HA extracted from soils amended with increasing doses of sludge showed a marked enhancement of aliphatic groups and of the peptide bands, concomitant with the amount of applied sludge (Piccolo et al., 1992). Adani and Tambone (2005) also reported more aliphatic HA in sludge-treated soils than in untreated ones. This result is remarkable as this modification of soil-HA occurred using a very low rate of applied sewage sludge. Soler Rovira et al. (2002) found that the HA extracted from sludge-amended soil is characterized by the prevalence of aliphatic character, a low content of oxygenated functional groups, a high content of sulfur, nitrogen, and polysaccharide components, and a low degree of humification.

Tarchitzky et al. (2007) found similar Fourier transform IR spectroscopy (FTIR) spectra in HA and FA samples extracted from FW- and TWW-irrigated brown alluvial soil. However, in the 2–4 and 4–6 cm layers, the HA of the TWW-irrigated soil exhibited a greater proportion of hydrophobic groups than that extracted from the FW-irrigated soil. According to ^{13}C-nuclear magnetic resonance spectroscopy with cross-polarization magic-angle spinning (CPMAS ^{13}C-NMR) spectra of HA extracted from different layers of FW- and TWW-irrigated soils, the aliphaticity in the TWW-irrigated soil did not change with depth and ranged from about 63% to 65%, whereas that of the FW-irrigated derived HA increased with depth from about 63% to 70% (Table 13.4). It appears that the enhanced degradation of SOM in TWW-irrigated soils is the result of a priming effect of OM degradation, thereby inducing decreased variability with depth. No distinctive differences were observed in the spectra between the distribution of the carbon-containing groups in FA extracted from the FW versus TWW treatments, or from different depths. A solvent mixture was applied with the aim of removing hydrophobic organic materials from the soil. The FTIR spectrum of the soil extract from the 0–2 cm layer of the TWW-irrigated soil exhibited the same peaks as the FW-irrigated soil, but their intensities and the ratio between them differed. For the 0–2 cm layer, the aliphatic peaks were more prominent and broad in the spectrum of the TWW extract, suggesting a greater proportion of aliphaticity in this material. There was also a relative increase in the polysaccharide peaks in the 0–2 cm layer of the TWW-irrigated soil, indicating the presence of recent, microbially derived OM.

Humic acid extracted from the topsoil in field-crop plots irrigated for almost 50 yrs with TWW exhibited higher H/C atomic ratios, and lower carboxyl and higher polysaccharide peaks in FTIR spectra relative to FW-irrigated plots. These results indicated more recent

Table 13.4 Aliphaticity as determined by cross-polarization magic-angle spinning (CPMAS) ^{13}C-nuclear magnetic resonance (NMR) spectra in humic and fulvic acids from different layers of freshwater- or treated-wastewater-irrigated brown alluvial soil (Gaaton)

Water quality	**Freshwater**			**Treated wastewater**		
Layer depth (cm)	0–2	2–4	4–6	0–2	2–4	4–6
Humic acid	63.4	66.3	69.9	63.4	65.7	64.6
Fulvic acid	65.7	63.4	72.3	66.3	63.4	71.8

Modified from Tarchitzky et al., 2007.

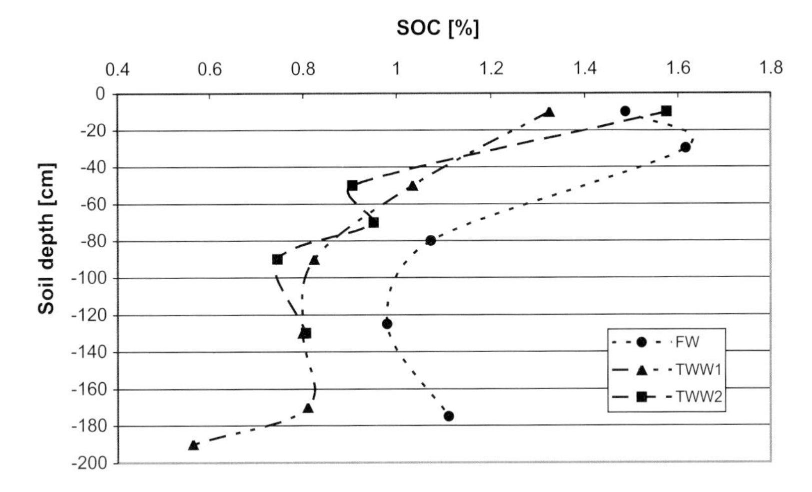

Figure 13.2 Soil organic carbon content (% of the soil profiles) in freshwater (FW)- and treated wastewater (TWW)-irrigated soils at the Yagur site. TWW1 and TWW2 represent two different plots irrigated with the same TWW. Reproduced, with permission, from Jüeschke, E., Marschner, B., Tarchitzky, J. and Chen, Y. (2008) Effects of treated wastewater irrigation on the dissolved on the dissolved and soil organic carbon in Israeli soils. *Water Science and Technology* **57**(5), 727–733.

and young HS in the TWW-irrigated soil. Soil organic matter seems to increase in the topsoil and tends to decrease after long-term irrigation with secondary TWW in the subsoil (Fig. 13.2) (Jüeschke et al., 2008).

Recent data obtained from clay and sandy clay soils indicate that both the frequency of tillage (field crop and orchard soils) and the irrigation period with TWW affect the content and characteristics of the SOM (Tarchitzky, Levy, and Chen, unpublished). The content of oxygen functional groups is presented in Table 13.5. Both tillage regime and water type had an impact on the SOM: the functional-group content was higher in the FW-irrigated soils, and decreased in the TWW plots according to the duration of TWW usage. The contents of carboxyl groups, phenol, and total acidity were higher in the non-tilled soils than in the tilled ones. The low content of functional groups was connected with the presence in the soil of young, highly aliphatic OM (Komonova and Alexandrova, 1973; Zhigunov and Simakov, 1977; Inbar, 1989).

Table 13.5 Functional group content in humic acids from the 0–10 cm layer of freshwater- or treated-wastewater-irrigated and tilled (field crops) or non-tilled (orchard) soil

Water quality	Irrigation period	Carboxyl		Phenol		Total acidity	
		Citrus	Field crop	Citrus	Field crop	Citrus	Field crop
Treated wastewater	Long-term[a]	2.6[c]	3.0	0.6[c]	0.4	3.2[c]	3.4
Treated wastewater	Short-term[b]	2.9	3.2	0.8	0.9	3.7	4.1
Freshwater		3.2	3.7	0.7	1.1	3.9	4.8

[a]Long-term: 15 yrs.
[b]Short-term: 4 yrs.
[c]10–30 cm layer.
Tarchitzky, Levy, and Chen, unpublished results.

13.5 The influence of treated-wastewater-borne organic matter on soil properties

Soils, whether in a natural or agricultural ecosystem, are exposed to mixtures of freshly decomposing OM, polysaccharides, and HS. This is obviously also true for soils irrigated with TWW, or to which biosolids compost has been added. The relative importance of each of these components is difficult to assess, and needs further research due to the complexity of the processes involved.

Swift (1996) concluded that polysaccharides are capable of producing stable aggregates, but their effect is transient and declines as the polysaccharides decompose. Adsorbed HS also produce stable aggregates and their effect is more persistent. The most stable and persistent reformed aggregates are produced by the combined action of polysaccharides and HS. According to Swift (1996), there is no doubt that the increased persistence results from the relative resistance of HS to biological decomposition in the soil compared to polysaccharides. In all instances, the involvement of soil microorganisms is an essential component of aggregate formation and stabilization.

The short-term effect of various types of OM addition on aggregate formation and stability were investigated by Muneer and Oades (1989), who showed that serial addition of glucose after wetting and drying cycles leads to substantial soil aggregation. Glucose plus calcium further amplified this process. In contrast, treating soils with $Na_4P_2O_7$, which is a powerful HS extractant (Stevenson, 1994), caused dispersion of most of the clay but did not disrupt large aggregates, probably because their main binding constituents were polysaccharides.

In an effort to corroborate the mechanism governing the macromolecular interactions of clays, HS, and polysaccharides under varying pH conditions and relative concentrations and charges of the organic components, Tarchitzky et al. (1993) performed flocculation measurements, which were later confirmed by the same group (Tarchitzky and Chen, 2002a, 2002b) using rheological measurements.

Studies on flocculation and dispersion characteristics of homo-ionic montmorillonite were performed as a function of exchangeable cations, HA, FA, and polygalacturonic acid concentrations, and pH. A matrix of experimental conditions was employed and the corresponding flocculation values were measured. Edge-face (E-F) and edge-edge (E-E) interactions at various pHs, as well as heteroflocculation and its response to pH and OM concentration were discussed in detail by Tarchitzky et al. (1993) and Tarchitzky and Chen (2002a, 2002b). In general, HA, FA, and polygalacturonic acid greatly increase the stability in suspension of clay particles (as reflected by increased flocculation values). This effect is pH-dependent due to changes in the edge charge of the clay and the increasing dissociation of the functional groups of organic macromolecules with pH. The relevance of these mechanisms to interaction of clays with DOM originating from TWW was also investigated (Tarchitzky et al., 1999). The previously presented concept was further employed for a study of the effects of humic-like substances originating from TWW on clay dispersion and hydraulic conductivity of soils. Maximum flocculation values were exhibited by Na-montmorillonite at the highest DOM concentration (Tarchitzky et al., 1999). In addition, these researchers tested the interactions of homo-ionic DOM originating from wastewater and sewage sludge with montmorillonite and a soil clay fraction (dark brown clayey grumosol and sandy loam hamra). Both HA and FA were isolated from the sewage sludge and purified (sewage sludge-HA and sewage sludge-FA, respectively). The sodium-saturated DOM isolated from wastewater caused dispersion of Na-montmorillonite and of the sandy loam clay fraction, but did not affect the clay fraction from the clayey grumosol (Fig. 13.3). The sodium-saturated sewage sludge-HA did not affect the stability of Na-montmorillonite suspensions. In contrast, sodium-saturated sewage sludge-FA stabilized the Na-montmorillonite suspension and an increase in the flocculation value was observed (reflecting a dispersive effect). The calcium-saturated DOM reduced the flocculation value of Ca-montmorillonite and that of the clay fraction of the clayey soil. The soil HA, however, had a very small dispersive effect on Ca-montmorillonite (Chen and Tarchitzky, 2008).

The effects observed at the microaggregate level were also tested in soil systems. Soil packed in columns and treated with wastewater effluent exhibited a sharp decrease in hydraulic conductivity, to only 20% of its initial value. The reduction in hydraulic conductivity is likely the result of decreased soil-pore sizes, which reflects two processes:

(i) retention of part of the DOM during water percolation; and
(ii) a change in pore-size distribution due to swelling and dispersion of clay particles.

The latter could result from a higher percentage of adsorbed sodium ions combined with the presence of HS originating from the wastewater effluent (Tarchitzky et al., 1999).

Several processes can cause the reduction in hydraulic conductivity – physical, biological, or chemical. Among the physical mechanisms, Vandevivere and Baveye (1992) include the release of entrapped air bubbles, filtration activity of solid particles suspended

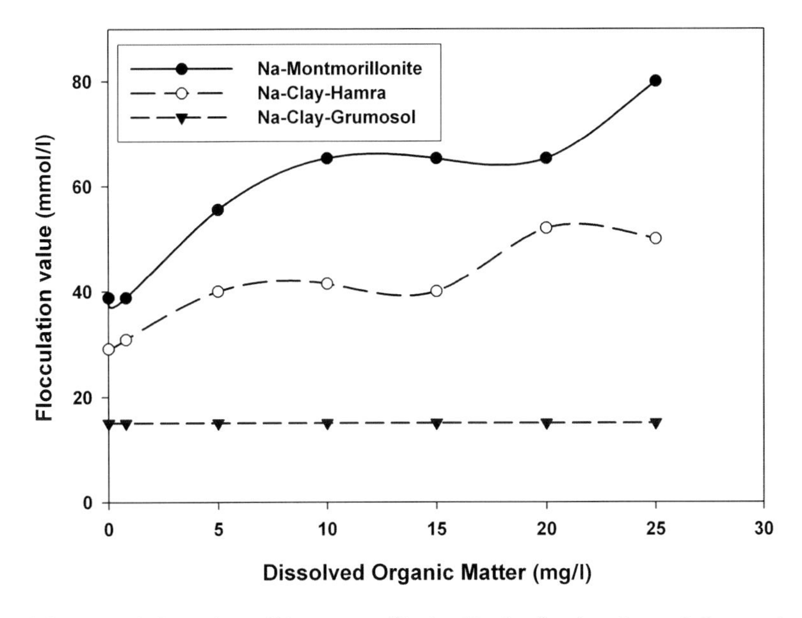

Figure 13.3 Flocculation values of Na-montmorillonite, Na-clay fraction of a sandy loam and a clayey soil as a function of dissolved organic matter (DOM) concentration (Tarchitzky and Chen, personal communication). (The values for Ca-montmorillonite did not change with the addition of the organic molecules.)

in the percolating liquid, and a progressive disintegration of soil aggregates. Clogging of the soil surface layer following TWW-irrigation has also been reported to result from the accumulation of suspended solids (de Vries, 1972; Rice, 1974; Metzger et al., 1983). According to Vinten et al. (1983), at least with respect to sandy and sandy loam soils, the coarse fraction of the effluent is responsible for most of the decrease in hydraulic conductivity. Balks et al. (1997) presented a microbial response and the formation of a semicontinuous film on the soil surface consisting of polysaccharide slime as the main factor reducing permeability in soils following the application of meat-processing effluent. The effect of TWW on decreasing soil hydraulic conductivity was connected to blockage of soil pores by microbial growth and extracellular carbohydrate production. According to some investigators, only high carbon concentrations and C/N ratios in the wastewater are able to cause a change in hydraulic conductivity (Magesan et al., (1999, 2002)).

Levy et al. (1999) concluded that in TWW with a high OM load, the OM fraction determines the hydraulic conductivity of the soils, whereas in low-OM-load TWW, it is the electrolyte concentration and composition in the water that appear to determine the hazard to the soil's hydraulic properties. In contrast, in a soil with low clay content and no carbonate, elevated SOM content from TWW-irrigation did not prevent a reduction in aggregate stability, which resulted in lower infiltration rates compared to a FW-irrigated plot (Lado et al., 2005). However, Agassi et al. (2003), in laboratory and field studies, did not find an increased influence of raindrop impact on aggregate disintegration, sealing processes, runoff or soil erosion in TWW-irrigated soils in comparison to FW-irrigated ones.

Dissolved organic carbon adsorption to, or coating of, soil particles can change their surface-area properties and consequently, their behavior. Farmers utilizing TWW in northern Israel reported a unique type of water distribution in drip-irrigated soils, consisting of:

(i) limited wetted area on the soil surface; and
(ii) small saturated areas around and below the dripper in TWW-irrigated soil, as opposed to the even, onion-like wet profile formed under FW-irrigation.

Following these observations in the field, it was hypothesized that TWW-irrigation introduces water-repellent organic constituents into the soil (Chen et al., 2003; Tarchitzky et al., 2007). Tests characterizing the water distribution showed the diameter of the saturated area on the soil surface and its water content (at a depth of 0–10 cm) to be smaller with TWW- than with FW-irrigation. Treated wastewater accumulated on the soil surface in small lenses and then flowed rapidly into the ground. The repellency of soils irrigated with FW and TWW was measured using the water-drop penetration time test, among others. Soils irrigated with FW were wettable, whereas those irrigated with TWW exhibited water-repellency.

Wallach et al. (2005) reported on water-repellency in the upper layer of a citrus orchard, and attributed this phenomenon to DOM added to the soil during TWW-irrigation. However, the very large content of OM in the surface layer of both the FW- and TWW-irrigated soils (about 6%) and the very small reported bulk density values (0.6 g/cm^3) lead us to think that the water-repellency should have been attributed to the litter layer under the trees. This observation is strengthened by the strong hydrophobicity reported in that paper for both FW- and TWW-irrigated orchard plots.

Fourier transform IR spectroscopy and ^{13}C-NMR analyses of organic components extracted from the soils using organic solvents indicated differences in composition only at a depth of 0–2 cm (Chen et al., 2003; Lerner, 2003; Lerner et al., 2003; Tarchitzky et al., 2007). It was also shown by Chen's group (Tarchitzky et al., 2007) that aliphatic components originating from the TWW are the reason for the induction of soil hydrophobicity in TWW-irrigated soils. Soil-repellency could be eliminated by solvent extraction of aliphatic carbon compounds. These were identified by FTIR, and a prominent ^{13}C-NMR peak at 32 ppm (a definitive peak of aliphatic carbon atoms).

13.6 Concluding remarks

As shown in this chapter, one of the major concerns of the use of TWW is its OM load. Wastewater originating from municipal, industrial, and agricultural sources contains varying concentrations of OM, much of which is removed during sewage water treatment. However, there is always some residual OM that is transferred to soils during TWW-irrigation. Little information has been reported regarding OM characteristics in TWW obtained from different levels of treatment. A classification and understanding of the OM based on particle size and molecular size as well as its chemical characteristics would greatly assist in the corroboration of SOM changes as a consequence of the incorporation of TWW-originated OM added to soils during TWW-irrigation.

Soils in both natural and agricultural ecosystems are exposed to mixtures of freshly decomposing OM, polysaccharides, and HS. The use of TWW can result in changes in the SOM content or in the ratios between its different components and their properties. Very little information is available on the changes in SOM with long-term use of TWW. In general, the input of particulate organic carbon, DOC, and/or colloidal OM will result in an increase or decrease in the SOM content according to the specific soil conditions and microbial activity. Further research should be conducted to assess the resultant quantitative and qualitative changes in SOM.

A result of SOM transformations along with different chemical conditions is a change in the physicochemical properties of the soil. Among the major effects of irrigation with OM-rich TWW (obtained following a low level of treatment) are:

(i) enhanced clay dispersion;
(ii) degradation of soil structure;
(iii) reduced hydraulic conductivity; and
(iv) hydrophobic (water-repellent) behavior of the soil.

The interactions between soil conditions and TWW quality, which determine the intensity of these phenomena, need further elaboration.

Acknowledgements

The authors wish to thank the DFG for their financial support of the project-Irrigation with municipal effluents: effects on physical soil properties, contaminant transport, soil carbon dynamics, soil microbial activity and crop quality; the Chief Scientist, The Israel Ministry of Agriculture and Rural Development, Wastewater Irrigation Projects; and the Glowa project, the BMBF, Germany.

References

Adani, F. and Tambone, F. (2005) Long-term effect of sewage sludge application on soil humic acids. *Chemosphere* **60**, 1214–1221.

Agassi, M., Tarchitzky, J., Keren, R., Chen, Y., Goldstein, D. and Fizik, E. (2003) Effects of prolonged irrigation with treated municipal effluent on runoff rate. *Journal of Environmental Quality* **32**, 1053–1057.

Alvarez-Bernal, D., Contreras-Ramos, S.M., Trujillo-Tapia, N., Olalde-Portugal, V., Frias-Hernández, J.T. and Dendooven, L. (2006) Effects of tanneries wastewater on chemical and biological soil characteristics. *Applied Soil Ecology* **33**, 269–277.

Amy, G.L., Bryant, C.W. and Belyani, M. (1987) Molecular weight distributions of soluble organic matter in various secondary and tertiary effluents. *Water Science and Technology* **19**, 529–538.

An, P.L., Hua, J.M., Franz, M., Winter, J. and Gallert, C. (2004) Changes of chemical and biological parameters in soils caused by trickling sewage. *Acta Hydrochimica et Hydrobiologica* **32**, 286–303.

Asano, T. (1998) Wastewater reclamation and reuse. *Water Quality Management Library*, Vol. 10 CRC Press, Boca Raton, FL.

Asano T., Burton, F.L., Leverenz, H.L., Tsuchihashi, R. and Tchobanoglous, G. (2007) *Water Reuse: Issues, Technologies and Applications*. Metcalf & Eddy|AECOM, McGraw Hill Inc., New York, NY.

Balks, M.R., McLay, C.D.A. and Harfoot, C.G. (1997) Determination of the progression in soil microbiological response, changes in soil permeability, following application of meat processing effluent to soil. *Applied Soil Ecology* **6**, 109–116.

Boyd, S.A., Sommers, L.E. and Nelson, D.W. (1980) Changes in the humic acid fraction of soil resulting from sludge application. *Soil Science Society of America Journal* **44**, 1179–1186.

Chefetz, B., Tarchitzky, J., Benny, N., Hatcher, P., Bortiatynski, J. and Chen, Y. (1998) Characterization and properties of humic substances originating from an activated sludge wastewater treatment plant. In: *Humic Substances: Structure, Properties and Uses* (eds. G. Davies and E. Gabbour), pp. 69–78. The Royal Society of Chemistry, Cambridge, UK.

Chen, Y., Lerner, O. and Tarchitzky, J. (2003) Hydraulic conductivity and soil hydrophobicity: effect of irrigation with reclaimed wastewater. In: *9th Nordic IHSS Symposium on Abundance and Functions of Natural Organic Matter Species in Soil and Water* (ed. U. Lundstrom), p. 19. Mid-Sweden University, Sundsvall, Sweden.

Chen, Y. and Tarchitzky, J. (2008). Organo-mineral complexes and their effects on the physico-chemical properties of soils. In: *Carbon Stabilization by Clays in the Environment*, CMS Workshop Lectures, Vol. 16 (eds D.A. Laird and J. Cervini-Silva), pp. 32–49. The Clay Minerals Society, Chantilly, VA.

Criquet, S., Braud, A. and Nèble, S. (2007) Short-term effects of sewage sludge application on phosphatase activities and available P fractions in Mediterranean soils. *Soil Biology and Biochemistry* **39**, 921–929.

De Nobili, M., Contin, M., Mondini, C. and Brookes, P.C. (2001) Soil microbiological biomass is triggered into activity by trace amounts of substrate. *Soil Biology and Biochemistry* **33**, 1163–1170.

Dignac, M.F., Ginestet, P., Rybacki, D., Bruchet, A., Urbain, V. and Scribe, P. (2000a) Fate of wastewater organic pollution during activated sludge treatment: natural of residual organic matter. *Water Research* **34**, 4185–4194.

Dignac, M.F., Derenne, S., Ginestet, P., Bruchet, A., Knicker, H. and Largeau, C. (2000b) Determination of structure and origin of refractory organic matter in bio-epurated wastewater via spectroscopic methods. Comparison of conventional and ozonation treatments. *Environmental Science and Technology* **34**, 3389–3394.

EPA (1992) *Guidelines for Water Reuse*. US Environmental Protection Agency, Washington DC, EPA/625/R-92/004.

Feigin, A., Ravina, I. and Shalhevet, J. (1991) Irrigation with sewage effluent. *Advanced Series in Agricultural Science* Vol. 17 Springer-Verlag, Berlin.

Filip, Z., Kanazawa, S. and Berthelin, J. (1999) Characterization of effects of a long-term wastewater irrigation on soil quality by microbiological and biochemical parameters. *Journal of Plant Nutrition and Soil Science* **162**, 409–413.

Fine, P., Hass, A., Prost, R. and Atzmon, N. (2002) Organic carbon leaching from effluent irrigated lysimeters as affected by residence time. *Soil Science Society of America Journal* **66**, 1531–1539.

Friedel, J.K., Langer, T., Siebe, C. and Stahr, K. (2000) Effects of long-term waste water irrigation on soil organic matter, soil microbial biomass and its activities in central Mexico. *Biology and Fertility of Soils* **31**, 414–421.

Fujita, Y., Ding, W. and Reinhard, M. (1996) Identification of wastewater dissolved organic carbon characteristics in reclaimed wastewater and recharged groundwater. *Water Environment Research* **68**, 867–876.

Gelsomino, A., Badalucco, L., Ambrosoli, R., Crecchio, C., Puglisi, E. and Meli, S.M. (2006) Changes in chemical and biological soil properties as induced by anthropogenic disturbance: a case study of an agricultural soil under recurrent flooding by wastewaters. *Soil Biology and Biochemistry* **38**, 2069–2080.

Gloaguen, T.V., Forti, M.C., Lucas, Y., Montes, C.R., Goncalves, R.A.B., Herpin, U., and Melfi, A.J. (2007) Soil solution chemistry of a Brazilian Oxisol irrigated with treated sewage effluent. *Agricultural Water Management* **88**, 119–131.

Hamer, U. and B. Marschner. (2005) Priming effects in different soil types induced by fructose, alanine, oxalic acid and catechol additions. *Soil Biology and Biochemistry* **37**, 445–454.

Her, N., Amy, G., McKnight, D., Sohn, J. and Yoon, Y. (2003) Characterization of DOM as a function of MW by fluorescence EEM and HPLC-SEC using UVA, DOC and fluorescence detection. *Water Research* **37**, 4295–4303.

Icekson-Tal, N., Michail, M., Kraitzer, T., Levanon, O., Sherer, D. and Shoham, G. (2007) *Groundwater Recharge with Municipal Effluent. Recharge Basins Soreq 1, Soreq 2, Yavne 1, Yavne 2, Yavne 3 and Yavne 4. Dan Reclamation Project*. Mekorot Water Company Ltd. Central District, Dan Region Unit, Israel.

Imai, A., Fukushima, T., Matsushige, K., Kim, Y. and Choi, K. (2002) Characterization of dissolved organic matter in effluents from wastewater treatment plants. *Water Research* **36**, 859–870.

Inbar, Y. (1989) *Formation of Humic Substances During The Composting of Agricultural Wastes and Characterization of their Physico-Chemical Properties*. Ph.D. thesis. Faculty of Agriculture, Hebrew University of Jerusalem, Rehovot, Israel.

Jüeschke, E., Marschner, B., Tarchitzky, J. and Chen, Y. (2004) Effects of irrigation with wastewater effluents on chemical and biological properties of organic soil components. In: *12th International Meeting of IHSS* (eds L. Martin-Neto, D. Milori and W.T.L. Da Silva). Embrapa Instrumentacao Agropecuaria, Sao Pedro, Brazil.

Jüeschke, E., Marschner, B., Tarchitzky, J. and Chen, Y. (2008) Effects of treated wastewater irrigation on the dissolved on the dissolved and soil organic carbon in Israeli soils. *Water Science and Technology* **57** (5), 727–733.

Kaushik, A., Nisha, R., Jagjeeta, K. and Kaushik, C.P. (2005) Impact of long term irrigation of a sodic soil with distillery effluent in combination with bioamendments. *Bioresources Technology* **96**, 1860–1866.

Komonova, M.M. and Alexandrova, I.V. (1973) Formation of humic acids during plant residue humification and their nature. *Geoderma* **9**, 157–164.

Kuzyakov, Y., Friedel, J.K. and Stahr, K. (2000) Review of mechanisms and quantification of priming effects. *Soil Biology and Biochemistry* **32**, 1485–1498.

Lado, M., Ben-Hur, M. and Assouline, S. (2005) Effects of effluent irrigation on seal formation, infiltration, and soil loss during rainfall. *Soil Science Society of America Journal* **69**, 1432–1439.

Leenheer, J.A. (1981) Comprehensive approach to preparative isolation and fractionation of dissolved organic carbon from natural waters and wastewaters. *Environmental Science and Technology* **15**, 578–587.

Lerner, O. (2003) *Hydraulic Conductivity and Soil Hydrophobicity: Effect of Irrigation with Reclaimed Wastewater*. (M.Sc. thesis). Ben Gurion University (in Hebrew with English abstract).

Lerner, O., Chen, Y., Shani, U., Tarchitzky, J., Arye, G. and Lowengart-Aycicegi, A. (2003) Changes in soil water distribution in soils irrigated with effluents. *Water Irrigation* **437**, 22–28 (in Hebrew with English abstract).

Levine, A.D., Tchobanoglous, G. and Asano, T. (1991) Size distributions of particulate contaminants in wastewater and their impact on treatability. *Water Research* **8**, 911–922.

Levy, G.J., Rosenthal, A., Tarchitzky, J., Shainberg, I. and Chen, Y. (1999) Soil hydraulic conductivity changes caused by irrigation with reclaimed wastewater. *Journal of Environmental Quality* **28**, 1658–1664.

Lingbo, L., Song, Y., Congbi, H. and Guangbo, S. (2005) Comprehensive characterization of oil refinery effluent-derived humic substances using various spectroscopic approaches. *Chemosphere* **60**, 467–476.

Ma, H., Ellen, H.E. and Yin, Y. (2001) Characterization of isolated fractions of dissolved organic matter from natural waters and a wastewater effluent. *Water Research* **35**, 985–996.

Magesan, G.N., Williamson, J.C., Sparling, G.P., Schipper, L.A. and Lloud-Jones, A.Rh. (1999) Hydraulic conductivity in soils irrigated with wastewaters of differing strengths: field and laboratory studies. *Australian Journal of Soil Research* **37**, 391–402.

Magesan, G.N., Williamson, J.C., Yeates, G.W. and Lloud-Jones, A.Rh. (2000) Wastewater C:N ratio effects on soil hydraulic conductivity and potential mechanisms for recovery. *Bioresources Technology* **71**, 21–27.

Manka, J. and Rebhun, M. (1982) Organic groups and molecular weight distribution in tertiary effluents and renovated waters. *Water Research* **16**, 399–403.

Manka, J., Rebhun, M., Mandelbaum, A. and Bortiger, A. (1974) Characterization of organics in secondary effluents. *Environmental Science and Technology* **5**, 606–609.

Meli, S., Porto, M., Belligno, A., Bufo, S.A., Mazzatura, A. and Scoppa, A. (2002) Influence of irrigation with lagooned urban wastewater on chemical and microbiological soil parameters in a citrus orchard under Mediterranean condition. *Science of the Total Environment* **285**, 69–77.

Metzger, L., Yaron, B. and Mingelgrin, U. (1983) Soil hydraulic conductivity as affected by physical and chemical properties of effluents. *Agronomie (Paris)* **3**, 771–778.

Muneer, M. and Oades, J.M. (1989) The role of Ca-organic interactions in soil aggregate stability. III. Mechanisms and models. *Australian Journal of Soil Research* **27**, 411–423.

Namour, Ph. and Müller, M.C. (1998) Fractionation of organic matter from wastewater treatment plants before and after a 21-day biodegradability test: a physical-chemical method for measurement of the refractory part of effluents. *Water Research* **32**, 2224–2231.

Nguyen, L.M. (2000) Organic matter composition, microbial biomass and microbial activity in gravel-bed constructed wetlands treating farm dairy wastewaters. *Ecological Engineering* **16**, 199–221.

Oved, T., Shaviv, A., Goldrath, T., Mandelbaum, R.T. and Minz, D. (2001) Influence of effluent irrigation on community composition and function of ammonia-oxidizing bacteria in soil. *Applied Environmental Microbiology* **67**, 3426–3433.

Painter, H.A. (1973) Organic compounds in solution in sewage effluents. *Chemistry and Industry* **17**, 818–822.

Peschel, G., and Wildt, T. (1988) Humic substances of natural and anthropogeneous origin. *Water Research* **22**, 105–108.

Piccolo, A., Zaccheo, P. and Genevini, P.G. (1992) Chemical characterization of humic substances extracted from organic-waste-amended soils. *Bioresources Technology* **40**, 275–282.

Ramirez-Fuentes, E., Lucho-Constantino, C., Escamilla-Silva, E. and Dendooven, L. (2002) Characteristics, and carbon and nitrogen dynamics in soil irrigated with wastewater for different lengths of time. *Bioresources Technology* **85**, 179–187.

Raunkjær, K., Hvitved-Jacobsen, T. and Nielsen, P.H. (1994) Measurement of pools of protein, carbohydrate and lipid in domestic wastewater. *Water Research* **28**, 251–262.

Rebhun, M. and Manka, J. (1971) Classification of organics in secondary effluents. *Environmental Science and Technology* **5**, 606–609.

Rice, R.C. (1974) Soil clogging during infiltration of secondary effluent. *Journal of the Water Pollution Control Federation* **46**, 708–716.

Rickert, D.A. and Hunter, J.V. (1971) General nature of soluble and particulate organics in sewage and secondary effluent. *Water Research* **5**, 421–436.

Saadi, I., Borisover., M., Armon, R. and Laor, Y. (2006) Monitoring of effluent DOM biodegradation using fluorescence, UV and DOC measurements. *Chemosphere* **63**, 530–539.

Sachedev, R., Ferris, J.J. and Clesceri, N.L. (1976) Apparent molecular weights of organics in secondary effluents. *Journal of the Water Pollution Control Federation* **48**, 570–578.

Shon, H.K., Vigneswaran, S. and Snyder, S.A. (2006) Effluent organic matter (EfOM) in wastewater: constituents, effects, and treatment. *Critical Reviews in Environmental Science and Technology* **36**, 327–374.

Soler Rovira, P.A., Brunetti, G., Polo, A. and Senesi, N. (2002) Comparative chemical and spectroscopic charcterization of humic acids from sewage sludges and sludge-amended soils. *Soil Science* **167**, 235–245.

Sophonsiri, Ch. and Morgenroth, E. (2004) Chemical composition associated with different particle size fractions in municipal, industrial, and agricultural wastewaters. *Chemosphere* **55**, 671–703.

Stevenson, F.J. (1994) *Humus Chemistry*. John Wiley & Sons, New York.

Swift, R.S. (1996) Organic matter characterization. In: *Methods of Soil Analysis. Part 3*. Chemical Methods. SSSA Book Series No. 5 (ed. D.L. Sparks), pp. 1011–1069. Soil Science Society of America, Madison, WI.

Tarchitzky, J. and Chen, Y. (2002a) Rheology of montmorillonite suspensions in the presence of humic substances. *Soil Science Society of America Journal* **66**, 406–412.

Tarchitzky, J. and Chen, Y. (2002b) Polysaccharides and pH effects on sodium-montmorillonite: flocculation, dispersion and rheological properties. *Soil Science* **167**, 791–801.

Tarchitzky, J., Chen, Y. and Banin, A. (1993) Humic substances and pH effects on sodium- and calcium-montmorillonite flocculation and dispersion. *Soil Science Society of America Journal* **57**, 367–372.

Tarchitzky, J., Golobati, Y., Keren, R. and Chen, Y. (1999) Wastewater effects on montmorillonite suspensions and hydraulic properties of sandy soils. *Soil Science Society of America Journal* **63**, 554–560.

Tarchitzky, J., Lerner, O., Shani, U., Arye, G., Lowengart-Aycicegi, A., Brenner, A. and Chen, Y. (2007) Water distribution pattern in treated wastewater irrigated soils: hydrophobicity effect. *European Journal of Soil Science* **58**, 573–588.

Vandevivere, P. and Baveye, P. (1992) Saturated hydraulic conductivity reduction caused by aerobic bacteria in sand columns. *Soil Science Society of America Journal* **56**, 1–13.

Vinten, A.J.A., Milgengrin, U. and Yaron, B. (1983) The effect of suspended solids in wastewater and soil hydraulic conductivity: II. Vertical distribution of suspended solids. *Soil Science Society of America Journal* **47**, 408–412.

de Vries, J. (1972) Soil filtration of wastewater effluent and the mechanism of soil clogging. *Journal of the Water Pollution Control Federation* **44**, 565–573.

Wallach, R., Ben-Arie, O. and Graber, E.R. (2005) Soil water repellency induced by long-term irrigation with treated sewage effluent. *Journal of Environmental Quality* **34**, 1910–1920.

Zhigunov, A.V. and Simakov, V.N. (1977) Composition and properties of humic acids separated from decomposing plant residues. *Soviet Soil Science* **9**, 687–693.

Chapter 14
Analysis of transport of mixed Na/Ca salts in a three-dimensional heterogeneous variably saturated soil

David Russo

14.1 Introduction

A shortage of rain and water resources exits in semiarid and arid zones, necessitating the use of treated wastewater (TWW) for irrigation, preserving scarce freshwater resources to fulfill the increasing demand of the urban sector. Inasmuch as TWW may contain relatively large quantities of soluble salts, predominantly Ca and Na ions, its use for irrigation may accelerate the contamination process of aquifers underneath the irrigated regions. Quantitative field-scale descriptions of chemical transport in the vadose (unsaturated) zone, therefore, are essential for improving the basic understanding of the transport of TWW in near-surface geological environments, and for providing predictive tools that, in turn, will be used to predict the future spread of TWW in these environments and to assess reliably the threat posed by the use of TWW for irrigation to the underlying water supplies.

The traditional approach for modeling chemical transport processes in the vadose zone has been to model water flow and solute transport by using soil macroscopic physical and chemical properties, which vary in a deterministic manner, obey physical and chemical laws, and are expressed in the form of partial differential equations. Using a deterministic approach, transport of TWW under cropped conditions was simulated in the past by Russo (1995a). Considering one-dimensional vertical flow, neglecting the spatial variability in the soil properties and the changes in the soil pore-size distribution due to soil solution–soil matrix interactions, two subsystems were considered in the simulations:

(i) the N-C-O subsystem describing the cycling of nitrogen and carbon compounds in the unsaturated zone; and
(ii) the major ions subsystem (Cl, Ca, Mg, K, Na) contributing to salinity.

Treated Wastewater in Agriculture, First Edition, edited by Guy J. Levy, Pinchas Fine and Asher Bar-Tal

The two subsystems were considered as loosely coupled, with limited interactions primarily through ammonia sorption, which competes with the major cations. As the focus of the simulations was on major ion and redox chemistry, the fate of other contaminants such as phosphate, boron, and trace metals was not considered in the simulations. The main findings of the study of Russo (1995a) suggest that under conditions of irrigation with TWW:

(i) the attenuation of the N-compounds (NH_3 and NO_3) in the unsaturated zone may be significant, particularly when the amounts of the applied TWW of a given composition are controlled by considering the consumption of N by the relevant crop; and
(ii) as in irrigation with freshwater, the unsaturated zone may provide a very limited capacity for the long-term attenuation of salinity loading to the watertable.

In reality, however, the relevant soil properties vary in space (e.g. Nielsen et al., 1973; Russo and Bresler, 1981; Jones and Wagenet, 1984; Russo and Bouton, 1992; Russo et al., 1997, among others). This spatial heterogeneity is generally irregular and occurs on a scale that is not captured by laboratory samples; it has a distinct effect on water flow and solute transport, as has been observed in field experiments (e.g. Schulin et al., 1987; Butters et al., 1989; Ellsworth et al., 1991; Roth et al., 1991; Flury et al., 1994; Forrer et al., 1999) and demonstrated by simulation (e.g. Russo, 1991; Russo et al., 1994, 1998, 2001; Tseng and Jury, 1994; Harter and Yeh, 1996; Roth and Hammel, 1996; Foussereau et al., 2001).

The spatial variability in the soil properties and the paucity of measurements that are available for site characterization preclude the deterministic description of the heterogeneous soil; this has led to a concerted effort to develop theories of flow and transport based on stochastic concepts. The rationale for using stochastic concepts is that they enable soil heterogeneity and data uncertainty to be treated quantitatively. Published stochastic treatments of vadose zone solute transport focused on either analytical (e.g. Dagan and Bresler, 1979; Jury, 1982; Destouni and Cvetkovic, 1989, 1991; Russo, 1993a,b, 1995b, 1998) or numerical (e.g. Russo, 1991; Russo et al., 1994, 1998, 2001, 2006; Tseng and Jury, 1994; Harter and Yeh, 1996; Roth and Hammel, 1996; Foussereau et al., 2001) analyses considering a single non-interacting solute. However, very little has been done to develop a suitable framework for the analysis of field-scale transport of mixed-ion salts, which contains ions that interact with the soil matrix (i.e. Ca and Na), and, consequently, may change the soil pore-size distribution (e.g. Russo, 1989a,b; Russo, 2004).

Changes in the soil pore-size distribution due to soil solution–soil matrix interactions may considerably affect the soil hydraulic conductivity and to a lesser extent may affect the soil water retention (see, e.g., Bresler et al. (1982), Shainberg (1984) and Russo (2005) for comprehensive reviews). Because the magnitude of the soil solution–soil matrix interactions depends on the concentration and composition of the soil solution, and on soil water content (Russo and Bresler, 1977a,b), the flow is coupled to the transport through the dependence of the hydraulic conductivity and water retention on solute concentrations; consequently, soil solution–soil matrix interactions must be considered in the analysis of field-scale transport of mixed-ion salts such as TWW. The analysis of field-scale transport of mixed-ion salts is further complicated by the fact that in unsaturated flow, the relevant flow parameters – the hydraulic conductivity and the water capacity, which, in turn, depend on few soil parameters – (see equation 14.7 below) depend also on the flow-controlled attributes (pressure head, water content) in a highly non-linear fashion. Consequently, in a spatially heterogeneous, variably saturated soil, the evaluation of the effects on water flow and chemical transport of the soil solution–soil matrix interactions is extremely complex.

A fundamental question, therefore, is how to develop mathematical models capable for predicting water flow and transport of mixed-ion salts in a realistic, spatially heterogeneous, variably saturated soil, for the case in which the flow is coupled to the transport through soil solution–soil matrix interactions. In the following, a few advances in this area will be presented and analyzed. The plan of the present chapter is as follows: the conceptual framework for modeling transport of interacting solutes in spatially heterogeneous, variably saturated soils is presented in 14.2; results of flow and transport simulations designed to assess the long-term effects of the soil solution–soil matrix interactions on the soils capability to transfer water and solutes under realistic conditions are presented and discussed in 14.3; 14.4 concludes the chapter.

14.2 Modeling of transport of mixed Na/Ca salts in spatially heterogeneous, variably saturated soils

Generally, the transport of TWW in the vadose zone should be treated as a multicomponent reactive transport of solutes (see, e.g., Russo, 1995a). In order to simplify matters and as the focus of this chapter is on the soil solution–soil matrix interactions, the simple but practical soil–water system containing only the major ions in the irrigation water (Na, Ca and Cl) is considered here.

14.2.1 The coupling-interaction parameters

The analysis of the transport of the interacting mixed Na/Ca salts in a realistic, three-dimensional, spatially heterogeneous, variably saturated flow system requires quantification of the effect of the soil solution concentration and composition on the spatial variability of soil properties pertinent to flow and transport. These properties (to be referred to hereafter as the coupling-interaction parameters) include the soil hydraulic conductivity, K, the soil water content, θ, the retardation factors for Na and Ca, R_f^{Na} and R_f^{Ca}, respectively, and the elution factor for Cl, E_f, all as functions of the pressure head and the concentration and the composition of the soil solution.

The modeling is based on a theoretical approach (Russo and Bresler, 1977b; Bresler, 1978; Russo, 1988), originally developed to evaluate the coupling-interaction parameters on the macroscopic (Darcy) scale, pertinent to flow and transport processes on the laboratory scale. The theoretical approach combines the mixed-ion diffuse double layer theory and experimental data of the structure and organization of the clay particles (relevant to the pore scale) with a conceptual model of the porous medium and experimental data of the soil pore-size distribution pertinent to the Darcy (laboratory) scale. To evaluate coupling-interacting parameters on a larger scale, more amenable to flow and transport processes on a realistic field-scale, the theoretical approach is coupled with measured spatial distributions of the relevant soil properties, which include the hydraulic conductivity and the water retention functions at an "inert" reference state, the soil cation exchange capacity (CEC), the soil-specific surface area, A_s, and the soil bulk density, ρ_b. The resultant spatial distributions of the coupling-interaction parameters, in turn, can be quantified in terms of their first two statistical moments (i.e. mean value and variance), expressed as functions of the soil solution concentration and composition and the pressure head (see, e.g., Russo, 1989a).

14.2.2 The governing partial differential equations

Following Russo et al. (2004), the transport problem is analyzed here by means of physically based flow and transport models and a stochastic presentation of the soil properties that affect water flow and solute transport. It is assumed that the water flow is described locally by the Richards equation, the physical parameters of which are visualized as realizations of stationary random space functions (RSFs). It is further assumed that the transport of each of the ions of the mixed Na/Ca salt is described locally by the classical, one-region, convection-dispersion equation (CDE). When the ions in the soil solution interact with the soil matrix, the properties of the spatially heterogeneous soil that affect the water flow and the solute transport are expressed as functions of the spatial coordinate vector, $\underline{x}$, the pressure head, ψ, the sum of the molar concentration of the cations, $C_0 = c_{Na} + c_{Ca}$ [mol/ℓ], and the ratio between the molar concentration of mono- and divalent cations, $R = c_{Na}/c_{Ca}$, in the soil solution.

In a Cartesian coordinate system (x_1, x_2, x_3), where x_1 [L] is directed vertically downwards, assuming local isotropy, considering water uptake by the plant roots and neglecting the effect of osmotic gradients on the flow, the "mixed" form of the Richards equation, governing saturated-unsaturated flow in a non-rigid, three-dimensional, spatially heterogeneous soil is:

$$\frac{\partial(\theta^r\theta)}{\partial t} = \sum_{i=1}^{3}\frac{\partial}{\partial x_i}\left[K^r K\frac{\partial\psi}{\partial x_i}\right] - \frac{\partial K^r K}{\partial x_1} - S_w \tag{14.1}$$

In (14.1) t [T] is time; $\psi = \psi(\underline{x}, t)$ [L] is the pressure head; $\theta = \theta(\psi, \underline{x})$ and the scalar $K = K(\psi, \underline{x})$ [L/T] are the volumetric water content and the hydraulic conductivity, respectively, of the soil in a rigid "stable" reference state in which the changes in the soil pore-size distribution (PSD) due to soil solution–soil matrix interactions are negligibly small (i.e., when $R \to 0$, and $C_0 \to \infty$); $\theta^r = \theta^r(C_0, R, \psi, \underline{x})$ and $K^r = K^r(C_0, R, \psi, \underline{x})$ are relative water content and relative conductivity, respectively, expressing the effect of the soil solution–soil matrix interactions on the soil PSD; and $S_w = S_w(\underline{x}, t)$ [T^{-1}] is a sink term representing water uptake by plant roots, given (Nimah and Hanks, 1973a,b) by:

$$S_w(\underline{x}, t) = -R_e(\underline{x}, t)K[\psi(\underline{x}, t)][\Psi_r(t) - \psi(\underline{x}, t) - \pi(\underline{x}, t)] \tag{14.2}$$

where R_e [L^{-2}] is the root effectiveness function, Ψ_r [L] is the total pressure head at the root-soil interface, and π [L] is the osmotic pressure head of the soil solution.

Similarly, neglecting ion precipitation or dissolution, the equation governing the transport of the mixed Na/Ca salt is:

$$\frac{\partial(\theta^*_m c_m)}{\partial t} = \sum_{i=1}^{3}\sum_{j=1}^{3}\frac{\partial}{\partial x_i}\left\{\theta D_{ij}\frac{\partial c_m}{\partial x_j}\right\} - \sum_{i=1}^{3}\frac{\partial(u_i\theta c_m)}{\partial x_i} \tag{14.3}$$

where m = 1 to 3 indicates the chloride, sodium, and calcium ions, respectively; c_m [mol/ℓ] is the resident solute concentration of the m-th ion, expressed as mass per unit volume of soil solution; θ^*_m is an "effective" volumetric water content, given (Russo, 1988) by

$$\theta^*_m(\psi, \underline{x}, R, C_0, c_m) = \theta(\psi, \underline{x}, R, C_0) - \theta_{ex}(\psi, \underline{x}, R, C_0, c_m) \tag{14.4a}$$

for the chloride (m = 1), and by

$$\theta^*_m(\psi, \underline{x}, R, C_0, c_m) = \theta(\psi, \underline{x}, R, C_0) + \theta_{adm}(\psi, \underline{x}, R, C_0, c_m) \tag{14.4b}$$

for the sodium (m = 2) or for the calcium (m = 3) cations, respectively; θ_{ex} and θ_{adm} are the exclusion and the adsorption volume fractions, respectively, given (Russo, 1988) by

$$\theta_{ex}(\psi, \underline{x}, R, C_0, c_m) = \Gamma^-(\psi, \underline{x}, R, C_0)A_{ex}(\underline{x})\rho_b(\underline{x})/c_m, \ m = 1 \tag{14.5a}$$

and

$$\theta_{adm}(\psi, \underline{x}, R, C_0, c_m) = \Gamma^+_m(\psi, \underline{x}, R, C_0)A_{ad}(\underline{x})\rho_b(\underline{x})/c_m, \ m = 2, 3 \tag{14.5b}$$

where A_{ex} [L^2/M], A_{ad} [L^2/M], and ρ_b [M/L^3] are the specific exclusion surface area, the specific adsorption surface area, and the bulk density of the soil, respectively, and Γ^-_{ex} [mol/P^2] and Γ^+_{ad} [mol/P^2] are the quantity of monovalent anionic charges repelled by, and the quantity of monovalent cationic charges adsorbed to a unit area of the solid surface, respectively; u_i (i = 1,2,3) [L/T] are the components of the pore water velocity vector; and D_{ij} (i, j = 1, 2, 3) [L/T^2] are the components of the pore-scale dispersion tensor. When molecular diffusion is small enough to be excluded from the analysis, D_{ij} is given (Bear, 1972) as:

$$D_{ij} = \lambda_T|\underline{u}|\delta_{ij} + (\lambda_L - \lambda_T)u_i u_j/|\underline{u}| \tag{14.6}$$

where λ_L and λ_T [L] are the longitudinal and the transverse components of the pore-scale dispersivity tensor; δ_{ij} is the Kronecker delta, and $|\underline{u}| = (u_1^2 + u_2^2 + u_3^2)^{1/2}$.

14.2.3 Characterization of the flow and transport parameters

The local-scale soil properties that are hypothesized to control the field-scale spread of the mixed Na/Ca salts are the soil hydraulic functions, K(ψ) and θ(ψ) at the "stable" reference state, the local soil coupling-interaction parameters K^r, θ^r, θ_{ex}, θ_{adm} (m = 2, 3), and the longitudinal, λ_L, and the transverse, λ_T, components of the pore-scale dispersivity tensor.

For a given soil material, the clay fraction of which is mostly montmorillonite, a procedure to estimate the local soil coupling-interaction parameters, all as functions of ψ, R and C_0, was outlined by Russo (1988). The inherent soil data required are the soil hydraulic functions, $K(\psi)$ and $\theta(\psi)$, at the "stable" reference state, the soil CEC, the soil-specific surface area, A_s, and the soil bulk density, ρ_b. To quantify the spatial variations in these properties, the soil is regarded as a continuum, the properties of which are continuous functions of the spatial coordinates. The elementary volume representing the relevant soil properties at a given spatial coordinate (that is, the support) is characterized by a macroscopic length-scale that is small in comparison with the characteristic length-scale of the field heterogeneities, but contains many pores. The soil properties vary in space in an irregular fashion; the characteristic length-scale of these variations is greater than the length-scale of the support, but small in comparison with the characteristic length-scale of the flow domain.

The main concern in this chapter is the effect of the spatial variations in the hydraulic properties and in the soil coupling-interaction parameters on the transport of the mixed Na/Ca salts; consequently, deterministic, constant values can be adopted for the components of the pore-scale dispersivity tensor. To simplify the analysis, it is assumed that the local K (ψ) and $\theta(\psi)$ relationships in the "stable" reference state obey the van Genuchten (1980) parametric expressions. Ignoring local hysteresis and local anisotropy, and considering the pressure head, ψ, as the dependent variable, they read:

$$\Theta(\psi,\underline{x}) = \left\{\frac{1}{1+[\beta(\underline{x})|\psi|]^{n(\underline{x})}}\right\}^{m(\underline{x})} \tag{14.7a}$$

$$K(\psi,\underline{x}) = K_s(\underline{x})\frac{\left\{1-[\beta(\underline{x})|\psi|]^{n(\underline{x})-1}\{1+[\beta(\underline{x})|\psi|]^{n(\underline{x})}\}^{-m(\underline{x})}\right\}^2}{\left\{1+[\beta(\underline{x})|\psi|]^{n(\underline{x})}\right\}^{m(\underline{x})/2}} \tag{14.7b}$$

where $\Theta = (\theta - \theta_{ir})/(\theta_s - \theta_{ir})$, is the effective water saturation; θ_s and θ_{ir} are the saturated and the irreducible water contents, respectively; K_s is the saturated conductivity; β and m are parameters related to the soil pore-size distribution, and $n = 1/(1 - m)$.

It is assumed further that each of the parameters of equation (14.7), as well as the soils CEC, A_s and ρ_b, denoted by $p(\underline{x})$, is a second-order stationary, statistically anisotropic RSF, characterized completely by a constant mean, $p(\underline{x})$, independent of the spatial position, and a two-point covariance, $C_{pp}(\underline{x}, \underline{x}')$, that, in turn, depends on the separation vector, $\underline{\xi} = \underline{x} - \underline{x}'$, and not on $\underline{x}$ and $\underline{x}'$ individually. The heterogeneous soil may be visualized as being composed of a three-dimensional structured arrangement of blocks of different soil materials that may exhibit specific sizes, but are not completely regular. The covariance, $C_{pp}(\underline{\xi})$, therefore, depends on the amount of overlap between these blocks. For example, when the overlapping blocks are allowed to vary randomly in size, the covariance is described by an exponential model, i.e.,

$$C_{pp}(\underline{\xi}) = \sigma_p^2 \exp[-\underline{\xi}'] \tag{14.8}$$

where $\underline{\xi}' = (\underline{x} - \underline{x}')/\underline{I}_p$ is the scaled separation vector, $\xi' = |\underline{\xi}'| \sigma^2_p$ and $\underline{I}_p = (I_{p1}, I_{p2}, I_{p3})$ are the variance and the correlation length-scales of $p(\underline{x})$, respectively.

14.3 Simulation of transport of mixed Na/Ca salts in spatially heterogeneous, variably saturated soils

Simulation is a powerful tool that can be regarded as a "numerical experiment", which, in turn, can provide detailed information (that is unaffected by uncertainty due to measurement errors or inadequate sampling) on the consequences of characteristics of soil, crop, weather, and the irrigation practice for the movement and spreading of water and solutes under realistic conditions. At the price of reduced generality, it circumvents most of the stringent assumptions of analytical studies and facilitates analysis of simplified yet realistic situations, at a fraction of the cost of physical experiments.

14.3.1 Results of the transport simulation of Russo et al. (2004)

Taking into account the coupling of the flow to the transport through the dependence of the hydraulic conductivity and water retention on solute concentrations, transport of mixed Na/Ca salts and a tracer solute in a realistic, three-dimensional, heterogeneous, variably saturated soil under cropped conditions was simulated by Russo et al. (2004). Assuming that locally the constitutive relationships for unsaturated flow are given by the van Genuchten, 1980van Genuchtens (1980) model equation 14.7, taking into account realistic atmospheric forcing conditions at the soil surface along with water extraction by plant roots, the simulations were performed by combining a stochastic generation method for producing realizations of the heterogeneous soil properties in sufficient resolution, with an efficient numerical method for solving the physically based, partial differential equations governing three-dimensional flow equation 14.1 and transport equation 14.3 in spatially heterogeneous, variably saturated soils.

Climatic conditions, crop (a citrus orchard with a complete cover of the soil surface), and agricultural practice typical of the central area of the coastal region of Israel, were considered in these simulations. Annual amounts of irrigation, rainfall and evapotranspiration were 700, 580 and 1005 mm/yr, respectively. The three-dimensional flow domain spanned 10 m in each of the horizontal directions and 2.5 m in the vertical direction. The flow scenario started at the beginning of the irrigation season (1 April) and proceeded for four successive years. For Cl, Na, and Ca, initial concentrations were $c_i = 10$, 2.5, and 3.75 mol/ℓ, respectively; the respective inlet concentrations during the irrigation and the rain seasons were $c_0 = 10$, 9, and 0.5 mol/ℓ, and $c_0 = 1.5$, 0.4, and 0.55 mol/ℓ, respectively. Two pulses ($t_0 = 0.05$ d) of the tracer solute (with initial concentration, $c_i = 0$, and inlet concentration, $c_0 = 10$ mol/ℓ), were applied during the second irrigation events in the first year and in the third year. For more details see Russo et al. (2004).

Cumulative distribution plots of the input soil coupling-interaction parameters, i.e. the relative conductivity, K^r, relative water content, θ^r, elution factor for Cl, $E_f = 1 - \theta_{ex}/\theta$, and the retardation factor for Na, $R_f = 1 + \theta_{ad2}/\theta$, are demonstrated in Figure 14.1 for selected values of the pressure head, ψ, the chloride concentration, $C = 10^{-3}C_0[1 + (R + 1)^{-1}]$

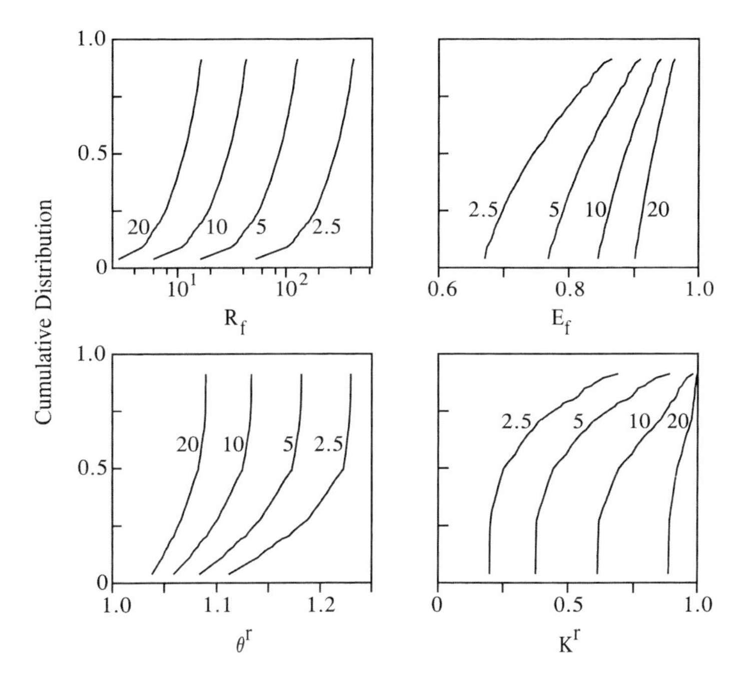

Figure 14.1 Cumulative distribution plots of relative conductivity, K_r, relative water content, θ_r, the elution factor for Cl, E_f, and the retardation factor for Na, R_f, for different chloride concentrations, C (denoted by the numbers labeling the curves, in meq/ℓ), for sodium adsorption ratio, SAR = 10, and pressure head, $\psi = -2$m. Reproduced from Russo et al., 2004, with permission.

[meq/ℓ], $SAR = c_{Na}/\sqrt{c_{Ca}}$, and the sodium adsorption ratio, $SAR = c_{Na}/\sqrt{c_{Ca}} = R = 10^{-3}C_0[1 + (R + 1)^{-1}]^{1/2}$ $[(meq/\ell)^{1/2}]$.

Simulated profiles of the horizontal averages, the mean values and the standard deviations of C and SAR are depicted in Figures 14.2 and 14.3. These figures suggest that during the rainy seasons, after the first few rain events, the displacement of the relatively high-sodium, concentrated soil solution by a relatively low-sodium diluted solution is confined to the upper part of the soil profile. After a series of rain events, the leached zone is extended to a deeper soil depth, but the decrease in the soil solution SAR is still confined to the few centimeters in the vicinity of the soil surface. This stems from the non-linear nature of the Na/Ca exchange isotherm, which increases with increasing SAR and decreasing C. The reverse tendency occurs during the irrigation season. The relatively low-sodium, diluted soil solution in the upper part of the soil profile is displaced by a relatively high-sodium concentrated solution. After the first few irrigation events, the increase in both soil solution concentration and SAR is confined to the upper part of the soil. After a series of irrigation events, however, the salinized/alkalinized zone is extended to a deeper soil depth. In other words, the characteristics of the Na/Ca exchange isotherm coupled with the seasonal variations of the inlet concentrations at the soil surface intensify the soil solution–soil matrix interactions during the rain events (soil leaching) and moderate them during the irrigation events (soil salinization).

The response of the concentration-dependent flow system (expressed in terms of profiles of the mean and the standard deviation of the log-hydraulic conductivity, logK and the longitudinal and the transverse components of the head-gradient and the velocity

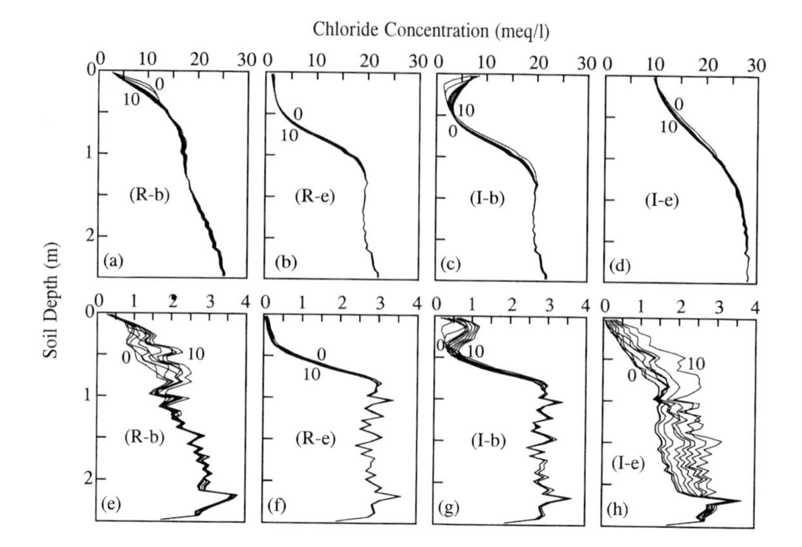

Figure 14.2 Profiles of the horizontal averages, the mean values (a, b, c, d), and the standard deviations (e, f, g, h), of the chloride concentration. The profiles in this figure represent different elapsed times (0, 1, 2, 3, 4, 6, 8, 10 d) following the cessation of a rain/irrigation event, at the beginning (a, e) and at the end (b, f) of the second rain season, and at the beginning (c, g) and at the end (d, h) of the third irrigation season. Reproduced from Russo et al., 2004, with permission.

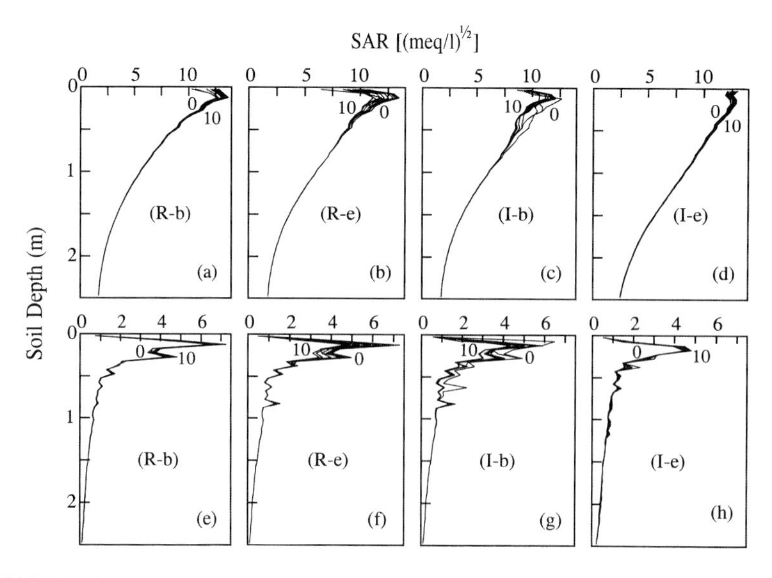

Figure 14.3 Profiles of the horizontal averages, the mean values (a, b, c, d), and the standard deviations (e, f, g, h) of the sodium adsorption ratio, SAR. The profiles in this figure represent different elapsed times (0, 1, 2, 3, 4, 6, 8, 10 d) following the cessation of a rain/irrigation event, at the beginning (a, e) and at the end (b, f) of the second rainy season and at the beginning (c, g) and at the end (d, h) of the third irrigation season. Reproduced from Russo et al., 2004, with permission.

vectors, $J_i = \delta(-\psi - x_1)/\delta x_i$, and u_i, $(i = 1, 2)$, respectively), to the seasonal changes in C and SAR is demonstrated in Figures 14.4, 14.5 and 14.6. Note that in Figures 14.5 and 14.6 only the transverse components of $\underline{J}$ and $\underline{u}$ in the direction parallel to the x_2 axis, J_2 and u_2, respectively, are shown; their counterparts, J_3 and u_3 (not shown here) generally exhibit the same pattern. The results in Figure 14.4(a, c) suggest a considerable decrease in both the mean and the standard deviation of log K during the rainy season, which, in turn, is compensated for by considerable increases in the mean value of J_1 (associated with a slight decrease in its SD), in the absolute deviation of the mean J_2 (and J_3, not shown here) from zero, and in their standard deviations (Fig. 14.5a, c). Consequently, the decrease in the mean value of u_1 and in the standard deviations of u_i $(i = 1, 2, 3)$, and the increase in the absolute deviation of the mean u_2 (and u_3, not shown here) from zero are moderate (Fig. 14.6a, c). The reverse tendency occurs during the irrigation season; the increase in the mean and the standard deviation of log K (Fig. 14.4b, d) is accompanied by a decrease in the mean value of J_1 and in the absolute deviation of the mean values of J_2 and J_3 from zero, and in their standard deviations (Fig. 14.5b, d). Consequently, the mean value of u_1 and the standard deviations of u_i, $(i = 1, 2, 3)$ increase, whereas the absolute deviation of the mean u_2 and u_3 from zero decrease during the irrigation season (Fig. 14.6b, d).

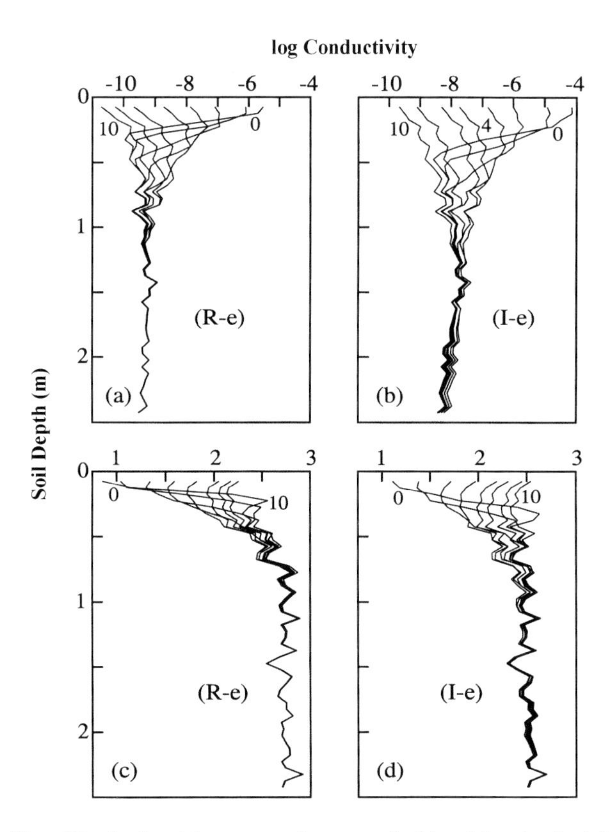

Figure 14.4 Profiles of the horizontal averages, the means (a, b) and the standard deviations (c, d) of y = log K at the end of the second rainy season (a, c), and at the end of the third irrigation season (b, d). The profiles in this figure represent different elapsed times (0, 1, 2, 3, 4, 6, 8, 10 d) following the cessation of a rain/irrigation event. Reproduced from Russo et al., 2004, with permission.

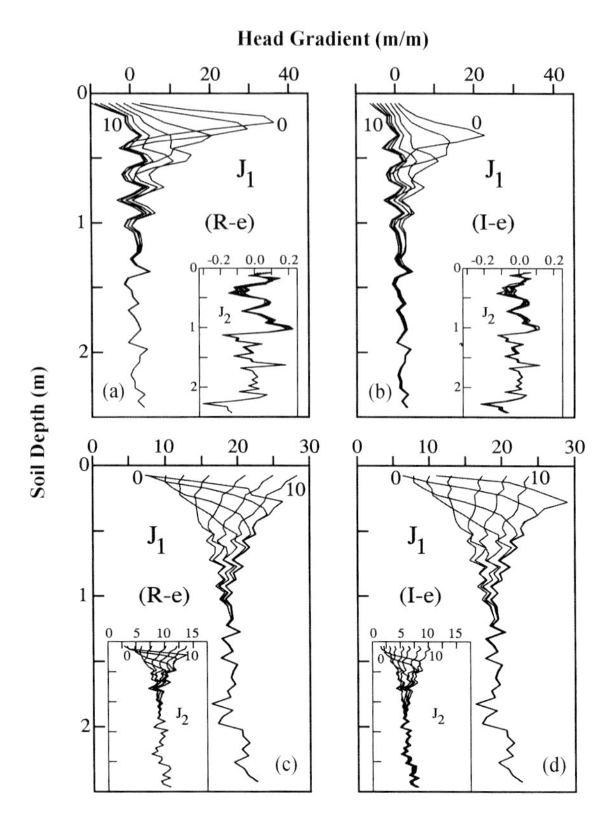

Figure 14.5 Profiles of the horizontal averages, the means (a, b) and the standard deviations (c, d) of the longitudinal and the transverse (parallel to the x_2 axis) components of the head-gradient vector, $\underline{J}$, at the end of the second rainy season (a, c) and at the end of the third irrigation season (b, d). The profiles in this figure represent different elapsed times (0, 1, 2, 3, 4, 6, 8, 10 d) following the cessation of a rain/irrigation event. Reproduced from Russo et al., 2004, with permission.

The movement and spread of the tracer solute in the heterogeneous soil, which, in turn, may be quantified in terms of integrated measures of the tracer solute transport, such as the moments of the spatial distribution of the tracer solute concentration at a given elapsed time, may serve as a measure of the capability of the soil to transfer water and solutes. The total mass of the tracer solute, M, the principal components of the coordinate vector of the centroid of the tracer solute plume, $\underline{R}$, and the principal components of the second spatial moment of the tracer solute plume, $S_{ij}(t)$ (i, j = 1, 2, 3), providing measures of the mass, location and spread of the tracer solute plume, respectively, are depicted in Figure 14.7 as functions of time, for the first and the third years.

The differences between the 2 yrs reflect the changes in the transport properties of the soil related to soil alkalinity, which, in turn, induce an increase in the strength of the soil solution–soil matrix interactions, and, concurrently, in the rearrangement of the soil PSD. These effects decrease the rate at which the tracer solute mass is lost by leaching through the lower boundary (Fig. 14.7a), decrease the rate of displacement of the centroid of the tracer solute plume (Fig. 14.7b), and diminish the tracer solute spread about its centroid (Fig. 14.7c), particularly in the longitudinal (vertical) direction.

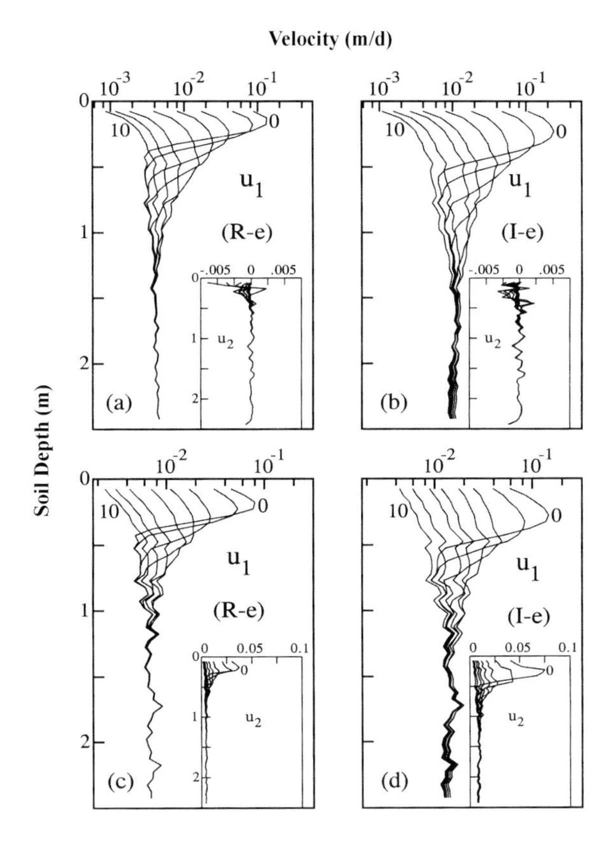

Figure 14.6 Profiles of the horizontal averages, the mean (a, b) and the standard deviation (c, d) of the longitudinal and the transverse (parallel to the x_2 axis) components of the velocity vector, $\underline{u}$, at the end of the second rain season (a, c), and at the end of the third irrigation season (b, d). The profiles in this figure represent different elapsed times (0, 1, 2, 3, 4, 6, 8, 10 d) following the cessation of a rain/irrigation event. Reproduced from Russo et al., 2004, with permission.

The decrease in R_1 in the third year compared with the first year, reflects the decrease in the tracer solute velocity, $V_1 = dR_1(t)/dt$, induced by the reduction in the hydraulic conductivity caused by the rearrangement of the soil PSD. Similarly, the changes in the structure of S_{11} with time reflect the changes of the tracer solute concentration distribution that occur because of the variability induced in the tracer solute velocity field by small-scale heterogeneity in the soil hydraulic properties. The decreases in both the magnitude of S_{11} and the temporal fluctuations of S_{11}, in the third year compared with the first year (Fig. 14.7c), therefore, suggest a reduction in the small-scale heterogeneity of the soil hydraulic properties, caused by the alkalinization process.

Identification of the process controlling the tracer solute transport, and the possible changes in this process due to soil solution–soil matrix interactions, can be accomplished by analyzing the tracer solute travel time PDF, $g_1(T; L)$, (where L is the vertical distance from the solute inlet zone at the soil surface, and $T = R_1(t)/L$), approximated by the mean solute breakthrough curves at a given horizontal control plane (CP) located at L.

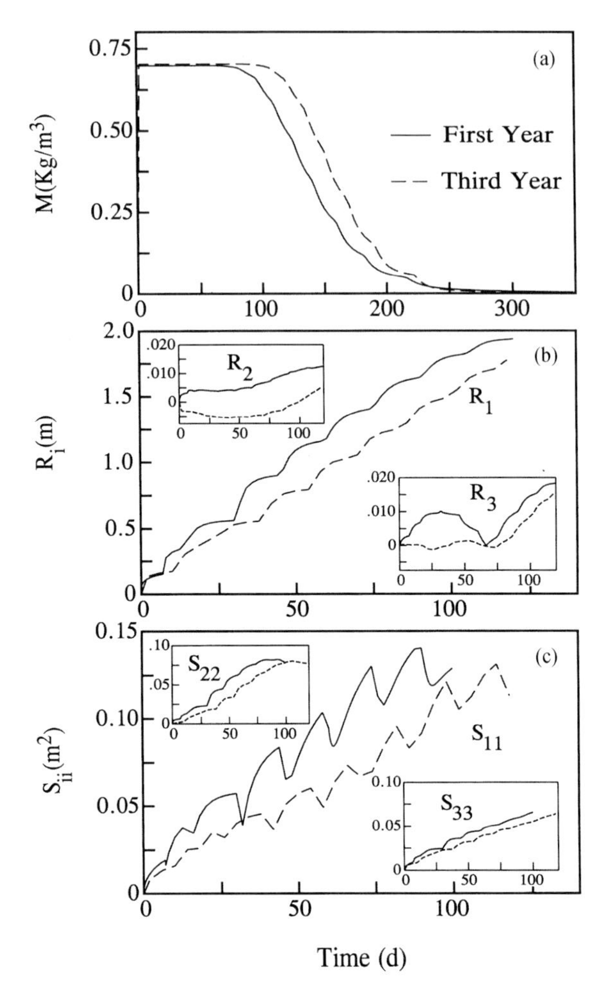

Figure 14.7 Total mass, M, of the tracer solute (a), the principal components of the coordinate location, $\underline{R}$, of the tracer center of mass (b), and the principal components of the spatial covariance tensor, S_{ij}, (c) as functions of time. Results are depicted for the first and the third years. Reproduced from Russo et al., 2004, with permission.

The goodness of fit of $g_1(T; L)$, as represented for a specific process, to the simulated $g_1(T; L)$, therefore, may be used in order to identify the process controlling the tracer solute transport. A Fickian PDF for the tracer solute travel time associated with the classical CDE, representing a transport process the travel-time variance of which grows linearly with travel distance, was selected by Russo et al. (2004) and is addressed here. The normalized solute travel time PDFs, g_1L/V_1, calculated from the simulated flux concentrations, and the fitted Fickian travel time PDFs are depicted in Figure 14.8 as functions of the scaled travel time, T, for horizontal CPs located at three different vertical distances, L, and for the first and the third years. For both years, the fitted travel time PDFs at the shallow CP located within the root zone, cannot reproduce the skewing in the simulated travel time PDFs, originating from the complex flow pattern within the root zone. For CPs located at greater depths below the root zone, however, the goodness of fit of the Fickian travel time

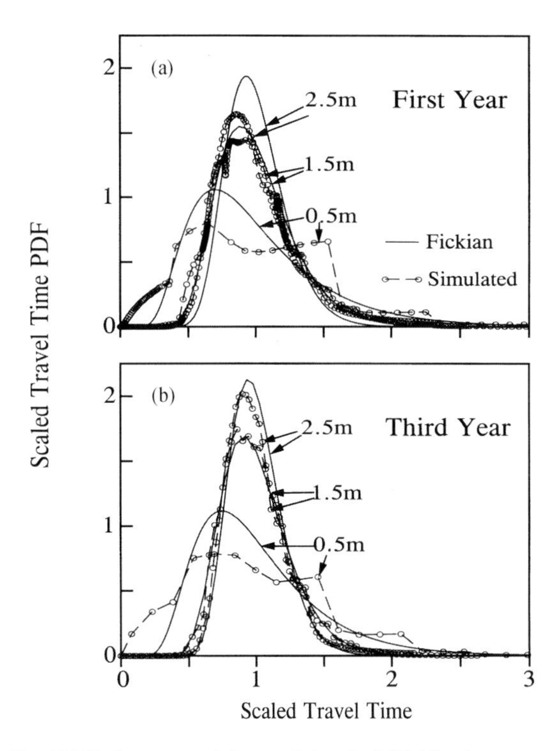

Figure 14.8 Normalized PDFs for one-particle travel time (g_1L/V_1) for the tracer solute, as a function of scaled travel time, $T = R_1(t)/L$, for control planes located three different vertical distances, L, from the injection zone. $R_1(t)$ is the longitudinal component of the coordinate location of the tracer center of mass and t is time. Results are depicted for the first (a) and the third (b) years. Reproduced from Russo et al., 2004, with permission.

PDFs (with parameters optimized at the shallow CP) to the simulated travel time PDFs improves, particularly in the third year. During the first year, the simulated travel time PDFs at the deeper CPs are characterized by earlier breakthroughs and smaller peaks than the fitted travel time PDFs. This suggests that when the process of soil alkalinization is in its initial stages, the travel-time variance associated with the simulated transport initially grows at a rate faster than linearly in travel distance, which decreases with increasing travel distance. In other words, in this case the transport is essentially convection-dominated, controlled by the disparity between regions of different velocities. On the other hand, during the third year, when the alkalinization process is developed, the Fickian travel time PDFs at the deeper CPs provide relatively good fit to the simulated travel time PDFs. This suggests that the reduction in the small-scale heterogeneity in the hydraulic properties, and, concurrently, in the spatial variability of the velocity vector, particularly in the vertical direction, caused by soil alkalinity, essentially damps out the extremely fast travel times, slows down the longitudinal spreading of the plume, and enhances the lateral dispersion of the tracer solute; these, in turn, promote mixing between regions of differing convection, leading to Fickian behavior. In other words, the development of soil alkalinity, associated with a reduction in the small-scale heterogeneity in the hydraulic properties, and, concurrently, in the spatial variability of the velocity vector, may transform the convection-dominated transport to a convection-dispersion transport.

14.3.2 Discussion of the results of the transport simulation of Russo et al. (2004)

The results presented in 14.3.1 suggest that on the field-scale, under realistic flow conditions over an extended period of time, including water extraction by plant roots, the adverse effects of low solute concentration and relatively high SAR on the flow and the transport may be moderate, much less than one would expect in light of the results of previous experimental observations and theoretical studies on the vulnerability of the Darcy-scale, soil hydraulic conductivity to low C and high SAR (e.g. McNeal, 1968; Russo and Bresler, 1977a, b). This discrepancy may be explained, based on three distinct features of a realistic flow system, taken into account in the simulations of Russo et al. (2004):

(i) the periodicity of the rain/irrigation events that are imposed on the soil surface;
(ii) the spatial heterogeneity in the hydraulic properties of the variably saturated soil; and
(iii) the three-dimensionality of the flow domain.

Consider a heterogeneous flow domain viewed as a mixture of different soil materials, i.e. relatively fine-textured materials with appreciable capillary forces and conductivity, $K^f(\psi)$, and relatively coarse-textured ones with relatively small capillary forces and conductivity $K^c(\psi)$. Close to saturation, $\psi \to 0$, $K^c(\psi) \gg oK^f(\psi)$, and the gap between the two decreases with decreasing ψ, approaching zero at $\psi = \psi_c$, where ψ_c is a critical pressure head that depends on parameters of both $K^f(\psi)$ and $K^c(\psi)$; for $\psi < \psi_c$; however, $K^f(\psi)$ is larger than $K^c(\psi)$ and the gap between the two increases as ψ continues to decrease.

During infiltration periods, which generally occupy a small proportion (typically less than 5%) of the rain/irrigation cycles, relatively high water content develops in the upper part of the soil profile. The most permeable zones of the heterogeneous soil, therefore, are those associated with relatively coarse-textured soil materials in which the effect of the soil solution–soil matrix interactions on conductivity is less significant. In other words, during infiltration, soil solution–soil matrix interactions mainly reduce the conductivities in the less permeable zones, and, consequently, decrease the overall conductivity and velocity and increase their spatial variability (Figures 14.4 and 14.6).

On the other hand, during redistribution periods, which are associated with diminishing water content because of water extraction by plant roots and downward flow, the most permeable zones of the heterogeneous soil are those associated with relatively fine-textured soil materials in which the effect of the soil solution–soil matrix interactions on the conductivity is most significant. In other words, for a substantial portion of the rain/irrigation cycles, soil solution–soil matrix interactions mainly reduce the conductivities in the most permeable zones; consequently, both the magnitude and the spatial variability in the conductivity (Fig. 14.4) and, concurrently, in the velocity (Fig. 14.6), are reduced. The decrease in the components of the velocity vector, particularly of its longitudinal component, slows down the tracer solute movement (Fig. 14.7b), decreases the tracer solute spread about its centroid, particularly in the vertical direction (Fig. 14.7c) and diminishes the skewing of the tracer solute travel time PDF (Fig. 14.8).

The three-dimensionality of the flow domain allows the development of lateral head-gradients (Fig. 14.5), and, concurrently, allows fluid particles to bypass zones of low conductivity laterally. Consequently, the substantial decrease in the conductivity during the rainy seasons (Fig. 14.4) may be compensated for by an increase in both the mean and the perturbations of the head-gradient vector (Fig. 14.5); therefore, the resultant velocity vector (Fig. 14.6) is much less affected by soil solution–soil matrix interactions than the hydraulic conductivity. Furthermore, because of the periodicity of the rain/irrigation events, a substantial proportion of the rain/irrigation cycle is occupied by redistribution periods during which the water content diminishes. Therefore, because soil solution–soil matrix interactions decrease with decreasing water content (Russo, 1988), their effect on the hydraulic conductivity and on the velocity is diminished during the redistribution periods.

14.4 Summary and concluding remarks

Experimental data and theoretical results published in the past (see, e.g., Bresler et al. (1982), Shainberg (1984) and Russo (2005) for comprehensive reviews), suggested that soil solution–soil matrix physicochemical interactions, enhanced by diluted soil solutions with relatively high sodium to calcium ratios, may considerably affect both water flow and solute transport. This is particularly so in fine-textured soils associated with substantial clay fractions, subjected to a rain storm associated with diluted solution and relatively high application rate (Russo, 1988).

The results of the flow and transport simulations (Russo et al., 2004) discussed in this chapter, suggest that on the field-scale, under realistic flow conditions and over an extended period of time, the adverse effects of low solute concentration and relatively high SAR on the flow and the transport are smaller compared with the case in which the transport takes place through a Darcy (laboratory) scale, one-dimensional vertical, spatially homogeneous flow domain, restricted to the infiltration stage only (Russo, 1988).

This finding has practical implications regarding the use of TWW for irrigation. For example, in the central area of the coastal region of Israel, TWW is extensively used for the irrigation of citrus groves planted on Hamra Red Mediterranean Soils. These waters are characterized by a similar chloride concentration (C.10 meq/P) to, and smaller sodium adsorption ratio [SAR $<$ 10(meq/P)$^{1/2}$] than, the values of C and SAR employed in the simulation study of Russo et al. (2004).The aforementioned finding suggests, therefore, that the use of the TWW for irrigation in this area, is not expected to significantly reduce the capability of the heterogeneous, variably saturated soil to transfer water and solutes. It should be emphasized, however, that generality is not claimed for this conclusion. Rather, it might be true for cropping conditions with a distinct rainy period during the winter and irrigations during the rest of the year, and for the group of soils (sandy clay), with properties similar to those of the Hamra soil employed in the simulation study of Russo et al. (2004).

An attempt to generalize the results of Russo et al. (2004), at least in a qualitative sense, is demonstrated in Figure 14.9, in which combinations of soil solution concentration (in terms of C) and composition (in terms of SAR) required to maintain K^r∃0.75 (that is, no more than a 25% reduction in K relative to the "stable" reference state) are given for various degrees of water saturation, and for soils of several textures. Figure 14.9 generalizes the concept of the "threshold concentration" (Quirk and Schofield, 1955) in terms

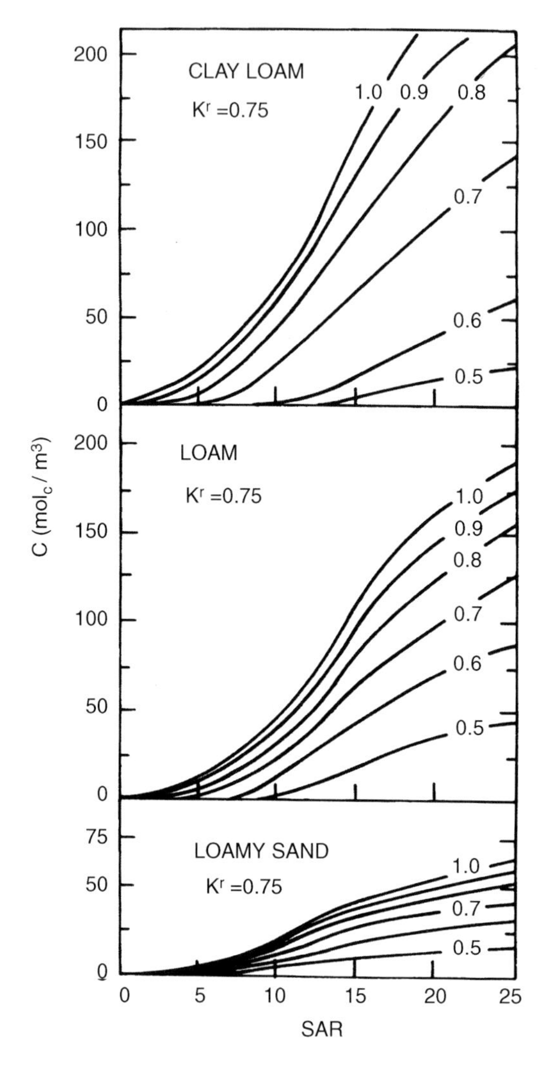

Figure 14.9 Combinations of the concentration, C, and the sodium adsorption ratio, SAR, of the equilibrium solution, and degree of water saturation, Θ (the numbers labeling the curves) at which 25% reduction in hydraulic conductivity occurs, for three soils of different texture, loamy sand (LS), loam (L), and clay loam (CL). Reproduced from Russo, 1988, by permission of the American Geophysical Union.

of soil type and soil water status. This figure clearly demonstrates that the vulnerability of the soil to the adverse effects of low C and high SAR depends on both soil texture and soil water status. Consequently, the curves in Figure 14.9 may be used for water quality classification in relation to soils of different textures, and for water and soil management. For a given SAR, the solution concentration required to maintain $K^r \exists 0.75$ (that is, the "threshold concentration") decreases as the soil water content decreases and as the soil texture becomes coarser. Figure 14.9 suggests that a decrease in the water application rate and, concurrently, in water saturation, may improve the efficiency of the reclamation of alkaline soils, especially the fine-textured ones.

Before concluding, it is worth emphasizing that the results presented in this chapter are relevant to soils having a clay fraction dominated by smectite minerals (e.g., montmorillonite), with solutions containing Na, Ca, and Cl ions only. The conclusions drawn from the present chapter should be considered with caution, inasmuch as the numerical results presented here are based on several simplifying assumptions regarding processes and entities at three different length scales: the micro (pore) scale, the local (Darcy) scale, and the macro (field) scale. These assumptions include:

(i) the description of the physicochemical interactions between the soil solution and the soil matrix at the micro scale (based on the mixed-ion diffuse double layer theory, the structure of the clay particles, the soil PSD, and hydrodynamic principles (Russo and Bresler, 1977b; Russo, 1988), assuming that their effects on the changes in the soil PSD, are reversible);
(ii) the description of the local-scale constitutive relationships for unsaturated flow (the van Genuchten model, ignoring local hysteresis and local anisotropy), and the local-scale transport (the classical, one-region, CDE, neglecting ion precipitation or dissolution); and
(iii) the description of the structure of the heterogeneous soil on the field scale (exponential covariance with axisymmetric anisotropy), and the statistics of the relevant soil properties and the flow-controlled attributes (statistically homogeneous).

The aforementioned simplifying assumptions might limit the applicability of the results presented in this chapter. Furthermore, it should be emphasized that the numerical results of (Russo et al., 2004) are based on analyses of single realizations of the relevant soil properties. Results of the analyses of transport of tracer and sorptive solutes under transient flow conditions in a variably saturated, three-dimensional, heterogeneous soil (Russo et al., 1998), however, suggest that because of the relatively large lateral extent of the solute input zone, the simulated results of Russo et al. (2004) might be sufficiently accurate to indicate appropriate trends.

References

Bear, J. (1972) *Dynamics of Fluids in Porous Media*. Elsevier, New York, NY.

Bresler, E. (1978) Theoretical modeling of mixed-electrolyte solution flows for unsaturated soils. *Soil Science* **125**, 196–203.

Bresler, E., McNeal, B.L. and Carter, D.L. (1982) *Saline and Sodic Soils: Principles-Dynamics-Modeling*. Springer, Berlin, Heidelberg and New York.

Butters, G.L., Jury, W.A. and Ernst, F.F. (1989) Field scale transport of bromide in an unsaturated soil. 1. Experimental methodology and results. *Water Resources Research* **25**, 1575–1581.

Dagan, G. and Bresler, E. (1979) Solute transport in unsaturated heterogeneous soil at field scale. 1. Theory. *Soil Science Society of America Journal* **43**, 461–467.

Destouni, G. and Cvetkovic, V. (1989) The effect of heterogeneity on large scale solute transport in the unsaturated zone. *Nordic Hydrology* **20**, 43–52.

Destouni, G. and Cvetkovic, V. (1991) Field scale mass arrival of sorptive solute into the groundwater. *Water Resources Research* **27**, 1315–1325.

Ellsworth, T.R., Jury, W.A., Ernst, F.F. and Shouse, P.J. (1991) A three-dimensional field study of solute transport through unsaturated layered porous media, 1. Methodology, mass recovery and mean transport. *Water Resources Research* **27**, 951–965.

Flury, M., Fluhler, H., Jury, W.A. and Leuenberger, J. (1994) Susceptibility of soils to preferential flow of water: A field study. *Water Resources Research* **30**, 1945–1954.
Forrer, I., Kasteel, R., Flury, M. and Fluhler, H. (1999) Longitudinal and lateral dispersion in an unsaturated field soil. *Water Resources Research* **35**, 3049–3060.
Foussereau, X, Graham, W.D., Akpoji, G.A., Destouni, G. and Rao, P.S.C. (2001) Solute transport through a heterogeneous coupled vadose-saturated zone system with temporally random rainfall. *Water Resources Research* **37**, 1577–1588.
Jones, A.J. and Wagenet, R.J. (1984) In-situ estimation of hydraulic conductivity using simplified methods. *Water Resources Research* **20**, 1620–1626.
Jury, W.A. (1982) Simulation of solute transport using a transfer function model. *Water Resources Research* **18**, 363–368.
Harter, T, and Yeh, T.-C.J. (1996) Stochastic analysis of solute transport in heterogeneous, variably saturated soils. *Water Resources Research* **32**, 1585–1595.
McNeal, B.L. (1968) Prediction of the effect of mixed salt solutions on soil hydraulic conductivity. *Soil Science Society of America Proceedings* **32**, 190–193.
Neuman, S.P., Feddes, R.A. and Bresler, E. (1975) Finite element analysis of two-dimensional flow in soils considering water uptake by roots. 1. Theory. *Soil Science Society of America Proceedings* **39**, 224–230.
Nielsen, D.R., Biggar, J.W. and Erh, K.T. (1973) Spatial variability of field-measured soil-water properties. *Hilgardia* **42**, 215–260.
Nimah, M.N. and Hanks, R.J. (1973a) Model for estimating soil water, plant and atmospheric relationships. 1. Description and sensitivity. *Soil Science Society of America Proceedings* **37**, 522–527.
Nimah, M.N. and Hanks, R.J. (1973b) Model for estimating soil water, plant and atmospheric relationships. 2. Field test of the model. *Soil Science Society of America Proceedings* **37**, 528–532.
Quirk, J.P. and Schofield, R.K. (1955) The effect of electrolyte concentration on soil permeability. *Journal of Soil Science* **6**, 163–178.
Roth, K. and Hammel, K. (1996) Transport of conservative chemical through an unsaturated two-dimensional Miller-similar medium with steady state flow. *Water Resources Research* **32**, 1653–1663.
Roth, K., Jury, A.W., Fluhler, H. and Attinger, W. (1991) Transport of chloride through an unsaturated field soil. *Water Resources Research* **27**, 2533–2541.
Russo, D. (1988) Numerical analysis of the nonsteady transport of interacting solutes through unsaturated soil: I. Homogeneous systems. *Water Resources Research* **24**, 271–284.
Russo, D. (1989a) Field-scale transport of interacting solutes through the unsaturated zone. I. Analysis of the spatial variability of the input parameters. *Water Resources Research* **25**, 2475–2485.
Russo, D. (1989b) Field-scale transport of interacting solutes through the unsaturated zone. II. Analysis of the spatial variability of the field response. *Water Resources Research* **25**, 2487–2495.
Russo, D. (1991) Stochastic analysis of vadose-zone solute transport in a vertical cross section of heterogeneous soil during nonsteady water flow. *Water Resources Research* **27**, 267–283.
Russo, D. (1993a) Stochastic modeling of macrodispersion for solute transport in a heterogeneous unsaturated porous formation. *Water Resources Research* **29**, 383–397.
Russo, D. (1993b) Stochastic modeling of solute flux in a heterogeneous partially-saturated porous formation. *Water Resources Research* **29**, 1731–1744.
Russo, D. (1995a) Contaminant transport below an orchard irrigated with sewage water. In: *The Raanana Sewage Purification Project: Recommendations of Criteria for the Sewage Water Quality for Different Purification Alternatives* (eds P. Fine, N. Haruvi and I. Shainberg), pp. 35–49. Preliminary Report, Institute of Soils, Water and Environmental Sciences, The Agricultural Research Organization, Bet Dagan, Israel (in Hebrew).
Russo, D. (1995b) Stochastic analysis of the velocity covariance and the displacement covariance tensors in partially saturated heterogeneous anisotropic porous formations. *Water Resources Research* **31**, 1647–1658.
Russo, D. (1998) Stochastic analysis of flow and transport in unsaturated heterogeneous porous formation: effects of variability in water saturation. *Water Resources Research* **34**, 569–581.
Russo, D. (2005) Physical aspects of soil salinity. In: *Encyclopedia of Soils in the Environment 3* (ed. D. Hillel), pp. 442–453. Elsevier Ltd., Oxford, U.K.
Russo, D. and Bouton, M. (1992) Statistical analysis of spatial variability in unsaturated flow parameters. *Water Resources Research* **28**, 1911–1925.

Russo, D. and Bresler, E. (1977a) Effect of mixed Na/Ca solutions on hydraulic properties of unsaturated soils. *Soil Science Society of America Journal* **41**, 713–717.

Russo, D. and Bresler, E. (1977b) Analysis of the saturated-unsaturated hydraulic conductivity in a mixed Na-Ca soil system. *Soil Science Society of America Journal* **41**, 706–710.

Russo, D. and Bresler, E. (1981) Soil hydraulic properties as stochastic processes: I. An analysis of field spatial variability. *Soil Science Society of America Journal* **45**, 682–687.

Russo, D., Russo, I. and Laufer, A. (1997) On the spatial variability of parameters of the unsaturated hydraulic conductivity. *Water Resources Research* **33**, 946–956.

Russo, D., Zaidel, J. and Laufer, A. (1994) Stochastic analysis of solute transport in partially saturated heterogeneous soil: I. Numerical experiments. *Water Resources Research* **30**, 769–779.

Russo, D., Zaidel, J. and Laufer, A. (1998) Numerical analysis of flow and transport in a three-dimensional partially saturated heterogeneous soil. *Water Resources Research* **34**, 1451–1468.

Russo, D., Zaidel, J. and Laufer, A. (2001) Numerical analysis of flow and transport in a combined heterogeneous vadose zone-groundwater system. *Advances in Water Resources* **24**, 49–62.

Russo, D., Zaidel, J. and Laufer, A. (2004) Numerical analysis of transport of interacting solutes in a three-dimensional unsaturated heterogeneous soil. *Vadose Zone Journal* **3**, 1286–1299.

Russo, D., Zaidel, J., Fiori, A. and Laufer, A. (2006) Numerical analysis of flow and transport from a multiple-source system in partially saturated heterogeneous soil under cropped conditions. *Water Resources Research* **42**, W06415, doi: .

Schulin, R., van Genuchten, M.Th., Fluhler, H. and Ferlin, P. (1987) An experimental study of solute transport in a stony field soil. *Water Resources Research* **23**, 1785–1794.

Shainberg, I. (1984) The effects of electrolyte concentration on the hydraulic properties of sodic soils. In: *Soil Salinity Under Irrigation: Processes and Management* (eds I. Shainberg and J. Shalhevet), pp. 49–64. Springer-Verlag, New York.

Tseng, P.H. and Jury, W.A. (1994) Comparison of transfer function and deterministic modeling of area-average solute transport in heterogeneous field. *Water Resources Research* **30**, 2051–2064.

van Genuchten, M.Th. (1980) A closed-form equation for predicting the hydraulic conductivity of unsaturated soils. *Soil Science Society of America Journal* **44**, 892–898.

Index